Tamarix: A CASE STUDY OF ECOLOGICAL CHANGE IN THE AMERICAN WEST

Tamarix

A CASE STUDY OF ECOLOGICAL CHANGE
IN THE AMERICAN WEST

Edited by
Anna Sher
and
Martin F. Quigley

OXFORD
UNIVERSITY PRESS

OXFORD
UNIVERSITY PRESS

Oxford University Press is a department of the University of Oxford.
It furthers the University's objective of excellence in research, scholarship,
and education by publishing worldwide.

Oxford   New York
Auckland   Cape Town   Dar es Salaam   Hong Kong   Karachi
Kuala Lumpur   Madrid   Melbourne   Mexico City   Nairobi
New Delhi   Shanghai   Taipei   Toronto

With offices in
Argentina   Austria   Brazil   Chile   Czech Republic   France   Greece
Guatemala   Hungary   Italy   Japan   Poland   Portugal   Singapore
South Korea   Switzerland   Thailand   Turkey   Ukraine   Vietnam

Oxford is a registered trademark of Oxford University Press
in the UK and certain other countries.

Published in the United States of America by
Oxford University Press
198 Madison Avenue, New York, NY 10016

Library of Congress Cataloging-in-Publication Data
Tamarix : a case study of ecological change in the American West / edited by
Anna Sher, Martin F. Quigley.
     p. cm.
Includes bibliographical references and index.
ISBN 978-0-19-989820-6 (hardcover : alk. paper)
1. Tamarisks—West (U.S.) 2. Tamarisks—Control—West (U.S.)   I. Sher, Anna.
II. Quigley, Martin F.
QK495.T35T36 2013
582.160978—dc23   2012019809

ISBN 978-0-19-989820-6

9 8 7 6 5 4 3 2 1
Printed in the United States of America
on acid-free paper

This book is dedicated to the memory of John P. Taylor, Ph.D. (1954–2004), mentor, scientist, colleague, and friend;

to all those in the field working for greater understanding of this complex genus and its ecology;

and to my wife Fran, without whom this book would not be a reality.

—AS

CONTENTS

PART III: **THE HUMAN ELEMENT**

PART IV: **MANAGEMENT**

FOREWORD

Juliet C. Stromberg

Three decades ago Benjamin Everitt, a physical geographer, published a paper with the imploring title "Ecology of Saltcedar: A Plea for Research" (Everitt 1980). Since that time, nearly 200 peer-reviewed journal articles have been published on the biology or ecology of the *Tamarix* species that grow along rivers of the American West (Stromberg et al. 2012). Some of these are reviews (Shafroth et al. 2005; Sogge et al. 2008; Stromberg et al. 2009; Hultine et al. 2010). The time is ripe for an edited volume that addresses the many fascinating, and often controversial, issues surrounding *Tamarix*.

Also three decades ago, I moved to Arizona to begin my doctoral research on the ecology of riparian vegetation. Little did I know this seemingly benign topic would become rife with debate. One provocative area revolves around differing human perceptions of native and introduced species (and, in general, of landscape change). My first jarring moment vis-à-vis this issue was when, then confident in my belief that "exotic species do not belong here" because "natives are good," I asked taxonomist Donald Pinkava whether a particular riparian species I had identified was exotic or native. He responded, and I paraphrase, "Why does it matter?" Over the decades my opinion on this topic has undergone a radical change. Gaining historical perspectives on people's changing perceptions of *Tamarix* and other introduced species, and on the long effort, first to increase and then to minimize the influence of tamarisk in riparian zones of the American West, is important for scientists and land managers alike.

Another controversy is intertwined with the basic ecological principle that competition for limited resources structures communities. In this case, the competitors are an odd pairing of a shrubby tree and a technologically enhanced primate, with the limiting resource being fresh water. Research and debate continue to address how much water is transpired by groundwater-dependent plants (phreatophytes), how much is used by local plants versus interlopers, the precision with which water use and stream depletion can be measured, and whether riparian plant water use is justified by the other ecosystem services the vegetation provides. These topics are important. Availability of water to sustain freshwater ecosystems and thriving human populations are issues of local, regional, and global concern.

Considerable research on the influence of *Tamarix* on other ecosystem properties, such as soil salinization, has now accumulated. Some of my research on riparian plant communities in the 1980s and 1990s focused on perennial, undammed reference streams, and I recall being surprised to *not* find high levels of salt in the soils beneath *Tamarix*. As I dove into the primary literature to discover why my expectation was not borne out, I found myself repeating Everitt's mantra about the need for research. I also began railing about the seemingly unfounded mythology that had developed around this once exclusively Old World and now New World taxon (of novel genetics). Syntheses that take into account multiple species and multiple environmental contexts are critical for laying to rest this issue of *Tamarix*, halophytes, and salinization.

Another of the myths about *Tamarix* that was oft-repeated (and is still heard today) is that the reason for its abundance is its ability to outcompete other riparian plants. I was delighted, then, to read a paper published by Anna Sher in 2000 in which she approached the subject of competition between riparian tree seedlings from a basic ecological perspective. Her findings, which revealed *Populus* (cottonwood) to be the better competitor in the strict sense of the word, were a refreshing contribution to the literature. Many people still are surprised to hear that *Tamarix* is not the better competitor, per se, although it certainly is more stress tolerant. Sutherland (1996) noted that one of the 20 commonest sins of biologists is not telling the world what you have found. Key research findings such as Sher's need many outlets to reach a wide audience.

A small songbird, the southwestern willow flycatcher, is a central (and controversial) player in the story of *Tamarix* and its management in the American West. In 1998 I was invited by the U.S. Fish and Wildlife Service to serve on a committee to develop a recovery plan for this riparian-obligate, federally endangered species. In parts of its range, the flycatcher builds most of its nests in *Tamarix*. As advocates for the flycatcher, we struggled to develop plans that would accommodate species recovery while not unduly restricting the many *Tamarix*-removal-based restoration actions, including the release of *Diorhabda* beetles for biocontrol. In retrospect, much of our effort was moot. Despite conclusions by entomologists regarding the limited capacity for spread of the biocontrol insects, the beetles are finding their way across the West into flycatcher breeding grounds, much as *Tamarix* did over the past century.

If you will indulge me one last personal memory: when I was in my early twenties and beginning to research an endangered plant in the wilds of Wisconsin, a neighborhood boy asked me with astonishment if I could really spend three years studying one species. Yes, and many people can spend lifetimes studying a single species and its interactions. The welcome result, in the case of this volume, is documentation of the intriguing tale of *Tamarix*. Populations of many species across the world are expanding their ranges (or, in an interesting semantic choice, "invading"; Crawford and Hoagland 2009), creating new species assemblages and novel ecosystems. Examination of vegetation change through time is a critical component of ecology (Mitchell 2011). Editions such as this, when combined with syntheses of other successful species, are important for reasoned ecological analyses of the causes and consequences of shifting plant dominance.

Literature Cited

Crawford, P. H. C., and B. W. Hoagland. 2009. Can herbarium records be used to map alien species invasion and native species expansion over the past 100 years? Journal of Biogeography 36:651–661.

Everitt, B. L. 1980. Ecology of saltcedar: A plea for research. Environmental Geology 3:77–84.

Hultine, K. R., J. Belnap, and C. van Riper III, et al. 2010. Tamarisk biocontrol in the western United States: Ecological and societal implications. Frontiers in Ecology and the Environment 8:467–474.

Mitchell, F. J. G. 2011. Exploring vegetation in the fourth dimension. Trends in Ecology and Evolution 26:45–52.

Shafroth, P. B., J. R. Cleverly, T. L. Dudley, J. P. Taylor, C. van Riper III, E. P. Weeks, and J. N. Stuart. 2005. Control of *Tamarix* spp. in the western U.S.: Implications for water salvage, wildlife use, and riparian restoration. Environmental Management 35:231–246.

Sher, A. A., D. L. Marshall, and S. A. Gilbert. 2000. Competition between *Populus deltoides* and invasive *Tamarix ramosissima* and the implications for reestablishing flooding. Conservation Biology 14:1744–1754.

Sogge, M., S. Sferra, and E. Paxton. 2008. Saltcedar as habitat for birds: Implications to riparian restoration in the southwestern United States. Restoration Ecology 16:146–154.

Stromberg, J. C., D. C. Andersen, and M. L. Scott. 2012. Riparian floodplain wetlands of the arid and semiarid Southwest. Pages 343–356, *in* D. Batzer and A. Baldwin, editors. *Wetland Habitats of North America: Ecology and Conservation Concerns*. University of California Press, Berkeley.

Stromberg, J. C., M. K. Chew, P. L. Nagler, E. P. Glenn. 2009. Changing perceptions of change: The role of scientists in *Tamarix* and river management. Restoration Ecology 17:177–186.

Sutherland, W.J. 1996. The twenty commonest surveying sins. Pages 408–410 *in* W. J. Sutherland, editor. Ecological Census Techniques: A Handbook. Cambridge University Press, Cambridge, UK.

PREFACE

Tamarix as a Teaching Tool

This book is designed to accomplish at least two objectives. The first is to be the first comprehensive treatment of *Tamarix* in the United States, a subject that is complex and politically charged. The second is to be a case study for a number of important subjects, including invasive species ecology, water issues in the West, interactions between humans and their environment, range management and restoration, and the philosophy of science. We understand that the book is likely to be used as a reference book, with chapters read in isolation; however, given the complexity of the subject matter, we urge the reader to read the introduction at the very least, or a section rather than a single chapter, to gain context. Because there are so many facets to this species (and by species we mean the hybrid swarm that represents *Tamarix* spp. in North America) and so many different perspectives about how it should be viewed and therefore managed, as editors we decided that it was important to allow each author's voice to be intact, as well as to present a complete thesis. This means that there is overlap between some chapters, but only when the subject is being treated in a different way, or used for a different (and sometimes contradictory) argument. It also means that this volume is in some ways more similar to a literary anthology than a science textbook. We explain how the chapters may be used in concert to stimulate critical thinking and understanding.

Invasive Species Ecology

At its core, this book confronts the fundamental question that is increasingly being asked: is the label "invasive" a useful ecological tool? Does such a designation even exist biologically, and if so, what are the implications for management? Using *Tamarix* as a case study, we can delve into the issues of how a single species can have an impact on an ecosystem, and the implications of this impact for other trophic levels. A reading of the entire book is appropriate to gain this full picture, with particular emphasis on the parts "Ecology," paired with "Management" and/or "The Human Element," depending on the focus. If only one of these sections is read, the introduction and the conclusion can also give a good overview and context, as can chapter 3, "Extent and Projections of *Tamarix* Distribution in North America."

Water Issues in the West

Tamarix is an ideal case study for investigating the complexity of water issues in the American West. To use this text, the following chapters from part I, "Biology and Range," and part III, "The Human Element" give several perspectives on *Tamarix* and water, including the implications for policy:

Chapter 4	Evapotranspiration by Tamarisk in the Colorado River Basin
Chapter 5	Tamarisk: Ecohydrology of a Successful Plant
Chapter 6	Tamarisk Water Use by Tamarisk
Chapter 7	*Tamarix*, Hydrology, and Fluvial Geomorphology
Chapter 16	Tamarisk Introduction, Naturalization, and Control in the United States, 1818–1952
Chapter 17	The Politics of a Tree: How a Species Became National Policy

Interactions between Humans and Their Environment

The fields of conservation biology and environmental studies both concern the intersection of humans and our environment, combining aspects of life and earth sciences with social and political fields. *Tamarix* is a genus with uncommonly rich stories to tell. Any invasive species will have interesting elements of ecology and management, but this is one that also involves an endangered species and has been the subject of lawsuits, national policy, and deep emotions both for and against it. The entire book is relevant to this topic, with particular emphasis on the part III, "The Human Element" paired with part IV, "Management." We suggest book-ending these with the introduction and concluding chapter.

Range Management and Restoration Ecology

Given that anyone doing restoration in the West is likely to encounter *Tamarix*, it is critical to understand how it affects its environment and how these affect one's ability to restore an area to a more desirable state. For those academic courses that seek to use *Tamarix* as a case study only to better understand the complex issues involved in restoration that go beyond biology, part IV on management will be useful, and one may consider following chapters from part I, "Biology and Range," and part III, "Ecology":

Chapter 7	*Tamarix*, Hydrology, and Fluvial Geomorphology
Chapter 8	Tamarisk and Salinity: An Overview
Chapter 10	*Tamarix* as Wildlife Habitat
Chapter 13	*Tamarix* and Soil Ecology
Chapter 14	Fire Ecology of *Tamarix*

Philosophy of Science and Environmental Ethics

Although as scientists we like to think of ourselves as being purely objective at all times, as humans this is of course rarely true to the extent we would like it to be. This book represents a dialogue among scientists who represent a continuum of attitudes about this single plant. As a tool for critical thinking and identifying the role of perspective in scientific inquiry, one may read the following chapters in juxtaposition to each other:

16. Tamarisk Introduction, Naturalization, and Control in the United States, 1818–1952	compared to	17. The Politics of a Tree: How a Species Became National Policy and 20. Tamarisk Management: Lessons and Techniques
5. Tamarisk: Ecohydrology of a Successful Plant	compared to	6. Water Use by *Tamarix* and 8. Tamarisk and Salinity: An Overview
11. Tamarisk in Riparian Woodlands: A Bird's Eye View	compared to	22. Bring on the Beetles! The History and Impact of Tamarisk Biological Control and 10. *Tamarix* as Wildlife Habitat

And one may read the following chapters in the context of attempting to reconcile the spectrum of opinions:

Chapter 1	Introduction to the Paradox Plant
Chapter 15	*Tamarix*: Passenger or Driver of Ecosystem Change?
Chapter 18	A Philosophical Framework for Assessing the Value of Tamarisk
Chapter 25	The Future of *Tamarix*

There are, of course, myriad other courses in various disciplines that may find this book to be a useful tool. We welcome your contacting us with your ideas. We hope that you find this text as interesting and stimulating to read and discuss as we have found it to write and edit. See www.oup.com/us/Tamarisk to contact the authors, for color figures, and for other teaching resources.

—AS and MFQ

See URL: www.oup.com/us/Tamarisk to contact the authors, for color figures, and other teaching resources

ACKNOWLEDGMENTS

The genesis of this project was two organized oral sessions for the annual conference of the Ecological Society of America in 2006 and 2007 that Julie Stromberg agreed to co-chair with me. Julie is nothing if not the pioneer in the field of tamarisk ecology in the United States, with more publications in the field than anyone. I am grateful to her for her collaborations with me. Quite simply, one cannot write a book about tamarisk without including her.

If Julie is the mother of tamarisk ecological research, for me, the field has two fathers: John P. Taylor and Patrick Shafroth. John did removal and restoration work that inspired land managers and scientists alike, including giving me field sites at Bosque del Apache Wildlife Refuge for my dissertation work. He was an important mentor to me, and our relationship, like so many of his connections with academia, would foreshadow the important bridges between managers and scientists that were to come. John's sudden passing in 2004 was a terrible loss. I first met Patrick at the Refuge during a very small research symposium in the 1980s. I was a student then and already a groupie of Pat's work on tamarisk. There weren't many papers on the ecology of the genus then, and I relied heavily on Julie's and Pat's research.

Many other people have helped and inspired me in my academic path, culminating in this book. These begin with Brent Smith at Earlham College, who first sparked my interest in invasive plants, my graduate advisor, Diane Marshall at the University of New Mexico, who gave me indepedence while supporting me 100%, my fellow grad students and postdocs who helped keep me going, and all the undergraduates who assisted me in the field. Data collection in tamarisk stands means heat in triple digits, swarms of mosquitoes, and fighting through the underbrush. I am grateful to my Fulbright mentors Deborah Goldberg and David Ward when I studied in Israel, and Joe DiTomaso and David Spencer at the University of California. I would not have stayed in the "tamarisk field" were it not for my research collaborations with Ken Lair and Scott Nissen and the excellent work of my graduate students, including Robin Bay, Stephanie Gieck, Stephanie Strudley, Michelle Ohrtman, Maggie Gaddis, Hisham El Waer, and Rob Anderson. The Denver Botanic Gardens leadership and my staff in Research and Conservation made it possible for me to continue my *Tamarix* work when I moved to Colorado, especially Michelle DePrenger-Levin. And I must thank the Tamarisk Coalition for its abundant support for this project, dedication to riparian health, and for allowing me to serve as its board president for four years.

We have been very fortunate to work with the authors who are represented in this book. They have been attentive, responsive, patient, and enthusiastic. Quite simply, they are the reason for this book: its quality and value is to their credit. Special thanks to Tom Dudley for his vision of what the book could be and for mentoring both Gail Drus and Stephanie Strudley in writing their chapters. I am grateful to Monty Sher, Joseph Kraus, Tim Carlson, Kevin Hultine, Pat Shafroth, John Gaskin, James Cleverly, and Mark Carny for information, ideas, and feedback on the first and final chapters. The support and encouragement of Ellen Bierhorst, Glo Harris, Gene Bierhorst, and many, many other friends, colleagues, and family members was invaluable. And I must thank Jeremy Lewis, our editor at Oxford University Press, who made working with them so easy.

Finally, and most importantly, I thank my wonderful coeditor Martin for being the kick-in-the-pants I needed him to be, and my loving partner Fran, who took care of our son Jeremy to give me time to work on the book. In the words of an Israeli colleague, "Publishing [a book] is like childbirth; it is a long, messy and difficult process." And, of course, it is so very worthwhile in the end.

—AS

CONTRIBUTORS

Daniel A. Auerbach, Ph.D.
Department of Biology
Colorado State University
Ft. Collins, Colorado

Heather L. Bateman, Ph.D.
Department of Applied Sciences and
    Mathematics
Arizona State University, Polytechnic
    campus
Mesa, Arizona

Robin F. Bay, M.S.
Senior Ecologist/Project Manager
Habitat Management, Inc.
Denver, Colorado

Dan W. Bean, Ph.D.
Director
Palisade Insectary, Biological Pest
    Control Program
Colorado Department of Agriculture
Palisade, Colorado

Vanessa B. Beauchamp, Ph.D.
Department of Biological Sciences
Towson University
Towson, Maryland

Mark K. Briggs, M.S.
Chihuahuan Desert Program
World Wildlife Fund
Tucson, AZ

Tim Carlson, P.E.
Former Director
Tamarisk Coalition
Current: Environmental Engineer
601 Rambling Rd
Grand Junction, Colorado

Matthew K Chew, Ph.D.
Center for Biology + Society
Arizona State University
Tempe, Arizona

James R. Cleverly, Ph.D.
Terrestrial Ecohydrology Research
    Laboratory
Plant Functional Biology and Climate
    Change Cluster
School of the Environment
University of Technology Sydney
Sydney, New South Wales

Peter Dalin, Ph.D.
Swedish University of Agricultural
    Sciences
Department of Ecology
Uppsala, Sweden

Gina Dello Russo
Coordinator
FWS New Mexico Invasive Species
    Strike Team
Fish and Wildlife Service
Bosque del Apache National Wildlife
    Refuge
San Antonio, New Mexico

Cameron H. Douglass, Ph.D.
    Candidate
Department of Bioagricultural
    Sciences and Pest Management
Colorado State University
Fort Collins, Colorado

Gail M. Drus, M.S., Ph.D.
Department of Ecology, Evolution and
    Marine Biology
University of California Santa Barbara
    Santa Barbara, California

Tom Dudley, Ph.D.
Marine Science Institute
University of California
Santa Barbara, California

Paul Evangelista, Ph.D.
Natural Resource Ecology Laboratory
Colorado State University
Fort Collins, Colorado

John F. Gaskin, Ph.D.
USDA Agricultural Research Service
Sidney, Montana

Catherine A. Gehring, Ph.D.
Department of Biological Sciences
Northern Arizona University
Flagstaff, Arizona

Edward P. Glenn, Ph.D.
Department of Soil, Water and
    Environmental Science
University of Arizona
Tucson, Arizona
eglenn@ag.arizona.edu

Jim Graham, Ph.D.
Oregon State University
College of Earth, Ocean, and
    Atmospheric Science
Corvallis, OR

Charles R. Hart, Ph.D.
Associate Department Head
Department of Ecosystem Science and
    Management
Texas AgriLife Extension Service
Stephenville, Texas

Kevin Hultine, Ph.D.
Department of Research
Desert Botanical Garden
Phoenix, AZ

Catherine S. Jarnevich, Ph.D.
U.S. Geological Survey
Fort Collins Science Center
Fort Collins, Colorado

Tyler D. Johnson, M.S.
US Forest Service
Bridger-Teton National Forest
Jackson, Wyoming

Kenneth D. Lair, Ph.D.
Former: Lockeford Plant Materials
    Center
Natural Resource Conservation
    Service-USDA
Lockeford, CA
Current: Restoration Ecologist/
    Revegetation Specialist
    Hesperia, CA

William S. Longland, Ph.D.
Great Basin Rangeland Research
    USDA
Agricultural Research Service
Reno, NV

Kelley A. Meinhardt, M.S.
Department of Civil and
    Environmental Engineering
University of Washington
Seattle, Washington

David M. Merritt, Ph.D.
Stream Systems Technology Center
Watershed, Fish, Wildlife, Air, and Rare
    Plants Staff, USDA Forest Service
and
Natural Resource Ecology
    Laboratory
Colorado State University
Fort Collins, Colorado

Pamela L. Nagler, Ph.D.
U.S. Geological Survey
Southwest Biological Science Center
Sonoran Desert Research Station
Tucson, AZ

Scott J. Nissen, Ph.D.
Department of Bioagricultural
    Sciences and Pest Management
Colorado State University
Ft. Collins, Colorado

Michelle Ohrtman, Ph.D.
Plant Science Department
South Dakota State University
Brookings, South Dakota

Eben H. Paxton, Ph.D.
U.S. Geological Survey,
Pacific Island Ecosystems Research
    Center
Kilanea Field Station
Hawaii National Park, HI

Martin F. Quigley, MLA, Ph.D.
Kurtz Professor of Botany and
    Arboretum Director
Department of Biological Sciences
University of Denver
Denver, Colorado

Naomi Reshotko, Ph.D.
Chair
Department of Philosophy
University of Denver
Denver, Colorado

Michael L. Scott, Ph.D.
Fort Collins Science Center
U.S. Geological Survey
Fort Collins, Colorado

Patrick B. Shafroth, Ph.D.
Fort Collins Science Center
U.S. Geological Survey
Fort Collins, Colorado

Anna Sher, Ph.D.
Department of Biological Sciences
University of Denver
Denver, Colorado

Mark K. Sogge, M.S.
U.S. Geological Survey
Southwest Region
Flagstaff, AZ

Juliet Stromberg, Ph.D.
School of Life Sciences
Arizona State University
Tempe AZ

Stephanie Strudley, M.S.
Formerly: Department of Biological
    Sciences
University of Denver

Charles van Riper III, Ph.D.
U.S. Geological Survey
Southwest Biological Science Center
Sonoran Desert Research Station
Tucson, AZ

Erika Zavaleta, Ph.D.
Environmental Studies Department
University of California
Santa Cruz, CA

1

Introduction to the Paradox Plant

Anna Sher

Tamarix spp. is the most successful exotic tree in the western United States and is an organism of contradictions (see figure 1.1). Rarely has a wild-growing plant attracted the same level of political, public, and scientific attention from such diverse perspectives. Is it is good or bad for wildlife? Is it the cause or consequence of ecosystem degradation? Is its growth and spread promoted by an increase or a decrease in ecosystem disturbance? Does it unify diverse stakeholder groups, spanning the political spectrum, or is it a lightning rod for disagreement? As this book will explore, the answer to all these questions is simply—or not so simply—yes.

From the late 1950s through the early 1980s, the group of species growing in the new world, referred to collectively as *Tamarix*, was considered an environment-destroying invasive, a view that received little argument from anyone. Although the significant political debates surrounding this plant would come later, by the mid-1980s, both the public and scientists had begun to reconsider the problem—and potential benefits—of *Tamarix*. The number of scientific publications on the plant has had exponential growth since 1985 (see figure 1.2). The largest cluster of these, although accounting for only 3% of the total, was published in the *Journal of Arid Environments*. This explosion of academic attention corresponded with drought in the American Southwest and increasing concern about water use by *Tamarix*. Between 1975 and 2010, there were 3,595 newspaper articles in over 600 US newspapers that mentioned *Tamarix* (or its common names, tamarisk and saltcedar), and 80.5% of these also included the word "water." These articles show a similar pattern of increase starting in the mid-1980s (see figure 1.3). The publications came from 48 states, with the greatest number appearing in California (722), New Mexico (577), Colorado (378), and Texas (353). Not surprisingly, these states have also led the way, both scientifically and in land management policy, concerning *Tamarix*.

FIGURE 1.1 A stand of mature *Tamarix* growing along the Rio Grande, in New Mexico, with scientist Dr. Ken Lair.

Source: Photo courtesy of Michelle Ohrtman.

Publications in the popular press even included editorials and fiction inspired by the plant, such as the short story "The Tamarisk Hunter," featured on the cover of *High Country News* (a magazine "For people who care about the West"). As in the real news stories, it focused on water use, beginning, "A big tamarisk can suck 73,000 gallons of river water a year" (Paolo 2006). Similar statements were appearing in mainstream news outlets as well, and the coverage was universally negative. However, in just a few years, a more complex picture of tamarisk began to emerge, as reflected in the editorial, "Tammy and Me: It's Complicated" (Wolcott 2010), which included the line, "I raise a toast to my sworn enemy and newfound friend: *Tamarix ramosissima*." The question was raised with more frequency in both the scientific and popular literature about whether invasive species were a significant problem or simply victims of xenophobia, an aversion to things foreign, even as landowners and managers struggled with real–world problems caused by *Tamarix* and other exotic weeds (Sher 2006).

There are no native species in the Tamaricaceae family in North America; there is, however, a sister family, Frankeneaceae, in which there are several species native to North America (see chapter 2, "Genetics of *Tamarix*"). As a group, members of the genus *Tamarix* are estimated to be the third most common woody species along rivers in the southwestern US, and are the second-most dominant in cover (Friedman et al. 2005). *Tamarix* has recently been estimated to cover several

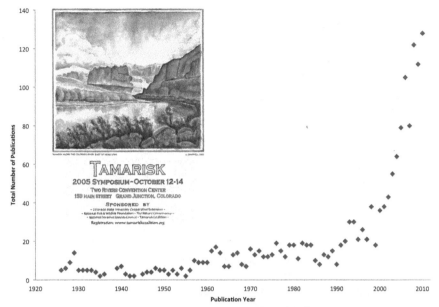

FIGURE 1.2 The number of academic publications from 1920 to 2011 that included the words "*Tamarix*," "tamarisk," "saltcedar," or "salt cedar." *Inset*: The increase of scientific interest helped fuel the first national symposia, such as the one shown on this 2005 promotional poster by the Tamarisk Coalition.

Source: Data compiled from a combined online Web of Science search that included the Science Citation Index Expanded (SCI-EXPANDED) (1899–present), the Social Sciences Citation Index (SSCI) (1900–present), and the Arts & Humanities Citation Index (A&HCI) (1975–present).

hundred thousand ha in North America, from northern Mexico to Montana and from Kansas to California (see chapter 3, "Measuring Extent and Projections of *Tamarix* Distribution in North America"). Although it has naturalized as far as the East Coast and North Dakota (see figure 1.4), *Tamarix* has only achieved ecosystem dominance in the warm, dry regions of the United States (National Resources Conservation Service [NRCS] 2010). The current geographic range may change under predicted scenarios of climate change, however, and there is evidence that the genus is evolving toward greater cold tolerance (Friedman et al. 2011).

It is believed that *Tamarix* was first introduced to North America as an ornamental in the early 1800s but may only have come to the Southwest some decades later (Robinson 1965; Chew 2009). It was planted along the sides of streams as a bank stabilizer in the late 1800s by the Army Corps of Engineers and by the early 1900s was well-established (Graf 1978). Between the 1920s and 1960s, it is estimated to have increased cover from 4,000 hectares to over 500,000 hectares (Robinson 1965). In the late 1960s it began to be considered a pest, often in the context of its water use (see chapter 16, "Tamarisk Introduction, Naturalization, and Control in the United States, 1818–1952); however, it continued to be sold and promoted for many years for windbreak, erosion control, and as an ornamental.

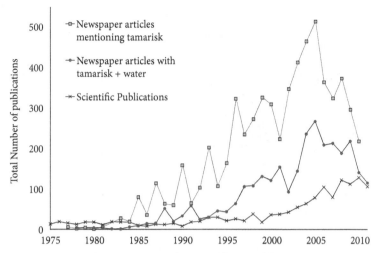

FIGURE 1.3 Total number of articles published between 1975 and 2010 that included the words "*Tamarix*," "tamarisk," "saltcedar," or "salt cedar" in newspapers as compared to scientific publications. To distinguish newspaper articles as those specifically about the plant (rather than place names), the search words included "tree," "bush," or "plant". Graph also shows those that also included the word "water". Newspaper search was of 600 US publications, using Newsbank® Access World News. Scientific publications were identified using Web of Science.

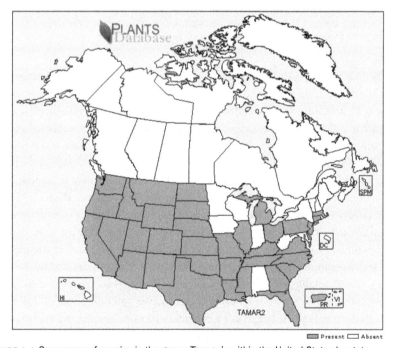

FIGURE 1.4 Occurrence of species in the genus *Tamarix* within the United States by state.

Source: Data and image from the United States Department of Agriculture Natural Resources Conservation Service (USDA-NRCS) PLANTS database.

The family Tamaricaceae is indigenous in Eurasia and North Africa (Baum 1978). Several species of *Tamarix* have dispersed to other continents and regions by anthropogenic means for shade, erosion control, and as an ornamental (see chapter 19, "Botany and Horticulture of Tamarisk"). New world countries where it has naturalized include Argentina (Gaskin and Schaal 2003), Australia (Griffin et al. 1989), Canada (Alberta Agricultural and Rural Development [AARD] 2008), Mexico (Glenn and Nagler 2005), South Africa (CARA 2001), and the United States (Robinson 1965). Most research on the ecology of invasive tamarisk has been conducted on populations in the United States, and these studies make up the scope of this book. Given the difficulty in distinguishing *Tamarix* species from each other in the field, and the apparent similarity in their niche and behavior, scientists and land managers have generally lumped this group together, although research is needed to validate this assumption (Natale 2010). Since nearly all published research on the genus in North America is on these US populations, *Tamarix* as used in this book refers to *T. ramosissima*, *T. chinensis*, and their hybrids unless otherwise specified. Further, this grouping of species is referred to in the singular, as "the species" rather than "the genus," to exclude otherwise specified members of the genus.

Biology and Ecology

Tamarix is a deciduous tree with a shrublike growth form (see figure 1.5). It grows most thickly in the high-moisture conditions of riparian fringes, but is highly drought tolerant once established and thus also occurs on upper terraces and far from perennial flows (Natale et al. 2010). It can attain heights of over 8 meters and can form very dense thickets. Near-monoculture populations of the tree can stretch for many kilometers along rivers (Hink and Ohmart 1984) (see figure 1.6). Such monospecific vegetation structure was unknown in North American riparian zones before the arrival of the species, and it is a dramatic departure from historic conditions along western rivers, which ranged from nearly empty (e.g., shifting sand bars) to mixed-species riparian forests with two and three canopy layers that were for the most part restricted to the riparian fringe (Webb et al. 2007).

The life history of *Tamarix* is paradoxical, given that it uses strategies of both the elephant and the mouse: as a long-lived tree that creates shade and other conditions inhospitable to other plants, it can be a dominant competitor (like the elephant), while at the same time producing multitudes of offspring (like the mouse) via wind-and-water-dispersed seeds that colonize disturbed areas quickly (see chapter 9, "*Tamarix* from Organism to Landscape"). It is tolerant of—and even benefits from—many forms of disturbance, including fire (see chapter 14, "Fire Ecology of *Tamarix*"). Not only this, but *Tamarix* is more stress tolerant than most riparian trees, with higher thresholds for salinity, low soil nutrients, heat,

FIGURE 1.5 Mature *Tamarix ramosissima* tree growing in Florence, Colorado.

Source: Photo courtesy of Stephanie Gieck.

FIGURE 1.6 Near-monoculture *Tamarix* population growing long the Colorado River in Utah.

Source: Photo courtesy of Anna Sher.

and drought (see chapter 5, "Tamarisk: Ecohydrology of a Successful Plant," and chapter 8, "Tamarisk and Salinity: An Overview"). Therefore it does not fit neatly into the classic model by Grime et al. (1988), which insists that there is a limited number of possible life history strategies, since environments that are both highly disturbed and stressful will prevent reestablishment of vegetation. However, the riverbank environment that *Tamarix* prefers can, at different times or locations, be both disturbed and stressful as well as the lushly stable, highly competitive environment one expects in the absence of disturbance or stress. Given the current success of *Tamarix* in riparian ecosystems in the western United States, it is no surprise that this plant is a master of all trades, representing a category of invasive weed that is rare but not unheard of (Sher and Hyatt 1999). That is, *Tamarix* can exploit increases and decreases in both chemical and physical flux, rapidly colonizing soil laid bare by flood or fire (when it acts as a ruderal and is not competitive or stress tolerant); then, as a sapling, it tolerates poor soils and drought (when it acts as a stress tolerator); finally, as a mature tree, it can competitively exclude most other species (as a "stress-tolerant competitor" sensu Grime). It can be both a weak and a strong competitor, and a weak and a strong drought tolerator, depending on age.

The relationship between *Tamarix* and water has been a contentious issue in the scientific literature (see chapter 5, "Tamarisk: Ecohydrology of a Successful Plant"; chapter 6, "Water Use by *Tamarix*"; and chapter 7, "*Tamarix*, Hydrology, and Fluvial Geomorphology"). It is a phreatophytic species, meaning that it is capable of putting down deep tap roots to exploit the water table; however, its dense and copious fine roots can draw moisture from the entire soil profile (Nippert et al. 2010). Despite wildly differing published estimates based on myriad methods of measuring water use, the current consensus appears to be that the range of water use of *Tamarix* is 0.7 to 1.4 meters per year (electrochemical time of flight [EToF] of 0.3 to 0.7, centering on a mean value of 0.5), although the extremes occur only under very specific conditions (see chapter 4, "Evapotranspiration by Tamarisk in the Colorado River Basin"). The higher amount has been observed for dense stands in mesic conditions, over a growing period of 300 days (EToF of 0.7; Devitt et al. 1998). It has been observed in many areas that presence of *Tamarix* is associated with the lowering of ground tables and drying up of streams; most likely this is in areas where other phreatophytic species (such as those of *Populus* and *Salix*) did not historically occur, or were sparse. Photographs from the early 1900s suggest that such areas are not uncommon, and that *Tamarix* often took advantage of such habitats (Webb et al. 2007).

Given the changes in the physical structure of the forest, soil chemistry, fire regimes, and hydrology associated with the species, it is not surprising that wildlife associated with *Tamarix* thickets tends to be markedly different than in stands where native *Populus* (cottonwood) and *Salix* (willow) occur (see chapter 10, "*Tamarix* as Wildlife Habitat"). For mammals, some observations suggest dramatically decreased diversity in *Tamarix* stands relative to other riparian areas (Hink and Ohmart 1984). Although goats will browse young saplings, few other vertebrate herbivores appear to eat it; beavers do, but they strongly prefer native species (Kimball and Perry 2008). Research on herpetofauna suggests that *Tamarix* control is likely to benefit, or at least not harm, native lizards (Bateman et al. 2008), and preliminary findings suggest that lizard diversity is greater in mixed stands over *Tamarix* monocultures (Bateman and Ostoja 2010).

However, whether *Tamarix* is considered good or bad for wildlife is still hotly debated, particularly with regard to bird communities (see chapter 11, "Tamarisk in Riparian Woodlands: A Bird's Eye View"). While simple species counts are sometimes similar or even greater when tamarisk is present than in purely native stands, bird species assemblages are consistently found to differ greatly between pure-native, mixed, and *Tamarix*-dominated stands (Ellis 1995; Brand et al. 2008; Walker 2008). Many species of birds can and do use tamarisk, but some studies have reported lower bird diversity in *Tamarix* stands, due in part to the complete absence of certain guilds (Hunter et al. 1988; Brand et al. 2008).

The southwestern willow flycatcher (*Empidonax traillii extimus*) is the animal species for which the most research has been done on habitat value of *Tamarix* (chapter 11). SWFL (as it is often abbreviated) is a listed species under the US Endangered Species Act, and its range overlaps areas of *Tamarix* dominance. There has been some disagreement as to whether *Tamarix* represents equivalent or inferior habitat for SWFL (Dudley and DeLoach 2004: Sogge et al. 2008: van Riper et al. 2008). Some argue that nesting habitat of SWFL appears to be most strongly related to water resources, rather than vegetation type (Hatten et al. 2010). However, it is clear that this species does use *Tamarix*, and that any restoration plan must consider how it will mitigate habitat loss (Zavaleta et al. 2001), even if that habitat is nonnative.

Invertebrate studies are few, but reflect patterns similar to those found for birds. Although a high overall density and richness of arthropods can be found in the branches and beneath tamarisk trees, community composition differs from native vegetation (see chapter 12, "*Tamarix* as Invertebrate Habitat"). This is likely due to the different structure and chemistry of leaf tissue and detritus of *Tamarix* (see chapter 9, "*Tamarix* from Organism to Landscape"). Nitrogen concentrations, carbon to nitrogen ratios (C:N), and salinity of decomposing leaves also appear to affect the soil microbial community under *Tamarix*, including greatly decreased density of beneficial mycorrhizal fungi (see chapter 13, "*Tamarix* and Soil Ecology").

Taken together, we can see that this highly adaptable plant has clearly made a place for itself in the western ecosystem, including having become the foundation of novel communities at all trophic levels.

Passenger or Driver?

Because so many of the ecological changes associated with *Tamarix* are also associated with increased human impact, another ironic aspect of the plant is whether it was the cause or consequence of these changes (Sher and Stromberg 2006). Using the passenger versus driver model first proposed by MacDougall and Turkington (2005), strong arguments have been made for *Tamarix* on both sides, but the most compelling may be the case for it being both (see chapter 15, "*Tamarix*: Passenger or Driver of Ecosystem Change?").

In the driver model, *Tamarix* is responsible for drought, elevated environmental salinity, increased fire frequency and intensity, and changed biota (see figure 1.7). It has been argued that overestimates of water use by *Tamarix*, as well as other traits of the genus, were used to vilify the species in the 1960s (see chapter 16), a clouding of the facts that would have reverberations for decades to come. Although many aspects of riparian degradation are associated with *Tamarix*, human use of water resources and reduced overbank flooding can also cause each of these effects (Stromberg et al. 2009). For example, the correlation of *Tamarix* presence with elevated soil salinity does not prove causation (chapters 5 and 8). Thus, it has been proposed that *Tamarix* is more a passenger of anthropogenic change, because it is one of the few species that has been able to adapt to the conditions created by man (see figure 1.8).

However, a third possibility exists: that the reality lies between these two; an acknowledgement that *Tamarix* certainly does take advantage of anthropogenic disturbance, but that, once established, it can be a powerful ecosystem engineer (Sher and Stromberg 2006; see also figure 1.9). Although the weight of the arrows connecting these environmental components has yet to be determined, thinking of the "*Tamarix* problem" in this way has important implications for management.

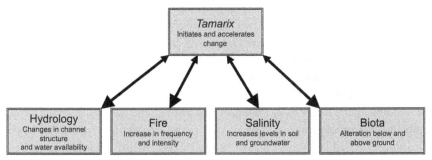

FIGURE 1.7 The driver model for understanding *Tamarix's* role in the ecosystem.

Source: Adapted from Stromberg (2007).

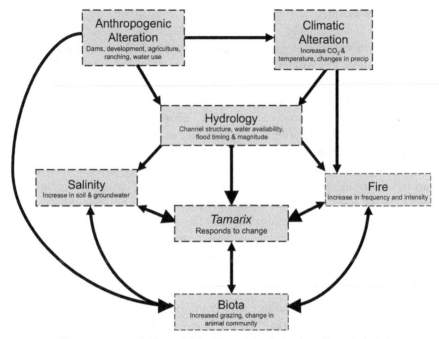

FIGURE 1.8 The passenger model for explaining *Tamarix's* relationship with ecological change, ultimately caused by human alteration of hydrology and the environment.
Source: Adapted from Stromberg (2007).

The History of Management

One can take either a bottom-up or top-down approach to the management of ecosystems where *Tamarix* occurs, focusing on hydrology (bottom up) or on the plant itself (top down). Those who argue that *Tamarix* is primarily a passenger will generally favor promoting historic hydrological regimes, where feasible. Issues such as elevated soil salinity, fire risk, and aridity will be directly mitigated by reinstatement of overbank flooding and gradual seasonal decline in the water table, and native species establishment will be more likely as well (Lovell et al. 2009). There is some evidence that where tamarisk is not overly dense, the addition of water can promote some native plant establishment (Nagler et al. 2005). However, where *Tamarix* occurs as a thick monoculture, native trees cannot establish, even after significant flooding (Taylor et al. 1999). Although soil salinity may be reduced by flooding, many native trees, including cottonwood and willow, are shade intolerant and so cannot grow in a *Tamarix* understory. In a monoculture of *Tamarix*, bird abundance and diversity are low (van Riper et al. 2008), and many ecosystem services provided by more diverse assemblages are missing. To regain them by creating an opportunity for native species to reestablish, at least some *Tamarix* must be removed.

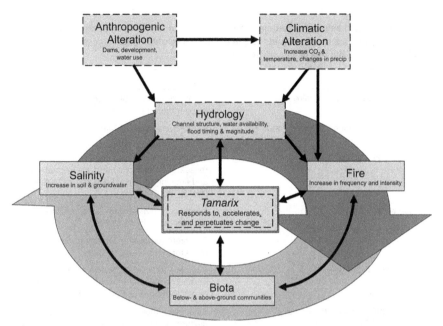

FIGURE 1.9 A combined passenger-driver model that incorporates elements of both. In this model, *Tamarix* is both the cause and the consequence of changes in salinity, hydrology, biota, and fire.

Tamarix has been targeted for removal by many states' management programs, and more so since the droughts of the late 1990s. At the time of this printing (2012), *Tamarix* species are included on the noxious weed lists of Colorado, Montana, Nebraska, Nevada, New Mexico, North Dakota, Oregon, South Dakota, Texas, Washington, and Wyoming (NRCS 2010). This means that the sale and planting of the species is prohibited or discouraged, and/or that state funding supports targeted control of the plant. In some states, such as Colorado, it is on "List B," meaning that statewide eradication isn't feasible but that efforts will be made to prevent the further spread of the plant by removing it in some places.

Financial gains associated with *Tamarix* removal have been estimated as $95–$7,000 per acre (Zavaleta 2000), figures that would help mobilize control efforts. This calculation was based on the values of lost water, increased flooding (from channel narrowing associated with sediment collection), and loss of wildlife computed for a 55–year period, with some of variance attributed to the degree to which future impacts are considered. Many issues are currently driving *Tamarix* control (see table 1.1). For example, dense stands of tamarisk along formerly open riverbanks have changed the dynamic of recreational rafting, making landing impossible and overgrowing limited camping areas. In one case, a river boatman trapped under a flipped raft drowned because his companions could not

TABLE 1.1

A list of restoration objectives that may warrant removal of *Tamarix* in locations where it is dense and/or a monoculture, and the general means by which these objectives may be reached, including the role of *Tamarix* removal

Restoration objective	Means
Create fire break/reduce wildfire risk	Complete removal of *Tamarix* without revegetation (see ch. 14)
Widen a channel that has been constricted	Complete removal of *Tamarix* without revegetation (see ch. 7)
Decrease soil salinity	Removal of *Tamarix* combined with revegetation to reduce surface evaporation and flushing to remove existing salts (see chs. 8 and 5)
Reduce water loss	Removal of *Tamarix* with revegetation by xeric species (see chs. 4 and 5)
Decrease groundwater salinity	Complete removal of *Tamarix* and other phreatophytes (see ch. 8)
Increase bird diversity	Selective removal of *Tamarix* to promote mixture of native and exotic trees (van Riper et al. 2008; but see ch. 11, this vol.)
Improve habitat for wildlife generally	Selective removal of *Tamarix* to promote mixture of native and exotic trees (see ch. 10)
Improve aesthetics	Removal of *Tamarix* wholly or to promote mixture of native and exotic trees
Create access point to water (for both humans and larger animals)	Selective removal of *Tamarix*
Improve recreational use (to include rafting, camping, fishing, and hunting)	Removal of *Tamarix* with revegetation using shade trees and versatile ground cover
Security concerns (visibility, e.g., on the border)	Removal of *Tamarix*

get through the dense tamarisk to reach him in time (M. Quigley, pers. comm.). In a survey of boaters and other river users along the Green and Colorado Rivers, 88% answered in the affirmative when asked if they would like tamarisk removed (Allred 2012).

In 2002, river rafters and other citizens alarmed by the impenetrable walls of *Tamarix* they had observed along their waterways founded the nonprofit organization the Tamarisk Coalition, a group that became a major actor in coordinating efforts and facilitating information sharing, with the goal of restoring riparian corridors. But they were not the only ones paying attention.

In 2004, the Departments of the Interior and Agriculture, along with the National Invasive Species Council, sponsored the first national conference on the species, where 14 other organizations and groups were represented. Over 300 stakeholders from federal, state, local, and tribal government offices, academic institutions, and nonprofits concerned with environmental issues met to discuss what was known, and to produce a set of "guiding principles" to foster collaborative efforts.

Concerns about the impacts of *Tamarix*, fueled by the droughts in the Southwest, led to the signing of the Salt Cedar and Russian Olive Control Demonstration Act in October 2006 (see chapter 17, "The Politics of a Tree: How a Species Became National Policy"). There was bipartisan support for the bill, and concern about *Tamarix* unified many disparate groups, including those in the ranching and agriculture sectors, nonprofit conservation organizations, such as The Nature Conservancy, and hunting groups, such as Ducks Unlimited. This legislation required the Department of Interior to assess the extent of *Tamarix* and Russian olive (*Elaeagnus angustifolia*) infestation and to carry out demonstration projects to manage it; however, federal spending on foreign wars and and the rising national debt crisis delayed funding until 2011. Although the role of manipulating hydrology in managing *Tamarix* invasion (bottom up) is fully acknowledged in this document and in the conferences happening at the time, most of the discussion focused on removal to achieve other ecosystem goals (top down).

How to kill *Tamarix* was well understood by that time, and relied on a combination of chemical and mechanical approaches (see chapter 20, "*Tamarix* Management: Lessons and Techniques"). Chemical control refers to the application of herbicides to the leaves or trunk; because this is a perennial species, effective agents are those that are systemic; they are translocated to the whole plant, including the roots, from the point of application. Because tamarisk can resprout from its root crown, mechanical removal of only the aboveground portion of the tree by chain sawing, burning, or mulching will not kill it. Thus, these control approaches must be followed by a systemic herbicide or root removal as well, often requiring heavy equipment (see figure 1.10). When these approaches to tamarisk removal have been followed by edaphic conditions that promote natives (such as well-timed overbank flows), restoration has been successful (see chapter 21, "Tamarisk Management at Bosque del Apache National Wildlife Refuge").

In the late 1990s, biological control for *Tamarix* focused on trials with species of *Diorhabda* (initially considered *D. elongata*, now reclassified as *D. carinulata*, (Tracy and Robbins 2009), a host-specific beetle native to Asia that feeds on the leaves of *Tamarix* in both its larval and adult stages (see chapter 22, "Bring on the Beetles! The History and Impact of Tamarisk Biological Control"; see also figure 1.11). Cage trials were run at 10 sites in 1999 with only mixed success because there were some problems with beetles going prematurely dormant (Lewis et al. 2003). In addition, it required repeated defoliations to kill the trees (Hudgeons et al. 2007). Nonetheless, the beetle showed promise and was deemed safe for the ecosystem, and in 2001 it was officially released in the wild. By 2010, *D. carinulata* could be found in *Tamarix* trees along the Colorado, Dolores, Green, Gunnison, San Juan, Virgin, and White Rivers and their tributaries (Tamarisk Coalition 2010).

FIGURE 1.10 A root rake mounted on a tractor being used to remove Tamarix from the soil.
Source: Photo courtesy of Scott Dressel-Martin, Denver Botanic Gardens.

Given that biological control by an insect is more difficult to control than other methods, concern began mounting that defoliation by *D. carinulata* would occur in areas where *Tamarix* had an important ecological role as nesting habitat for birds (chapter 11). This concern came to a head in March of 2009 when the Center for Biological Diversity and the Maricopa Audubon Society filed a notice of intent to sue the Animal and Plant Health Inspection Service (APHIS) and

FIGURE 1.11 Larvae of *Diorhabda* spp., the biological control introduced for *Tamarix* in the United States, on a branch of *Tamarix*.
Source: Photo courtesy of Andrew Norton.

US Fish and Wildlife Service (USFWS) for violation of the Endangered Species Act (Kenna 2009). The charge was that the release of the beetle threatened the southwestern willow flycatcher. APHIS revoked permits for interstate movement of *D. carinulata* soon after, and in the following year an agreement was reached in which APHIS no longer participated in or funded biocontrol releases, and penalties with large fines were established for transporting beetles over state lines (Dudley and Bean 2012). However, the beetle is already well established in many watersheds and its effects are being integrated into management plans for tamarisk across the west (chapter 22). Whether mortality from beetle defoliation results in improved habitat and/or other ecosystem services will depend very heavily on continued management.

Critical in any management plan of tamarisk will be the establishment of goals that go beyond killing tamarisk (see chapter 23, "Riparian Restoration in the Context of *Tamarix* Control"). It is not unusual for land managers to state their goal as "tamarisk removal"; thus, project success has often been measured in terms of acres treated, rather than improvement in biodiversity or other measure of ecosystem function (Sher et al. 2010). For instance, tamarisk removal may be an important component of reducing fire risk, creating river access, or promoting habitat for eagles.

The success of restoration after *Tamarix* removal has been highly variable, and depends in part on how success is defined (see chapter 24, "Revegetation after Tamarisk Removal"). It is clear that in many cases, removal of *Tamarix* without active revegetation does not result in replacement by native species (Harms and Hiebert 2006). Even with plantings of native vegetation by seeds, plants, or poles, the availability of water and soil chemistry plays a major role in determining restoration success (Bay and Sher 2008). It is also apparent that much of the highly variable response of native vegetation to *Tamarix* removal may be rooted in regional differences in biotic and abiotic conditions (see chapter 25, "The Future of *Tamarix*").

Conclusion

Tamarix is nothing if not a highly successful plant in its new world range. The ways in which this genus differs from native plants in growth form, chemistry, and physiology have allowed it to proliferate, changing the appearance and ecology of rivers across the American West. Whether these changes are good or bad, and the degree to which they are attributable to human-caused effects is both a matter of science and perspective (see chapter 18, "A Philosophical Framework for Assessing the Value of Tamarisk"). Regardless, there are currently many active *Tamarix* control programs throughout its range, with the primary goals of returning ecosystems to native species dominance and recovering lost ecosystem services. There is evidence that the calls for land managers and policy makers to have a larger perspective when it comes to land occupied by *Tamarix*, to focus instead on broader goals that may or may not include complete *Tamarix* removal, have been successful (chapter 25).

Tamarix was an ideal poster child to focus attention and funds on riparian restoration for the last decade. That this era may be drawing to a close is evidenced by a decline in public attention, while increases in partnerships and in scientific research continue, to the benefit of ecosystem management in the West. Thus, paradoxically, the end of exclusively *Tamarix*-focused management may in fact begin real restoration of the habitats where *Tamarix* occurs.

Literature Cited

AARD. 2008. Weed Alert: *Tamarix ramosissima*. Government of Alberta Agriculture and Rural Development. June 26, 2008. [Online]. http://www1.agric.gov.ab.ca/$department/deptdocs.nsf/all/prm12239 Accessed January 22, 2011.

Allred, E. C. 2012. Knowledge, norms and preferences for tamarisk management in the Green and Colorado River corridors of the Colorado Plateau. Master's thesis. Utah State University, Logan.

Bateman, H. L., A. Chung-MacCoubrey, and H. L. Snell. 2008. Impact of non-native plant removal on lizards in riparian habitats in the Southwestern United States. *Restoration Ecology* 16:180–190.

Bateman, H. L. and S. M. Ostoja. 2010. Saltcedar, herpetofauna, and small mammals: evaluating the impacts of non-native plant biocontrol. The Wildlife Society's 17th Annual Conference. Snowbird, UT.

Baum, R. 1978. *The Genus Tamarix*. Israel Academy of Sciences and Humanities. Jerusalem, Israel.

Bay, R. F., and A. A. Sher. 2008. Success of active revegetation after *Tamarix* removal in riparian ecosytems of the southwestern United States: A quantitiative assessment of past restoration projects. *Restoration Ecology* 16:113–128.

Brand, L. A., G. C. White, and B. R. Noon. 2008. Factors influencing species richness and community composition of breeding birds in a desert riparian corridor. *Condor* 110:199–210.

CARA (Conservation of Agricultural Resources Act). 2001. Regulations 15 and 16 (regarding problem plants). South Africa Conservation of Agricultural Resources Act, 1983 (Act No. 43).

Chew, M. K. 2009. The monstering of tamarisk: How scientists made a plant into a problem. Journal of the History of Biology 42:231–266.

Devitt, D. A., A. Sala, D. B. Smith, J. R. Cleverly, L. Shaulis, and R. Hammett. 1998. Bowen ratio estimates of evapotranspiration for *Tamarix ramosissima* stands on the Virgin River in southern Nevada. Water Resources Research 34:2407–2414.

Dudley, T. and D. Bean. 2012. Tamarisk biocontrol, endangered species risk and resolution of conflict through riparian restoration. BioControl 57:331–347.

Dudley, T. L. and C. J. DeLoach. 2004. Saltcedar (*Tamarix* spp.), endangered species, and biological weed control: Can they Mix? Weed Technology 18:1542–1551.

Ellis, L. M. 1995. Bird use of saltcedar and cottonwood vegetation in the middle Rio Grande valley of New Mexico, USA. Journal of Arid Environments 30:339–349.

Friedman, J. M., G. T. Auble, P. B. Shafroth, M. L. Scott, M. F. Merigliano, M. D. Freehling, and E. R. Griffin. 2005. Dominance of non-native riparian trees in western USA. Biological Invasions 7:747–751.

Friedman, J. M., J. E. Roelle, and B. S. Cade. 2011. Genetic and environmental influences on leaf phenology and cold hardiness of native and introduced riparian trees. International Journal of Biometeorology 55:775–787.

Gaskin, J. F., and B. A. Schaal. 2003. Molecular phylogenetic investigation of U.S. invasive *Tamarix*. Systematic Botany 28:86–95.

Glenn, E. P., and P. L. Nagler. 2005. Comparative ecophysiology of *Tamarix ramosissima* and native trees in western US riparian zones. Journal of Arid Environments 61:419–446.

Graf, W. L. 1978. Fluvial adjustments to the spread of tamarisk in the Colorado Plateau region. Geological Society of American Bulletin 89:1491–1501.

Harms, R. S., and R. D. Hiebert. 2006. Vegetation response following invasive tamarisk (*Tamarix* spp.) removal and implications for riparian restoration. Restoration Ecology 14:461–472.

Hatten, J. R., E. H. Paxton, and M. K. Sogge. 2010. Modeling the dynamic habitat and breeding population of Southwestern Willow Flycatcher. Ecological Modeling 221:1674–1686.

Hink, V. C., and R. D. Ohmart. 1984. Middle Rio Grande biological survey. U. S. Army Corps of Engineers. Albuquerque, NM.

Hudgeons, J. L., A. E. Knutson, K. M. Heinz, C. J. DeLoach, T. L. Dudley, R. R. Pattison, and J. R. Kiniry. 2007. Defoliation by introduced *Diorhabda elongata* leaf beetles (Coleoptera: Chrysomelidae) reduces carbohydrate reserves and regrowth of *Tamarix* (Tamaricaceae). Biological Control 43:213–221.

Hunter, W. C., R. D. Ohmart, and B. W. Anderson. 1988. Use of exotic saltcedar (*Tamarix chinensis*) by birds in arid riparian systems. *Condor* 90:113–123.

Kenna, M. 2009. Center for Biological Diversity and Maricopa Audubon Society v. Animal and Plant Health Inspection Service and U.S. Fish and Wildlife Service. Court for the District of Arizona. Case 4:09-cv-00172-FRZ.

Kimball, B., and K. Perry. 2008. Manipulating Beaver (*Castor canadensis*) Feeding Responses to Invasive Tamarisk (*Tamarix* spp.). Journal of Chemical Ecology 34:1050–1056.

Lewis, P. A., C. J. DeLoach, A. E. Knutson, J. L. Tracy, and T. O. Robbins. 2003. Biology of *Diorhabda elongata deserticola* (Coleoptera: Chrysomelidae), an Asian leaf beetle for biological control of saltcedars (*Tamarix* spp.) in the United States. Biological Control 27:101–116.

Lovell, J. T., J. Gibson, and M. S. Heschel. 2009. Disturbance regime mediates riparian forest dynamics and physiological performance, Arkansas River, CO. American Midland Naturalist 162:289–304.

Nagler, P. L., O. Hinojosa-Huerta, E. P. Glenn, J. Garcia-Hernandez, R. Romo, C. Curtis, A. R. Huete, and S. G. Nelson. 2005. Regeneration of native trees in the presence of invasive saltcedar in the Colorado River delta, Mexico. Conservation Biology 19:1842–1852.

Natale, E., S. M. Zalba, A. Oggero, and H. Reinoso. 2010. Establishment of *Tamarix ramosissima* under different conditions of salinity and water availability: Implications for its management as an invasive species. Journal of Arid Environments 74:1399–1407.

Nippert, J. B., J. J. Butler Jr., G. J. Kluitenberg, D. O. Whittemore, D. Arnold, S. E. Spal, and J. K. Ward. 2010. Patterns of *Tamarix* water use during a record drought. Oecologia 162:283–292.

NRCS, U. 2010. PLANTS Profile: *Tamarix ramosissima* Ledeb. saltcedar. PLANTS Database. http://plants.usda.gov/java/nameSearch?keywordquery=Tamarix+ramosissima+&mode=sciname&submit.x=0&submit.y=0.

Paolo, B. 2006. The Tamarisk Hunter. *High Country News*, June 26. Paonia, Colorado.

Robinson, T. W. 1965. Introduction, spread and areal extent of saltcedar (*Tamarix*) in the western states. U.S. Geologic Service. Professional Paper 491-A. Washington DC.

Sher, A. A. 2006. The Invasion of Our Rockies: Hype or Management Priority? 2006 State of the Rockies Report Card, 57–63, Colorado College, Colorado Springs, CO.

Sher, A. A., and J. Stromberg. 2006. ESA Special Session: Passengers versus drivers of ecosystem change: Current debate on *Tamarix* and riparian invasion. Ecological Society of America 91st Annual Meeting, Memphis, TN.

Sogge, M. K., S. J. Sferra, and E. H. Paxton. 2008. *Tamarix* as habitat for birds: Implications for riparian restoration in the southwestern United States. Restoration Ecology 16:146–154.

Stromberg, J., M. K. Chew, P. L. Nagler, and E. P. Glenn. 2009. Changing perceptions of change: The role of scientists in *Tamarix* and river management. Restoration Ecology 17:177–186.

Taylor, J. P., D. B. Wester, and L. M. Smith. 1999. Soil disturbance, flood management, and riparian woody plant establishment in the Rio Grande floodplain. Wetlands 19:372–338.

Tracy, J. L., and T. O. Robbins. 2009. Taxonomic revision and biogeography of the Tamarix-feeding *Diorhabda elongata* (Brullé, 1832) species group (Coleoptera: Chrysomelidae: Galerucinae: Galerucini) and analysis of their potential in biological control of *Tamarisk*. Zootaxa 2101:1–152.

van Riper, C., K. L. Paxton, C. O'Brien, P. B. Shafroth, and L. J. McGrath. 2008. Rethinking Avian Response to *Tamarix* on the lower Colorado River: A Threshold Hypothesis. Restoration Ecology 16:155–167.

Walker, H. A. 2008. Floristics and Physiognomy determine migrant landbird response to tamarisk (*Tamarix ramosissima*) invasion in riparian areas. Auk 125:520–531.

Webb, R. H., S. A. Leake, and R. M. Turner. 2007. *The Ribbon of Green: Change in Riparian Vegetation in the Southwestern United States*. Arizona University Press. Tuscon.

Wolcott, M. 2010. Tammy and me: It's complicated. Page 27, *Inside/Outside Southwest*, Durango, CO.

Zavaleta, E. 2000. Valuing ecosystem services lost to *Tamarix* invasion in the United States. Pages 462–467 *in* H. Mooney and R. J. Hobbs, editors. Invasive Species in a Changing World. Island Press, Washington, DC.

Zavaleta, E. S., R. J. Hobbs, and H. A. Mooney. 2001. Viewing invasive species removal in a whole-ecosystem context. Trends in Ecology & Evolution 16:454–459.

PART I

Biology and Range

2

Genetics of *Tamarix*

John F. Gaskin

Genetic studies can be used to help us understand more about invasion identities and mechanisms, and to provide information that helps in the management of invasive species. Molecular-based approaches such as DNA sequencing or DNA fragment analyses can clarify taxonomic and evolutionary relationships, uncover evidence of hybridization events and distinguish closely related species that cannot easily be differentiated morphologically (cryptic species), elucidate methods of reproduction, and determine population structure and origins of introduced plant species (Gaskin et al. 2011). Molecular methods have their limitations; they are most useful when good taxonomic, morphological, ecological, historical, and demographic information is available to complement them and provide the context for understanding results. Molecular methods can also be costly and time-consuming, and at times, gaps in understanding plant speciation can be addressed without resorting to them. But DNA– based molecular markers have the advantage that they are not normally influenced by current environmental stimuli, unlike most morphological and other phenotypic data. In addition, the variety of molecular methods is continually expanding so that some questions can now be more easily answered, and improvements in protocols and equipment are making these tools cheaper and more accessible to those who do not specialize in their use.

Tamarix, due to its notoriety, has been subject of several genetic studies in the last decade. These studies have been done to determine which species are in the United States, how they are interacting with each other (hybridization), and why they have been such successful invaders. The use of genetic studies has been required partially because *Tamarix* species are not always easy to identify. Since the genus was first named by Linnaeus in 1753, the identification of species has necessarily been based primarily on morphology (the shape and arrangement of plant parts) and distribution. Baum (1978) produced a treatment of the genus in which he gave detailed information for 54 species found worldwide. The genus *Tamarix*, and its family Tamaricaceae, are not native to the new world; all species in the Americas have been introduced, usually intentionally (Horton 1964). For North America, McClintock (1951) performed the first study of the identity of six species found in California, which was followed by a detailed description of eight species naturalized in the United States and Canada by Baum (1967). The treatment of Tamaricaceae in *Flora of North America North of Mexico* (Gaskin in press) recognizes eight species that are able to spread and persist without help from humans (i.e., are naturalized): *Tamarix africana* and *T. canariensis*, found commonly on the Gulf Coast and less commonly on southern coastal areas of the Atlantic; *T. gallica*, found in the same areas as *T. africana* and *T. canariensis* but also inland in southwestern states; *T. aphylla*, found commonly in horticultural settings in the southwestern United States, and rarely as an invasive; *T. chinensis* and *T. ramosissima*, found commonly and forming the large invasion in western states; *T. parviflora*, not as common but also invasive in the Southwest and Pacific Northwest; and *T. tetragyna*, a rare naturalized species in Georgia.

Identification of *Tamarix* in North America is relatively simple for some species such as *T. aphylla*, with its distinct overlapping leaves, and *T. parviflora* with its four petals, sepals, and anthers compared to the typical five (see figure 2.1). However, even a character such as petal number can vary within an individual plant in this genus. The remaining common North American species are more difficult to identify. For example, Baum (1978) contends that *T. ramosissima* and *T. chinensis* are easily distinguished by the type of margin on the sepals and how the anther filaments are inserted into the nectar disc, which is found just above the petals (figure

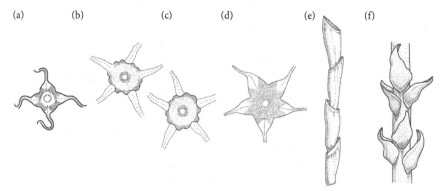

(a)   (b)   (c)   (d)   (e)   (f)

FIGURE 2.1 Flower nectary disc and filament insertion morphology of (a) *Tamarix parviflora*, (b) *T. chinensis*, (c) *T. ramosissima*, and (d) *T. canariensis* and *T. gallica*. Leaf morphology and attachment of (e) *T. aphylla* and (f) *T. ramosissima*.

Source: Figure by S. Parsons, adapted from Gaskin and Schaal (2003). Printed with permission of the American Society of Plant Taxonomists.

2.1b vs. c). Crins (1989) noted that *T. chinensis* and *T. ramosissima* are not so easily distinguishable and also questioned Baum's (1978) placement of the two species in different sections of the genus, believing them to be much more closely related. The taxonomic confusion continued with some botanists understandably suggesting that *T. ramosissima* and *T. chinensis* are the same species (Welsh et al. 1993; Allred 2002). This confusion is obviously problematic for those attempting to understand the ecology of a species. For example, there are populations in New Mexico (at the Bosque del Apache National Wildlife Refuge) which were researched and published under *T. chinensis*, but later determined to be *T. ramosissima* (Sher et al. 2002). In Asia, *T. chinensis* and *T. ramosissima* are considered distinct species, originating from different areas, and with putative differences in characteristics, such as salt tolerance (Baum 1967). Any misidentification can be especially important when the species in question are invasive and the targets of a biological control program. Biological control for *Tamarix* is the use of its natural enemies from Eurasia to control the numbers and spread of invasive species of this genus. These control programs rely on correct taxonomic identity to determine origins of the invasive organism, which is where potential natural enemies (biological control agents) that evolved with *Tamarix* can be found. Early *Tamarix* biological control efforts did not have accurate identification of the invasive *Tamarix* species.

Even *Tamarix* taxonomic experts may argue over the identification of a plant specimen, especially when characteristics used to distinguish species are inconsistent or change because of the environment the plant was grown in. In such cases, DNA analysis can provide an objective species identification, since molecular markers can potentially distinguish species and individuals that are morphologically similar. To help clarify which *Tamarix* species are found in the United States, Gaskin and Schaal (2002) sequenced DNA of plants from United States and Asia. They found that not only were the major invasive species, *T. ramosissima* and *T. chinensis*, distinguishable using DNA, but that these two species often formed hybrids in the United States. Hybridization is clearly a source of taxonomic confusion, with many plants having intermediate morphologies that are not easy to identify with a botanical key.

Hybridization events are thought to increase invasiveness in some cases (Ellstrand and Schierenbeck 2000) by providing a relatively rapid mechanism for increasing genetic diversity, which enables species to adapt and evolve in new and changing environments. Hybrids can also contain novel gene combinations, further material for adapting to the new range (Anderson 1949). Any increase in genetic diversity can be especially important for successful invasion when an introduced species has lost genetic variation due to a founder effect, where only a small portion of the species' genetic diversity from its native range makes it to the new invasive range. Founding effect is a common pre-invasion occurrence because a limited number of individuals are the ones to colonize a new range.

Tamarix ramosissima and *T. chinensis* appear to have distinct ranges in Asia, thus the hybrids found in the United States are novel, and may have never been encountered by potential biological control agents. Will these beetles attack the invasive hybrids as much as the parent species? An ecological question of concern is, are hybrids better invaders than their parents? These are important questions still being researched, and studies on genetic diversity of the invasive *Tamarix* in their native range have only recently been performed (e.g., *T. chinensis*; Jiang et al. 2011). Also notable is that, according to DNA analyses, the two species appear to be close relatives (Gaskin 2003), and thus not from different sections of the genus as proposed by Baum (1978). This result would suggest that a biological control agent would more likely attack both *Tamarix* species than if the species were distantly related (Wapshere 1974).

The number of hybrid plants in the *Tamarix* United States invasion was estimated at 23% by Gaskin and Schaal (2002). However, this estimate was based on a single fragment of diploid DNA (i.e., with two copies in each plant, one coming from each parent), which is only a small portion of the plant's genome. When examining only one fragment of DNA it may not be possible to distinguish a pure parental type from a plant that is the product of many generations of crossing with hybrids or backcrossing with parental species (introgression) since the initial hybridization event. This is because after many generations of hybridization, a plant could have randomly ended up with its two copies of the fragment of DNA being from one parental species. Thus, it would falsely appear to be one of the parent species even though most of its genome is a combination of both parental species DNA. To determine more about the hybrid status of a plant, it is better to look at many fragments of DNA from an individual and compare that information to known, pure, parental species. To do this, Gaskin and Kazmer (2009) used amplified fragment length polymorphisms ([AFLPs]; similar to a DNA fingerprint), which is another type of genetic analysis that represents many more fragments of DNA from an individual (in this case, 148 fragments). From AFLP analysis they determined that approximately 85% of the invasion in the United States was made up of hybrids of *T. ramosissima* and *T. chinensis*, and these hybrids produce ample amounts of viable seed. Clearly, while both *T. ramosissima* and *T. chinensis* are capable of being invasive, their hybrid progeny have been even more successful. In addition, Gaskin and Kazmer (2009) found a strong correlation between latitude and species, with northern plants being similar to *T. ramosissima* from Asia, southern plants being similar to *T. chinensis* from Asia, and mid-latitude plants consisting of hybrid combinations of these two species (see figure 2.2). Also, recent studies of *Tamarix* from the Green River in Utah showed that even the original plants that established in the 1930s were hybrids, and the level of hybridization has been stable in that region for more than 70 years (Gaskin et al. 2011).

In 2003, researchers found *Tamarix* with strange morphologies growing on the shore of Lake Mead (Barnes 2003). The leaves were not similar to any of the species known in the United States; they were neither vaginate, as in *T. aphylla*, nor sessile, as in *T. ramosissima* and *T. chinensis* (figure 2.1e vs. f). At the same time, similar morphologies were noted in southwest Arizona (P. Shafroth, unpublished

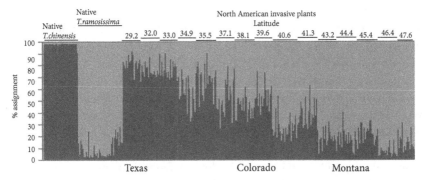

FIGURE 2.2 Percentage assignment scores of North American invasive *Tamarix* plants as compared to their native parental species *T. chinensis* and *T. ramosissima* (*left side* of graph). Note how percentage assignment to *T. chinensis* (*dark grey*) drops, and percentage assignment to *T. ramosissima* (*light grey*) increases with an increase in latitude.
Source: Many US samples provided by J. Friedman, US Geological Survey.

data). Both *T. aphylla* and *T. ramosissima* were also noted growing in each area. Gaskin and Shafroth (2005), using both nuclear and chloroplast DNA markers, determined that the plants were hybrids of *T. aphylla* and either *T. ramosissima* or *T. chinensis*. Though numbers of these hybrids were low, it is of interest that hybridization was not at first suspected since *T. aphylla* is a very distant relative of the other two species involved (Baum 1978); this genus is reported to have 2n = 24 chromosomes (Baum 1978), making hybridization more feasible due to chromosome compatibility, even among more distantly related species of the genus. It is also suspected, but not well confirmed, that there exist hybrid combinations between *T. gallica* and *T. canariensis*, and with each of these and *T. ramosissima* and *T. chinensis* (Gaskin and Schaal 2003; see figure 2.3).

In addition to concerns over species identification and biological control, there was a question of how the *Tamarix* invasion spread to new areas. Was it from natural travel of seed via wind or animals? Or did clonal propagation result from

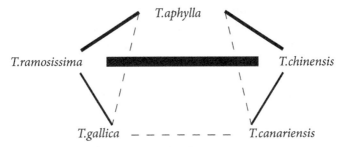

FIGURE 2.3 Diagram of hybridization potential between invasive *Tamarix* spp. in North America. Thicker lines between species indicate stronger evidence for hybridization and more common occurrence in North America. Dashed lines indicate a lack of good evidence of hybridization.

the movement of pieces of plants, which are capable of rooting and growing, traveling along waterways? Or was spread due to intentional movement of plants for ornamental use, with subsequent seed spreading from gardens to natural areas? To test this, Gaskin and Kazmer (2006) compared nuclear and chloroplast DNA sequences of ornamental and nearby naturalized plants from Montana, North Dakota, and Wyoming. They found that ornamental and naturalized genotypes were mostly dissimilar, and that the majority of naturalized genotypes originated from other naturalized plants, not from nearby ornamental plants. However, ornamental plants could not be excluded as contributors to invasive populations because all chloroplast and nuclear haplotypes found in the ornamental plants were found at low frequencies in the wild. These findings suggest that while ornamental tamarisk plants are not the sole source of the tamarisk invasion, they do have potential to contribute genetic material to an invasion or reestablish populations after existing invasive individuals are removed.

Genetic studies can also be in the form of comparison of growth and reproductive performance of individuals from different origins, to provide insight into how plants are successful as invaders. Any differences in individuals when grown in a common setting (same soil, temperature, light, water, etc.), such as seed output or plant size, may be attributed to differences in heritable genetic material, or maternal effects (e.g., enhanced seedling growth due to larger seed size). Sexton et al. (2002) performed such a study, comparing *T. ramosissima* from the southern and northern latitudinal extremes of the US invasion (Arizona and Montana). Plants were grown from seed in climate-controlled growth chambers. They found genetic differences in plants from the two latitudes; plants from the north allocated more growth to root biomass when grown at colder temperatures (perhaps an adaptation to surviving the first winter as a young plant). Genetic differences between northern and southern plants may explain how the invasion is able to succeed in such a range of environments. The plants also showed high levels of phenotypic plasticity (i.e., the tolerance of a wide variation of environmental factors by a single genotype) for gas exchange and morphological traits. Phenotypic plasticity is claimed to be a helpful trait for plants when they first arrive in a new environment, before they have had time to adapt through natural selection (Richards et al. 2006; but see Hulme 2008).

A latitudinal study, performed by Friedman et al. (2008), examined cold hardiness in both *Tamarix* and cottonwoods (*Populus deltoides* subsp. *molinifera*) from latitudes spanning from Texas to Montana. The temperatures needed to kill stems of *Tamarix* plants from Montana were 5 to 21 degrees centigrade lower than those needed for Texas plants. Cottonwood, a native to North America, also showed similar latitudinal variation in withstanding freezing temperatures. The research used the *Tamarix* SSR molecular markers of Gaskin et al. (2006) to compare the genetic makeup of the plants collected and found that northern plants were genetically similar to *T. ramosissima*, and southern plants more similar to *T. chinensis*. Mid-latitude plants again corresponded to hybrid combinations of the two species, as in Gaskin and Kazmer (2006). The authors suggest that quick adaptation of *Tamarix* to the climate extremes in North America could be due either to multiple introductions from Asia of plants adapted to different latitudes or to adaptation after introduction facilitated by hybridization and the subsequent increased genetic variability necessary for rapid evolution. In either case, it seems as if human activity in introducing many plants or allowing the hybridization of previously isolated species has contributed to creating one of the most successful invaders in the United States.

In conclusion, these genetic studies have helped us gain basic knowledge of the *Tamarix* invasion. We now have a better understanding of the species involved in the invasion, their evolutionary relationships, and the contribution of hybridization to the invasion. This information can be used to enhance the efficacy of biological control, which relies on close pairwise coevolutionary relationships of the target weed and the specialist biological control agent. We also have a better understanding of the relationship between horticultural and invasive *Tamarix*, which can help advise noxious weed policies. Genetic studies on invasive *Tamarix* tolerance of different latitudes in the United States helps us understand how the invasion became successful, how far it might spread, and how quickly it might adapt to a changing climate. Without this basic knowledge, invasion management programs, especially biological control, may suffer lower efficacy when trying to reduce the ecological and economic effects of this infamous invasive plant.

Literature Cited

Allred, K. 2002. Identification and taxonomy of *Tamarix* (Tamaricaceae) in New Mexico. Desert Plants 18:26–32.

Anderson, E. C. 1949. *Introgressive Hybridization*. John Wiley, New York, NY.

Barnes, P. 2003. Reproductive and population characteristics of *Tamarix aphylla* at Lake Mead National Recreation Area, Nevada. Master's thesis. University of Nevada. Las Vegas.

Baum, B. R. 1978. *The Genus Tamarix*. Israel Academy of Sciences and Humanities. Jerusalem, Israel.

Baum, B. R. 1967. Introduced and naturalized tamarisks in the United States and Canada. Baileya 15:19–25.

Crins, W. J. 1989. The Tamaricaceae of the southeastern United States. Journal of the Arnold Arboretum 70:403–425.

Ellstrand, N. C, and K. A. Schierenbeck. 2000. Hybridization as a stimulus for the evolution of invasiveness in plants? Proceedings of the National Academy of Science 97:7043–7050.

Friedman, J. M., J. E. Roelle, J. F. Gaskin, A. E. Pepper, and J. R. Manhart. 2008. Latitudinal variation in cold hardiness in introduced *Tamarix* and native *Populus*. Evolutionary Applications 1:598–607.

Gaskin, J. F. 2003. Molecular systematics and the control of invasive plants: A case study of *Tamarix* (Tamaricaceae). Annals of the Missouri Botanical Garden 90:109–118.

Gaskin, J.F. In press. Tamaricaceae. In *Flora of North America North of Mexico*. Flora of North America Editorial Committee, eds. 1993+. Flora of North America North of Mexico. 12+ vols. Oxford University Press, New York, NY, and Oxford, UK, Vol. 6.

Gaskin, J. F., A. E. Pepper, J. R. Manhart. 2006. Isolation and characterization of 10 polymorphic microsatellites in saltcedars (*Tamarix chinensis* and *T. ramosissima*). Molecular Ecology Notes 6:1147–1149.

Gaskin, J. F., A. S. Birken, D. J. Cooper. 2012. Levels of novel hybridization in the saltcedar invasion compared over seven decades. Biological Invasions 14:693–699.

Gaskin, J. F., and B. A. Schaal. 2002. Hybrid *Tamarix* widespread in U.S. invasion and undetected in native Asian range. Proceedings of the National Academy of Sciences of the United States of America 99:11256–11259.

Gaskin, J. F., and B. A. Schaal. 2003. Molecular phylogenetic investigation of us invasive *Tamarix*. Systematic Botany 28:86–95.

Gaskin, J. F., and D. J. Kazmer. 2006 Comparison of ornamental and wild saltcedar (*Tamarix* spp.) along eastern Montana, USA, riverways using chloroplast and nuclear DNA sequence markers. Wetlands 26:939–950.

Gaskin, J., and D. Kazmer. 2009. Introgression between invasive saltcedars (*Tamarix chinensis* and *T. ramosissima*) in the USA. Biological Invasions 11:1121–1130.

Gaskin, J. F., M. C. Bon, M. J. W. Cock, M. Cristofaro, A. D. Biase, R. De Clerck-Floate, C. A. Ellison, H. L. Hinz, R. A. Hufbauer, M. H. Julien, and R. Sforza. 2011. Applying molecular-based approaches to classical biological control of weeds. Biological Control 58:1–21.

Gaskin, J. F., and P. B. Shafroth. 2005. Hybridization of *Tamarix ramosissima* and *T. chinensis* (saltcedars) with *T. aphylla* (athel) (family Tamaricaceae) in the southwestern USA determined from DNA sequence data. Madroño 52:1–10.

Horton, J. S. 1964. Notes on the introduction of deciduous *Tamarix*. U.S. Forest Service Research Note RM-16. Ft. Collins, CO, p. 7.

Hulme, P. E. 2008. Phenotypic plasticity and plant invasions: Is it all jack? Functional Ecology 22:3–7.

Jiang Z., Y. Chen, and Y. Bao. 2012. Population genetic structure of *Tamarix chinensis* in the Yellow River Delta, China. Plant Systematics and Evolution 298:147–153.

McClintock, E. 1951. Studies in California ornamental plants. 3. The tamarisks. Journal of California Horticultural Society 12:76–83.

Richards, C. L., O. Bossdorf, N. Z. Muth, J. Gurevitch, and M. Pigliucci. 2006. Jack of all trades, master of some? On the role of phenotypic plasticity in plant invasions. Ecology Letters 9:981–993.

Sexton, J. P., J. K. McKay, and A. Sala. 2002. Plasticity and genetic diversity may allow saltcedar to invade cold climates in North America. Ecological Applications 12:1652–1660.

Sher, A. A., D. L. Marshall, and T. P. Taylor. 2002. Establishment patterns of native *Populus* and *Salix* in the presence of invasive nonnative *Tamarix*. Ecological Applications 12:760–772.

Wapshere, A. J. 1974. A strategy for evaluating the safety of organisms for biological weed control. Annals of Applied Biology 77:201–211.

Welsh, S. L., N. D. Atwood, S. Goodrich, and L. C. Higgins. 1993. *A Utah Flora*. Brigham Young University. Provo, UT.

3

Measuring Extent and Projections of Tamarix Distribution in North America

Catherine S. Jarnevich, Paul Evangelista, and Jim Graham

Tamarisk (*Tamarix* spp.) has successfully invaded riparian ecosystems throughout the western United States (Jarnevich et al. 2011). Occurring from Argentina to Canada, it continues to invade new environments that lie well beyond the latitudinal confines in its native range. It was once believed to have reached its geographic limits within the United States; however, new invasions continue to outpace treatment and control. The absence of spatially explicit inventories has impeded management strategies (e.g., detection, assessment, mitigation) and limits the ability to forecast potential spread and risk of new invasions. Furthermore, tamarisk's ecological impacts on native systems cannot fully be realized without a quantitative assessment of its occurrence and distribution. In this chapter, we review some of the methods that have been used to map tamarisk infestations and predict distributions, using field surveys, online databases, remote sensing, and spatial models. The methods presented have had varying degrees of success, and some may have better applications under specific site conditions or at specific geographic locations.[1]

Field surveys, where persons familiar with a species physically visit locations to search for that species, are still the most common means of mapping tamarisk distributions for land managers (Tamarisk Coalition 2008). For early detection of new invasions, small geographic areas or species-poor environments, this

---

[1] Any use of trade, product, or firm names in this chapter is for descriptive purposes only and does not imply endorsement by the US Government.

may still be a practical approach for inventories. However, this is rarely the case. The need for tamarisk inventories spans from watersheds to states, across multiple land ownerships and jurisdictions (e.g., private, state, federal). Under these circumstances, field surveys are often too costly and time-intensive to justify land managers' investing limited resources in them. Not only are they inefficient and sometimes incomplete, but the data are often obsolete within a few years due to tamarisk's rapid growth and spread. This is not to imply that field survey data do not have value or are low priority, only that costs may exceed allocated budgets or outweigh the benefits to land management agencies.

There is, in fact, a shortage of tamarisk distribution data, and updated field surveys provide a critical foundation for some of the new methods and techniques being used to predict tamarisk distribution. Data are often readily available at a county or state scale (e.g., the Plants database, www.plants.usda.gov), but this is generally too coarse to inform effective management. Other regional and national online data repositories are becoming increasingly popular for land managers, researchers and the general public. For example, T-map (www.tamariskmap.org) was specifically designed for data compilation and dissemination of tamarisk occurrences throughout the United States These data can be collected and shared by anyone, professionals or citizen scientists, to create an integrated data set of tamarisk locations at local, regional, or national scales (see figure 3.1).

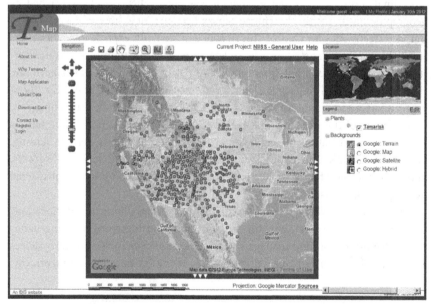

FIGURE 3.1 Tamarisk locations compiled via the Tamarix Cooperative Mapping Initiative (T-Map) website (http://www.tamariskmap.org).

The types of tamarisk data collected from the field can range from a location coordinate taken from a global positioning system (GPS) to a mapped polygon of a stand with detailed site descriptions. The more information collected from a field survey, the greater the utility of the data. A location coordinate may simply reveal that tamarisk exists at a certain location, whereas more detailed surveys may include information on stand size, density, percent cover, and age-class (Barnett et al. 2007). For larger stands, some land managers rely on aerial photography to inventory tamarisk. These may be from actual photographs or from aerial images in a digital format. In either case, polygons are manually drawn or digitized around identified stands, such as tamarisk mapping by the Tamarisk Coalition near Grand Junction, Colorado, utilizing high resolution aerial and satellite photos (see figure 3.2). Everitt and Deloach (1990) used aerial photographs to map tamarisk in Texas and Arizona. They found that in late fall and early winter months, tamarisk foliage turned a yellow-orange color that was distinguishable from other vegetation. This may be a practical approach if tamarisk phenology is predictable and synchronous, and the associated vegetation is limited to a few species; however, the timing of foliage change of tamarisk and other present species is not synchronous at all latitudes. Still, the increased availability of aerial photography, such as products offered by National Agriculture Imagery Program ([NAIP]; http://www.fsa.usda.gov/FSA/), warrants further consideration and testing.

Another type of data that can be useful in determining species distributions are interviews or surveys with resource managers and other local experts (Marvin et al. 2009) responsible for the management and control of invasive species within a local area. For example, the Western Weed Coordinating Committee conducted a survey of county weed coordinators, asking them to estimate acreage of specific weed species for each quarter quadrangle in their county (Thoene 2002). However, although surveys of this type provide finer resolution data than county level data, they are based on estimates and are geographically incomplete (without a 100% response rate) and inconsistent because the way people estimate acreage varies (Nagler et al. 2011).

We can, however, use this information to try to estimate the current abundance of tamarisk. In 2004, the Western Weed Coordinating Committee asked county weed coordinators from 17 western states to provide infested acreage estimates of tamarisk for quarter quadrangles within their county (see figure 3.3). Estimated acreage of tamarisk across these 17 western states totals 880,291 acres. However, not all counties responded to the survey, resulting in estimates for 98,394 quarter quadrangles of the 112,727 in the study area. Many nonresponse counties bordered those with high acreage estimates. We also have field data for quarter quadrangles where weed coordinators estimated zero acres, indicating presence locations missed by the survey. Also, this sum of acreage across quarter quadrangles is lower than that of Robinson (1965), who estimated 900,000 acres. Thus, we can assume this number is an underestimate and that there are still at least this many acres of tamarisk.

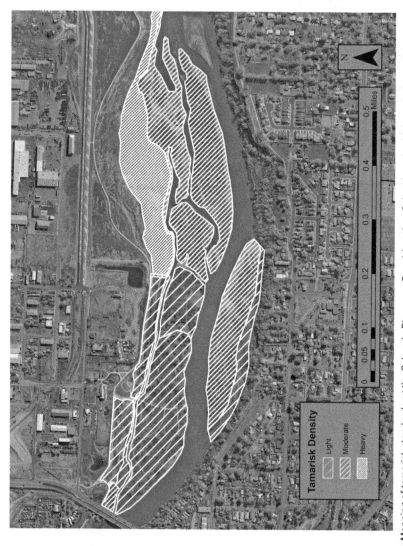

FIGURE 3.2 Mapping of tamarisk stands along the Colorado River near Grand Junction, Colorado, using high-resolution aerial and satellite photos.

Source: Tamarisk Coalition.

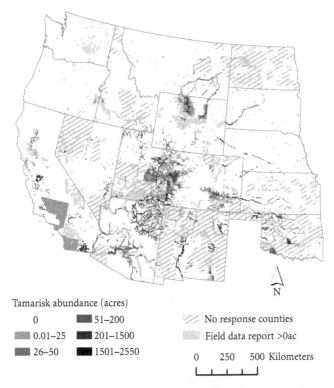

Tamarisk abundance (acres)

0	▬ 51–200	▨ No response counties
▬ 0.01–25	▬ 201–1500	▨ Field data report >0ac
▬ 26–50	▬ 1501–2550	0 250 500 Kilometers

N

FIGURE 3.3 Quarter-quadrangle acreage estimates for tamarisk based on a 2004 survey of counties in the western United States. Counties that did not respond to the survey are highlighted along with counties with a zero acreage response where we have field data indicating tamarisk presence. Acreage estimates produced by the Western Weed Coordinating Committee with funding from the Center for Invasive Plant Management.

Source: Adapted from Nagler et al. (2011).

Detecting and Mapping Tamarisk Using Remote Sensing

Remote sensing has played an important, but limited, role in mapping and detecting invasive plants (Huang and Asner 2009; He et al. 2011). Aerial photography is the simplest form of remote sensing. It was used to map tamarisk infestations as early as 1947 (Robinson 1965) and probably remains the most commonly used method today (see Tamarisk Coalition 2008). Generally, aerial photography is acquired in natural color, grey-scale, or color-infrared at high resolutions. Depending on the floral heterogeneity, time of year, and size of the infestation, this may be sufficient for mapping tamarisk (Everitt and Deloach 1990; Everitt et al. 1996). Detection of tamarisk and other species from aerial photography can be enhanced with new software and methods. For example, Ge et al. (2006) analyzed color aerial photographs at 1-m² resolution using a texture analysis for tamarisk in Northern California. They found that color (grey tones) alone could not distinguish tamarisk from associated vegetation; however, the use of textural classifiers greatly improved the ability to classify cover types.

Today, remote sensing applications include a suite of high resolution (i.e., spatial, spectral) sensors and instrumentation ranging from those that can be flown on order by small aircraft to satellites that orbit the earth regularly. Despite new technologies, detecting a specific plant species in forest, range, or riparian landscapes can be extremely challenging. Most approaches attempt to discriminate a species' spectral reflectance from other vegetation and environmental features. Several small-scale field studies using portable spectroradiometers, devices that can record the wave lengths being reflected off objects including plant species, have demonstrated that tamarisk can be spectrally distinguished from other plants (Griffith et al. 2005; Tree and Slusser 2005). Griffith et al. (2005) was also able to show that phenological attributes of tamarisk and color differences among its flowers had different spectra (see figure 3.4). However, remote sensing of tamarisk has had limited success, and conventional processing methods (e.g., supervised and unsupervised classification) have not proved to be reliable (Carter et al. 2009). Akasheh et al. (2008) used an iterative classification procedure to map riparian species with high-resolution multispectral airborne sensors in July on the Rio Grande, New Mexico. Although they were able to achieve 86% classification accuracy, it should be noted that only four cover types were included in the study (i.e., tamarisk, cottonwood, willow, and Russian olive). In southern California, Hamada et al. (2007) used discriminant analyses of presence/absence data and hierarchical clustering with hyperspectral imagery collected in October. Overall accuracy of their research varied by scene and minimum patch size, and results tended to over overestimate tamarisk distribution.

Time-series analyses of remotely sensed data are increasingly being used for detecting broad-scale invasions (Pavri and Aber 2004; Robinson et al. 2008) and monitoring the impacts of mitigation treatments (Anderson et al. 2005; Everett et al. 2007). The distinctiveness of any phenological attribute, including temporal characteristics, can vary widely with regional climate, latitudinal gradients, and species

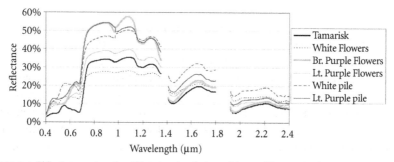

FIGURE 3.4 Reflectance values for different parts of a tamarisk tree. "Tamarisk" refers to a branch from a tree; "flowers" refers to a branch with flowers of the specified color; and "pile" refers to a pile of flowers of the specified color. These different phenological stages for tamarisk have different values.
SOURCE: FROM GRIFFITH ET AL. (2005)

richness within an ecosystem. As a result, the timing of acquiring remotely sensed data is critical and may be difficult to predict. Evangelista et al. (2009) tested six Landsat satellite scenes and derived vegetation indices from different months of the growing season in the lower Arkansas River in southeastern Colorado. Scenes were acquired for April, May, June, August, September, and October. The researchers used a habitat modeling technique to relate remote sensing derived products to tamarisk presence. The specific technique was Maxent (v3.2.1; Phillips et al. 2006), and the remote sensing imagery included bands 1 through 5 and 7, which were analyzed with normalized difference vegetation index ([NDVI]; Rouse et al. 1974), ratio vegetation index ([RVI]; Jordan 1969), soil-adjusted vegetation index ([SAVI]; Huete 1988) and tasseled cap transformations (i.e., soil brightness, vegetation greenness, and soil/vegetation wetness; Kauth and Thomas 1976). Each scene and its associated vegetation indices were analyzed independently and collectively as a time series. The time-series analyses outperformed the single-scene analyses (see figure 3.5), and top predictors were June tasseled cap wetness, September tasseled cap wetness, and October band 3, respectively. Using the same methods to quantify tamarisk encroachment on southwestern willow flycatcher (*Empidonax traillii extimus*) habitat, York et al. (2011) had similar results for study areas in Utah, Nevada, and Arizona. In both cases, time-series analyses outperformed all of the single-scene analyses, while imagery from October had the best single-scene results.

FIGURE 3.5 Tamarisk infestations detected along the Arkansas River and irrigation ditches near the town of Riverdale, Colorado. The results shown here are from a time-series analysis from Landsat 7 ETM+ over the course of a single growing season. Heavy tamarisk infestations are indicated in black; a color example of this type of map can be found at www.oup.com/us/Tamarisk.

Source: Evangelista et al. (2009).

Although the role of remote sensing in mapping tamarisk invasions remains limited, new high-resolution sensors, analytical tools, and methods are rapidly improving our ability to detect infestations. In many cases, mapping large infestations of tamarisk through aerial or satellite imagery can be easily done, especially in semiarid regions of the Southwest where floral heterogeneity is minimal and tamarisk can be more easily distinguished from other plants. The imagery required to conduct simple remote surveys is often freely available from state and federal government websites, and data are provided in commonly used geographic information system (GIS) formats. For more challenging study sites (e.g., high floral diversity, low tamarisk abundance) more elaborate methods may be required (Evangelista et al. 2009; York et al. 2011). Even these methods, however, are relatively easy to use for most GIS practitioners and many of the analytical tools needed are included with software (e.g., Esri ArcGIS) or downloadable from the internet (e.g., Maxent model).

Predicting Tamarisk Current and Potential Distribution

In addition to mapping tamarisk distributions, resource managers also want to know where new invasions are most likely to occur. By understanding the environmental conditions that may facilitate or prohibit infestations and the invasive potential of a new species, resource managers can make more informed decisions. Predictive models have become important tools for managing invasive species because they can fill knowledge gaps (Jarnevich et al. 2006), guide early detection strategies (Robinson et al. 2010), and help prioritize monitoring and control efforts (Hoffman et al. 2008). For tamarisk, predictive models have been especially useful at local (e.g., Evangelista et al. 2008), regional (Jarnevich et al. 2011) and national scales (e.g., Morisette et al. 2006) and under future climate scenarios (Kerns et al. 2009). However, because all these models use different algorithms to relate tamarisk presence to the environment, different environmental predictors, and different spatial scales, they are not directly comparable. Predictive models are not new to science or management; however, today's models can better integrate field observations with geospatial data (e.g., GIS data, remote sensing) using spatial statistics to produce probability maps. Known as habitat suitability models, they are designed to relate species occurrence data to environmental conditions to define habitat parameters and predict suitability in unsampled locations (see figure 3.6; Elith and Leathwick 2009).

LOCAL-SCALE MODELING

Models developed at local spatial scales can provide insights into factors related to a species' distribution at a management unit scale. Local spatial scales include those for an area such as a national park or a county and are at a fine spatial resolution, including predictions at 10m^2 or 30m^2. These models are site specific and can

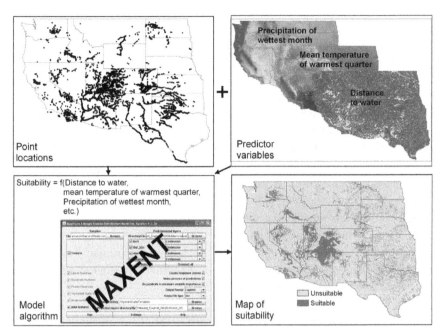

FIGURE 3.6 Habitat suitability models relate presence locations such as those for tamarisk to the environment using some sort of algorithm such as the Maxent program. These relationships can then be applied to maps of the environmental attributes to generate predicted habitat suitability such as that for tamarisk in the western United States.

assist in local management planning, such as prioritizing among species locally for monitoring and control efforts. Different factors control a species' distribution at different scales (Luoto et al. 2007), and different areas within the species range for broad-ranging species such as tamarisk (Murphy and Lovett-Doust 2007).

Models at these local management scales can also take into account anthropogenic features under management control, such as hiking trails and fine scale land cover/land use information. Soils may also play an important role at finer scales. These finer scale models (with lower resolutions of 30m^2) may be able to pick up more localized habitats, such as springs, that may be missed at coarser resolutions (e.g., 1-km^2). Additionally, they may pick up micro-scale differences, such as phenological relationships that may change with latitude.

Evangelista et al. (2008) developed a model of tamarisk (*T. chinensis*) distribution in the Grand Staircase Escalante Monument, Utah. Predictors include variables derived from a ten-meter digital elevation model representing topographic features and resource availability for tamarisk, including overland distance to water, degree slope, insolation, soil wetness index, and aspect (e.g., facing north or south). The models were evaluated with a test data set consisting of known presence locations not used in developing the model and performed well in predicting these test locations as presence locations.

LARGE-SCALE MODELING

Models developed across large spatial extents such as the western United States, North America, or the globe can help identify general patterns of invasion. These models are at a coarser resolution such as 1km². They can help set management priorities, identifying areas for early detection efforts or to begin control efforts. They can inform how to best allocate limited resources for these activities, such as prioritizing areas for federal funding of control efforts. For example, with tamarisk it may be best to concentrate control efforts upstream, moving downward, so that controlled areas are not constantly being bombarded with new propagules.

Predictors at these broader spatial scales include climate data, such as temperature and precipitation. These factors affect the distribution of tamarisk at larger spatial extents, illustrated by the somewhat distinct pattern of tamarisk presence and absence at stream gauges in relation to mean annual minimum temperature. In the western United States, tamarisk was much more common in areas that did not get as cold (Friedman et al. 2005). Remote sensing imagery can also play an important role in models at these scales, such as the national tamarisk map produced using three metrics related to timing of vegetation indexes derived from moderate resolution imaging spectroradiometer (MODIS) satellite time series data, consisting of values collected every 16 days for three years, along with remote sensing derived land cover data (Morisette et al. 2006). In this case, the southwestern United States was confirmed to be most suitable for tamarisk.

Since publication of Morisette et al. (2006), several new locations were reported (Jarnevich et al. 2011). Jarnevich and colleagues updated the national modeling of Morisette et al. (2006) for the western United States using the new location data and a different modeling algorithm (see figure 3.6). Distance to water was the most important contributor to the model by far, followed by mean temperature of the warmest quarter, and precipitation of the wettest month. This model did not include remotely sensed predictors as the other large-scale model did.

The differences between the 2006 national model and the 2011 western United States model that included these new data locations underscore the utility of iterative modeling for mapping invasive species such as tamarisk, as Stohlgren and Schnase (2006) suggested. The initial model triggered individuals and organizations with data not available for the first round to contribute locations, leading to the improved model.

The Hyper-Envelope Modeling Interface (HEMI) is a new modeling approach that allows automatic placement of the environmental envelopes for a species, visualization of the occurrence data and the envelope in environmental space, and the ability to modify the model based on expert opinion. The HEMI models for tamarisk at relatively small extents (e.g., Grand Staircase Escalante National Monument) used wetness index and distance to water, finding most tamarisk locations in relatively low wetness areas at a range of relatively moderate distances from water (0 to 4 km) in the monument (Graham et al. unpublished data). For the western United States the model included minimum temperature of the coldest month and annual precipitation, finding tamarisk in lower precipitation areas.

There are also regional models within the western United States including the northwest and southwest. Predictions for the northwestern United States were concerned with potential distribution changes with changing climate, and thus used only climate data as predictors (Kerns et al. 2009). This type of study can be useful to determine priority locations for control and early detection by distinguishing between locations currently suitable now but not in the future, suitable now and in the future, and unsuitable now but suitable in the future. They predicted tamarisk in warm, dry areas in the northwestern United States, and point out that although tamarisk has not been perceived as a problem in the northwest, it has been there for a long time.

Another study predicted tamarisk habitat in the southwestern United States and Mexico using climate and remote sensing derived plant phenology predictors (Cord et al. 2010). Similar to other large scale models, their predictions suggested that there is still considerable uninvaded suitable habitat. They again found tamarisk associated with warm and arid climates in their study area.

Farther northward and higher elevation expansion could occur as a result of climate warming. *Tamarix* appears to be limited by its sensitivity to extremely low temperatures, as supported by field observations (Friedman et al. 2005). Distribution models applied to climate change scenarios in the northwestern United States further support this by indicating a two- to tenfold increase in suitable *Tamarix* habitat (Kerns et al. 2009). However, we need to be able to produce more accurate models of current distribution before forecasts can be trusted.

Future Needs

Current remote sensing tools combined with distribution modeling can provide useful information for the management of tamarisk. There are some easily identified areas for future work. Accurate predictive models of tamarisk distribution require data covering the entire environmental range of the species. Tamarisk has invaded large regions of the southwestern United States and Mexico, but these data have not been aggregated and made available as a set. Similarly, little occurrence data for tamarisk within its native range in northern Africa and Asia are available. These data gaps can led to underpredictions of the potential distribution of invasive species (Beaumont et al. 2009).

Large-scale collaborative efforts are needed to aggregate location data for species such as tamarisk to make large-scale distribution predictions possible. The large extent models described above were reliant on data aggregation efforts. There are hundreds of databases containing data on invasive species in the United States alone (Crall et al. 2006). Data for many invasive species, including tamarisk, are available from both their native and invasive range. This makes invasive species data management a global problem. There are also other types of databases that may contain information important to mapping the distribution of a species, such as the National Phenology Network's data. This type of information recording the phenology—time of leafing and flowering—may aid remote sensing efforts in mapping tamarisk in different regions of the country. Attempts to integrate data into large centralized databases, such as the Global Biological Information Facility (GBIF) have had some success. GBIF does contain occurrence data on tamarisk (see the GBIF data portal at http://data.gbif.org/search/tamarix) and provides maps of the occurrence data. The map for *Tamarix ramosissima* shows that global data aggregation has the advantages of worldwide data but often at the cost of quality, exemplified by the occurrences in the Atlantic Ocean (at the time of writing). GBIF and other aggregators are also not collecting data on the management and control status of an invasive species. The Global Invasive Species Information Network ([GISIN]; www.gisin.org) was created to integrate and disseminate data on invasive species from all over the world (Simpson et al. 2009). In the future, GISIN could provide a single-source for occurrence data for modeling invasive species and information for managing invasive species. Additional efforts to fill gaps for tamarisk and other species throughout the world may lead to more comprehensive models.

Abundance information rather than mere presence is a better indication for determining effects of tamarisk, such as observed avian response to tamarisk on the lower Colorado River (Van Riper et al. 2008). We would like to move from models predicting occurrence to models predicting abundance, but abundance information needs to be generated in a systematic way. For example, Evangelista et al. (2007) developed a model to predict tamarisk biomass based on field data of estimated cover and height. Consistently collected cover and height data over a landscape scale could be used to derive biomass estimates that could then be used to predict biomass.

Conclusions

Compiled information from different studies, including surveys, remote sensing, and fieldwork, can provide an estimate of tamarisk distribution and abundance. However, because of lack of consistency within the approaches, the compiled data can have gaps spatially (e.g., county not responding to survey) or methodologically (e.g., remote sensing mapping used different methods). Modeling can help fill in these gaps and provide scenarios of what distribution and abundance may be. These models will never be a perfect representation of actual distribution and abundance, but they can provide resource managers and others with useful information that can be used in efforts to control invasive species such as tamarisk.

Literature Cited

Akasheh, O. Z., C. M. U. Neale, and H. Jayanthi. 2008. Detailed mapping of riparian vegetation in the middle Rio Grande River using high resolution multi-spectral airborne remote sensing. Journal of Arid Environments 72:1734–1744.

Anderson, G. L., R. I. Carruthers, S. Ge, and P. Gong. 2005. Cover: Monitoring of invasive *Tamarix* distribution and effects of biological control with airborne hyperspectral remote sensing. International Journal of Remote Sensing 26:2487–2489.

Barnett, D. T., T. J. Stohlgren, C. S. Jarnevich, G. W. Chong, J. A. Ericson, T. R. Davern, and S. E. Simonson. 2007. The art and science of weed mapping. Environmental Monitoring and Assessment 132:235–252.

Beaumont, L. J., R. V. Gallagher, W. Thuiller, P. O. Downey, M. R. Leishman, and L. Hughes. 2009. Different climatic envelopes among invasive populations may lead to underestimations of current and future biological invasions. Diversity and Distributions 15:409–420.

Carter, G., K. Lucas, G. Blossom, C. Lassitter, D. Holiday, D. Mooneyhan, D. Fastring, T. Holcombe, and J. Griffith. 2009. Remote sensing and mapping of tamarisk along the Colorado River, USA: A comparative use of summer-acquired hyperion, thematic mapper and QuickBird data. Remote Sensing 1:318–329.

Cord, A., D. Klein, and S. Dech. 2010. Remote sensing time series for modeling invasive species distribution: A case study of *Tamarix* spp. in the US and Mexico. Proceedings of the International Congress on Environmental Modelling and Software: Modelling for Environment's Sake, Ottawa, Canada.

Crall, A. W., L. A. Meyerson, T. J. Stohlgren, C. S. Jarnevich, G. J. Newman, and J. Graham. 2006. Show me the numbers: What data currently exist for nonnative species in the USA? Frontiers in Ecology and the Environment 4:414–418.

Elith, J., and J. R. Leathwick. 2009. Species distribution models: Ecological explanation and prediction across space and time. Annual Review of Ecology Evolution and Systematics 40:677–697.

Evangelista, P., S. Kumar, T. J. Stohlgren, A. W. Crall, and G. J. Newman. 2007. Modeling aboveground biomass of *Tamarix ramosissima* in the Arkansas River basin of southeastern Colorado, USA. Western North American Naturalist 67:503–509.

Evangelista, P., S. Kumar, T. J. Stohlgren, C. S. Jarnevich, A. W. Crall, J. B. Norman III, and D. Barnett. 2008. Modelling invasion for a habitat generalist and a specialist plant species. Diversity and Distributions 14:808–817.

Evangelista, P. H., T. J. Stohlgren, J. T. Morisette, and S. Kumar. 2009. Mapping invasive tamarisk (*Tamarix*): A comparison of single-scene and time-series analyses of remotely sensed data. Remote Sensing of Environment 1:519–533.

Everett, J., C. Yang, C. Fletcher, C. J. Deloach, and M. R. Davis. 2007. Using remote sensing to assess biological control of saltcedar. Southwest Entomologist 32:93–103.

Everitt, J. H., and C. J. Deloach. 1990. Remote-sensing of Chinese tamarisk (*Tamarix-Chinensis*) and associated vegetation. Weed Science 38:273–278.

Everitt, J. H., D. E. Escobar, M. A. Alaniz, M. R. Davis, and J. V. Richerson. 1996. Using spatial information technologies to map Chinese tamarisk (*Tamarix chinensis*) infestations. Weed Science 44:194–201.

Friedman, J. M., G. T. Auble, P. B. Shafroth, M. L. Scott, M. F. Merigliano, M. D. Preehling, and E. K. Griffin. 2005. Dominance of non-native riparian trees in western USA. Biological Invasions 7:747–751.

Ge, S., R. Carruthers, P. Gong, and A. Herrera. 2006. Texture analysis for mapping *Tamarix parviflora* using aerial photographs along the Cache Creek, California. Environmental Monitoring and Assessment 114:65–83.

Griffith, J., R. McKellip, and J. Morisette. 2005. Comparison of multiple sensors for identification and mapping of tamarisk in western Colorado: Preliminary findings. ASPRS Annual Conference, March 7–11. Baltimore, MD.

Hamada, Y., D. A. Stow, L. L. Coulter, J. C. Jafolla, and L. W. Hendricks. 2007. Detecting tamarisk species (*Tamarix* spp.) in riparian habitats of Southern California using high spatial resolution hyperspectral imagery. Remote Sensing of Environment 109:237–248.

He, K. S., D. Rocchini, M. Neteler, and H. Nagendra. 2011. Benefits of hyperspectral remote sensing for tracking plant invasions. Diversity and Distributions 17:381–392.

Hoffman, J. D., S. Narumalani, D. R. Mishra, P. Merani, and R. G. Wilson. 2008. Predicting potential occurrence and spread of invasive plant species along the North Platte River, Nebraska. Invasive Plant Science and Management 1:359–367.

Huang, C. Y., and G. P. Asner. 2009. Applications of remote sensing to alien invasive plant studies. Sensors 9:4869–4889.

Huete, A. R. 1988. A soil-adjusted vegetation index (Savi). Remote Sensing of Environment 25:295–309.

Jarnevich, C. S., P. Evangelista, T. J. Stohlgren, and J. Morisette. 2011. Improving national-scale invasion maps: Tamarisk in the Western United States. Western North American Naturalist 71:164–175.

Jarnevich, C. S., T. J. Stohlgren, D. Barnett, and J. Kartesz. 2006. Filling in the gaps: Modelling native species richness and invasions using spatially incomplete data. Diversity and Distributions 12:511–520.

Jordan, C. F. 1969. Derivation of leaf-area index from quality of light on the forest floor. Ecology 50:663–666.

Kauth, R., and G. Thomas. 1976. The tasselled cap-a graphic description of the spectral-temporal development of agricultural crops as seen in Landsat. Pages 41–51 *in* Proceedings of the Symposium on Machine Processing of Remotely Sensed Data, West Lafayette, IN.

Kerns, B. K., B. J. Naylor, M. Buonopane, C. G. Parks, and B. Rogers. 2009. Modeling tamarisk (*Tamarix* spp.) habitat and climate change effects in the northwestern United States. Invasive Plant Science and Management 2:200–215.

Luoto, M., R. Virkkala, and R. K. Heikkinen. 2007. The role of land cover in bioclimatic models depends on spatial resolution. Global Ecology and Biogeography 16:34–42.

Marvin, D. C., B. A. Bradley, and D. S. Wilcove. 2009. A novel, web-based, ecosystem mapping tool using expert opinion. Nature Areas Journal 29:281–292.

Morisette, J. T., C. S. Jarnevich, A. Ullah, W. J. Cai, J. A. Pedelty, J. E. Gentle, T. J. Stohlgren, and J. L. Schnase. 2006. A tamarisk habitat suitability map for the continental United States. Frontiers in Ecology and the Environment 4:11–17.

Murphy, H. T., and J. Lovett-Doust. 2007. Accounting for regional niche variation in habitat suitability models. Oikos 116:99–110.

Nagler, P. L., E. Glenn, C. S. Jarnevich, and P. B. Shafroth. 2011. Distribution and abundance of saltcedar and Russian olive in the western United States. Critical Reviews in Plant Sciences 30:508–523.

Pavri, F., and J. Aber. 2004. Characterizing wetland landscapes: A spatiotemporal analysis of remotely sensed data at Cheyenne Bottoms, Kansas. Physical Geography 25:86–104.

Phillips, S. J., R. P. Anderson, and R. E. Schapire. 2006. Maximum entropy modeling of species geographic distributions. Ecological Modeling 190:231–259.

Robinson, T. 1965. Introduction, spread and areal extent of saltcedar (*Tamarix*) in the western United States. US Geological Survey Professional Paper 491-A.

Robinson, T. P., R. D. van Klinken, and G. Metternicht. 2010. Comparison of alternative strategies for invasive species distribution modeling. Ecological Modeling 221:2261–2269.

Robinson, T. P., R. D. v. Klinken, and G. Metternicht. 2008. Spatial and temporal rates and patterns of mesquite (*Prosopis* species) invasion in Western Australia. Journal of Arid Environments 72:175–188.

Rouse, J. W., R. H. Haas, J. A. Schell, and D. W. Deering. 1974. Monitoring vegetation systems in the Great Plains with ERTS. Proceedings of the Third Earth Resources Technology Satellite-1 Symposium.

Simpson, A., C. Jarnevich, J. Madsen, R. Westbrooks, C. Fournier, L. Mehroff, M. Browne, J. Graham, and E. Sellers. 2009. Invasive species information networks: Collaboration at multiple scales for prevention, early detection, and rapid response to invasive alien species. Biodiversity 10:5–13.

Stohlgren, T. J., and J. L. Schnase. 2006. Risk analysis for biological hazards: What we need to know about invasive species. Risk Analysis 26:163–173.

Tamarisk Coalition. 2008. Colorado Tamarisk Mapping and Inventory Summary Report. Tamarisk Coalition, Grand Junction, CO. [Online]. http://www.tamariskcoalition.org/PDF/Colorado's%20Inventory%20&%20Mapping%20summary%20REVISED%20 2–08.pdf.

Thoene, J. 2002. Implementation of a GIS for Regional Management of Leafy Spurge (*Euphorbia esula*) and Yellow Starthistle (*Centaurea solstitialis*) in the Western United States. MS GIS capstone project. University of Denver, Denver, CO.

Tree, R., and J. Slusser. 2005. Measurement of spectral signatures of invasive plant species with a low cost spectrometer *in* Ultraviolet Ground- and Space-based Measurements, Models and Effects. SPIE Proceedings.

Van Riper, C., K. L. Paxton, C. O'Brien, P. B. Shafroth, and L. J. McGrath. 2008. Rethinking avian response to *Tamarix* on the Lower Colorado River: A threshold hypothesis. Restoration Ecology 16:155–167.

York, P., P. Evangelista, S. Kumar, J. Graham, C. Flather, and T. Stohlgren. 2011. A habitat overlap analysis derived from Maxent for tamarisk and the Southwestern Willow Flycatcher. Frontiers of Earth Science 5:120–129.

4

Evapotranspiration by Tamarisk in the Colorado River Basin

Erika Zavaleta

The stewardship of river ecosystems is central to any region's long-term future, and removal of exotic tamarisk trees has played a role in this stewardship in the American West. Restoration of tamarisk-invaded rivers and riverbanks could generate many benefits, but debate has persisted over the question of whether increased water savings would be among them. A panel of experts was convened in November 2008 to try to answer this question, specifically for the Colorado River basin.[1]

Background

Over the past 50 years, tamarisk has gained a reputation as an aggressive invasive species that uses large quantities of water (see chapter 16, this volume). Though no known projects document water recovery following Russian olive removal, several projects have observed recovering springs and wetlands and rebounding ground-water levels following tamarisk removal. These examples are widely used to support

---

[1] This chapter is a synthesis of an independent peer review of the current literature and knowledge, coordinated by the Tamarisk Coalition as a product of a commissioned report. The parties to that MOU were the Central Arizona Water Conservation District, Colorado Water Conservation Board, New Mexico Interstate Stream Commission, Six Agency Committee, Southern Nevada Water Authority, Utah Division of Water Resources, and Wyoming State Engineer's Office. It is excerpted with permission by the Tamarisk Coalition from the Independent Peer Review of Tamarisk and Russian olive Evapotranspiration for the Colorado River Basin, as a part of the Colorado River Basin Tamarisk and Russian Olive Assessment.

the assertion that tamarisk exploits valuable water resources. The most notable of these projects were completed at Spring Lake in Artesia, New Mexico; Coachella Creek, California; and Eagle Borax Springs Works, Death Valley National Park, California (Rowlands 1990; DiTomaso 2004).

At Spring Lake, tamarisk had invaded and covered a 13-acre spring-fed lake, eliminating its surface water by 1968. Tamarisk was effectively controlled with herbicides in 1989, and by 1992 the water table had resurfaced. The tamarisk infestation at the Nature Conservancy's Coachella Valley Preserve had replaced approximately 80% of the area's vegetation. This dense stand depressed the groundwater table, which was suspected to have decreased the output of local springs. The stand was removed over a five-year period, after which oasis springs in the area rapidly recovered. At Eagle Borax Works Springs, historical records described a natural spring and associated ponds, progressively drying as tamarisk spread began in 1950. In 1971 the park staff conducted a controlled burn of 10 acres. Eight weeks later, the water table had risen 1.2 feet and a one-acre pond had reappeared. These examples provide anecdotal evidence that tamarisk control provides water savings. As a result numerous control efforts have occurred to increase water supplies. However, few projects have collected the data necessary to show that actual savings have occurred.

The only high-quality field study we could identify of tamarisk control resulting in an increase in water supplies was performed on the Gila River upstream from San Carlos Reservoir in Arizona. The report stated:

> During the first few years of the 10-year study, the natural hydrologic system was monitored using observation wells, streamflow gauges, and meteorological instruments. Following this initial monitoring period, the phreatophytes were removed from the flood plain and the effects on streamflow were evaluated. The average effect of vegetation removal over the entire study reach was that the Gila River changed from a continually losing river for most years before clearing to a gaining stream during some months for most years following clearing. Specifically, average monthly values of gain or loss from the stream indicated that before clearing, the river lost water to ground water during all months for most years. After clearing, the river gained ground-water inflow during March through June and during September for most years. (Culler et al. 1982)

Although the research clearly indicated that tamarisk control improved stream flow in the Gila River, the project failed to include a revegetation component. Revegetation is essential to prevent extreme erosion or the reinvasion of non-natives such as tamarisk and Russian olive. Because of the lack of revegetation, this study did not address the question of whether water savings would occur if replacement vegetation were established.

The only other major study of this type was performed on the Pecos River in New Mexico. This study evaluated water salvage following the removal of 18,800 acres of tamarisk from the floodplain of the Pecos River near Artesia. Tamarisk was almost entirely removed, excluding some thickets on wildlife refuges and ten-meter strips along each bank left for erosion control. No discernible streamflow gain was observed. Several explanations were provided for this lack of increase in the Pecos River base flow, including error in streamflow measurements, masking of salvage by variations in climate, and capture of salvaged water by groundwater pumping (Shafroth et al. 2005). Although not specifically stated in the citation, another important reason for no discernable gain in stream flow was that cleared tamarisk sites were often allowed to immediately undergo secondary encroachment by extremely dense, near-100% canopy closure, monotypic stands of kochia (*Kochia scoparia*). This is a common occurrence on many land ownerships (public and private) along the broader riparian corridors of the Pecos River in southeastern New Mexico when no revegetation is conducted following tamarisk removal. This is compounded by the widespread distribution of an aggressive, high leaf area index (LAI), herbicide-resistant ecotype of kochia in this locale for many years. Under this scenario, and perhaps for this reason primarily, no water salvage is likely to occur when dense secondary invasions fill the ecological void, essentially replacing tamarisk ET in the absence of natural or artificial revegetation (Lair unpublished data 2006; pers. comm. 2009).

All these anecdotal examples and studies have led to some confusion about the potential for water savings. The objective of this study was to review and focus discussion of ET rates and tamarisk management in order to inform decisions by the Colorado River basin states about the potential benefits and cost-effectiveness of removing tamarisk and restoring riparian lands. Details about the selection of panel members, its final composition, and the specific task put before it can be found in Appendix A at the end of this chapter. What follows is an excerpt from the original report, which is divided into sections that address major questions posed to the panel by the parties that commissioned the study.

Panel Questions

1) What does existing research tell us about the use of water by tamarisk in different ecological settings and what information gaps require additional research?
2) Can ET measurements from lower latitude states be used to infer potential ET rates in higher latitude states? What about elevation differences?
3) What is known about ET rates for replacement vegetation, including both riparian and upper terrace floodplain species? What is an appropriate palette of replacement species for each ecosystem within the Colorado River watershed?
4) What role does infestation density play in overall ET rates?
5a) Can the Panel agree on a narrower range of tamarisk ET than is described in the literature?

5b) Can a range of water savings per acre be agreed to? Can a relative range of water savings between tamarisk and replacement plant communities be agreed to?

5c) Can water be saved and stream flows increased in the Colorado River system by implementing tamarisk control and restoration actions?

6) What are the benefits or impacts if tamarisk management within the Colorado River basin states does not occur?

Panel Questions Answered

Question 1: What does existing research tell us about the use of water by tamarisk in different ecological settings and what information gaps require additional research?

A key conclusion of the panel is that *native vegetation can use either more or less water than tamarisk*, depending on the identity of the native species, stand densities, and environmental and site conditions. Panel members expressed two distinct perspectives on how ET rates can be predicted or extrapolated from one site to another: (1) ET depends on several factors that vary by site, making extrapolation relatively complex; and (2) ET is relatively well predicted from canopy characteristics and reference ET (ET_0) for the site in question. We describe both.

Variation in evapotranspiration from tamarisk systems: Within-species ET rates of tamarisk or Russian olive can vary widely under the same general climatic conditions because of wide variations in vegetation density, vegetation health, height and age of the vegetation, nature of understory vegetation, and access to groundwater. Often, stand characteristics and access to groundwater vary widely with distance from a stream, soil type, local topography, geology, and stream-groundwater interactions. Therefore, it is unreasonable to expect or to use a single value for ET from tamarisk systems or even a narrow range (Johns 1989). It is possible, however, to express water consumption in relatively narrow ranges for specific classes of vegetation stand characteristics. For example, old, sparse stands of tamarisk are expected to have much lower water consumption rates than densely populated five-year-old stands that have ready access to groundwater. Vegetation growing in salinized soil or substantially above a water table, or that is infested with insects, is expected to fall into a lower range of water consumption.

Allen et al. (2007) sampled ET derived by satellite-based energy balance along a 150-kilometer river corridor of the Middle Rio Grande Valley of New Mexico and derived frequency distributions of estimated seasonal ET from cottonwood and tamarisk. These frequency distributions, shown in figure 4.1, exhibit wide variation in ET from both cottonwood and tamarisk. ET from tamarisk exhibited larger variance due to its tendency to grow across a wider range of water availability, water table depth, soil types, and salinity conditions, whereas cottonwoods are usually found close to stream channels.

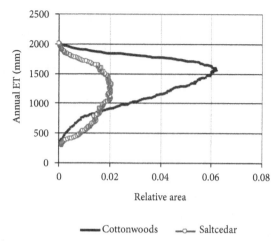

FIGURE 4.1 Sampled ET derived by satellite-based energy balance along a 150-kilometer river corridor of the middle Rio Grande.

Source: Adapted from Allen et al. (2007).

Because ET rates from tamarisk vary widely, ET models have not yet been developed for these species that are readily transferred from one local area to another. Models common to agriculture and other "uniform" systems do not transfer well to tamarisk because of the complex surface energy and aerodynamic characteristics and heterogeneity typical of these riparian species.

Most information and estimates of water consumption by riparian systems are therefore based on in-place measurements. These measurements show a wide range of peak and annual water consumption within each species, as described later in this report. Some of the variation in reported values is real and is caused by differences in weather (location, year) and stand characteristics such as density, age, health, and soil conditions such as salinity. Still more of the variation, however, likely results from biases or error in the measurements themselves. This bias or error can arise from the method employed, the care and quality exercised by the data collector, or the context of the measurement, where physical conditions required by the measurement system can be violated. Error can also arise from extrapolating short-term measurements (over hours, days, or weeks) to the entire growing season or year. Likely error ranges for several common ET measurement methods are summarized in table 4.1. Underreporting of methodological details in the literature means that it is not always possible to evaluate the accuracy and precision of a reported value.

Evapotranspiration measurements: ET measurements include soil water balance, lysimeter container study, eddy covariance, Bowen ratio, remote sensing, scintillometry, and sap flow. Improperly designed experiments or measurements will give erroneous data (Anderson and Idso 1985). Further, estimates of water savings from water salvage projects are often optimistic (Johns 1989). Many panel members felt that values from tank/container studies should not be used in developing ET estimates, but that other evaporation measurements were valid.

TABLE 4.1

Expected error (one standard deviation) for various types of ET measurements in riparian systems.*

Method	Error relative to actual ET, %	Additional error caused by equipment and operator malfunction, %
Lysimeter container studies	20–100	5–40
Soil water balance	10–30	10–40
Bowen ratio	10–20	5–40
Eddy covariance	15–40	10–40
Remote sensing energy balance	10–20	10–30
Remote sensing using vegetation indices	15–40	10–20
Sap flow	15–40	20–100
Scintillometers	10–35	5–30

*Errors can be much larger if measurements are made by people without specific expertise in the method and underlying theory used.

Source: R.G. Allen (pers. comm.)

Comparing and extrapolating across sites: To normalize across sites with different weather and climate, ET or water consumption measurements should not only be reported in terms of absolute units, such as millimeters or acre-feet, but should also be normalized by dividing by a reference ET (ET_o) to create the "fraction of reference ET," ET_oF. This is because ET demand is a strong function of weather and climate. ET rates can change by a factor of five or more day to day as weather changes. For the same month, average ET can change by as much as 20% from year to year due to differences in weather systems. Annual ET for the same location can change by as much as 15% to 20% for the same reasons. Because ET is an energy-governed process, its demands generally decrease with increased elevation and latitude and increase with the regional dryness. For example, annual ET demands in western Colorado may be only one-half to two-thirds those for the Mojave.

Reference ET, ET_o, is a standardized representation of climatic evaporative demand at a given site. ET by a particular species such as tamarisk, when expressed as a fraction of reference ET, or an ET_oF, are more transferrable across locations and weather. The ET_oF is equivalent to the commonly used "crop coefficient" and effectively describes the physiological and local physical characteristics (such as water availability or stand density) that impact the ET rate while the ET_o describes the impact of weather and climate on the consumptive rate.[2]

---

[2] ET_o is best calculated using the ASCE-EWRI (2005) standardization of the Penman-Monteith equation if dependable data for solar radiation, wind speed, air temperature, and humidity are available. If these data are not available, then a simpler but dependable method such as Hargreaves and Samani (1985) is recommended.

Use of a stand density function to estimate EToF: Vegetation density functions have been used to predict expected values for ET_oF as a function of stand appearance and characteristics. A basic model for this is given in the FAO-56 publication on evapotranspiration (Allen et al. 1998). Refinements of the method are described in ASABE (2007). Similar methods should be explored for riparian systems to provide more narrowed ranges of expected water consumption. These vegetation-characteristic-based methods can be combined with groundwater elevation models, precipitation, and ET_o models to produce more accurate estimates of water consumption.

Question 2: Can ET measurements from lower latitude states be used to infer potential ET rates in higher latitude states? What about elevation differences?

Most studies to date on ET rates have taken place in the lower basin states of the Colorado River. The full latitudinal range of tamarisk is therefore not well represented by current ET studies. This should be addressed by future ET studies, as should the possibility that other tamarisk species besides *T. ramossissima* (the most-studied one) could have different ET dynamics. In the meantime, ET rates can reasonably be re-scaled to new latitudes and elevations by expressing tamarisk ET as a fraction of reference ET (ET_oF) for each site (see Question 1). We estimate that error from this approach can be within 10% to 20% of ET rates measured on-site, provided that differences between sites in water availability and precipitation patterns are taken into account.

Question 3: What is known about ET rates for replacement vegetation, including both riparian and upper terrace floodplain species? What is an appropriate palette of replacement species for each ecosystem within the Colorado River watershed?

Riparian restoration and prevention of reinvasion both require the promotion of replacement vegetation following tamarisk removal. Replacement vegetation is also critical for bank stabilization and erosion control, wildlife habitat enhancement, forage production for wildlife and livestock, recreation, aesthetics, and other ecosystem services. Replacement species need to be selected foremost for suitability to target sites, since without their successful establishment none of these goals will necessarily be met. However, characteristics such as their effects on ET and wildlife habitat may also guide their selection.

Generally speaking, ET rates for replacement vegetation are not as well studied as for tamarisk, particularly for facultative or nonphreatophytic species that tend to dominate plant composition on upper floodplain terraces. There are very few data for herbaceous cover species other than selected sacaton grasses (*Sporobolus* spp.) and inland saltgrass (*Distichlis spicata*). It is clear that replacement species exhibit a very wide range of ET values, from values higher than tamarisk to values significantly lower (see table 4.2).

In the riparian zone phreatophytic tree communities, mainly cottonwood-willow (*Populus-Salix*) vegetation typical of former floodplains in Western river systems, exhibit ET rates comparable to tamarisk at maturity and full canopy closure. Many other replacement species (e.g., shrubs and grasses) that have been studied exhibit lower ET rates. In the mesic riparian fringe, phreatophytic trees are likely an appropriate revegetation choice for both site suitability and the recovery of habitat, aesthetic and other values associated with this zone. There may be a period of several (5–15) years following cottonwood/willow planting when their ET rate, although increasing as they mature, is still significantly less per unit land area than the tamarisk stands they replaced. In some cases this may

TABLE 4.2

Daily and/or annual evapotranspiration (ET) estimates for native, nonphreatophytic vegetation types (individual species and plant community associations) occurring in upper terrace floodplain sites, western United States.

Veg Type	ET [1] mm d^{-1} (m yr^{-1}) [ac-ft yr^{-1}]	Study Location	Method	Citation
Mesquite	(0.4)	San Pedro River, AZ	BR	Scott et al. 2000
(*Prosopis* spp.)	1.6–2.4	San Pedro River, AZ	BR	Scott et al. 2000
	(0.6–0.7)	San Pedro River, AZ	EC	Scott et al. 2004
	(1.02)	Arizona	??	Gatewood et al. 1950
	[2.1–2.3]	San Pedro River, AZ	EC/BR	Scott et al. 2006
Honey mesquite (*Prosopis glandulosa/P. juliflora*)	(0.47)	Lower Colorado River (near Blythe, CA)	BR ??	Wiesenborn 1995
	[2.9]	Acme-Artesia area, NM		USBR 1979
Velvet mesquite (*Prosopis velutina*) woodland	(0.64–0.69)[++]	San Pedro River, AZ	SF/EC	Leenhouts et al. 2006
Velvet mesquite shrubland	(0.57)[++]	San Pedro River, AZ	SF/EC	Leenhouts et al. 2006
Mixed saltcedar/Honey mesquite	(1.0)	Lower Colorado River (near Blythe, CA)	BR	Wiesenborn 1995
Mixed saltcedar/Screwbean mesquite (*Prosopis pubescens*)	(0.37)	Lower Colorado River (near Blythe, CA)	BR	Wiesenborn 1995
Savannah woodland: velvet mesquite/big sacaton mixed stand	3.5	Tucson, AZ	EC	Yepez et al. 2003
Arrowweed (*Pluchea sarothroides*)	(0.37)	Lower Colorado River (near Blythe, CA)	BR	Wiesenborn 1995
Quailbush (*Atriplex lentiformis*)	(0.69)	Lower Colorado River (near Blythe, CA)	BR	Wiesenborn 1995

(Continued)

TABLE 4.2 (CONTINUED)

Veg Type	ET [1] mm d⁻¹ (m yr⁻¹) [ac-ft yr⁻¹]	Study Location	Method [2]	Citation
Inland saltgrass (Distichlis spicata)	(0.3–1.2)	Various sites	LYS	Weeks et al. 1987
	1.1–4.5	Sonora, NM	LYS	Miyamoto et al. 1996
Inland saltgrass	(0.45–1.15)	Owens Valley, CA	LYS	Young and Blaney 1942
	(0.4–0.9)	Santa Ana, CA	LYS	Young and Blaney 1942
	(0.25–1.25)	Los Griegos, NM	LYS	Young and Blaney 1942
Inland saltgrass/alkali sacaton	[1.2]	Acme-Artesia area, NM	??	USBR 1979
Big sacaton (Sporobolus wrightii)	0.3–1.6	San Pedro River, AZ	BR	Scott et al. 2000
	(0.55) ++	San Pedro River, AZ	SF/EC	Leenhouts et al. 2006
	[1.8]	San Pedro River, AZ	EC/BR	Scott et al. 2006
	50% of mesquite shrubland site; 25% of cottonwood site	San Pedro River, AZ	SF/EC??	Qi et al. 1998
Alkali sacaton/desert seepweed (Suaeda suffrutescens)	(1.05–1.2)	Carlsbad, NM	LYS	Blaney and Hanson 1965
	(0.57–0.67)	Artesia & Bitter Lakes NWR, NM	BR	Weeks et al. 1987
	(0.40)	Artesia & Bitter Lakes NWR, NM	EC	Weeks et al. 1987
"Grassland" (saltgrass/alkali sacaton??)	[0–1.99]	Los Lunas, NM	??	USACE/USBR 2002

[1] Values without parentheses or brackets are reported in mm d⁻¹ units; values within parentheses are reported in (m yr⁻¹) units; values within brackets are reported in [ac-ft yr⁻¹] units.

[2] Methods include Bowen ratio (BR), Eddy Covariance (EC), Sap Flow (SF), and Lysimeter (LYS).

"??" symbol indicates that it was unclear what the specific ET measurement technique or plant species was

++ Growing season only.

Source: Adapted from Shafroth et al. (2005), with additions.

not hold true as in older tamarisk stands, which may use less water than younger, more vigorous stands.

Tamarisk also invades well beyond the mesic riparian fringe; upper floodplain terraces comprise the vast bulk of tamarisk-infested area. Arid- to xeric-adapted shrubs and grasses are appropriate choices for replacement species in this zone. Replacement of tamarisk with diverse facultative or nonphreatophytic native vegetation that is adapted to upper floodplain terraces (with their deeper water tables and/or higher soil salinity) may reduce ET. Although studies were conducted under different site conditions, these native species and plant community associations exhibit ET values ranging roughly from 50% to 75% of mean tamarisk stand values often cited in the literature (table 4.2).

The choice of appropriate replacement species can be based on three driving factors at nested scales: climate (regional-scale), hydrology/water table characteristics (reach-scale), and salinity (site-scale). Examples of candidate revegetation species recommended for use in riparian restoration in the Colorado River basin, rated according to hydrologic regime and salinity tolerance, are described Shafroth et al. (2008). Appropriate replacement species could also depend on whether target sites are on free-flowing or regulated river reaches.

The upper Colorado River basin contains tamarisk at elevations up to 9,000 feet (though it is mapped only up to 6,500 feet). In the upper basin, replacement vegetation differs between two principal substrates: Colorado Plateau sandstones (with native vegetation such as bunchgrasses and shrubs), and marine shales (with native vegetation such as salt-tolerant alkali sacaton, saltgrass, saltbush, and greasewoods). In the lower Colorado River basin, replacement vegetation outside of the mesic riparian zone (which supports cottonwood/willow and associated species) includes a range of xeric-adapted and salt-tolerant shrubs, grasses, and forbs. Salinity is a pervasive challenge in the lower basin for revegetation. As a result, restoration of these highly saline, xeric sites may be extremely difficult. Panel members had divergent views about the practicality, feasibility, and cost of trying to restore infested areas on such sites in the lower Colorado basin. Native plant community restoration is achievable on many these sites, but economic feasibility rests with value of the restored habitat as perceived or assigned by the managing agency or landowner. In comparison, several areas in the upper basin will likely experience passive revegetation after tamarisk control because infestations are less dense, there is good presence of native vegetation, and periodic over-bank flooding occurs.

Question 4: What role does infestation density play in overall ET rates?

ET rates vary positively with canopy cover for similar age class and ecological setting. Many existing studies do not report canopy cover. However, it is probably sufficient to treat canopy cover categorically, that is, high, medium, and low categories that can be used to assign ET rates or ranges to particular stands. One study, at the Cibola National Wildlife Refuge, explicitly compared three nearby stands of different canopy cover and found that ET ranged from 0.55 to 0.88 to 1.34 m/year in low, medium, and high canopy stands, respectively (Christopher Neale, pers. comm.).

On tamarisk stands that have reached higher canopy cover, with near-monotypic or full monotypic composition and full canopy closure (the middle Rio Grande, Pecos, and lower Colorado being prime examples), age class will make a definite difference, yielding a wide spectrum of ET rates within the same canopy cover class. This is especially true as plants lose vigor under deeper water table conditions (i.e., drought). Two scenarios illustrate this specifically:

1. A young, monotypic tamarisk stand with higher green foliage: woody-stem biomass ratio (often maintained by high fire frequency) will exhibit higher ET rates than an old, decadent, monotypic stand (undisturbed for decades) that is 95%-plus woody biomass—with both states exhibiting 100% canopy closure.
2. Similarly, a young and active, but patchy or mixed tamarisk stand with relatively low plant (or stem) density may still have higher ET rates than an older, decadent, monotypic stand with higher densities and full canopy closure.

Question 5a: Can the Panel agree on a narrower range of tamarisk ET than is described in the literature?

Tamarisk evapotranspiration varies depending on many interacting factors, such as climate; canopy cover, age, and health; water table depth; water quality and salinity. Based on all available evidence, the Panel assembled by the Tamarisk Coalition reached consensus that the range of tamarisk ET on western rivers is 700 mm to 1400 mm/year, (ET_oF of 0.3 to 0.7, centering on a mean value of 0.5). Actual area-wide ET rates might be lower than this average, as most of the flux tower studies were set in denser stands to provide homogeneous measurement conditions (see table 5.1 in chapter 5, this volume).

The extremes of this ET range occur in distinct settings. In the southwestern United States along the Colorado River, a healthy, dense,[3] tamarisk forest well supplied with groundwater can use up to 1,400 millimeters of water per year over a 300-day growing period, suggesting an ET_oF of 0.7. A similar stand experiencing water and salinity stress, such as on upper floodplain terraces, would likely have significantly lower evapotranspiration (Hattori 2004). Similarly, lower stand densities result in lower ET rates.

Recent measurements of ET using Bowen ratio systems at the Cibola Refuge (unpublished data) by the US Bureau of Reclamation have shown a wide variation of tamarisk ET within the same area and illustrate how variation in stand canopy

---

[3] with green leaf area index (LAI) values of 4 or above.

cover, water table depth, soil properties, and salinity levels interact to influence ET.[4] The implication is that remote sensing, which is based on stand canopy cover along with reference ET, will provide a good estimate of the upper limit of ET over a broad area for a particular stand but not of actual or average ET on a plant basis.

Finally, tamarisk often occurs in mixed stands with other species such as cottonwood and Russian olive (Cleverly et al. 2002, Cleverly et al. 2006), such as in the upper reaches of the Colorado River system with narrow riparian zones in canyons. Based on the similar ET rates observed for tamarisk and mesic riparian tree species (see Question 3) and the consistency of ET rates across mixed stands with varying tamarisk cover (Nagler et al. 2005),[5] these mixed stands would have ET rates within the middle range of monotypic tamarisk stands.

Question 5b: Can a range of water savings per acre be agreed to? Can a relative range of water savings between tamarisk and replacement plant communities be agreed to?

The range of water savings (or loss) is large and depends on site ecology, hydrology, and the identity of replacement vegetation. Potential water savings depends upon the replacement of tamarisk with species that require less water. As described in Question 3, this can only occur on sites appropriate for more xeric replacement vegetation. The selection of replacement vegetation should also take into account other desired values, such as wildfire management and wildlife habitat.

In general, potential water savings will range from 50% to 60% to less than zero (if replacement vegetation uses more water than tamarisk). Water salvage will occur only for a few years (during early growth) in areas where riparian species such as cottonwood and willow are the appropriate replacement vegetation for tamarisk. Cottonwoods and willows (CW) can have ET similar to tamarisk.[6] For other replacement vegetation besides CW, potential water savings are higher but vary among species and depend strongly on site factors. Across many studies conducted since 1985 (and excluding tank studies as per the panel's recommendation

---

[4] The annual average ET_oF values varied between 0.3 (peak ET of 2.5 mm/day, 540m m/year) for a location 1.5 kilometers away from the river with a green LAI of around 2 and high salinity level in the ground water, to 0.7 (peak ET of 7.5 mm/day, 1.3 m/year) at a location 750 meters from the river with average salinity levels and similar water depths to the first site but with green LAI values of 4. A third site only 200 meters from the Colorado River presented a measured annual ET_oF value of 0.45 with average peak ET in the summer of 5 mm/day and a total seasonal ET of 870m m/year. This latter site had the best water quality, lowest depth to the water table, but a stand density resulting in a green LAI of 2.6.

[5] Nagler et al. (2005) compared wide-area ET rates on the middle Rio Grande and the upper San Pedro and lower Colorado Rivers using flux tower and remote sensing data and found that riparian vegetation used about 800mm to 900mm/year on all three rivers, despite differing from less than 5% to greater than 80% in tamarisk cover.

[6] Consensus estimate for CW: 0.9–1.4 m/yr [ET_oF = 0.45–0.7]; consensus estimate for tamarisk 0.7–1.4 m/yr [ET_oF = 0.3–0.7]. CW rates do not range as low as tamarisk because CW cannot colonize more xeric, upper floodplain terrace sites that tamarisk can. However, rates as low as 0.6 m/yr have been observed for stressed CW stands.

in Question 1, ET by xeric replacement species such as rabbitbrush, saltgrass, and other xeric-adapted herbs and shrubs averaged 0.5 m/yr (45%) lower than ET by tamarisk (Zavaleta et al. unpublished data).

The greatest opportunity for meaningful water savings will thus be on upper terraces located within the floodplain away from the river, where the water table is deep, the replacement native species more xeric in character, and reduced opportunity exists for reestablishment of tamarisk due to reduced frequency or absence of recurrent overbank flooding. However, the greatest opportunities for recovery of other ecosystem service values may occur in the mesic riparian fringe where water savings are lower.

Finally, the structure and composition of replacement vegetation communities will not be entirely under the control of managers and will vary—planted species can decline and/or other species can possibly colonize over time. It is therefore important to consider ET associated with a range of communities that could come to inhabit the site.

Question 5c: Can water be saved and stream flows increased in the Colorado River system by implementing tamarisk control and restoration actions?

Stream flows can be increased in the Colorado River system through appropriate and well-planned tamarisk control and restoration measures that include

- revegetation as a critical component;
- replacement vegetation for tamarisk on upper floodplain terraces composed of more xeric native vegetation suitable for site-specific precipitation, soils, salinity, and groundwater depths; and
- long-term maintenance of the restoration action occurs.

More conclusive and quantitative answers to the questions of whether and how much water savings will likely occur are not yet available. Well-planned restoration experiments coupled with good ET and hydrologic monitoring and modeling would help provide a more conclusive and quantitative answer. As of now, we have no direct, long-term, before/after studies of tamarisk removal and restoration to the point of mature native/replacement vegetation to conclusively answer this question. The best we have are a few before/after studies that detected short-term reductions in ET or increases in surface water, but these do no not address the long-term course of water recovery as replacement vegetation matures.

Whether water makes it to the channel and increases surface flow or enters groundwater depends on the hydrology of the system. Ground-based measurement, aerial extrapolation, and modeling would help identify where recovered water will go on a reach-by-reach basis. In losing reaches, water recovered from tamarisk will likely go to groundwater stores rather than surface flows and will not be measured as in-stream increases. Hydrologic conductivity also must be sufficient for the salvaged water to enter the channel. Because river systems tend to have very coarse-textured sediments, hydrologic conductivities are almost always sufficient to permit this.

Reaches can be assessed as gaining or losing by several methods, including ground-based measurement of the water table gradient away from the channel, ground-based measurement of rates of upward or downward water flux through the channel bottom, and aerial photography to extrapolate from point measurements on the ground. Where a reach is gaining, the water table rises (i.e., has a higher elevation) away from the channel. In gaining reaches the riparian corridor tends to be much wider with gradation from vigorous vegetation at the channel margins to lower and lower vegetation away from the channel. By contrast, in losing reaches the water table and the phreatophyte riparian vegetation that depend on it can both can decline relatively quickly with lateral distance from the channel. These generalities are, of course, subject to local terrain and geology.

Any increases in Colorado River basin flow due to salvage will be difficult to measure even if they are considerable. Even on gaining reaches, potential water salvage is often a small part of the entire river discharge, would mainly occur during the summer when plants use water, and will be difficult to measure because of large natural variation in flows and stream gauge accuracy. Warming trends and climate variability will also contribute background noise against which long-term water recovery must be measured. Modeling and remote sensing approaches, described more in Question 9, can help overcome these challenges on a reach-by-reach basis.

Question 6: What are the benefits or impacts if tamarisk management within the Colorado River basin states does not occur?

Proactive management has always produced better results, for lower costs, than reactive steps taken in crisis mode. It is reasonable to expect that without tamarisk management, this invasive species will continue to spread, especially in the upper basin, and likely with invasive Russian olive in its understory. Because ecosystems are dynamic, we cannot forecast how rapidly spread will occur in the future or what other surprises might arise. However, continued expansion into new areas will most likely increase ET. Continued expansion will also heighten concerns about fire, because tamarisk-fueled fires kill native cottonwood and willow forests. Continued expansion will likely produce further negative effects on federally listed upriver fishes because of infilling of critical backwater habitat. Proactive management of these exotic species and of the river and riparian ecosystems they affect is necessary to address not only water salvage but also biodiversity conservation, the stability of river beds and banks, and many other values. One panel member disagrees that removing or controlling tamarisk will be beneficial even if there is water savings. Recent research on the lower Colorado River shows that some birds do use tamarisk and that when it is cleared, habitat value decreases (Hinojosa-Huerta 2008; van Riper 2008; Sogge et al. 2008).

The other critical point is that tamarisk management has already occurred; the *Diorhabda* beetle has been released and is spreading on a regional scale within the upper basin (see chapter 22, this volume). At this stage, we must consider what benefits and impacts will accrue if biocontrol proceeds without any additional management measures. First, biocontrol by itself will not finish the job of controlling tamarisk. Without follow-up action, areas where tamarisk has been defoliated by the biocontrol agent may experience reinvasion, resprouting, and new invasions. Second, the chance to reclaim and restore tamarisk-invaded sites controlled by beetles is best when it is proactive rather than reactive. Active revegetation of these sites could save water in the end because it would allow managers to direct or accelerate reestablishment and successional trajectories of competitive native plant communities, rather than allowing equally competitive invasive species to colonize these sites. Finally, biocontrol will reduce ET in the short term by reducing tamarisk ET. However, monitoring after biocontrol will be essential for adaptive management responses such as detection of the need to move in and control secondary invasions.

Conclusion

A key conclusion of the panel is that *native vegetation can use either more or less water than tamarisk*, depending on the identity of the native species, stand densities, and environmental and site conditions such as depth to groundwater and salinity. Panel members expressed two distinct perspectives on how ET rates can be predicted or extrapolated from one site to another: (1) ET depends on several factors that vary by site, making extrapolation relatively complex, and (2) ET is relatively well predicted from canopy characteristics and reference ET (ET_o) for the site in question. Most studies to date on ET rates have taken place in the lower basin states of the Colorado River, but it was agreed that these results could be extrapolated to northern latitudes.

In general, water savings by removing tamarisk will range from 50% to 60% to less than zero (if replacement vegetation uses more water than tamarisk). Water salvage will usually occur only for a few years (during early growth) in areas where riparian species such as cottonwood and willow are the appropriate replacement vegetation for tamarisk. The greatest opportunity for meaningful water savings will likely be on upper terraces located within the floodplain. However, the greatest opportunities for recovery of other ecosystem service values may occur in the mesic riparian fringe where water savings are lower.

More conclusive and quantitative answers to the questions of whether and how much water savings will likely occur are not yet available. Well-planned restoration experiments coupled with good ET and hydrologic monitoring and modeling would help provide a more conclusive and quantitative answer. Whether water makes it to the channel and increases surface flow or enters groundwater depends on the hydrology of the system.

Tamarisk management is already occurring—as described in this chapter, an effective biocontrol agent for tamarisk has been released and is spreading on a regional scale within the upper basin. At this stage, we must consider what benefits and impacts will accrue if biocontrol proceeds without any additional management measures. First, biocontrol by itself will not finish the job of controlling tamarisk. Second, the chance to reclaim and restore tamarisk-invaded sites controlled by beetles is best when it is proactive rather than reactive. Finally, biocontrol will reduce ET in the short term by reducing tamarisk ET. However, monitoring after biocontrol will be essential for adaptive management responses such as the need to control secondary invasions.

All but one of the panel members viewed tamarisk as a negative component of the system overall, one whose continued spread will be a detriment to the river system and whose control is desirable regardless of whether water savings can be demonstrated.

Literature Cited

Allen, R., L. Pereira, D. Rais, and M. Smith.1998. Crop evapotranspiration—guidelines for computing crop water requirements—FAO irrigation and drainage paper 56. Food and Agriculture Organization of the United Nations, Rome, Italy.

Allen, R. G., M. Tasumi, A. T. Morse, R. Trezza, W. Kramber, I. Lorite, and C. W. Robison. 2007. Satellite-based energy balance for mapping evapotranspiration with internalized calibration (METRIC)—Applications. ASCE J. Irrigation and Drainage Engineering 133(4):395–406.

Anderson, M., and S. B. Idso. 1985. Evaporative Rates of Floating and Emergent Aquatic Vegetation: Water Hyacinths, Water Ferns, Water Lilies and Cattails. Proceedings of the 17th Conference on Agriculture and Forest Meteorology and 7th Conference on Biometeorology and Aerobiology, 1985 May 21–24, Scottsdale, Arizona. American Meteorological Society, Boston, MA.

ASABE. 2007. *Design and Operation of Farm Irrigation Systems*. 2nd ed. American Society of Agricultural and Biological Engineers, p. 863.

ASCE-EWRI. 2005. The ASCE Standardized Reference Evapotranspiration Equation. Report 0-7844-0805-X, ASCE Task Committee on Standardization of Reference Evapotranspiration. American Society of Civil Engineers. Reston, VA.

Blaney, H. F., and E. G. Hanson. 1965. Consumptive use and water requirements in New Mexico. New Mexico State Engineering Technical St. Eng. Tech. Report 32, p. 82.

Cleverly, J., C. Dahm, J. Thibault, D. McDonnell, and J. Coonrod. 2006. Riparian ecohydrology: Regulation of water flux from the ground to the atmosphere in the middle Rio Grande, New Mexico. Hydrological Processes 20:3207–3225.

Cleverly, J., C. Dahm, Thibault, D. Gilroy, and J. Coonrod. 2002. Seasonal estimates of actual evapo-transpiration from *Tamarix ramosissima* stands using three-dimensional eddy covariance. Journal of Arid Environments 52:181–197.

Culler, R. C., R. L. Hanson, R. M. Myrick, R. M. Turner, and F. P. Kipple. 1982. Evapotranspiration before and after clearing phreatophytes, Gila River flood plain, Graham County, Arizona. U.S. Geological Professional Paper 655-P.

DiTomaso, J. 2004. Biology and Ecology of *Tamarix*, Presented at the 2004 Western Society of Weed Science meeting. Colorado Springs, CO.

Gatewood, J. S., T. W. Robinson, B. R. Colby, J. D. Helm, and L. C. Halpenny. 1950. Use of water by bottom-land vegetation in lower Safford Valley, Arizona. US Geological Survey Water-Supply Paper 1103: 210.

Hargreaves G. H., and Z. A. Samani. 1985. Reference crop evapotranspiration from temperature. Applied Engineering in Agriculture 1(2):96–99.

Hattori, K. 2004. The Transpiration Rate of Tamarisk Riparian Vegetation. MS thesis. Utah State University, Logan.

Hinojosa-Huerta, O., H. Iturribarria-Rojas, E. Zamora-Hernandez, A. Calvo-Fonseca. 2008. Densities, species richness and habitat relationships of the avian community in the Colorado River Delta, Mexico. Studies in Avian Biology 37:74–82.

Johns, E. L. (ed). 1989. *Water Use by Naturally Occurring Vegetation Including an Annotated Bibliography*. Report prepared by the Task Committee on Water Requirements of Natural Vegetation, Committee on Irrigation Water Requirements, Irrigation and Drainage Division, American Society of Civil Engineers, p. 216. [Online]. http://www.kimberly.uidaho.edu/water/WaterUseNaturalVegetation.pdf.

Lair, K. D. 2006. Summary of Pecos River Revegetation Research—Unpublished Data. Bureau of Reclamation, Denver Technical Service Center, Albuquerque Area Office, Brantley Dam Field Office, Elephant Butte Division Office. July 3.

Leenhouts, J. M., J. C. Stromberg, and R. L. Scott. 2006. Hydrologic requirements of and evapotranspiration by riparian vegetation along the San Pedro River, Arizona. U.S. Geological Survey Fact Sheet 2006–3027. U.S. Geological Survey, Department of the Interior, Washington DC.

Miyamoto, S., E. P. Glenn, and M. W. Olsen. 1996. Growth, water use and salt uptake of four halophytes irrigated with highly saline water. Journal of Arid Environments 32:141–159.

Nagler, P., R. Scott, C. Westenburg, J. Cleverly, E. Glenn, and A. Huete. 2005. Evapotranspiration on western U.S. rivers estimated using the Enhanced Vegetation Index from MODIS and data from eddy covariance and Bowen ratio flux towers. Remote Sensing of Environment 97:337–351.

Qi, J., M. S. Moran, D. C. Goodrich, R. Marsett, R. Scott, A. Chehbouni, S. Schaeffer, J. Schieldge, D. Williams, T. Keefer, D. Cooper, L. Hipps, W. Eichinger, and W. Ni. 1998. Estimation of evapotranspiration over the San Pedro riparian area with remote and in situ measurements. Paper 1.13. Session 1: Integrated observations of semi-arid land-surface-atmosphere interactions, American Meteorological Society Special Symposium on Hydrology. Phoenix, AZ.

Rowlands, P. G. 1990. History and treatment of the saltcedar problem in Death Valley National Monument. Pages 46–56 *in* M. R. Kunzmann, R. R. Johnson, and P. S. Bennett, editors. Tamarisk Control in southwestern United States. U.S. Dept. of Interior, National Park Service, Cooperative National Park Resources Studies Unit. University of Arizona, Tucson.

Scott, R. L., D. Goodrich, L. Levick, R. McGuire, W. Cable, D. Williams, R. Gazal, E. Yepez, P. Ellsworth, and T. Huxman. 2006. Determining the riparian groundwater use within the San Pedro Riparian National Conservation Area and the Sierra Vista Sub-Basin, Arizona. U.S. Geological Survey Scientific Investigations Report, no. 2005–5163.

Scott, R. L., E. A. Edwards, W. J. Shuttleworth, T. E. Huxman, C. Watts, and D. C. Goodrich. 2004. Interannual and seasonal variation in fluxes of water and carbon dioxide from a riparian woodland ecosystem. Agricultural and Forest Meteorology 122(1–2):65–84.

Scott, R. L., W. J. Shuttleworth, D. C. Goodrich, and T. Maddock III. 2000. The water use of two dominant vegetation communities in a semiarid riparian ecosystem. Agricultural and Forest Meteorology 105:241–256.

Shafroth, P. B., J. R. Cleverly, T. L. Dudley, J. P. Taylor, C. van Riper III, E. P. Weeks, and J. N. Stuart. 2005. Profile: Control of *Tamarix* in the western United States: Implications for water salvage, wildlife use, and riparian restoration. Environmental Management 35(3): 231–246.

Shafroth, P. B., V. B. Beauchamp, M. K. Briggs, K. Lair, M. L. Scott, and A. A. Sher. 2008. Planning riparian restoration in the context of *Tamarix* control in western North America. Restoration Ecology 16(1):97–112.

Sogge, M. K., S. J. Sferra, and E. H. Paxton. 2008. *Tamarix* as habitat for birds: Implications for riparian restoration in the southwestern United States. Restoration Ecology 16(1):146–154.

U.S. Army Corps of Engineers (USACE), U.S. Bureau of Reclamation (USBR). 2002. Final environmental assessment and finding of no significant impact for Rio Grande habitat restoration project, Los Lunas, New Mexico. U.S. Army Corps of Engineers District Office and U.S. Bureau of Reclamation Area Office, Albuquerque, NM.

U.S. Bureau of Reclamation (USBR). 1979. Final environmental statement: Pecos River Basin water salvage project, New Mexico-Texas. Regional Office Southwest Region, Bureau of Reclamation, Amarillo, TX.

van Riper III, C., K. L. Paxton, C. O'Brien, P. B. Shafroth, and L. J. McGrath. 2008. Rethinking avian response to *Tamarix* on the lower Colorado River: A threshold hypothesis. Restoration Ecology 16(1):155–167.

Weeks, E. P., H. L. Weaver, G. S. Campbell, and B. D. Tanner. 1987. Water use by saltcedar and replacement vegetation in the Pecos River floodplain between Acme and Artesia, New Mexico: Studies of evapotranspiration. U.S. Geological Survey Professional Paper 491-G.

Wiesenborn, W. D., editor. 1995. Vegetation management study, Lower Colorado River. Phase II Final Report. Bureau of Reclamation, p. 72. Lower Colorado Region, Boulder City, NV.

Yepez, E. A., D. G. Williams, R. L. Scott, and L. Guanghui. 2003. Partitioning overstory and understory evapotranspiration in a semiarid savannah woodland from the isotopic composition of water vapor. Agricultural and Forest Meteorology 119:53–68.

Young, A. A., H. F. Blaney.1942. Use of water by native vegetation. California Department of Public Works, Divisiono of Water Resources Bulletin 50:160.

Appendix A: Peer Panel Process

A peer review selection process was developed to ensure that panel members had the appropriate experience and expertise to participate, represented various aspects on the issue, had no vested interest in the outcome of the discussion, and were available. The selection process was open with key stakeholders' reviews requested to identify and approve panel members.

The Tamarisk Coalition identified a list of approximately 25 experts from its contact network and from stakeholder recommendations. These individuals were identified based on their knowledge and expertise in the areas of riparian and floodplain upper terrace ecosystems ecology; ET rate measurement of vegetation associated with these ecosystems; hydrologic interaction between vegetation, groundwater and surface water; and tamarisk control and revegetation approaches. Every effort was made to bring together a balanced panel with diverse experiences and opinions.

The Peer Review

The purpose of the Peer review was to synthesize information on tamarisk evapotranspiration (ET) rates as well as those of potential replacement vegetation. The literature on tamarisk and native riparian vegetation describes a wide range of ET rates for each vegetation type. Partially as a result of this incongruence there is no consensus among the scientific community if a reduction in tamarisk will increase the availability of water resources.

The Panel's tasks were to

1. reach a consensus, not necessarily unanimity, on a narrower range of tamarisk and replacement vegetation ET rates in various ecosystems and climates, considering elevation and latitude;
2. identify areas of additional research needs; and
3. reach a consensus, not necessarily unanimity, on how groundwater and surface water may respond to changes in vegetation over time.

The selected panel was composed of the following 10 individuals:

- Richard Allen, University of Idaho, remote sensing, ET measurement, and modeling and hydrology
- Dan Bean, Colorado Department of Agriculture, biological control entomologist
- Dan Cooper, Los Alamos National Laboratory, Remote sensing—LIDAR
- Ed Glenn, University of Arizona, ecophysiology and remote sensing
- David Groeneveld, Hydrobio, remote sensing
- Ken Lair, H. T. Harvey and Associates, restoration vegetation
- Christopher Neale, Utah State University, remote sensing and mapping, ET measurement and modeling
- Richard Niswonger, US Geological Survey, hydrology and modeling
- Anna Sher, University of Denver and Denver Botanic Gardens, restoration ecology
- Erika Zavaleta, University of California, Santa Cruz, restoration ecology

Dr. Zavaleta served as the Panel chair, and Tim Carlson of the Tamarisk Coalition served as moderator. The Panel met at the University of California, Santa Cruz, over a two-day period dictated by the schedules of the Panel members (November 11–12, 2008). This chapter is a shortened version of the official report made from this meeting, with contributions from each of the participants and edited by Dr. Zavaleta.

5

Tamarisk: Ecohydrology of a Successful Plant

Pamela L. Nagler and Edward P. Glenn

The title of this chapter is adapted from an influential 1987 paper "Tamarix: Impacts of a Successful Weed" (Brotherson and Field 1987). That paper made the case for tamarisk (*Tamarix* spp.) removal, as a high-water-use, invasive species that out-competed and displaced native vegetation on western US rivers. The paper declared that tamarisk was capable of drying up water courses, using as much as 4 meters of water per year (i.e., a water table depth of 4 meters of water over an indefinite area, the same way that rainfall is expressed), water that could otherwise support environmental or human water needs. It was also thought to out-compete native trees for establishment sites due to prolific seed production. The paper concluded: "A firm commitment must be made concerning the control of saltcedar because of its unparalleled aggressiveness." This conclusion became the reigning paradigm in riparian restoration efforts for the next decade.

Much has been learned since 1987. The intervening years have produced new methods of unobtrusively measuring water use in natural stands of riparian vegetation, and numerous new studies have been published on the physiology and ecology of tamarisk and other riparian species (see chapter 1, this volume). In the five years prior to 1987, six papers on the biology or ecology of tamarisk in North America were published; by contrast, from 2006 to 2010, 117 papers were published, covering all aspects of tamarisk physiology, water use, interactions with native plants, habitat value for wildlife, abundance, distribution, and potential for control (see reviews in Glenn and Nagler 2005; Shafroth et al. 2009, and chapters therein).

We modified the title of the 1987 paper to reflect the more nuanced view of tamarisk that has emerged from these studies. We replaced the word "impacts" (which were mostly negative due to its assumed high water use) with "ecohydrology" to denote a more objective assessment of tamarisk water use and ecological interactions with native plants. Similarly, we substituted the word "plant" for "weed." A weed is an undesirable, troublesome, unattractive plant, growing where it is not wanted. It is no longer clear that tamarisk fits that definition under all circumstances on western US rivers (Stromberg et al. 2009).

In this chapter we review tamarisk water use, salt tolerance, potential for salinizing flood plains, rooting depths, and ecohydrological interactions with native plants on western rivers. The working hypothesis emerging from our review is that tamarisk is a stress-adapted species with low to moderate water use that tends to replace mesic vegetation when conditions on flow-regulated rivers become unsuitable for those species, rather than as an invasive species that displaces and out-competes native species under all conditions.

Tamarisk Water Use

EARLY STUDIES AND THE DEMONIZING OF TAMARISK

A fascinating history of tamarisk research (see chapter 16, this volume) describes the origins of the perception that tamarisk has unusually high water use. The first concerted efforts to determine tamarisk water use were conducted by US Geological Survey (USGS) hydrologists trying to determine how much water could be salvaged for human use by clearing tamarisk from western rivers. Starting in the 1930s, the Phelps Dodge Company in Arizona began attempting to obtain water credits to open a new open-pit copper mine and mill in Morenci. It proposed removing tamarisk from the middle Gila River, whose surface waters were already fully apportioned for agriculture. In theory, clearing tamarisk could reduce evapotransporation (ET) losses, thereby increasing surface flows, and the increases could be used to operate the new mine and mill. Copper production became a national priority during World War II, and the USGS was charged with measuring phreatophyte water use. They established study plots on the Gila River, the Pecos River, and at other sites where tamarisk was considered a problem.

ET requires a source of energy to convert water from the liquid to the vapor phase. This energy is supplied by net radiation (R_n); a simplified equation for the surface energy balance is:

$$\lambda ET = R_n - G - H \qquad (1)$$

where λ is the latent heat of evaporation of water; R_n is net radiation flux (R_s minus outgoing shortwave and longwave radiation); and G is soil heat flux (units are W m^{-2}). ET is now measured on the ground across the continent (AmeriFlux) and globe (FluxNet) using an array of eddy covariance moisture and carbon flux towers. Bowen ratio towers are also common. They measure fluxes of moisture, H, and CO_2 in and out of a canopy, and additional instruments measure windspeed,

R_n (by net radiometers), and G (by soil heat flux plates) so that all terms in equation (1) are measured.

When the USGS scientists began their studies, these methods were not yet well developed. Nevertheless, several innovative approaches to ET estimation were attempted, some of which helped pave the way for methods used today. For example, Weeks et al. (1987) deployed a prototype eddy covariance flux tower and obtained ET estimates of 600 to 1100 mm yr^{-1} on the Pecos River in New Mexico, similar to estimates obtained today. Similar moderate estimates of tamarisk ET were obtained by others. Ball et al. (1994) measured rates of 700 to 800 mm yr^{-1} by Bowen ratio tower measurements at Blythe on the lower Colorado River, within the range of native vegetation (*Atriplex lentiformis* and *Prosopis glandulosa*) measured at the same location.

On the other hand, some high-end estimates of tamarisk water use were also generated. These came mainly from lysimeter tanks installed in tamarisk stands (van Hylckama 1974; Horton and Campbell 1974; Gay and Hartman 1982; Davenport et al. 1982). Shallow concrete tanks were filled with soil and planted with tamarisk, and then water was added to create a shallow water table. Evapotransporation was estimated by the rate at which water had to be added to the tank to maintain the water level. These tanks produced ET rates at or above the local rate of potential ET (ET_o), and ranged from 1,500 to 2,500 mm yr^{-1} (Gay and Hartman, 1982). A few produced rates as high as 3,000–4,000 mm yr^{-1} (Davenport et al. 1982)

We now understand the reasons for these high rates. First, the tanks had an unnaturally high water table compared to natural stands, and, for all phreatophytes, ET tends to decline as depth to water increases. Second, the tanks tended to be isolated from surrounding plants, creating an "oasis" or "clothesline" effect that can increase ET to well above ET_o (Allen et al. 1998). Warm, dry air surrounding the tank will flow toward the transpiring, cooler plants in the tank, greatly increasing their capacity for ET by increasing atmospheric water demand and the net energy available to evaporate water. In fact, the authors of the lysimeter studies urged caution in concluding that large water savings could be obtained by clearing tamarisk (Davenport et al. 1982; van Hylckama 1974; Horton and Campbell 1974). Davenport et al. (1982) pointed out that tamarisk water use declined with increasing depth to water, and van Hylckama (1974) pointed out that natural stands often grow on terraces with water depths of four meters, and that potential water savings based on lysimeters were overestimates of what could actually be achieved. Horton and Campbell (1974) noted the high variability of ET among different stands of tamarisk in lysimeters and called for measurements in wild stands of plants. In 1982 Anderson conducted field and laboratory experiments showing that tamarisk did not have unusually high rates of either transpiration or photosynthesis compared to other plants. This study was important in that it directly compared transpiration and photosynthesis of tamarisk and native plants on a leaf area basis over a range of temperatures and light intensities in controlled studies, and then extended the findings to natural stands of plants. It showed that tamarisk downregulated transpiration rapidly in response to stress, reducing water use, and concluded: "Failure to treat stomatal resistance as a variable in attempts to predict ET from meteorological data and stand characteristics may result in significant overestimates" (Anderson 1982).

Nonetheless, subsequent reviews of tamarisk water use tended to emphasize the high-end estimates and to ignore the caveats. Brotherson and Field (1987) summarized the evidence as indicating that water use by tamarisk ranged from 1300 to 4700 mm yr^{-1}, with typical values of 1700 mm yr^{-1}, approximately equal to ET_o. Often, tamarisk water use was described metaphorically to emphasize the amount of water lost for human use. Di Tomaso (1998) stated that tamarisk water use was among the highest measured for any phreatophyte in the southwestern United States, equal to twice the water use of the Southern California cities. An analysis of the potential economic savings that could result from tamarisk control (Zavaleta 2000) cited an unpublished metasurvey of 20 ET studies showing that tamarisk used 300 to 460 mm yr^{-1} more water than native vegetation, which could produce water salvage worth \$233 million if tamarisk were to be replaced with native species throughout its range in the western United States. Hence, it became a priority to estimate actual ET by natural stands of tamarisk, something that was difficult to do in the 1980s.

Perhaps the most egregious overestimate of tamarisk water use was the often-cited statement that a single tree can transpire 757 L day^{-1} (e.g., Di Tomaso 1998). Owens and Moore (2007) attributed the estimate to an unsupported statement in a review article on tamarisk control methods (Holdenbach 1987). They showed that this estimate is two standard deviation units beyond the daily ET of other woody species, and is likely to be five times too high based on their own measurements of tamarisk ET.

CONTEMPORARY ESTIMATES OF TAMARISK

New technological developments over the past 20 years have produced operational methods for unobtrusively measuring ET in natural stands of plants at temporal and spatial scales relevant to hydrological and ecological studies. The methods include moisture flux towers, plant sap-flow sensors, and satellite imagery (Glenn et al. 2007). Results from the different measurement systems are shown in table 5.1 and are discussed in more detail below.

Moisture flux towers measure either the gradients in temperatures and moisture content of air at two heights over a canopy (Bowen ratio towers) or fluxes of heat and moisture in and out of canopies at a single point over the canopy (eddy covariance towers; Glenn et al. 2007). Both methods produce estimates of ET over a footprint area of several thousand square meters. These towers have been employed on western rivers having varying amounts of tamarisk, including the upper San Pedro River (Scott et al. 2008), the lower Colorado River (Westenberg et al. 2006; Chatterjee, 2010), the middle Rio Grande (Cleverly et al. 2002, 2006;

TABLE 5.1

Annualized rates of evapotranspiration (ET) by tamarisk stands measured by different methods on western US rivers.*

Location	Method	ET (mm yr-1)	Comments	References
Virgin River, NV	Bowen ratio flux tower	700–1400	Same site in two consecutive years.	Devitt et al. 1998
Lower Colorado River, CA	Bowen ratio flux tower	400–1400	Three sites at Cibola National Wildlife Refuge.	Chatterjee 2010
Lower Colorado River, CA	Bowen ration flux tower	780–980	Three sites at Havasu National Wildlife Refuge.	Westenberg et al. 2006
Middle Rio Grande, NM	Eddy covariance flux tower	800–1200	Two sites over multiple years	Cleverly et al. 2002, 2006
Lower Colorado River, CA	Sap-flow sensors	307–1460	Six sites over two years at Cibola National Wildlife Refuge	Nagler et al. 2009b, c
Pecos River, NM	Sap-flow sensors	700	One site in dense stands	Owens and Moore 2007
Dolores River, UT	Sap-flow sensors	220	Typical stand on river terrace	Hultine et al. 2010
Pecos River, TX	Diurnal ground water fluctuations	420–1180	Dense stands on riverbanks	Hatler and Hart 2009
Lower Colorado River, CA and AZ	Remote sensing	447–1155	Wide area estimates for six wildlife refuges, Lake Mead to US/Mexico border	Murray et al. 2009
Upper Basin of Colorado River, NV, WY, UT	Remote sensing	220–580	Tributary rivers to Colorado	Nagler et al. 2011

*The table includes all studies that used unobtrusive measurement techniques on natural stands of plants and had sufficient data to estimate annual rates of ET.

Dahm et al. 2002), and the Virgin River (Devitt et al. 1998). Over saltcedar monocultures, ET rates ranged from 250 to 1400 mm yr^{-1}, with a mean value of 850 mm yr^{-1}. Expressed as a fraction of ET_o (ET_oF), the mean is 0.47 of potential ET for the flux tower studies in table 5.1. For perspective, ET_o for a grass reference crop ranges from 1,400 to 2,000 mm yr^{-1} in the southwestern United States, and alfalfa (the top irrigated crop in the region) uses 1.0 to 1.3 times ET_o (Allen et al. 1998; Murray et al. 2009).

Sap-flow sensors measure the rate of water flow through plant stems to the leaves. They attach to the trunk or branches of a plant and introduce a source of low grade heat into either the woody tissue or the transpiration stream. Heat pulse methods introduce a pulse of heat into the stem and measure the velocity of the sap by the time it takes the heat pulse to reach a heat sensor placed upstream from the heat source (Green et al. 2003). Granier sensors introduce a constant source of heat into the transpiration stream through a needle penetrating the sapwood layer, and measure transpiration by the temperature at the heat source and at a similarly placed needle downstream from the heat source (Granier 1987). Both methods must be empirically calibrated for each plant type and application. Tissue heat balance methods wrap a heating wire around small stem segments, and uniformly heat the stem segment through a constant-voltage power source (Kjelgaard et al. 1997; Grime and Sinclair 1999). This method solves an energy balance equation and does not require calibration for different plant species.

All these methods have sources of error or uncertainty, and they require ancillary measurements to scale them from stem measurements to plant stand estimates (see chapter 4, this volume). However, they represent an advance over earlier intrusive methods for obtaining plant-level transpiration, such as porometers, which enclose the leaves of plants in measurement chambers and cannot be easily scaled to stand-level plant water use.

Sap flux sensors have been deployed on tamarisk shrubs on the lower Colorado River (Nagler et al. 2009b, c); the Virgin River (Sala et al. 1996; Devitt et al. 1997); the Dolores River, a tributary of the Colorado River (Hultine et al. 2010); the Pecos River (Owens and Moore 2007); and the middle Rio Grande (Moore et al. 2008). In general, results are similar to moisture flux tower estimates, ranging from 220 to 1500 mm yr^{-1}. Sala et al. (1996) measured rates for individual shrubs that were 1.6 to 2.0 times ET_0 at a nonsaline site on the Virgin River, Nevada, but rates were only half as great at a nearby saline site. Similar rates were found for other riparian species. They did not extrapolate their results to whole stands, but speculated that tamarisk could have high water use if it maintained higher leaf area index ([LAI]; defined as the one-sided surface area of leaves on a plant in square meters per square meter of ground area) than other plants.

Differences in tamarisk ET among sites are due to differences in stand structure (e.g., Sala et al. 1996; Nagler et al. 2008, 2009a,b), depth to the water table (Horton et al. 2001a,b, and 2003), salinity of the water table (Cui et al. 2010), and soil texture differences among sites (Nagler et al. 2008, 2009a,b). For example, five of six tamarisk sites measured on a terrace of the Cibola National Wildlife Refuge on the lower Colorado River had marked midday depression of ET (Xu and Shen, 2005) that reduced water loss per leaf (see figure 5.1) (Nagler et al. 2009a,b). Leaf area index of tamarisk stands generally ranges from 1 to 4 (Nagler et al. 2004), with higher ET recorded for stands with higher LAI (Sala et al. 1996; Nagler et al. 2008, 2009a,b). Salinities above about 10 g L^{-1} in the aquifer, or water tables below 3 to 4 meters depth, tend to lower both LAI and ET. ET can also be reduced by daytime cavitation of the soil-plant water continuum in sandy soils (Sperry et al. 1998; Nagler et al. 2009a, b). As a phreatophyte, tamarisk usually obtains its water from the wetted capillary fringe of soil above the aquifer, and transport through this rather narrow zone of water can become a constraint on plant water uptake in sandy soils (Hultine et al. 2006).

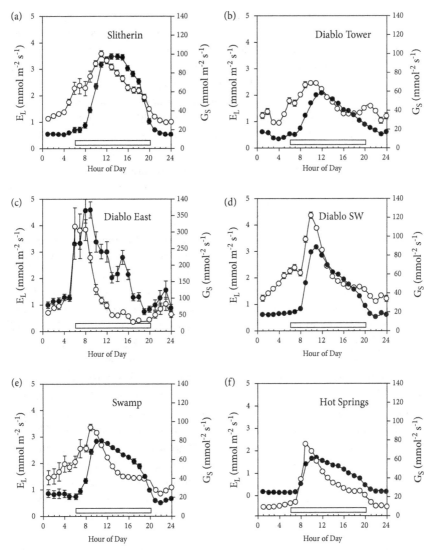

FIGURE 5.1 Diurnal curves of transpiration (*closed circles*) and stomatal conductance (*open circles*) at six tamarisk stands at the Cibola National Wildlife Refuge on the lower Colorado River, measured by sap-flow sensors. Note the severe midday depression of stomatal conductance and transpiration at all sites except Slitherin.

Source: Adapted from Nagler et al. (2009b, c).

Other nonobtrusive ground methods have also been employed to estimate tamarisk ET. Hatler and Hart (2009) used diurnal fluctuations in groundwater depths to project ET rates of 420 to 1180 mm yr⁻¹ for dense stands on the Pecos River, Texas.

The above results are all point estimates of ET, and most were conducted in dense stands that might not be representative of typical stand densities along a river. Most river terraces are mosaics of sparse vegetation, areas of bare soil, and stands of dense vegetation, those near the river channel. Remote sensing with satellite imagery offers a means to scale point estimates over larger areas using vegetation indices such as the Normalized Difference Vegetation Index (NDVI) or the Enhanced Vegetation Index ([EVI]; (Glenn et al. 2007, 2010). Typically, a satellite VI is combined with ground ET and meteorological data from flux towers or sap-flow sensor stations to create an algorithm by which ET can be scaled over whole river systems. The error or uncertainty of these estimates is 10% to 30%, determined largely by the error or uncertainty in the ground estimates of ET.

Nagler et al. (2005) used remote sensing to compare ET on three rivers differing in tamarisk cover: the upper San Pedro in Arizona (less than 5% tamarisk); the middle Rio Grande (ca. 30%–50% tamarisk); and the lower Colorado River (85% tamarisk). All three rivers had similar, modest rates of ET, ranging from 851 to 874 mm yr^{-1}. The upper San Pedro was dominated by honey mesquite trees (*Prosopis glandulosa*) at varying densities, with cottonwoods (*Populus fremontii*) and willows (*Salix* spp.) along the river banks and giant sacaton grass (*Sporobulus wrightii*) in patches on the terraces. Cottonwoods and willows were codominant with tamarisk and Russian olive (*Elaeagnus angustifolia*) trees in the understory on the middle Rio Grande, while tamarisk and arrowweed (*Pluchea sericea*) were the dominant species on the lower Colorado River. Thus, ET was remarkably unaffected by species composition, and ET_oF was only 0.5 across rivers.

Evapotranspiration rates of tamarisk in the upper basin tributary streams were much lower than on the lower Colorado River below Lake Mead, averaging 330 mm yr^{-1} in the upper basin (Nagler et al. 2011) and 854 mm yr^{-1} in the lower basin (Murray et al. 2009). Tributary rivers such as the Dolores River (Hultine et al. 2010) tend to have lower plant cover and greater water stress than tamarisk stands on the lower Colorado River (Nagler et al. 2009a,b). As with flux tower and sap-flow studies, results for individual plant stands were much more variable than for larger river reaches. On the upper basin streams, most stands had rates below 350 mm yr^{-1}, but those at Lovelock, Nevada, on the Carson River, had an ET rate of 580 mm yr^{-1}, and midsummer rates equaled or exceeded ET_o (Nagler et al. 2011). On the lower Colorado, riparian ET (dominated by tamarisk) ranged from 447 mm yr^{-1} along the river in the Mohave Irrigation District south of Las Vegas, Nevada, to 1155 mm yr^{-1} at the Imperial National Wildlife Refuge north of Yuma, Arizona (Murray et al. 2009).

COMPARISON WITH NATIVE SPECIES AND PROSPECTS FOR WATER SALVAGE

Several reviews have attempted to compare ET rates by tamarisk to those of native plants, to estimate how much water could theoretically be saved by replacing tamarisk with a given native plant (Hughes, 1970; Shafroth et al. 2005; Carlson et al. 2006; Zavaleta 2000; Nagler et al. 2009c; see also chapter 4, this volume). However, all riparian species show a wide range in actual ET rates, determined as much by environmental circumstances as by species. As an example, Shafroth et al. (2005) proposed that saltgrass (*Distichlis spicata*) had lower water use than tamarisk and might result in a net salvage of water. An early attempt to salvage water along the Pecos River in New Mexico was also predicated on replacing tamarisk with saltgrass (Hughes 1970). However, the literature shows that saltgrass can have low ET rates (ca. 200–300 mm yr^{-1}; Shafroth et al. 2005) at some locations, but higher rates at other locations, for example 450 to 650 mm yr^{-1} in natural stands in Nevada (Nichols et al. 1997). Cottonwoods and willows generally have ET in the range of 450 to 1400 mm yr^{-1} (Dahm et al. 2002; Gazal et al. 2006; Schaeffer et al. 2000; Nagler et al. 2007; Hartwell et al. 2010), higher on average than tamarisk, while mesquites range from 400 to 600 mm yr^{-1} (Ansley et al. 2002; Scott et al. 2004, 2008), similar to tamarisk stands. Nagler et al. (2009b) compared tamarisk to four co-occurring native riparian species (mesquite, arrowweed, tamarisk, and cotton-wood) with respect to ET, LAI, and other physiological characteristics, based on published values and measurements at Cibola National Wildlife Refuge on the lower Colorado River. Evapotranspiration and LAI rates overlapped among species. A definitive experiment would need to actually remove tamarisk and replace it with a native species at a given location, with ET rates measured before and after replacement, but such experiments have rarely been reported.

Some studies have attempted to remove tamarisk from riparian areas without replacement vegetation to document a potential water savings. Cleverly et al. (2006) removed tamarisk from the understory of cottonwood trees at a site on the middle Rio Grande, and were able to document a one-time reduction in ET of 260 mm yr^{-1}, but the understory plants quickly grew back. More recently, Moore et al. (manuscript in preparation) removed tamarisk from a mixed cottonwood/tam-arisk canopy and found no net reduction in ET, because cottonwood ET increased when tamarisk was removed. Hatler and Hart (2009) treated tamarisk monocul-tures with herbicide along the banks of the lower Pecos River, Texas, and found an 83% reduction in ET after two years, but only 31% after four years, due to regrowth of tamarisk and replacement vegetation. They documented a temporary rise in the shallow aquifer in response to clearing. Barz et al. (2009) analyzed the costs and benefits of a program that eradicated tamarisk from a 370-kilometer stretch of the middle Pecos River in New Mexico. Using satellite imagery, they projected that ET decreased by 205 mm yr^{-1} from 2002 (pretreatment) to 2005 (posttreat-ment) over the 2,412-hectare treated area. The purpose was to increase the base flow in the river to fill downstream reservoirs. However, no net increase in base flow was recorded. Instead, it was assumed that the reduced ET was recovered in groundwater pumping for agriculture along the river. Total direct costs for herb-icide application were $2.2 million, and the (assumed) salvaged water was worth $63,000, making water salvage through herbicide treatment uneconomic. More importantly, they documented increased erosion of the channel and sediment run-off into downstream reservoirs following tamarisk removal, which would result in additional costs to maintain reservoir storage capacity in years to come. Further, avian nesting habitat was destroyed, and the high salinity of the aquifer precluded revegetation with native trees as replacement habitat on that river stretch.

As a historical note, none of the early proposals to salvage water by tamarisk removal led to a sustainable recovery of additional water. For example, from 1967 to 1982 the US Bureau of Reclamation conducted a tamarisk control program on a portion of the Pecos River in New Mexico in which tamarisk was replaced with saltgrass (Weeks et al. 1987). Despite projections of an increase in base flow and positive benefits/cost ratios (Great Western Research 1989), the project was not successful. Water table levels rose in some areas of the river, indicating that some salvage had been realized, but the expected doubling in river base flow did not occur (Welder 1988; Duncan et al. 1993). The investigators concluded that any water salvage appeared as a rise in the local water table rather than the desired increase in base flow to fill downstream reservoirs. Similarly, Hart et al. (2005) did not find a net increase in base flow or decrease in salinity in the lower Pecos River, Texas, following widescale eradiation of tamarisk. As for the Morenci copper mine that started the "water salvage" programs, Phelps Dodge eventually contributed to the construction of Horseshoe Dam on the Verde River, earning them rights to nine times as much water than could be optimistically recovered by tamarisk clearing on the Gila River.

Tamarisk and Salinity

SALT TOLERANCE OF TAMARISK AND NATIVE RIPARIAN PLANTS

Western US rivers tend to have moderate base levels of salts in their surface flows. These salts originate from natural sources in the watersheds, some of which were formed from the uplift of ancient sea beds. Salinity in surface flows is increased by the diversion of water for agriculture and the discharge of agricultural return flows back to the rivers. Surface flows range in salinity from low (e.g., 0.2 g L^{-1} in the upper San Pedro) to moderate (1 g L^{-1} in the lower Colorado River at the United States/Mexico border) to high (5 g L^{-1} in the Pecos River in New Mexico and Texas). On flow-regulated rivers without seasonal overbank flooding, salinities can reach 5 to 25 g L^{-1} in the aquifer and even higher in the vadose zone above the aquifer because of lack of flood flows to leach salts from the soils and aquifers (Nagler et al. 2009a, b). Hence, many flow-regulated rivers and associated wetlands have been converted from mesic to highly saline environments (Jolly et al. 2008), decreasing mesic native trees from the riverbanks and terraces, and dominance by salt-tolerant shrubs and grasses.

Tamarisk is a halophyte, falling on the salinity tolerance spectrum above other salt-tolerant riparian species such as arrowweed (*Pluchea sericea*) but below iodine bush *Allenrolfia occidentalis* (see figure 5.2; Glenn et al. 1998; Vandersande et al. 2001) or other euhalophytes that are actually stimulated by salinity (Miyamoto

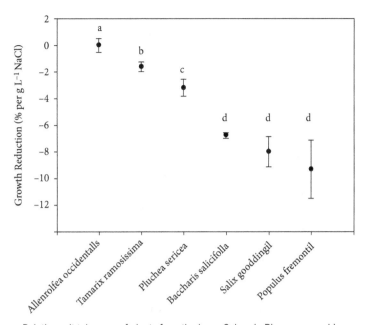

FIGURE 5.2 Relative salt tolerance of plants from the lower Colorado River measured in a greenhouse study in Tucson, Arizona. Different letters of values denote significant differences at P < 0.05.

Source: Glenn et al. (1998).

et al. 1996). In greenhouse trials, tamarisk growth was reduced by 2% per g L^{-1} salinity increase (figure 5.2), with growth potential reduced by 50% on a salinity of 25 g L^{-1} (Glenn et al. 1998). By contrast, the most common riparian pioneer trees, cottonwood and willow, are glycophytes, with little innate salinity tolerance (Rowland et al. 2004). In greenhouse trials, growth of these plants was reduced by 50% at a salinity of 5 to 6 g L^{-1} (Glenn et al. 1998), and field observations show that ET rates of cottonwoods are reduced by 50% when groundwater salinities are in the range of 2 to 5 g L^{-1} (Pataki et al. 2005). In riparian settings, cottonwoods and willows are usually not found where groundwater salinity exceeds 2.5 g L^{-1} (Pataki et al. 2005), whereas tamarisk grows on terraces where the groundwater ranges from 2.5 to 20 g L^{-1} (Cui et al. 2010; Natale et al. 2010). Mesquites tend to have higher salt tolerance than cottonwoods or willows, with growth reductions of 50% on 6 to 12 g L^{-1} in greenhouse trials (Felker et al. 1981). Field observations suggest that screwbean mesquite (*Prosopis pubescens*) has greater tolerance than honey mesquite (*P. glandulosa*) (authors' unpublished observations). On the lower Colorado River, stunted screwbean mesquites are found growing as occasional plants in tamarisk and arrowweed stands where groundwater salinities are in the range of 2.5 to 7.5 g L^{-1} (Nagler et al. 2008, 2009b, c).

Tamarisk does not usually grow on seawater salinities in coastal salt marshes; for example, it is absent from the estuaries of the northern Gulf of California although it grows in adjacent lower-salinity agricultural and riparian ecosystems (Glenn et al. 2006). However, tamarisk has become established at the backs of some Southern California salt marshes subject to periodic flooding by fresh water from storm runoff (Whitcraft et al. 2007).

TAMARISK AND THE SALINIZATION OF RIVERBEDS

Tamarisk has salt glands and actively excretes salts from it leaves (Storey and Thomson 1994). As a result, salts from tamarisk leaf litter can accumulate on the soil surface of nonflooding river terraces. An interesting chicken-and-egg dispute is whether native trees are excluded from riverbanks due to deposition of salts by tamarisk, or whether riverbanks became saline due to lack of overbank flooding, facilitating replacement of native trees by tamarisk when their mesic niches were no longer available (see chapter 8, this volume). Several reviews (e.g., Brotherson and Field 1987; Di Tomaso 1998; Zavaleta 2000) concluded that tamarisk salt deposition results in the competitive exclusion of cottonwood and willows from tamarisk-dominated riverbanks. A corollary assumption is that removing saltcedar might allow native vegetation to reestablish on affected sites (DeLoach et al. 2000; DeLoach and Carruthers 2004).

Only a few studies have attempted to quantify the role of tamarisk in salt deposition, and these present a more complicated picture than was initially assumed. Ladenburger et al. (2006) reported that soil salinity levels were higher under tamarisk shrubs compared to between shrubs, but that salt levels were not high enough to disrupt the germination of native shrubs and trees at riparian sites in Wyoming. Lesica and DeLuca (2004) also reported elevated salt levels under tamarisk shrubs, but germination of a native grass was actually higher on these soils than on control soils, and they concluded that saltcedar is not allelopathic. Cederborg (2009) reported elevated salt levels under tamarisk plants in both flooding and nonflooding sites on the middle Rio Grande, New Mexico, but paradoxically, salt levels were lowest under the oldest tamarisk plants, and were not so high that restoration with native vegetation would be inhibited by salinity. Yin et al. (2009), working in the native range of tamarisk in Asia, reported that salt levels were higher in tamarisk mounds than between mounds, but that fertility factors such as potassium, organic matter, and phosphorous were also higher, and that tamarisk mounds were overall a net positive factor in rehabilitating soils for restoration.

A second source of salts in riparian soils is capillary rise of water and salts from the aquifer into the vadose zone (Silliman et al. 2002). In fine-textured soils the capillary fringe can extend to 1 to 3 meters more above the water table (Gerla 1992) and salts can accumulate to high levels because of the evaporation of water from the soil surface and deposition of salts in the vadose zone. In fact, this is one mechanism by which salt flats are formed (Kinsman 1969). Vegetation tends to counter this effect, because plants can lower the water table through ET (Gerla 1992). Hence, it is not clear that simply removing tamarisk from a riverbank will improve salinity or allow native mesic vegetation to return. Very likely, a return of overbank flooding would be necessary to return mesic species to the riparian zone of many regulated rivers.

Three recent studies have shed further light on this question (Glenn et al. 2012; Merritt and Shafroth 2012; Ohrtman et al. 2012). These studies showed that salts on the terraces of the lower Colorado River were high primarily because of capillary rise of salts from the saline aquifer; salts accumulated in the soils over time due to lack of flooding. On the other hand, on the less-disturbed upper Colorado River, salt levels were higher in surface soils under tamarisk stands compared to native stands and accumulated to levels that could inhibit the germination of native plants (Merritt and Shafroth 2012). Ohrtman et al. (2012) conducted a detailed assessment of factors controlling soil salinity on the Middle Rio Grande. They found that lack of overbank flooding was the most important factor controlling salinity but other factors such as distance from the river and vegetation species composition and age were also significant. In general, soil salinity tended to be higher under *Tamarix* compared to native plants, but soil salinity was highest in bare areas compared to either *Tamarix* or native plants. They concluded that physical factors such as capillary rise and high evaporation rates increased salt deposition in bare areas, but that *Tamarix* salt exudates also contributed to soil salinity with *Tamarix* stands. Taken together, these studies show that although *Tamarix* appears to contribute to surface salinity, the role of capillary rise should not be underestimated.

Tamarisk and Water Stress

DROUGHT TOLERANCE AND ROOTING DEPTHS OF TAMARISK AND NATIVE RIPARIAN SPECIES

Regulated western rivers often have deeper aquifers than occur on unregulated rivers, due to diversion of water away from the river for human use and lack of overbank flooding to recharge aquifers. Even unregulated rivers are subject to low-flow conditions during droughts. Numerous physiological studies report greater drought tolerance for tamarisk compared to cottonwood and willow trees under field conditions (Sala et al.1996; Devitt et al.1997). In response to drydowns, tamarisk is able to maintain tighter stomatal control than mesic native species, limiting water loss during dry conditions (Anderson 1982; Devitt et al.1997). Tamarisk and mesquite are also less susceptible to cavitation of xylem elements during water stress than cottonwood and willow, a trait that can define maximum drought tolerance of a species.

Tamarisk also reportedly has greater rooting depth, faster root growth in response to a declining water table, ability to use vadose zone moisture as well as groundwater, and higher physiological tolerance to low water potential (Busch et al. 1992; Busch and Smith 1993; Horton et al. 2001a and b, 2003; Devitt et al. 1997) compared to cottonwood and willows, which are generally considered to be unable to access groundwater deeper than 2 to 3 meters below the surface (Pataki et al. 2005). Horton et al. (2003) found that tamarisk, willow, and cottonwood were all able to extract water from as deep as 4 meters on the Bill Williams River, but willow and cottonwood showed decreasing predawn water potential, indicating water stress, with increased depth to groundwater, whereas tamarisk was unaffected. Other studies show that tamarisk can utilize groundwater as deep as 6 meters (Horton and Campbell 1974).

However, tamarisk is not the deepest-rooted riparian tree on western floodplains. Mesquites can extract water from 10 meters (Scott et al. 2004) or deeper (Canadell et al. 1996). Fourwing saltbush (*Atriplex canescens*) and black greasewood (*Sarcobatus vermiculatus*), frequently found on upper riparian terraces in the Great Basin province, can root into groundwater 10 to 20 meters deep (Jordan et al. 2008), and have salt tolerance equal to tamarisk (Glenn and O'Leary 1984).

The analysis of stable isotopes of water (deuterium and ^{18}O) in stem tissues and in possible sources of water from the environment allows an accurate determination of where a plant obtains its water supply. Stable isotope studies have revealed complicated temporal and spatial patterns of water use by riparian plants across what appear to be uniform floodplains (Smith et al. 1998). At Bishop Creek in the Sierra Nevada, cottonwood (*Populus trichocarpa*) used mainly shallow soil moisture during the high water period in the spring, but by late summer it used only groundwater when surface soils were dry and river flow was low (Smith et al. 1998). On the other hand, Busch et al. (1992) found that large, mature cottonwoods and willows on the Bill Williams River (tributary to the lower Colorado River) used only groundwater, whereas tamarisk used both groundwater and shallow soil moisture at the same location. Horton et al. (2003), working at the same location but in different years, found that all of the trees used mainly groundwater. Mesquite uses mainly groundwater along the San Pedro River in Arizona, but it can also utilize vadose zone moisture during the rainy season, and it can redistribute water within the root zone as well (Scott et al. 2004). These studies are not contradictory, but they show that riparian species have much more flexibility in their use of water sources in the environment than was suspected before stable isotope analyses were possible.

Vandersandae et al. (2001) reported an interaction between salinity tolerance and drought tolerance of riparian plants from the Colorado River delta in Mexico. Tamarisk, willow, and cottonwood plants grown in 0.5 g L^{-1} NaCl were all able to extract water to about 2% soil moisture (−22 to −26 bar) before wilting, but on 4,000 ppm NaCl, willow and cottonwood wilted at 10% soil moisture, whereas tamarisk did not wilt until soil moisture was reduced to 3%. This experiment reflects the differences in salt tolerance among the species, since the soil solution became increasingly saline as water was extracted by the plants, but does not indicate a difference in drought tolerance per se. In actuality, riverbanks often become salinized as they dry, so tamarisk can be expected to persist longer than native trees on drying river benches. In controlled outdoor experiments, tamarisk was also able to maintain low stomatal resistance under both water stress and salt stress, whereas cottonwood and willow had high stomatal resistance under stress (Vandersandae et al. 2001; Nagler et al. 2003, 2004). These studies all support the conclusion that the drying and salinizing of river banks, and the lowering of water tables, due either to natural causes or flow regulation, are the primary mechanisms by which tamarisk has come to dominate western riparian corridors (Busch and Smith 1995; Cleverly et al. 1997; Everitt 1998; Shafroth et al. 1998).

CAN TAMARISK INVADE THE UPLANDS?

Several reviews have suggested that tamarisk can extend the vegetation zone in riparian corridors into adjacent upland areas because of its greater rooting depth compared to native trees (e.g., Morisette et al. 2006; Carlson et al. 2006). However, as shown earlier, tamarisk is not always the deepest-rooted species within a riparian corridor. On lower elevation streams, the higher terraces of the floodplain tend to be dominated by mesquites, which can cover vast areas away from the main river channel. On upper—elevation streams, the higher terraces tend to be dominated by phreatophyte shrubs such as black greasewood, along with facultative phreatophytes, such as fourwing saltbush, rabbitbrush (*Chrysothamnus nauseosus*), and deep-rooted bunch grasses such as giant sacaton (*Sporobulus wrightii*). Tamarisk can indeed become a dominant or codominant plant on these higher terraces, especially if it becomes salinized because of a lack of overbank flooding. However, this does not necessarily extend the area of the riparian corridor or increase riparian water use.

Conclusion

It is clear that the earlier view of tamarisk as a uniformly invasive species is not supported by more recent studies. A recent review of tamarisk distribution and abundance concluded that there are three types of riparian zones that support tamarisk (Nagler et al. 2011). In unaltered streams that still experience a seasonal pulse-flood regime, tamarisk tends to be a minor species. For example, it has less than 5% cover on the upper San Pedro River in the United States and the Sonoran headwater streams in Mexico. On the other hand, highly altered rivers such as the lower Colorado River are flow-regulated and channelized to prevent overbank flows. These rivers have developed saline aquifers and terraces, and no longer support extensive stands of mesic trees. Tamarisk and other halophytes dominate these riparian areas. Perhaps the most common condition are altered rivers such as the middle Rio Grande and the Colorado River delta in Mexico, that still have a seasonally variable hydroperiod, although large floods are now rare. These rivers support apparently stable associations of tamarisk and native trees in a patchwork mosaic of mesic and saline habitats within their riparian corridors.

All these stream types support at least some valuable riparian habitat for wildlife. Furthermore, the present review shows that prospects for salvaging large quantities of water for human use by removing tamarisk are probably illusory, as 50 years of failed demonstration projects show. Therefore, riparian management and restoration programs should emphasize strategies that maximize environmental, recreational, or aesthetic values of each riparian corridor on a case by case basis, recognizing that tamarisk can be a positive factor in maintaining riparian ecosystems, especially on flow-regulated rivers.

Literature Cited

Allen, R., L. Pereira., D. Rais, and M. Smith. 1998. Crop evapotranspiration—Guidelines for computing crop water requirements—FAO irrigation and drainage paper 56. Food and Agriculture Organization of the United Nations. Rome, Italy.

Anderson, J. 1982. Factors controlling transpiration and photosynthesis in *Tamarix chinensis* Lour. Ecology 63:8–56.

Ansley, R., W. Dugas, M. Heuer, and B. Kramp. 2002. Bowen ratio/energy balance and scaled leaf measurements of CO_2 flux over burned *Prosopis savanna*. Ecological Applications 12:948–961.

Ball, J., J. Picone, P. Ross. 1994. Evapotranspiration by riparian vegetation along the lower Colorado River. U.S. Bureau of Reclamation. Boulder City, NV. Final Report J-CP-3-08910.

Barz, D., R. Watson, J. Kanney, J. Roberts, and D. Groevenveld. 2009. Cost/benefits considerations for recent saltcedar control, middle Pecos River, New Mexico. Environmental Management 43:282–298.

Brotherson, J., and D. Field. 1987. *Tamarix*: Impacts of a successful weed. Rangelands 9:110–112.

Busch, D., N. Ingraham, and S. Smith. 1992. Water-uptake in woody riparian phreatophytes of the southwestern United States: A stable isotope study. Ecological Applications 2:450–459.

Busch, D., and S. Smith. 1993. Effects of fire on water and salinity relations of riparian woody taxa. Oecologia 94:186–194.

Busch, D., and S. Smith. 1995. Mechanisms associated with the decline of woody species in riparian ecosystems of the southwestern U.S. Ecological Monographs 65:347–370.

Canadell, J., R. Jackson, J. Ehleringer, H. Monney, O. Sala, and E. Schulze. 1996. Maximum rooting depth of vegetation types at the global scale. Oecologia 108:583–595.

Carlson, T., T. Stohlgren, and G. Newman. 2006. Arkansas River Tamarisk Mapping and Invenory Summary, Tamarisk Coalition. Grand Junction, CO. [Online] www.tamariskcoalition.org/PDF/Colorado%20River%20summary.pdf.

Cederborg, M. K. 2009. Quantifying soil and groundwater chemistry in areas invaded by *Tamarix* spp. along the middle Rio Grande, New Mexico. Doctoral dissertation, University of Denver. Denver, CO.

Chatterjee, S. 2010. Estimating evapotranspiraton using remote sensing: A hybrid approach between MODIS derived Enhanced Vegetation Index, Bowen Ratio System, and ground based micro-meteorological data. Doctoral dissertation, Wright State University.

Chew, M. 2009. The monstering of tamarisk: How scientists made a plant into a problem. Journal of the History of Biology 42:231–266.

Cleverly, J., C. Dahm, J. Thibault, D. Gilroy, and J. Coonrod. 2002. Seasonal estimates of actual evapo-transpiration from *Tamarix ramosissima* stands using three-dimensional eddy covariance. Journal of Arid Environments 52:181–197.

Cleverly, J., C. Dahm, J. Thibault, D. McDonnell, and J. Coonrod. 2006. Riparian ecohydrology: Regulation of water flux from the ground to the atmosphere in the Middle Rio Grande, New Mexico. Hydrological Processes 20:3207–3225.

Cleverly, J., S. Smith, A. Sala, and D. Devitt. 1997. Invasive capacity of *Tamarix ramosissima* in a Mohave Desert floodplain: The role of drought. Oecologia 111:12–18.

Cui, B., Q. Yang, K. Zhang, X. Zhao, and Z. You. 2010. Responses of saltcedar (*Tamarix chinenesis*) to water table depth and soil salinity in the Yellow River delta, China. Plant Ecology 209:279–290.

Dahm, C., J. Cleverly, J. Coonrod, J. Thibault, D. McDonnell, and D. Gilroy. 2002. Evapotranspiration at the land/water interface in a semi-arid drainage basin. Freshwater Biology 47:831–843.

Davenport, D., P. Martin, and R. Hagan. 1982. Evapotranspiration from riparian vegetation: Water relations and irrecoverable losses for saltcedar. Journal of Soil and Water Conservation 37:233–236.

DeLoach, C., and R. Carruthers. 2004. Biological control programs for integrated invasive plant management. Page 17 *in* Proceedings of Weed Society of America Meeting (CD-ROM). Weed Science Society of America. Kansas City, MO.

DeLoach, C., R. Curruthers, J. Lovich, T. Dudley, and S. Smith. 2000. Ecological interactions in the biological control of saltcedar (*Tamarix* spp.) in the United States: Toward a new understanding. Pages 819–873 *in* N. R. Spencer (ed.), Proceedings of the X International Symposium on Biological Control of Weeds. Montana State University. Bozeman, MT.

Devitt, D., A. Sala, K. Mace, and S. Smith. 1997. The effect of applied water on the water use of saltcedar in a desert riparian environment. Journal of Hydrology 192:233–246.

Devitt, D., A. Sala, S. Smith, J. Cleverly, L. K. Shaulis, and R. Hammett. 1998. Bowen ratio estimates of evapotranspiration for *Tamarix ramosissima* stands on the Virgin River in southern Nevada. Water Resources Research 34:2407–2414.

Di Tomaso, J. 1998. Impact, biology, and ecology of saltcedar (*Tamarix* spp.) in the southwestern United States. Weed Technology 12:326–336.

Duncan, K., S. Schemnitz, M. Suzuki, Z. Homesley, and M. Cardenas. 1993. Evaluation of saltcedar control: Pecos River, New Mexico. General Technical Reports, Rocky Mountain Forest Range Experiment Station, Forest Service, USDA 226:207–210.

Everitt, B. 1998. Chronology of the spread of tamarisk in the central Rio Grande. Wetlands 18:658–668.

Felker, P., P. Clark, A. Laag, and P. Pratt. 1981. Salinity tolerance of the tree legumes: Mesquite (*Prosopis glandulosa* var. *Torreyana*, *P. velutina* and *P. articulata*), algarrobo (*P. chinensis*), kiawe (*P. pallida*) and tamarugo (*P. tamarugo*) grown in sand culture on a nitrogen-free medium. Plant and Soil 61:311–317.

Gay, L., and R. Hartman. 1982. ET measurements over riparian saltcedar on the Colorado River. Hydrology and Water Resources in Arizona and the Southwest 12:9–15.

Gazal, R., R. Scott, D. Goodrich, and D. Williams. 2006. Controls on transpiration in a semiarid riparian cottonwood forest. Agricultural and Forest Meteorology 137:56–67.

Gerla, P. 1992. The relationship of water-table changes to the capillary fring, evapotranspiration, and precipitation in intermittaqnt wetlands. Wetlands 12:91–98.

Glenn, E., A. Huete, P. Nagler, K. Hirschboek, and P. Brown. 2007. Integrating remote sensing and ground methods to estimate evapotranspiration. Critical Reviews in Plant Sciences 26:139–168.

Glenn, E., K. Morino, P. Nagler, R. Murray, S. Pearlstein, and K. Hultine. 2012. Roles of saltcedar (*Tamarix* spp.) and capillary rise in salinizing a non-flooding terrace on a flow-regulated desert river. Journal of Arid Environments 79:56–65.

Glenn, E., and J. O'Leary. 1984. Relationship between salt accumulation and water content of dicotyledenous halophytes. Plant Cell and Environment 7:253–261.

Glenn, E., and P. Nagler. 2005. Comparative ecophysiology of *Tamarix ramosissima* and native trees in western US riparian zones. Journal of Arid Environments 61:419–446.

Glenn, E., P. Nagler, and A. Huete. 2010. Vegetation index methods for estimating evapotranspiration by remote sensing. Surveys in Geophysics 31:531–555.

Glenn, E., P. Nagler, R. Bursca, and O. Hinojosa-Huerta. 2006. Coastal wetlands of the northern Gulf of California: Inventory and conservation status. Aquatic Conservation—Marine and Freshwater Ecosystems 16:5–28.

Glenn, E., R. Tanner, S. Mendez, T. Kehret, D. Moore, J. Garcia, and C. Valdes. 1998. Growth rates, salt tolerance and water use characteristics of native and invasive riparian plants from the delta of the Colorado River delta, Mexico. Journal of Arid Environments 40:281–294.

Granier A. 1987. Evaluation of transpiration in a Douglas-fir stand by means of sap flow measurements. Tree Physiology 3:309–320.

Great Western Research. 1989. Economic analysis of harmful and beneficial aspects of saltcedar United States Reclamation Report, Contract No. 8-CP-0580.

Green S., B. Clothier, and B. Jardine. 2003. Theory and practical application of heat pulse to measure sap flow. Agronomy Journal 95:1371–1379.

Grime V., F. Sinclair. 1999. Sources of error in stem heat balance sap flow measurements. Agricultural and Forest Meteorology 94:103–121.

Hart, C., L. While, A. McDonald, and Z. Sheng. 2005. Saltcedar control and water salvage on the Pecos River, Texas, 1999–2003. Journal of Environmental Management 75:299–409.

Hartwell, S., K. Morino, P. Nagler, and E. Glenn. 2010. On the irrigation requirements of cottonwood (*Populus fremontii* and *Populus deltoides* var. *wislizenii*) and willow (*Salix gooddingii*) grown in a desert environment. Journal of Arid Environments 74:667–674.

Hatler, W., and C. Hart. 2009. Water loss and salvage in saltcedar (*Tamarix*) stands on the Pecos River, Texas. Invasive Plant Science and Management 2:309–317.

Holdenbach, G. 1987. Tamarix control. Pages 116–123 *in* M. R. Kunzmann, R. R. Johnson, and P. Bennett, editors. *Tamarix Control in the United States.* University of Arizona, Tuscon.

Horton, J., and C. Campbell. 1974. *Management of Phreatophyte and Riparian Vegetation for Maximum Multiple Use Values.* U.S. Forest Service, Research Paper RM-117.

Horton, J., S. Hart, and T. Kolb. 2003. Physiological condition and water source use of Sonoran Desert riparian trees at the Bill Williams River, Arizona, USA. Isotopes in Environmental and Health Studies 39:69–82.

Horton, J., T. Kolb, and S. Hart. 2001a. Physiological response to groundwater depth varies among species and with river flow regulation. Ecological Applications 11:1046–1059.

Horton, J., T. Kolb, and S. Hart. 2001b. Responses of riparian trees to interannual variation in ground water depth in a semi-arid river basin. Plant Cell and Environment 24:293–304.

Hughes, W., 1970. Economic Feasibility of Increasing Pecos Basin Water Supplies through Reduction of Evaporation and Evapotranspiration. Water Resources Research Institute Report No. 9. Colorado State University. Ft. Collins.

Hultine, K., D. Koepke, W. Pockman, A. Fravolini, J. Sperry, and D. Williams. 2006. Influence of soil texture on hydraulic properties and water relations of a dominant warm-desert phreatophyte. Tree Physiology 26:313–323.

Hultine, K., P. Nagler, K. Morino, S. Bush, K. Burtch, P. Dennison, E. Glenn, and J. Ehleringer. 2010. Sap flux-scaled transpiration by tamarisk (*Tamarix* spp.) before, during and after episodic defoliation by the saltcedar leaf beetle (*Diorhabda carinulata*). Agricultural and Forest Meteorology 150:1467–1475.

Jolly, I., K. McEwan, and K. Holland. 2008. A review of groundwater-surface water interactions in arid/semi-arid wetlands and the consequences for wetland ecology. Ecohydrology 1:43–58.

Jordan, F., W. Waugh, E. Glenn, L. Sam, T. Thompson, and T. L.Thompson. 2008. Natural bioremediation of a nitrate-contaminated soil-and-aquifer system in a desert environment. Journal of Arid Environments 72:748–763.

Kinsman, D. J. 1969. Modes of formation, sedimentary associations, and diagnostic features of shallow-water and supratidal evaporates. American Association of Petroleum Geologists Bulletin 53. doi:10.1306/5D25C801-16C1-11D7-8645000102C1865D.

Kjelgaard, J., C. Stockle, R. Black, and G. Campbell. 1997. Measuring sap flow with the heat balance approach using constant and variable heat inputs. Agricultural and Forest Meteorology 85:239–250.

Ladenburger, C., A. Hild, and D. Kazmer. 2006. Soil salinity patterns in *Tamarix* invasions in the Bighorn Basin, Wyoming, USA. Journal of Arid Environments 65:111–128.

Lesica, P., and T. DeLuca. 2004. Is *Tamarix* allelopathic? Plant and Soil 267:357–365.

Merritt, D., and P. Shafroth. 2012. Edaphic, salinity and stand structural trends in chronosequences of native and non-native dominated riparian forests along the Colorado River, U.S. Biological Invasions. doi:10.1007/s10530-012-0263-4.

Miyamoto, S., E. Glenn, and M. Olsen. 1996. Growth, water use and salt uptake of four halophytes irrigated with highly saline water. Journal of Arid Environments 32:141–159.

Moore, G., J. Cleverly, and M. Owens. 2008. Nocturnal transpiration in riparian *Tamarix* thickets authenticated by sap flux, eddy covariance and leaf gas exchange measurements. Tree Physiology 28:521–528.

Morisette, J., C. Jarnevich, A. Ultah, W. Cai, J. Pedetty, J. Gentile, T. Stohlgren, and J. Schnase. 2006. A tamarisk habitat suitability map for the continental United States. Frontiers in Ecology and Environment 4:11–17.

Murray, S., P. Nagler, K. Morino, and E. Glenn. 2009. An empirical algorithm for estimating agricultural and riparian evapotranspiration using MODIS Enhanced Vegetation Index and ground measurements. II. Application to the Lower Colorado River, U.S. Remote Sensing 1:1125–1133.

Nagler, P., A. Jetton, J. Fleming, K. Didan, E. Glenn, J. Erker, K. Morino, J. Milliken, and S. Gloss. 2007. Evapotranspiration in a cottonwood (*Populus fremontii*) restoration plantation estimated by sap flow and remote sensing methods. Agricultural and Forest Meteorology 144:95–110.

Nagler, P., E. Glenn, C. Jarnevich, and P. Shafroth. 2011. Distribution and abundance of saltcedar and Russian olive in the western United States. Critical Reviews in Plant Sciences 30:508–523.

Nagler, P., E. Glenn, K. Didan, J. Osterberg, F. Jordan, and J. Cunningham. 2008. Wide-area estimates of stand structure and water use of *Tamarix* spp. on the lower Colorado River: Implications for restoration and water management projects. Restoration Ecology 16:136–145.

Nagler, P., E. Glenn, and T. Thompson. 2003. Comparison of transpiration rates among saltcedar, cottonwood and willow trees by sap flow and canopy temperature methods. Agricultural and Forest Meteorology 116:73–89.

Nagler, P., E. Glenn, T. Thompson, and A. Huete. 2004. Leaf area index and Normalized Difference Vegetation Index as predictors of canopy characteristics and light interception by riparian species on the Lower Colorado River. Agricultural and Forest Meteorology 116:103–112.

Nagler, P., K. Morino, K. Didan, J. Erker, J. Osterberg, K. Hultine, and E. Glenn. 2009a a. Wide-area estimates of saltcedar (*Tamarix* spp.) evapotranspiration on the lower Colorado River measured by heat balance and remote sensing methods. Ecohydrology 2:18–33.

Nagler, P., K. Morino, S. Murray, J. Osterberg, and E. Glenn. 2009b. An empirical algorithhm for estimating agricultural and riparian evapotranspiration using MODIS Enhanced Vegetation Index and ground measurements. I. Description of method. Remote Sensing 1:1273–1297.

Nagler, P., P. Shafroth, J. LaBaugh, K. Snyder, R. Scott, D. Merritt, and J. Osterberg. 2009c. The potential for water savings through the control of saltcedar and Russian olive. Pages 33 *in* P. Shafroth, C. Brown, and D. Merritt, editors. Saltcedar and Russian Olive Control Demonstration Act Science Assessment. U.S. Geological Survey Scientific Investigations Report 2009-5247.

Nagler, P., R. Scott, C. Westenburg, J. Cleverly, E. Glenn, and A. Huete. 2005. Evapotranspiration on western US rivers estimated using the Enhanced Vegetation Index from MODIS and data from eddy covariance and Bowen ratio flux towers. Remote Sensing of Environment 97:337–351.

Nagler, P., T. Brown, K. Hultine, C. van Riper III, D. Bean, R. Murray, and E. Glenn. 2011. Regional-scale impacts of the *Tamarix* leaf beetles (*Diorhabda carinulata*) on the leaf phenology and water use of *Tamarix* spp. Remote Sensing of Environment (in review).

Natale, E., S. M. Zalba, A. Oggero, and H. Reinoso. 2010. Establishment of *Tamarix ramosissima* under different conditions of salinity and water availability: Implications for its management as an invasive species. Journal of Arid Environments 74:1399–1407.

Ohrtman, M., K. Lair, and A. A. Sher. 2012. Quantifying soil salinity in areas invaded by *Tamarix spp.* Journal of Arid Environments 85:114–121.

Owens M., and G. Moore. 2007. Saltcedar water use: Realistic and unrealistic expectations. Rangeland Ecology and Management 60:553–557.

Pataki, D., S. Bush, P. Gardner, D. Solomon, and J. Ehleringer. 2005. Ecohydrology in a Colorado River riparian forest: Implications for the decline of *Populus fremontii*. Ecological Applications 15:1009–1018.

Rowland, D., A. A. Sher, and D. L. Marshall. 2004. Cottonwood population response to salinity. Canadian Journal of Forest Research 34: 1458–1466.

Sala, A., S. Smith, and D. Devitt. 1996. Water use by *Tamarix ramosissima* and associated phreatophytes in a Mojave Desert floodplain. Ecological Applications 6:888–898.

Schaeffer, S., D. Williams, and D. Goodrich. 2000. Transpiration of cottonwood/willow forest estimated from sap flux, Agricultural and Forest Meteorology 105: 257–270.

Scott, R. E., W. Edwards, T. Shuttleworth, T. Huxman, C. Watts, and D. Goodrich. 2004. Interannual and seasonal variation in fluxes of water and carbon dioxide from a riparian woodland ecosystem. Agricultural and Forest Meterology 122:65–84.

Scott, R., W. Cable, T. Huxman, P. Nagler, M. Hernandez, and D. Goodrich. 2008. Multiyear riparian evapotranspiration and groundwater use for a semiarid watershed. Journal of Arid Environments 72:1232–1246.

Shafroth, P., C. Brown, and D. Merritt (eds.). 2009. Saltcedar and Russian Olive Control Demonstration Act Science Assessment. U.S. Geological Survey Scientific Investigations Report 2009-5247.

Shafroth, P., G. Auble, J. Stromberg, and D. Patten. 1998. Establishment of woody riparian vegetation in relation to annual patterns of streamflow, Bill Williams River, Arizona. Wetlands 18:577–590.

Shafroth, P., J. Cleverly, T. Dudley, J. Taylor, C. van Riper, E. Weeks, and J. Stuart. 2005. Control of *Tamarix* in the western United States: Implications for water salvage, wildlife use, and riparian restoration. Environmental Management 35:231–246.

Silliman, S., B. Berkowitz, J. Simunek, and M. van Genuchten. 2002. Fluid flow and solute migration within the capillary fringe. Ground Water 40:76–84.

Smith, S., D. Devitt, A. Sala, J. Cleverly, and D. Busch. 1998. Water relations of riparian plants from warm desert regions. Wetlands 18:687–696.

Sperry, J., F. Adler, G. Campbell, and J. Comstock. 1998. Limitation of plant water use by rhizosphere and xylem conductance: Results from a model. Plant, Cell and Environment 21:347–359.

Storey, R., and W. Thomson. 1994. An X-ray microanalysis study of the salt-glands and intracellular calcium crystals of *Tamarix*. Annals of Botany 73:307–313.

Stromberg, J., M. Chew, P. Nagler, and E. Glenn. 2009. Changing perceptions of change: The role of scientists in *Tamarix* and river management. Restoration Ecology 17:177–186.

van Hylckama, T. 1974. Water use by Saltcedar as Measured by the Water Budget method, USGS Professional Paper 491-E, Washington, DC.

Vandersande, M., E. Glenn, and J. Walworth. 2001. Tolerance of five riparian plants from the lower Colorado River to salinity, drought and inundation. Journal of Arid Environments 49:147–159.

Weeks, E., G. Weaver, G. Campbell, and B. Tanner. 1987. Water Use by Saltcedar and by Replacement Vegetation in the Pecos River Floodplain between Acme and Artesia, NM, USGS Professional Paper 491-G, Washington, DC.

Welder, G. 1988. Hydrologic Effects of Phreatophyte Control, Acme-Artesia Reach of the Pecos River, New Mexico, 1967–1982. U.S. Geological Survey Water-Resources Investigations Report 87-4148.

Westenberg, C., D. Harper, and G. DeMeo. 2006. Evapotranspiration by Phreatophytes along the Lower Colorado River at Havasu National Wildlife Refuge, Arizona. U.S. Geological Survey Scientific Investigations Report, 2006–5043, Henderson, NV.

Whitcraft, C., D. Talley, J. Crooks, J. Boland, and J. Gaskin. 2007. Invasion of tamarisk (*Tamarix* spp.) in a Southern California salt marsh. Biological Invasions 9:875–879.

Xu, D., and Y. Shen. 2005. External and internal factors responsible for midday depression of photosynthesis. Pages 297–298 *in* M. Pessarakli, editor. *Handbook of Photosynthesis.* 2nd ed. Taylor & Francis, Boca Raton, FL.

Yin, C. H., G. Feng, F. Zhang, C. Y. Tian, and C. Tang. 2009. Enhancement of soil fertility and salinity by tamarisk in saline soils on the northern edge of the Taklamakan Desert. Agricultural Water Management 97:1978–1986.

Zavaleta, E. 2000. The economic value of controlling an invasive shrub. Ambio 29:462–467.

6

Water Use by *Tamarix*

James R. Cleverly

Water use by *Tamarix* has been the focus of study, anecdote, and controversy for at least a half century. Over that time, a wealth of data from numerous methods has been collected to characterize evapotranspiration (ET) in *Tamarix* and invaded ecosystems. High rates of water use were expected from the beginning because *Tamarix* was used for erosion control and to de-water riparian ecosystems, thus opening space for and preventing uncontrolled flooding of irrigated agriculture (Everitt 1998). Furthermore, it had been commonly observed that communities occupied by *Tamarix* became increasingly xeric since invasion (Brotherson et al. 1984). These sorts of observations resulted in such statements as, "Saltcedar is a highly water-consuming naturalizing shrub that has escaped from cultivation and spread rapidly from one stream valley to another. An aggressive plant, it has not only invaded but has entirely replaced the native vegetation in many areas" (Robinson 1965).

Early studies were unable to reject the hypothesis that *Tamarix* is a profligate water waster. Evaluating previous studies of *Tamarix* transpiration, Cleverly et al. (2002) observed that these early studies were limited to short periods during ideal conditions. As longer and more extensive studies have been pursued, extraordinary water use by *Tamarix* has been called into question (e.g., see chapter 5, this volume). It has become clear that water use in *Tamarix* is highly variable in time as well as in space, making it difficult to scale from single measurements to a general understanding (see chapter 5, this volume). Fortunately, with the wealth of literature published in journals and reports, *Tamarix* has become an ideal study organism for investigating natural variability in ecosystem water flux.

The difficulty with interpreting short-term studies arises from an issue with scaling to longer periods, given that these studies occurred over a span of a week to a couple of months while attempting to quantify *Tamarix* water use over the course of years. This is where the wealth of studies is important. The object is to identify under which conditions *Tamarix* uses a great deal of water, under which conditions *Tamarix* water use is restricted, how that might change over time, and what managers and restoration practitioners can do to favor the balance. The bottom line is, of course, whether water salvage can be achieved through *Tamarix* restoration, that is, whether removing the tree will increase the water available for other uses. By carefully considering hydroclimatological conditions, and by carefully choosing replacement vegetation to fit into the hydroclimatological milieu, the answer is a cautious affirmation of moderate proportions (Shafroth et al. 2005; see also chapter 4, this volume).

While *Tamarix* has been observed on lake and reservoir shores and higher-order upstream habitats, invasion has focused on floodplains, where *Tamarix* presence has been unmistakable for decades. This chapter will therefore focus on *Tamarix* floodplain ecosystems. Also, the nomenclature used throughout this chapter refers to genera, for example, *Tamarix* or *Populus*. This is done to recognize many native species in genera found in the American West. Common genera found in groundwater-dependent ecosystems throughout the American Southwest include *Tamarix* (*Tamarix* or saltcedar), *Populus* (cottonwood), *Prosopis* (mesquite), *Distichlis* (saltgrass), *Salix* (willow), *Acer* (box elder), and *Baccharis* (seepwillow). *Populus* is a large genus containing both cottonwoods and noncottonwoods (e.g., aspen); here, *Populus* will be used purely for reference to cottonwoods. Likewise, *Acer* contains common nonphreatophytic species to which this chapter does not refer. *Tamarix* in this chapter comprises two species, *T. chinensis* and *T. ramosissima*, and their hybrids.

Riparian Hydrology

Riparian hydrological change and *Tamarix* invasion have long been the explanation one for the other. Whereas some attribute altered hydrology and geomorphology to the presence of *Tamarix* (Robinson 1965), others argue that anthropogenic hydrological modifications provide *Tamarix* with opportunities to invade that would not otherwise be available (Everitt 1998). While it does appear likely that *Tamarix* improves habitat quality for itself in the American Southwest, anthropogenic hydrological modifications provide the opportunity for *Tamarix* to dominate. Geomorphic and hydrologic modifications caused by *Tamarix* include decreased flow rate, increased deposition, and channel incision (Blackburn et al. 1982). These same modifications in riparian corridors are implemented through anthropogenic devices, such as placement of dams or jetty-jacks (Grassel 2002), each of which controls flooding and is linked to channel incision. Jetty-jacks are large iron tripods that were placed in the middle Rio Grande floodplain by the US Army Corps of Engineers to reduce flow rates, thereby increasing sediment deposition. Dam building has led to channel narrowing in addition to restriction of the flood pulse and summer minimum flow (Shafroth et al. 2002). Thus, the cause of changes in flow and the shape of a waterway is not always clear, either being the result of the presence of the plant, human alterations, or both.

Water inputs from just outside the riparian corridor, such as from irrigation, are characteristic of a "gaining reach." This means that the water table outside of the riparian corridor is at a higher elevation than the groundwater in the riparian corridor (see figure 6.1). A "losing reach," on the other hand, is characterized by groundwater flow out of the riparian corridor down into a water table that gets deeper with distance from the riparian corridor. Because recharge is scant in the tableland region between mountains and riparian corridors, a reach can be gaining or losing depending on the time lag of flow from the base of the mountains to the river. In the case of the middle Rio Grande, it can take about 50 years for water that falls as precipitation to reach the riparian corridor (Duffy 2004). In practical terms, the depression in recharge caused by the 1950s drought reached the Rio Grande during the 2000s drought, turning the reach into a losing reach, magnifying drought impacts, and ultimately favoring *Tamarix* (Duffy 2004; Cleverly et al. 2006).

The most astounding characteristic of *Tamarix* related to hydrology is the range of water levels in which it can persist and thrive. At the dry extreme, *Tamarix* is well known to occupy upper terraces, which, in semiarid landscapes, are nearly as dry as surrounding nonriparian areas (Reichenbacher 1984). The explanation for *Tamarix*'s success in marginal habitats is related to its physiology. *Tamarix* has plant–water relations that are similar to upland species, particularly

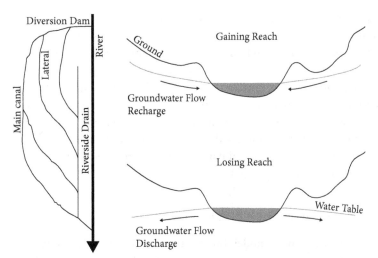

FIGURE 6.1 *Left side*: Schematic diagram of irrigation support structures in shallow river valleys. Water is withdrawn at the diversion dam, channeled into canals, and returned to the river via the riverside drain. *Right side*: Groundwater levels relative to river elevation in gaining and losing reaches.

FIGURE 6.2 Photograph of adventitious roots on an inundated *Tamarix* at Bosque del Apache National Wildlife Refuge, New Mexico.

Source: Photograph courtesy of J. R. Cleverly.

in physiological sensitivity to water stress (Pockman and Sperry 2000). At the opposite extreme, mature *Tamarix* is able to withstand long periods of flooding: 70 days fully submerged and 98 days partially submerged (Warren and Turner 1975), but inundation for "several months" is followed by plant death (Lehnhoff et al. 2011). *Tamarix*, like *Populus*, *Salix*, and *Baccharus* (among others), forms adventitious roots into floodwater to minimize the effects of anoxia in floodwater (see figure 6.2), but there is a toll in the form of delayed greening in the spring and lower water use. These multifaceted responses to groundwater level and flooding help explain the wide range in habitat and depth to groundwater where *Tamarix* is found (Lite and Stromberg 2005).

GROUNDWATER–VEGETATION INTERACTIONS

Hydrologists frequently model the relationship between groundwater depth and vegetation water use. A model in widespread use is MODFLOW 2000, a finite-difference groundwater flow model (McDonald and Harbaugh 1988; Wilcox et al. 2007). Evapotranspiration can be modeled in MODFLOW with a simple relationship between groundwater depth and ET. This is done by determining the "surface depth," that is, the minimal depth at which ET is not limited by access to water, and the "extinction depth," where the water table is below the root zone, ET becomes negligible, and ultimately plants die. This model has been improved over time, including incorporating functions specific for plant functional types (Baird and Maddock 2005). To apply this sort of model to *Tamarix*, the relationship between groundwater depth and transpiration rate must be determined.

In the early years of *Tamarix* water use studies, *Tamarix* water use was assumed to decline with depth to groundwater (Davenport et al. 1976a; Davenport et al. 1976b; Davenport et al. 1982). However, each of these papers cite Robinson (1965) who speculates on a possible relationship between groundwater depth and ET. Two studies, one in Arizona and another in China on native *Tamarix*, investigated transpiration responses to groundwater depth. In Arizona, depth to the water table was varied to 10 meters without having an effect on transpiration (Horton et al. 2001). In the Northwest China study, variation in groundwater depth to 24 meters did not have an effect on plant–water relations. In both cases, soil water storage contributed to maintaining transpiration rates and plant–water status regardless of water table depth.

Independence between *Tamarix* ET and groundwater depth allows *Tamarix* to inhabit sites of varying groundwater depth as well as sites with the driest conditions and terraces (Lite and Stromberg 2005; Stromberg et al. 2006; Stromberg et al. 2007). Consistent with the distribution of *Tamarix* in dry sites and sites with a variable water table, it was recently discovered that *Tamarix* water use is dictated not by static water table depth but by changes in that depth. For example, at Bosque del Apache in central New Mexico, *Tamarix* growing in a dense, monospecific thicket increased its water use by 50% when experiencing a drop in the water table of up to 7.5 cm/day (Cleverly et al. 2006). It appears that the xerophytic water relations of *Tamarix* allows these plants to maintain root function in recently desaturated soil layers while extending root growth to maintain connection with a declining water table.

HISTORIC VEGETATION DYNAMICS

What was the nature of riparian vegetation before the arrival of *Tamarix*? How has water use changed since ecosystems came to be dominated by *Tamarix*? Are native *Populus* and willow forests of today representative of historic riparian forests? These are some of the significant questions when considering the restoration of riparian ecosystems and water salvage. In the Paleocene period, formerly known as the early Tertiary period, mixed mesophytic forests were common throughout the American Southwest, but they have since become restricted to riparian corridors (e.g., *Populus* and *Salix*; Reichenbacher 1984). On the other hand, *Prosopis* and *Acacia* are derived from neotropical flora and, like *Tamarix*, are pre-adapted to uplands and riparian terraces in the semiarid Southwest (Reichenbacher 1984).

There is evidence that *Populus* and *Salix* were never dominant in recent prehistory. Along the Virgin River, Hughes (1993) questioned whether *Populus* and

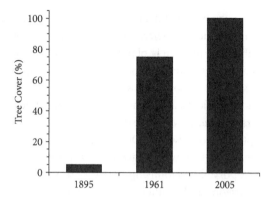

FIGURE 6.3 Historical estimates of tree cover along the Middle Rio Grande. Estimates are from 1895 (Scurlock 1998), 1961 (Campbell and Dick-Peddie 1964), and 2005 (Cleverly et al. 2006).

Salix had ever been dominant. In the late nineteenth century, shrubs occupied the higher margins and combined with extensive bare-soil patches in the Green River riparian corridor (Birken and Cooper 2006). Likewise, along the Rio Grande in New Mexico, late nineteenth-century vegetation was described as a few widely dispersed *Populus* copses, individual *Populus* trees along the riverside, riverside *Salix* thickets, and large expanses of *Distichlis* and xeric shrubs (e.g., *Prosopis* and *Atriplex*; see figure 6.3) (Scurlock 1998). Since then, canopy cover by *Populus* has increased along the Middle Rio Grande, nearing 75% of the 1961 total ground cover at some locations (Campbell and Dick-Peddie 1964). Since the 1960s, vegetation cover (*Populus* in the north, *Tamarix* in the south) has surpassed 100% (Cleverly et al. 2006).

The late nineteenth century saw an extreme drought, an event that recurs in the Rio Grande every 30 to 70 years (Gray et al. 2003). These "mega-droughts" are responsible for massive dieback of riparian vegetation (Johnson 2001). However, because of *Tamarix*'s drought tolerance, drought and deluge cycles serve to only enhance *Tamarix* dominance as drought sensitive native species die, opening more habitat for colonization by exotic species (Cleverly et al. 1997; Cleverly et al. 2006). Since ET increases proportionally with leaf area index (LAI) and cover (Cleverly et al. 2006), water use has certainly increased over the last 100-plus years along with vegetation density.

TAMARIX ESTABLISHMENT AND ECOSYSTEM RESTORATION

Water and climate come together to regulate *Tamarix* establishment in the same way that recruitment density dictates ecosystem dominance and *Tamarix* water use. *Tamarix* cover tends to be higher in reaches with late-summer peak flow because river regulation and flow alteration favors the species with matching life-history characteristics (Beauchamp and Stromberg 2007; Mortenson and Weisberg 2010). Flooding that occurs outside the narrow seed-viability windows for native *Populus* and *Salix* can favor opportunistic species like *Tamarix* if the out-of-season flooding opens suitable habitat (Stromberg et al. 2007). Some of the densest *Tamarix* stands are found along river reaches where flooding from monsoon rains occurs during the summer (Stromberg 1997; Shafroth et al. 2002; Cleverly et al. 2006; Akasheh et al. 2008). Recruitment of *Tamarix* is favored when high-flow years are followed by two to three years of drought (Birken and Cooper 2006), capitalizing on *Tamarix*'s improved stress tolerance relative to native species (Cleverly et al. 1997).

Conversely, *Tamarix* can be managed on regulated rivers by timing controlled floods and staged drawdown to favor native species by matching the historical hydrograph, which features spring flooding and groundwater decline during late May or early June (Sher et al. 2002; Bhattacharjee et al. 2006). *Tamarix* seedlings can be out-competed by native species because of their shade intolerance. For example, mortality in *Tamarix* has been observed following one to two seasons of shade under an *Acer negundo* (box elder) canopy in northern Colorado (Dewine and Cooper 2008). Native *Populus* and *Salix* also show a competitive advantage over *Tamarix*, a relationship that is density dependent and facilitated through imposition of a flood pulse and/or shallow water table (Sher et al. 2000; Sher and Marshall 2003; Bhattacharjee et al. 2009).

It is also important to understand the effects of restoration on *Tamarix* establishment and water use. Poorly controlled restoration efforts can result in hydrological changes that are favorable to the reestablishment of *Tamarix,* such as creating exposed soil (see chapter 7, this volume). In one case along the Rio Puerco in New Mexico, spraying *Tamarix* from a helicopter caused extreme erosion and channel widening (Vincent et al. 2009). On the other hand, carefully planned *Tamarix* removal can result in improved ecosystem function and even water salvage. In general, reduction of *Tamarix* water use is best accomplished by replacing it with upland and xeroriparian species along with a minimum of native trees (Shafroth et al. 2005; Cleverly and Dello Russo 2007; see also chapter 4, this volume). The achievements of the late John Taylor in leading successful conversions of monospecific *Tamarix* to xeric species like *Chrysothamnus nausesosus* (Chamisa or rabbitbrush) leave a legacy that continues to work toward habitat improvement and water use reduction in the Rio Grande (see chapter 21, this volume). In semiarid areas where water resources are becoming increasingly dear, conversion of *Tamarix* to herbaceous plants in riparian areas has greater potential for water salvage than does restoration in upland areas (Wilcox et al. 2006).

Potential and Actual Evapotranspiration

The primary energy source for ET derives from the sun. The solar constant (1,366 W/m²) represents the quantity of solar energy received by the outer surface of the atmosphere. The atmosphere absorbs some of that energy, and the quantity of solar energy absorbed by aerosols and clouds depends upon latitude and weather.

In the southwestern United States, on a cloud-free day 1,000 to 1,100 W/m² insolates the surface at midday in the summer. Occasionally, spikes of 1,200 W/m² can be observed, caused by reflection from a cloud's leading edge. Of the solar radiation reaching the ground, 11% to 30% is reflected back into the atmosphere by vegetated and soil surfaces, respectively (Warner 2004).

Potential ET is a baseline measurement of water loss from the environment when water is unlimited; it is derived from measurements of radiation, temperature, wind speed, or humidity, in various combinations. The most used semiempirical methods used in the estimation of crop and riparian potential ET are the Penman and Penman-Monteith combination equations (Brutsaert 1982; Jensen et al. 1990). These methods combine energy balance, wind speed, and vapor pressure measurements to predict potential ET. Understanding these measures is important for interpreting the highly variable measurements of *Tamarix* water use that have been published.

A SENSE OF REALISM

All measurements of ET have assumptions, one of which is the growing season. Most the riparian species in the American Southwest, including *Tamarix*, are winter deciduous. During the winter, ET rates are negligible (0.8 mm day⁻¹) even though potential ET can be two to five times that amount (see figure 6.4). By obviating these small wintertime rates and by assuming a growing season of 250 days, the absolute physical maximum for annual ET is less than 3,000 millimeters. Observed growing seasons are shorter (190–240 days), further reducing annual physical maximal ET below 3000 millimeters.

Within this framework, evaluation of early ET estimates of *Tamarix* do not seem to be as far out of the realm of possibility as one might expect. The early studies in the Bernardo lysimeters, for example, found a maximum rate near 11 mm day⁻¹, with common values between 7.9 mm day⁻¹ and 8.2 mm day⁻¹ (Gay and Fritschen 1979; Davenport et al. 1982). The limitation of these studies is their limited time scale; short-term measurements of a week to a month were standard at that time because instrumentation was less reliable and required constant monitoring and maintenance. The most significant finding from the Bernardo lysimetry study is the variability that was observed between trees. Davenport et al. (1982) observed differences of 45% in *Tamarix* water use among lysimeters. This level of variability in *Tamarix* ET has been recently observed in long-term measures of *Tamarix* water use (Cleverly et al. 2002). Knowledge of the conditions and habitats associated with thrifty or profligate water use is important to inform our understanding of *Tamarix* water use and to identify of appropriate locations and strategies for riparian restoration.

Even though many of these early, short-term measurements seem reasonable, Owens and Moore (2007) point out that there are yet other estimates that are well beyond physical reality. For example, it has been widely reported that *Tamarix* uses 757 liters (200 gallons) of water tree⁻¹ day⁻¹ (see chapters 5 and 16, this volume).

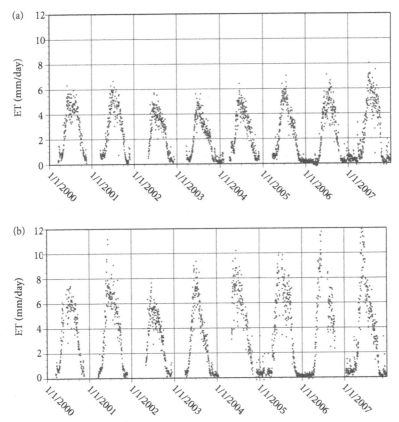

FIGURE 6.4 Time series of daily evapotranspiration (ET) in *Tamarix* (a) at Sevilleta National Wildlife Refuge, New Mexico, and (b) at Bosque del Apache National Wildlife Refuge, New Mexico.

The problem with this sort of estimate is that the density of *Tamarix* can be very high: over 2,400 plants ha⁻¹ (6,000 plants acre⁻¹) in a mature, monospecific thicket (Cleverly and Dello Russo 2007). If each plant were consuming 757 L day⁻¹ in such a dense stand, *Tamarix* ET would be approximately 1065 mm day⁻¹, which exceeds the physical maximum by two orders of magnitude (Cleverly and Dello Russo 2007).

OBSERVATIONS OF LONG-TERM *TAMARIX* WATER-USE PATTERNS

Tamarix water use is highly variable, but that variability is coming to be better understood. At a given location, *Tamarix* ET varies with distance from the river between open and dense stands (Devitt et al. 1997). *Tamarix* water use responds immediately to flooding or precipitation, is highest near the river's edge, and declines the most rapidly in response to drought in an open stand (Devitt et al. 1997). Similar observations have been made on the Rio Grande, where regional drought leads to an initial decline in *Tamarix* water use and LAI followed by rapid increases in ET and LAI when wet conditions returned (Cleverly et al. 2006). In both of these studies, *Tamarix* water use returned to predrought rates or higher whether pre-drought rates were high or low (Devitt et al. 1997; Cleverly et al. 2006).

Even today, long-term studies of *Tamarix* ET are rare, especially studies that directly measure ecosystem ET. The only methods currently available for directly measuring ET are lysimeters (if they're large enough) and eddy covariance. Eight years of eddy covariance measurements along the Rio Grande are presented in figure 6.4 a and b. The two *Tamarix* sites that are represented were established in 1999 and represented the same range in variability that was observed in the original Bernardo lysimeters (Davenport et al. 1982; Cleverly et al. 2002). Minimal water use was observed at the xeroriparian *Tamarix* woodland that never floods, has a constant water table, and relatively low LAI (figure 6.4a; Cleverly et al. 2006). Due to the absence of evaporative cooling, water-conservative *Tamarix* ecosystems are easily identified by higher air temperature. Removal of *Tamarix* at such a site should include revegetation with upland and low water-using native species. It would be foolhardy to plant early pioneer trees like *Populus* in these locations. In these xeric landscapes, modest water salvage could be achieved by restoring dryland species but additional water usage would be realized if restoration to *Populus* and *Salix* were the goal (Shafroth et al. 2005).

At the other extreme, ET in dense monospecific *Tamarix* thickets can approach the physical limit (see figure 6.4b). Even if a full *Populus* canopy is established at this site, ET would likely be reduced, although water salvage could be negligible depending on planting or establishment density of these large trees. Rather than planting dense stands of high-water-using native trees, restoration activities that will maximize water salvage operate with the goal of regenerating the prehistoric sparse mosaic of individual and widely spaced *Populus* trees within a grass and shrub matrix.

Water use in both woodlands (figure 6.4a) and thickets (figure 6.4b) has been increasing since the extreme drought of 2002/2003. Increasing ET at the mesic site began before the conclusion of the drought in response to groundwater decline (Cleverly et al. 2006). Observed annual water use in a mixed *Populus* and *Tamarix* forest is higher than water use at either of these *Tamarix* sites (Cleverly et al. 2006). Removal of *Tamarix* from the understory resulted in reduced ecosystem ET (Martinet et al. 2009). Daily ET rates in *Populus* seldom if ever reach the physical maximum. Instead, sensible heat advection and extended season length in this Rio Grande *Populus* forest leads to equally high annual water use, around 1 m year^{-1}, in *Populus* and *Tamarix* (Cleverly et al. 2006).

Conclusions

In general, those who take the view that *Tamarix* should be eradicated cite *Tamarix*'s spendthrift use of water; the opposing viewpoint is that native species are profligate water users and that *Tamarix* is frugal. Considering only water use, both opinions are correct because of the extraordinary variability in *Tamarix* ET. Evapotranspiration along riparian corridors in the American Southwest depends upon vegetation density, hydrology, groundwater dynamics, historical constraints and release from those restraints, history and dynamics of plant establishment, and restoration activities and level of success. Where conditions are favorable for *Tamarix*, daily ET can be near the physical limit. The foremost condition that favors profligate water use by *Tamarix* is summer flooding because native species that colonize bared sandbars, riverbanks, and lakeshores do not shed seeds that late into the year. Next, *Tamarix* is favored by deep water tables and variable groundwater depth. Where some of the favorable conditions are met but not others, *Tamarix* tends to form woodlands of restricted water use, although there is an indication that ET from such ecosystems may not remain restricted indefinitely.

Water salvage can be achieved with *Tamarix* control only when its removal doesn't create conditions that promote reinvasion by *Tamarix* and when it is replaced by appropriate xeric plants. *Tamarix* stands should be replaced by xeroriparian or upland species. Conversion of *Tamarix* thickets into sparse mosaics that mimic prehistoric plant communities will likewise benefit the water budget. Any change in ET due to restoration activities will be modest in either case, but carefully implemented management strategies are an improvement over continuously increasing water use by the current vegetation or by restoration activities that increase *Tamarix* presence or water consumption (Shafroth et al. 2005).

Literature Cited

Akasheh, O. Z., C. M. U. Neale, and H. Jayanthi. 2008. Detailed mapping of riparian vegetation in the middle Rio Grande River using high resolution multi-spectral airborne remote sensing. Journal of Arid Environments 72:1734–1744.

Baird, K. J., and T. Maddock. 2005. Simulating riparian evapotranspiration: A new methodology and application for groundwater models. Journal of Hydrology 312:176–190.

Bay, R. F., and A. A. Sher. 2008. Success of active revegetation after *Tamarix* removal in riparian ecosytems of the southwestern United States: A quantitiative assessment of past restoration projects. Restoration Ecology 16:113–128.

Beauchamp, V. B., and J. C. Stromberg. 2007. Flow regulation of the Verde River, Arizona, encourages *Tamarix* recruitment but has minimal effect on *Populus* and *Salix* stand density. Wetlands 27:381–389.

Bhattacharjee, J., J. P. Taylor, and L. M. Smith. 2006. Controlled flooding and staged drawdown for restoration of native cottonwoods in the middle Rio Grande valley, New Mexico, USA. Wetlands 26:691–702.

Bhattacharjee, J., J. P. Taylor, L. M. Smith, and D. A. Haukos. 2009. Seedling competition between native cottonwood and exotic saltcedar: Implications for restoration. Biological Invasions 11:1777–1787.

Birken, A. S., and D. J. Cooper. 2006. Processes of *Tamarix* invasion and floodplain development along the lower Green River, Utah. Ecological Applications 16:1103–1120.

Blackburn, W. H., R. W. Knight, and J. L. Schuster. 1982. Saltcedar influence on sedimentation in the Brazos River. Journal of Soil and Water Conservation 37:298–301.

Brotherson, J. D., J. G. Carman, and L. A. Szyska. 1984. Stem-diameter age relationships of *Tamarix ramosissima* in Central Utah. Journal of Range Management 37:362–364.

Brutsaert, W. 1982. *Evaporation into the Atmosphere: Theory, History, and Applications.* D. Reidel Publishing Company, Dordrecht, Holland.

Campbell, C. J., and W. A. Dick-Peddie. 1964. Comparison of phreatophytic communities on the Rio Grande in New Mexico. Ecology 45:492–502.

Chew, M. K. 2009. The monstering of Tamarisk: How scientists made a plant into a problem. Journal of the History of Biology 42:231–266.

Cleverly, J. R., C. N. Dahm, J. R. Thibault, D. E. McDonnell, and J. E. A. Coonrod. 2006. Riparian ecohydrology: Regulation of water flux from the ground to the atmosphere in the middle Rio Grande, New Mexico. Hydrological Processes 20:3207–3225.

Cleverly, J. R., C. N. Dahm, J. R. Thibault, D. J. Gilroy, and J. E. A. Coonrod. 2002. Seasonal estimates of actual evapo-transpiration from *Tamarix ramosissima* stands using three-dimensional eddy covariance. Journal of Arid Environments 52:181–197.

Cleverly, J. R., and G. Dello Russo. 2007. Salt cedar control: Exotic species in the San Acacia reach. Pages 76–79 *in* L. G. Price, P. S. Johnson, and D. Bland, editors. Water resouces of the middle Rio Grande. New Mexico Bureau of Geology and Mineral Resources, Socorro.

Cleverly, J. R., S. D. Smith, A. Sala, and D. A. Devitt. 1997. Invasive capacity of *Tamarix ramosissima* in a Mojave Desert floodplain: The role of drought. Oecologia 111:12–18.

Davenport, D. C., P. E. Martin, and R. M. Hagan. 1976a. Aerial spraying of phreatophytes with antitranspirant. Water Resources Research 12:991–996.

Davenport, D. C., P. E. Martin, E. B. Roberts, and R. M. Hagan. 1976b. Conserving water by antitranspirant treatment of phreatophytes. Water Resources Research 12:985–990.

Davenport, D. C., P. E. Martin, and R. M. Hagan. 1982. Evapotranspiration from riparian vegetation: Water relations and irrecoverable losses for saltcedar. Journal of Soil and Water Conservation 37:233–236.

Devitt, D. A., J. M. Piokowski, S. D. Smith, J. R. Cleverly, and A. Sala. 1997. Plant water relations of *Tamarix ramosissima* in response to the imposition and alleviation of soil moisture stress. Journal of Arid Environments 36:527–540.

Dewine, J. M., and D. J. Cooper. 2008. Canopy shade and the successional replacement of Tamarisk by native box elder. Journal of Applied Ecology 45:505–514.

Duffy, C. J. 2004. Semi-discrete dynamical model for mountain-front recharge and water balance estimation, Rio Grande of southern Colorado and New Mexico. Groundwater Recharge in a Desert Environment: The Southwestern United States 9:255–271.

Everitt, B. L. 1998. Chronology of the spread of Tamarisk in the central Rio Grande. Wetlands 18:658–668.

Gay, L. W., and L. J. Fritschen. 1979. An energy budget analysis of water use by saltcedar. Water Resources Research 15:1589–1592.

Grassel, K. 2002. Taking out the jacks: Issues of jetty jack removal in bosque and river restoration planning. WRP-6, Water Resources Program, University of New Mexico, Albuquerque.

Gray, S., J. Betancourt, C. Fastie, and S. Jackson. 2003. Patterns and sources of multidecadal oscillations in drought-sensitive tree-ring records from the central and southern Rocky Mountains. Geophysical Research Letters 30:1316.

Horton, J. L., T. E. Kolb, and S. C. Hart. 2001. Physiological response to groundwater depth varies among species and with river flow regulation. Ecological Applications 11:1046–1059.

Hughes, L. E. 1993. "The devil's own"—Tamarisk. Rangelands 15:151–155.

Jensen, M. E., R. D. Burman, and R. G. Allen, editors. 1990. Evapotranspiration and irrigation water requirements. American Society of Civil Engineers, New York.

Johnson, M. R. 2001. Historical change in woody riparian vegetation in the Republican River basin and impact on evapotranspiration. University of Nebraska, Lincoln, NE.

Lehnhoff, E. A., F. D. Menalled, and L. J. Rew. 2011. Tamarisk (*Tamarix* spp.) establishment in its most northern range. Invasive Plant Science and Management 4:58–65.

Lite, S. J., and J. C. Stromberg. 2005. Surface water and ground-water thresholds for maintaining *Populus-Salix* forests, San Pedro River, Arizona. Biological Conservation 125:153–167.

Martinet, M. C., E. R. Vivoni, J. R. Cleverly, J. R. Thibault, J. F. Schuetz, and C. N. Dahm. 2009. On groundwater fluctuations, evapotranspiration, and understory removal in riparian corridors. Water Resources Research 45: W05425. doi:10.1029/2008WR007152.

McDonald, M. G., and A. W. Harbaugh. 1988. A modular three-dimensional finite-difference ground-water flow model US Geological Survey, Department of the Interior.

Mortenson, S. G., and P. J. Weisberg. 2010. Does river regulation increase the dominance of invasive woody species in riparian landscapes? Global Ecology and Biogeography 19:562–574.

Owens, M. K., and G. W. Moore. 2007. Saltcedar water use: Realistic and unrealistic expectations. Rangeland Ecology and Management 60:553–557.

Pockman, W., and J. Sperry. 2000. Vulnerability to xylem cavitation and the distribution of Sonoran Desert vegetation. American Journal of Botany 87:1287–1299.

Reichenbacher, F. W. 1984. Ecology and evolution of southwestern riparian plant communities. Desert Plants 6:15–22.

Robinson, T. W. 1965. Introduction, spread and areal extent of saltcedar (*Tamarix*) in the Western States. Professional Paper 491-A, U.S. Geological Survey, Washington DC.

Sammis, T. W., C. L. Mapel, D. G. Lugg, R. R. Lansford, and J. T. McGuckin. 1985. Evapotranspiration crop coefficients predicted using growing-degree-days. Transactions of the Asae 28:773–780.

Scurlock, D. 1998. From the Rio to the Sierra: An Environmental History of the Middle Rio Grande Basin. General Technical Report RMRS-GTR-5. USDA Forest Service, Rocky Mountain Research Station, Fort Collins, CO.

Shafroth, P. B., J. C. Stromberg, and D. T. Patten. 2002. Riparian vegetation response to altered disturbance and stress regimes. Ecological Applications 12:107–123.

Shafroth, P. B., J. R. Cleverly, T. L. Dudley, J. P. Taylor, C. Van Riper, E. P. Weeks, and J. N. Stuart. 2005. Control of *Tamarix* in the western United States: Implications for water salvage, wildlife use, and riparian restoration. Environmental Management 35:231–246.

Shah, J. J. F., C. N. Dahm, S. P. Gloss, and E. S. Bernhardt. 2007. River and riparian restoration in the southwest: Results of the national river restoration science synthesis project. Restoration Ecology 15:550–562.

Sher, A. A., and D. L. Marshall. 2003. Competition between native and exotic floodplain tree species across water regimes and soil textures. American Journal of Botany 90:413–422.

Sher, A. A., D. L. Marshall, and J. Taylor. 2002. Spatial partitioning within southwestern floodplains: Patterns of establishment of native *Populus* and *Salix* in the presence of invasive, non-native *Tamarix*. Ecological Applications 12:760–772.

Sher, A. A., D. L. Marshall, and S. A. Gilbert. 2000. Competition between native *Populus deltoides* and invasive *Tamarix ramosissima* and the implications for reestablishing flooding disturbance. Conservation Biology 14:1744–1754.

Stromberg, J. C. 1997. Growth and survivorship of Fremont cottonwood, Goodding willow, and salt cedar seedlings after large floods in central Arizona. Great Basin Naturalist 57:198–208.

Stromberg, J. C., S. J. Lite, T. J. Rychener, L. R. Levick, M. D. Dixon, and J. M. Watts. 2006. Status of the riparian ecosystem in the upper San Pedro River, Arizona: Application of an assessment model. Environmental Monitoring and Assessment 115:145–173.

Stromberg, J. C., V. B. Beauchamp, M. D. Dixon, S. J. Lite, and C. Paradzick. 2007. Importance of low-flow and high-flow characteristics to restoration of riparian vegetation along rivers in and south-western United States. Freshwater Biology 52:651–679.

Vincent, K. R., J. M. Friedman, and E. R. Griffin. 2009. Erosional consequence of saltcedar control. Environmental Management 44:218–227.

Warner, T. T. 2004. *Desert Meteorology*. Cambridge University Press, Cambridge UK.

Warren, D. K., and R. M. Turner. 1975. Saltcedar (*Tamarix chinesis*) seed production, seedling establishment, and response to inundation. Journal of the Arizona Academy of Sciences 10:135–144.

Wilcox, B. P., M. K. Owens, W. A. Dugas, D. N. Ueckert, and C. R. Hart. 2006. Shrubs, streamflow, and the paradox of scale. Hydrological Processes 20:3245–3259.

Wilcox, L. J., R. S. Bowman, and N. G. Shafike. 2007. Evaluation of Rio Grande management alternatives using a surface-water/ground-water model. Journal of the American Water Resources Association 43:1595–1600

7

Tamarix, Hydrology, and Fluvial Geomorphology

Daniel A. Auerbach, David M. Merritt,
and Patrick B. Shafroth

Flowing water, sediment, and vegetation interact to form diverse riverine land-scapes. In arid and semiarid regions such as the southwestern United States, this varied biophysical regime shapes banks and channels (Poff et al. 2006a; Miller and Friedman 2009), alters the flux of nutrients (Molles et al. 1998; Fisher et al. 2007), and governs the composition and population dynamics of biological communities (Poff et al. 1997; Merritt et al. 2010). In turn, the structure of riparian vegetation communities mediates hydrologic and fluvial conditions, such as water temperature and turbidity, as well as the formation and destruction of fluvial forms (Naiman et al. 2005; Sandercock et al. 2007; Merritt 2012).

The spread of trees and shrubs in the introduced genus *Tamarix* (commonly referred to as tamarisk or saltcedar) has provided an opportunity to increase our understanding of the relationships between plants and fluvial processes. Native to Eurasia, *Tamarix* was first planted as an ornamental shrub and a cost-effective means of stabilizing stream banks, but it naturalized and spread rapidly during the twentieth century, concurrent with a period of large-scale river alteration due to dam and reservoir construction (Robinson 1965; Nagler et al. 2011). Research by Ringold et al. (2008) corroborated the finding of Friedman et al. (2005) that species of *Tamarix* are among the most common woody plants across all western floodplains and are especially abundant in valley bottoms within more xeric areas. The role of *Tamarix* in western US riparian ecosystems has prompted research on various aspects of its ecology and management, including its relationship to native riparian vegetation and wildlife, how it affects recreational values, and how it influences and is influenced by water availability (Shafroth et al. 2005, 2010b; Stromberg et al. 2009).

Here, we describe the hydrologic and geomorphic controls on *Tamarix* distribution and abundance as well as the reciprocal effects of *Tamarix* on hydrologic and geomorphic conditions. These relationships bear on key questions for river science and management. Does flow-regime alteration favor establishment of *Tamarix* over native taxa? How do *Tamarix* stands modify processes of channel narrowing and floodplain formation? After providing an overview of the basic geomorphic and hydrologic character of rivers in the American West, we examine how this setting has facilitated the regional success of *Tamarix* and review the influence of *Tamarix* on the form and function of these systems. We conclude by discussing the relevance of a shifting climate, vegetation management, and continued water-resource development to the future role of *Tamarix* in these ecosystems.

Riparian Vegetation in Relation to Hydrogeomorphic Form and Function in Western US Rivers

Rivers in the western United States progress from their headwaters, which are often in montane settings with high topographic relief, through intermediate-elevation foothills and valleys before reaching their terminus in a larger river, inland basin, or sea. The lower gradients and wider valleys typical of many downstream (higher-order) reaches permit the development of extensive and topographically complex floodplains (Leopold et al. 1964). The largest riparian forests occur along reaches subject to weaker lateral geologic constraints and to greater lateral hydrologic connectivity between channels and floodplains (Naiman et al. 2005; Cooper et al. 2012). Human actions such as grazing and agricultural clearing also influence the amount and type of vegetation in riparian ecosystems, and the widespread construction of dams and irrigation networks has altered patterns of discharge and sediment movement critical to riparian plants (Patten 1998; Graf 1999; Nilsson and Berggren 2000; Poff et al. 2007).

At the watershed scale, the amount, timing, and form of precipitation (e.g., rain vs. snow) combine with terrestrial land cover, geology, and human infrastructure to control the magnitude and seasonal distribution of flows reaching a particular location (Poff et al. 2006a, b). Flow variation shapes the fluvial landscape in arid and semiarid western rivers by affecting the delivery and removal of the sediment that forms bottomland surfaces such as floodplains, islands, bars, splay deposits, oxbows, and terraces (see figure 7.1; Schumm and Lichty 1963; Leopold et al. 1964; Hereford 1984; Tooth 2000). The energy available for erosion and transport of sediment changes as a function of discharge magnitude (Baker and Ritter 1975), parent material (e.g., percentage of sand, clay, or silt), and landscape setting (e.g., valley slope and constraint). The probability that a specific magnitude flow will occur on average defines a recurrence interval, and larger magnitude flows (floods) typically have longer recurrence intervals due to the inverse relation

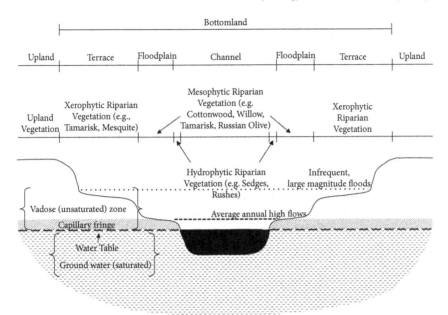

FIGURE 7.1 Major channel-floodplain land forms showing characteristic vegetation classes, positions where they grow within the bottomland, and key surface and groundwater elements. Relationships are simplified for clarity and do not illustrate the complex topography, stratigraphy, hydrology, and plant-community composition typical of most bottomlands. Bottomlands are the areas within alluvial valleys primarily influenced by stream flow and sediment transport, and are distinguished from higher-elevation uplands that do not consist of geologically recent river-affected sediment. Channels are linear depressions that contain continually or periodically flowing water and sediment. Floodplains border channels at an elevation approximately corresponding to the annual flood, and are constructed of river-transported sediment. Terraces are elongated surfaces that parallel channels above the floodplain, and are rarely inundated other than by very large-magnitude floods. Groundwater occurs as subsurface saturation, and its upper level is often called the water table. Above this but below ground, the capillary fringe forms a zone in which groundwater is drawn upward into interstitial spaces within sediment. More generally, the vadose zone extends from the soil surface to the water table, and is sometimes termed the "unsaturated zone." Hydrophytic riparian vegetation includes plants that are well adapted to growth in saturated or very wet conditions are termed "hydrophytic," whereas mesophytic riparian vegetation is better adapted to intermediate moisture levels. Xerophytic riparian vegetation consists of plants adapted to low moisture that occurs within bottomlands.
Source: Modified from figure 1 (p. 5) and figure 2 (p. 37) in Shafroth et al. (2010b).

between flow size and frequency. In some systems, intermediate-magnitude floods with recurrence intervals of less than five years may drive floodplain formation by depositing new sediment both in-channel and over banks (e.g., Miller and Friedman 2009). In contrast, the largest floods, with recurrence intervals greater than five to ten years, may be responsible for the most mobilization of channel and floodplain material (Wolman and Gerson 1978; Miller and Friedman 2009). Such "channel-resetting" floods act as a negative feedback on the channel-narrowing processes associated with increased sediment supply or decreased peak flows, and may maintain wider active channels in which perennial vegetation is scarce or absent (Sigafoos 1964; Dean et al. 2011).

These geomorphic processes and the water that flows laterally over banks and below ground as alluvial groundwater are primary determinants of the structure and species composition of riparian plant communities (Naiman et al. 2005). Variation in physiological, morphological, and phenological (timing-related) traits causes individual riparian species to differ in their responses to flow and sediment regimes. Flows are characterized by their magnitude, duration, frequency, timing, and rate of change (i.e., the speed of transition to flood stage), and a characteristic series of flows measured over an extended period constitutes a flow regime on a particular river (Poff et al. 1997). The compatibility of species traits with site sediment and flow regimes influences establishment, growth, survival, and dispersal processes through a variety of mechanisms (Karrenberg et al. 2002; Nilsson and Svedmark 2002; Poff et al. 2006a, b; Renofalt et al. 2007; Merritt et al. 2010). Sediment stratigraphy and texture influence vegetation by controlling moisture availability and the retention and cycling of nutrients (Merigliano 2005). High flows drive the formation and destruction of surfaces suitable for riparian tree seedling establishment (Scott et al. 1996; Friedman and Lee 2002). Floods also kill existing vegetation through scour, burial, or prolonged inundation. Riparian species have evolved a suite of adaptations in response to varied flows. The formation of aerenchyma (tissue with large intercellular air spaces) and active transport of oxygen to roots enable plants to tolerate flooded, submerged, and anoxic (oxygen–depleted) conditions (Merritt et al. 2010). Flexible stems, the ability to resprout after disturbance, and thick, furrowed bark enable many riparian species to survive shear, abrasion, ice scour, and burial. Production of numerous small seeds and ability to reproduce from branch and stem fragments enable some plants to persist in frequently disturbed river bottomlands. The interplay between these traits and the physical forces in rivers controls the spread and persistence of riparian plants such as *Tamarix*.

Reciprocal Feedbacks between Hydrology, Fluvial Form, and *Tamarix*

Tamarix possesses a set of traits that have allowed it to naturalize widely in North America (Friedman et al. 2005; Ringold et al. 2008), but the mechanisms responsible for its establishment and survival vary with position in the drainage network and with the chronology of hydrologic changes resulting from climate shifts and human modification of flow and sediment regimes (Glenn and Nagler 2005; Stromberg et al. 2007b; Merritt and Poff 2010; Mortenson and Weisberg 2010; Nagler et al. 2011). Similarly, the effect of *Tamarix* on fluvial geomorphic processes varies with factors such as local sediment supply and history of high and low flows.

ESTABLISHMENT

Tamarix has expanded through the river corridors of the arid western United States as larger-scale climate shifts, land-use changes, and dam construction altered local flow and sediment regimes to create establishment opportunities. Within the riparian zone, *Tamarix* may establish across a fairly broad range of soil texture and chemistry, but seedlings require plentiful light and moisture (Shafroth et al. 1995; Cooper et al. 1999; Taylor et al. 1999; Sher et al. 2002; Sher and Marshall 2003; Glenn and Nagler 2005). Accordingly, *Tamarix* has recruited on former active-channel surfaces exposed after peak flow reduction, on bare floodplain surfaces formed by large-magnitude floods, and relative to the spatiotemporal availability of alluvial groundwater (Graf 1982; Shafroth et al. 1998; Glenn and Nagler 2005; Merritt and Poff 2010).

Floods that create bars and islands through lateral and vertical accretion (gradual, layered accumulation) or deposit fresh sediment over banks within more confined valleys generate the bare, moist sites that provide the plentiful light and water required by germinating *Tamarix* seeds (Scott et al. 1996, Taylor et al. 1999; Merritt and Cooper 2000). However, reduced flood magnitude can also support *Tamarix* establishment by exposing portions of the former active channel to colonization (Graf 1982). Recruitment on previously flood-mobilized surfaces was especially widespread during the early stages of *Tamarix* spread in the mid-twentieth century (Campbell and Dick-Peddie 1964; Graf 1978, 1982; Hereford 1984). Comparing sites on the Yampa and Upper Green Rivers below Flaming Gorge Dam, Cooper et al. (2003) described *Tamarix* recruitment on surfaces both within and peripheral to channels, relative to flow regulation and valley confinement (park vs. canyon). On these rivers *Tamarix* established both on higher-elevation floodplains following high flows and on lower-elevation surfaces within the active channel in association with several years of lower peak flow. *Tamarix* also increased along the Bill Williams River following completion of Alamo Dam, reaching higher cover compared to the unregulated Santa Maria River upstream from the dam (Shafroth et al. 2002). Several flow-regime changes may have contributed to *Tamarix* expansion in this system, such as increased summer flows that have supported riparian vegetation in previously moisture-limited areas. The dam also reduced peak discharges, likely enabling *Tamarix* recruitment by exposing germination surfaces and reducing the frequency and magnitude of channel scouring flows (Shafroth et al. 2002). Despite such favorable effects in the period following dam construction, newly colonized patches were narrower along the Bill Williams after 25 years of regulation, suggesting that diminished floods could also limit further expansion by decreasing overall channel disturbance.

The timing of flood peaks and the rate of associated alluvial groundwater recession can favor or disfavor *Tamarix* germination relative to native species (see figure 7.2; Warren and Turner 1975; Graf 1982; Rood and Mahoney 1990; Shafroth et al. 1998; Horton and Clark 2001; Merritt and Poff 2010; Mortenson and Weisberg

2010). Like native *Populus* and *Salix, Tamarix* produces abundant small seeds widely dispersed by wind and water and able to germinate immediately. However, whereas *Populus* and *Salix* typically disperse seed earlier in the growing season in synchrony with the declining stage of spring snowmelt run-off floods (Mahoney and Rood 1998), *Tamarix* releases seeds somewhat later and over a longer period (figure 7.2; Stromberg 1993, 1997; Shafroth et al. 1998; Cooper et al. 1999; Sprenger et al. 2001). In some river systems, this seed-release timing may allow *Tamarix* to take advantage of establishment opportunities generated by floods occurring outside the late winter and spring, particularly if lethal high flows or rapid groundwater declines do not ensue (Warren and Turner 1975; Shafroth et al. 1998; Cooper et al. 1999; Beauchamp and Stromberg 2007). A longer period of seed release could afford *Tamarix* an advantage over some native colonizers along intermittent or flashy streams, where patches of bare sediment provide germination sites after late-summer rainfall-driven floods, or along regulated rivers, where floods occur later in the growing season (Sher et al. 2002; Beauchamp and Stromberg 2007; Mortenson and Weisberg 2010). Conversely, later seed release may disadvantage *Tamarix* if plants that establish earlier in the season have already occupied suitable sites or if subsequent floods remove *Tamarix* seedlings established at lower-elevation positions on the floodplain (Stromberg 1993, 1997; Sprenger et al. 2001; Merritt and Wohl 2002).

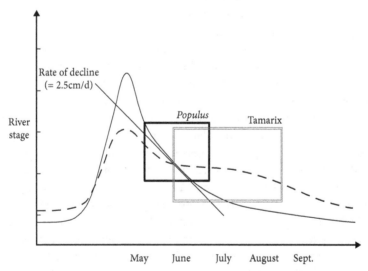

FIGURE 7.2 Recruitment box model comparing *Populus* and *Tamarix* establishment requirements relative to regulated (solid) and unregulated (dashed) snowmelt-flood hydrographs. Both species establish on bare, moist surfaces such as those created after floods scour existing vegetation or deposit new sediment. Note that the illustrated flow regime alteration is not representative of all regulated snowmelt rivers, and diversions for consumptive use such as agricultural irrigation may result in lowered summer flows as well as reduced peak–flow magnitude and duration.

Source: Based on concepts in Mahoney and Rood (1998).

Regulation of flood magnitude and timing has been proposed as an important driver of *Tamarix* spread (e.g., Stromberg et al. 2007a, b), although relatively few studies have closely examined how variability in flood timing and magnitude has affected the colonization of unregulated rivers by *Tamarix*. Merritt and Poff (2010) studied current *Tamarix* establishment, abundance, and dominance along both regulated and unregulated rivers relative to a composite index of flow alteration that included flood timing and magnitude. Distinguishing establishment from abundance and dominance revealed important differences in the relationship between flow alteration and the success of *Tamarix*. Contrary to the expectation that flow regulation promotes *Tamarix* establishment, this research showed equal or greater probability of recruitment along free-flowing reaches as compared to heavily regulated reaches. The authors noted that periodic disturbance on less regulated rivers likely provides more frequent establishment opportunities than are typical of systems with reduced or eliminated floods, and contributed to the much higher probability of native *Populus* establishment at sites experiencing less regulated flows. By comparison, dominance of *Tamarix* was positively related to the intensity of alteration, and the abundance of *Tamarix* increased up to an intermediate level of flow modification before declining with maximum hydrologic alteration. Similarly, Mortenson and Weisberg (2010) found that *Tamarix* canopy cover was positively related to flow alteration, but was highest at intermediate levels of hydrologic change. Furthermore, both studies indicated strong declines in *Populus* abundance and dominance (Merritt and Poff 2010) and cover (Mortenson and Weisberg 2010) with even moderate flow modification. This relationship suggested that different mechanisms limit *Tamarix* at low and high levels of regulation. Merritt and Poff (2010) proposed that overall low recruitment associated with the absence of fresh establishment surfaces may constrain *Tamarix* on intensively modified reaches, but that ecological interactions between *Tamarix* and other riparian trees may play a greater role on reaches where less regulated flow conditions support the recruitment of native species. Patterns of abundance and dominance are affected by growth and survival of maturing and adult individuals, and these results also highlight the degree to which flow and sediment regimes interact with multiple life history stages to determine the local ecological significance of *Tamarix*.

PERSISTENCE

In addition to influencing recruitment, flood intensity and duration affect juvenile and adult survival through inundation, scour, and burial. Although *Tamarix* displays inundation tolerance (Brotherson and Field 1987), prolonged flooding can cause mortality (Gladwin and Roelle 1998; Sprenger et al. 2001; Tallent-Halsell and Walker 2002) and has been suggested as a control method, particularly if the submergence of target plants is coordinated with native seed release timing (Vandersande et al. 2001; Lescia and Miles 2004; see Sprenger et al. 2001 for discussion of the risk to cottonwood seedlings). In a greenhouse study, Levine and Stromberg (2001) found that, relative to native species, *Tamarix* seedlings required a longer period of growth to reach sizes sufficient to survive sediment burial treatments. These authors proposed that restoring pre-regulation sediment regimes associated with occasional large floods might play a role in limiting the establishment and survival of *Tamarix*. On the Bill Williams River, *Tamarix* seedlings had much higher mortality rates than Goodding's willow (*Salix gooddingii*) due to both burial and scour associated with flooding (Shafroth et al. 2010a), but adult *Tamarix* in ephemeral channels in Spain displayed significant resistance to high flood velocities and shear stresses (Sandercock and Hooke 2010). Extrinsic factors such as local substrate cohesion and the density of neighboring vegetation combine with intrinsic features such as age and stem morphology to determine whether individual plants withstand the abrasive forces generated by floods (Sandercock et al. 2007). In contrast to flexible *Salix* stems and branches, the relatively dense, inelastic wood of mature *Tamarix* (Gerry 1954) tends to confer resistance rather than resilience, either supporting continued growth after burial and scour or withstanding damage under intense flood forces (Friedman et al. 2005b).

Decreased flood frequency and magnitude may favor *Tamarix* survival over native species by reducing the removal of plant litter from floodplains (Glenn and Nagler 2005; Stromberg et al. 2007a; Shafroth et al. 2010b). The accumulation of combustible *Tamarix* litter (leaves and dead branches), due in part to reduced rates of litter decomposition and removal by floods (Ellis et al. 1998; Molles et al. 1998; Stromberg and Chew 2002; Ellis 2001), has contributed to an increased intensity and frequency of fire in some western riparian zones (Busch 1995). The degree to which fire favors *Tamarix* over natives remains unclear (Ellis 2001; Glenn and Nagler 2005), but in contrast to some native species, *Tamarix* resprouts readily after fire and has increased in abundance following fires in some locations (see chapter 14, this volume; Busch 1995).

Decreased flooding may also reduce flushing of floodplain soils and thereby contribute to high levels of floodplain soil salinity (Jolly et al. 1993; see also chapter 8, this volume; Merritt and Shafroth 2012). Numerous studies have documented the high salt tolerance of both establishing and mature *Tamarix* (Busch and Smith 1995; Shafroth et al. 1995, Glenn et al. 1998; Vandersande et al. 2001). Floodplains with high soil salinity relative to regularly flooded ones may favor *Tamarix* over *Populus* and other glycophytic (salt-intolerant) native species (see chapter 8, this volume; Merritt and Shafroth 2012). Although *Tamarix* may contribute to salinization of floodplains by exuding salt from its leaves, Merritt and Shafroth (2012) found that other causes of salinity (e.g., upward transport to near-surface soils due to evaporation of capillary water) overshadow the salinizing effect of *Tamarix* along the flow-regulated lower Colorado River. Further, salt concentrations in *Tamarix*-dominated stands along the relatively free-flowing upper Colorado River only occasionally reached levels that would significantly affect survival of native glycophytes, despite being elevated relative to native dominated stands (Merritt and Shafroth, 2012). Thus, presence of *Tamarix* in more saline floodplain habitats may reflect its tolerance of relatively high soil salt concentrations more than its direct modification of floodplains (Merritt and Cooper 2000; Glenn et al. 2012).

Supplemental soil moisture provided by stream flow and alluvial groundwater during dry times of year is a key factor promoting the growth of distinct riparian plant assemblages in arid and semiarid regions. Low or ephemeral flows may affect tree density and composition by exposing portions of the active channel to colonization (Scott et al. 1996) or by producing water stress that limits the extent of riparian stands away from the active channel (Graf 1978, 1982; Stromberg et al. 2007b). Among riparian trees and shrubs, *Tamarix* is well adapted to low-moisture conditions (Cleverly et al. 1997; Horton and Clark 2001; Glenn and Nagler 2005; see also chapters 5, 6, and 19, this volume). *Tamarix* tends to dominate sites with intermittent surface flows, in contrast to native cottonwood and willow species that can dominate *Tamarix* along reaches characterized by perennial surface water, relatively stable and shallow ground water tables, or high and low flows ranging within the rooting depths of these species (Busch and Smith 1995; Shafroth et al. 2002; Lite and Stromberg 2005). Functional trade-offs between drought tolerance and tolerance of inundation and disturbance constrain bottomland vegetation (Stromberg et al. 2008). *Tamarix* is well adapted to both these limiting factors, however, having traits that enable it to occupy drier sites that still experience disturbance and inundation (Pockman and Sperry 2000; see also chapter 1, this volume).

Comparing portions of the San Pedro River in Arizona that spanned a hydrologic gradient from perennial to highly intermittent flow, Stromberg (1998a) observed an increase in the relative abundance of *Tamarix* corresponding to less perennial flow and declining groundwater levels. At sites along the Bill Williams River, juvenile *Tamarix* survived greater interannual groundwater declines than native saplings (Shafroth et al. 2002). At a regional scale (southern Arizona), Stromberg et al. (2007b) found more *Tamarix* patches and greater *Tamarix* basal area along reaches with intermittent surface flow and dam-altered flow regimes compared to perennial and free-flowing reaches. Nonetheless, despite its drought tolerance, continued groundwater declines can eventually result in decreased *Tamarix* abundance and replacement by species with even greater tolerance of xeric conditions, as noted by Graf (1982) for sections of the Salt and Gila Rivers.

CHANGES IN CHANNEL FORM

While flow and sediment regimes regulate *Tamarix* recruitment and survival, *Tamarix* and other floodplain plants can simultaneously affect channel form. Riparian vegetation mediates sediment erosion and deposition through surface and subsurface mechanisms. Stems and other aboveground biomass create drag that increases floodplain roughness, reduces boundary shear stress, and slows the passage of water (Griffin et al. 2005; Sandercock et al. 2007; Merritt 2012). These changes increase sediment deposition, reduce erosion relative to unvegetated surfaces, and increase flow depth and turbulence (e.g., Griffin et al. 2005). Roots may reinforce banks, with the degree of increased cohesion depending on factors such as plant size and species, root depth, and the percentages of sand and clay within banks (Pollen-Bankhead et al. 2009; Merritt 2012).

Tamarix is not unique in its influence on fluvial geomorphic processes (Friedman et al. 1996; Friedman et al. 2005b; Sandercock et al. 2007). Stromberg (1998b) found comparable rates of sedimentation in patches dominated by *Tamarix* and native cottonwood and willow (*Populus* and *Salix*) along the free-flowing San Pedro River. Numerical modeling suggested that flexible stems of sandbar willow (*Salix exigua*), which may flatten against the ground at high flows, contributed to preventing significant floodplain erosion during a major flood (Griffin and Smith 2004). Intra- and interspecific differences in stem density, diameter, and rigidity likely generate variation in the hydraulic effects of riparian vegetation (Kean and Smith 2004; Griffin et al. 2005). Nonetheless, high stem density in stands of *Tamarix* may increase its influence on sediment deposition and retention relative to low-density, mature cottonwood gallery forests, for instance (Hereford 1984; Pollen-Bankhead et al. 2009).

Observations of channel narrowing along western American rivers, concurrent with the spread of *Tamarix*, have prompted debate about the degree to which *Tamarix* controls or contributes to decreased width (Graf 1982; Hereford 1984; Everitt 1998). Hydrologic changes that reduce flood intensity and thereby decrease erosion and sediment transport are sufficient to narrow channels even in the absence of new stands of riparian vegetation. However, dam construction or periods of drought that generate such hydrologic change were concurrent with *Tamarix* establishment on many rivers, complicating attribution of causality. Narrowing and dense floodplain vegetation are of particular concern when they increase hydraulic roughness, flood stage, and overbank inundation. Such patterns have been observed in association with *Tamarix* establishment along western rivers (Blackburn et al.1982; Graf 1982). Thus, a number of studies have sought to disentangle whether narrowing occurred largely independently of *Tamarix*, primarily because of *Tamarix* establishment, or due to hydrologic changes combined with *Tamarix* establishment.

Several sections of the Green River in Utah have been the focus of research into the relationship between changing flow regimes, fluvial geomorphology, and *Tamarix*. Graf (1978) proposed that *Tamarix* cohorts established on islands, marginal bars, and alluvial fans when several years of below-average peak flow followed a flood year. He suggested that a period of drought during the early 1930s likely promoted stabilization of previously dynamic channel and floodplain features by *Tamarix*, leading to subsequent sediment deposition and aggradation (figure 7.3). The finer temporal resolution of research by Allred and Schmidt (1999) provided additional insight into the extent of narrowing that followed construction of the upstream Flaming Gorge Dam ([FGD]; completed in 1963). Analysis of hydrologic records combined with excavations of *Tamarix* root crowns and mapping of

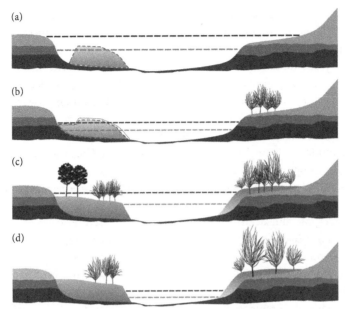

FIGURE 7.3 Channel form changes in response to flow alteration and *Tamarix* establishment. Shifts in form for a hypothetical river channel in the southwestern United States are illustrated over several decades, relative to a combination of reduced flooding and the establishment of riparian vegetation such as *Tamarix*. The heavy dashed lines indicate large floods (*dark gray*) and annual high flows (*lighter gray*). (A) Large floods mobilized sediments across a wide cross-sectional area, including transient bars and islands within the channel and at its margins (*gray polygon*). Riparian vegetation may occur along the river, but its persistence within this area is limited by intense floods. (B) A period of drought, upstream dams, or a combination of the two reduce flood magnitudes, thereby decreasing the movement of formerly mobile surfaces and exposing previously inundated and scoured areas to colonization by riparian vegetation, such as *Populus*, *Salix*, or *Tamarix*. (C) Subsequently higher flows narrow the overall channel width, as reduced velocity and shear stress within new stands of vegetation increase sediment deposition and reduce erosion. Infilling and accretion may occur on surfaces that stabilized independently of vegetation establishment, with colonization by native trees and *Tamarix* extending into these areas. Riparian trees such as *Tamarix* are often partially buried by sediment, so that stems at the present ground surface are well above the surface on which the tree originally established. In conjunction with extensive root networks, these below-ground portions of the plant may reinforce banks and floodplains, further reducing erosion. (D) A scenario in which drought and dams eliminate any significant floods and reduce baseflow. This might result in floodplains dominated by mature *Tamarix* after native trees with less tolerance of drought and lowered groundwater levels are lost.

floodplain stratigraphy enabled these authors to associate the stabilization and vertical accretion of previously transient channel formations with increased roughness and reduced velocity near *Tamarix*. In agreement with the hypothesis that *Tamarix* acted as a feedback on underlying hydrogeomorphic changes, they concluded that a large reduction in bankfull channel width (the distance encompassed by typical high flows) had occurred during the 1930s, coincident with drought and initial *Tamarix* colonization of the study area, and that dam-induced flow alteration contributed to a second phase of narrowing after 1963, when *Tamarix* was already abundant. Grams and Schmidt (2002, 2005) noted that vegetation, including *Tamarix*, played a role in the development of the post-dam land forms below FGD. They concluded that *Tamarix* had contributed to narrowing in some reaches, but that peak flow reductions, sediment loads, and reach-scale variation in width and sediment input (e.g., alluvial fans) ultimately moderated the type and magnitude of channel change. Also on the upper Green River below FGD, Birken and Cooper (2006) demonstrated a clear relationship between *Tamarix* recruitment and an interannual flow pattern consisting of a large magnitude flood followed by several years of lower peak discharge.

Stratigraphic and dendrochronologic records (based on dating tree growth rings) of *Tamarix* establishment both before and after filling of FGD supported the conclusion that *Tamarix* contributed to narrowing prior to dam-caused flow alteration (figure 7.4). Nonetheless, these authors observed a post-dam shift in the position of *Tamarix* establishment from higher, more marginal sites to lower surfaces nearer the contemporary active channel (Birken and Cooper 2006). A fine-scale dendrogeomorphic analysis of floodplains in Brown's Park and Lodore Canyon (below FGD) corroborated this observation, indicating that flow alteration by FGD promoted the formation of lower, inset deposits suitable for *Tamarix* establishment (Alexander 2008). Most of the post-FGD accretion on these surfaces resulted from controlled flood releases that have further isolated stands of established vegetation from the river rather than removing them. Consequently, Alexander (2008) attributed post-dam narrowing to regulation of high flows, with the resulting alteration of geomorphic processes enhanced by *Tamarix*.

Studies in the Rio Grande watershed have also clarified interactions between *Tamarix* spread, hydrologic alteration, and fluvial geomorphic dynamics. Along the central Rio Grande (New Mexico and western Texas), Everitt (1998) described a series of human modifications (dams, channelization, levees, and floodplain agriculture) that reduced flows and decreased sediment movement both prior to and during the spread of *Tamarix* in the 1930s. This chronology suggested that *Tamarix* did not play a primary role in channel narrowing and aggradation, though it did not exclude *Tamarix* as a factor in these processes. Friedman et al. (2005b) observed that *Tamarix* establishment and concurrent decreases in peak flow promoted narrowing and hindered lateral channel migration along the unregulated lower Rio Puerco, a tributary to the Rio Grande. Modeling and observation of erosion along reaches where *Tamarix* was removed have strengthened the conclusion that *Tamarix* significantly reduces bank and floodplain susceptibility to erosion and increases sediment deposition in this system (Griffin et al. 2005, Vincent et al. 2009). In the Big Bend (southwestern Texas) section of the Rio Grande, recent research has analyzed how reduced peak and mean flows generated a series of changes in channel morphology that were enhanced by *Tamarix* establishment

Historic flow regime intact

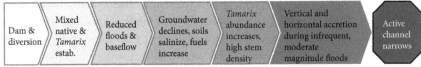

| Historic flows persist | Native seed release adapted to flood timing | Scour & deposition create & maintain bare, moist sites | *Tamarix* establishes on late season, low flow channel surfaces | Subsequent floods remove *Tamarix* on low flow surfaces | Mixed native & *Tamarix* stands on channel margins | Historic active channel width |

Unregulated river and floodplain, flow alteration

| Period of drought | Reduced flood magnitude & frequency | Former channel surfaces exposed | *Tamarix* seedlings establish | Drought ends, moderate floods return | Stabilized bars & islands accrete with overbank deposition | Active channel narrows |

Regulated river and floodplain, flow alteration

| Dam & diversion | Mixed native & *Tamarix* estab. | Reduced floods & baseflow | Groundwater declines, soils salinize, fuels increase | *Tamarix* abundance increases, high stem density | Vertical and horizontal accretion during infrequent, moderate magnitude floods | Active channel narrows |

FIGURE 7.4 Conceptual pathways of channel alteration. The historic natural flow regime maintains abundant native vegetation and historic active channel width, with *Tamarix* present but not dominant. In unregulated rivers, climate shifts alter flood magnitude and frequency and lower late-season flow (not shown). *Tamarix* seedlings establish in areas previously subjected to higher-intensity flooding, and subsequent peak flows result in accelerated aggradation in *Tamarix* stands. This produces higher, drier, and less-connected surfaces on which *Tamarix* may persist and dominate. In regulated rivers, damming reduces flood magnitudes and facilitates the establishment of *Tamarix* and native species within former active-channel habitats. Diminished flows favor *Tamarix* persistence; riparian vegetation reduces erosion and promotes sediment deposition during occasional high flows to enhance the narrowing of the active channel.

(Dean and Schmidt 2011; Dean et al. 2011). Periodic large floods historically maintained a wide, sandy channel in this system, similar to other arid and semiarid rivers (Schumm and Lichty 1963; Graf 1982; Martin and Johnson 1987). Upstream dams and diversions, however, reduced overall discharge volume as well as flood frequency during the early and mid-twentieth century. Subsequent channel aggradation and narrowing resulted in over-bank flooding despite diminished flows, and *Tamarix* establishment contributed to narrowing by promoting the vertical accretion of channel bars and inset surfaces (Dean and Schmidt 2011). River regulation was important to the underlying alteration of flow and sediment regimes on the Rio Grande (Dean and Schmidt 2011), but Dean et al. (2011) proposed that comparable processes might occur in unregulated rivers, particularly following a rapid climate change toward warmer and drier regimes.

Collectively, these studies support the idea that *Tamarix* and other species of riparian vegetation have acted as a positive feedback on channel-narrowing processes in western rivers, building on climate- and dam-induced hydrologic changes by stabilizing in-channel forms and increasing vertical accretion (figures 7.3 and 7.4; Friedman et al. 1996; Allred and Schmidt 1999; Friedman et al. 2005b; VanLooy and

Martin 2005; Sandercock et al. 2007; Dean and Schmidt 2011, Dean et al. 2011). As a caveat to this conclusion, these interactions vary with watershed and site-specific differences in a range of factors, including local climate, valley constraint, and human floodplain uses. Furthermore, the long-term interplay between *Tamarix* and channel form continues to develop with ongoing changes in the river environment.

Tamarix and Future Floodplain Dynamics

Extensive research in varied riverine settings has given us a clearer understanding of the interacting hydrogeomorphic factors promoting *Tamarix* naturalization and the mechanisms by which *Tamarix* influences channel morphology. Such insights are critical to addressing future reciprocal relationships as western river ecosystems continue to change (Auerbach et al. 2012). Climate warming and drying, biological interactions, and *Tamarix* control and riparian restoration efforts are likely to mediate the role of *Tamarix*.

Changes in CO_2 concentration, temperature, precipitation, and watershed hydrology may combine in complex ways to affect the extent and abundance of *Tamarix*, but current projections suggest that warming and altered drought frequency may facilitate further expansion throughout the western United States (Perry et al. 2012). Increases in minimum winter temperatures, which currently limit *Tamarix* latitudinal and elevation extent, could enable *Tamarix* to extend its range northward or into higher elevation reaches (Sexton et al. 2002; Friedman et al. 2008). Climate change is expected to influence stream flow timing and variability as well as mean annual runoff in the western United States (Barnett et al. 2008; Luce and Holden 2009; Perry et al. 2012). Earlier spring snowmelt is projected to affect the timing and magnitude of peak flow, resulting in earlier and possibly smaller snowmelt runoff floods (Mote et al. 2005). Summer low flows are also expected to decrease in rivers throughout the West (Dettinger and Cayan 1995; Cayan et al. 2001; Luce and Holden 2009). The water demands of the rapidly growing human population in this region may increase the intensity of regulation on some rivers, particularly if droughts stress the existing supply infrastructure (Palmer et al. 2009; Sabo et al. 2010). Thus, increasing intermittency (Stromberg et al. 2007b), later non-snowmelt high flows, or reduced growing season flows (Merritt and Poff 2010), could favor greater *Tamarix* abundance and dominance throughout the region it already occupies, while disfavoring native riparian species with earlier seed release or less drought tolerance.

Management decisions may exacerbate or mitigate such outcomes as future river regulation affects the flow regime to influence riparian plant community composition (Merritt et al. 2010; Auerbach et al. 2012). For instance, the elimination of physical disturbance and inundation from flooding may lead to more establishment opportunities for shade-tolerant species, such as Russian olive (*Elaeagnus angustifolia*), than for disturbance-adapted *Tamarix* or other pioneer trees and shrubs (Katz and

Shafroth 2003; Reynolds and Cooper 2010; but see Mortenson and Weisberg 2010 for an alternative view). Alternatively, seasonal dam releases and rates of recession timed to favor cottonwood establishment may promote its recruitment along some reaches (Shafroth et al. 1998; Rood et al. 2005; Merritt and Poff 2010). The assessment and implementation of environmental flows, designed to accommodate the hydrologic life-history requirements of species of concern, has emerged as an important frontier in river research and management (Poff et al. 2010). While it is unlikely that even very aggressive flow management would extirpate or eliminate *Tamarix* from western US rivers (Merritt and Poff 2010), this strategy may substantially improve conditions for native species and thereby restrict dominance by *Tamarix* within riparian zones (Nagler et al. 2005; Stromberg et al. 2007a; Shafroth et al. 2010a).

A variety of other *Tamarix* control and riparian restoration efforts may also reduce *Tamarix* abundance within floodplains it already occupies or hinder its colonization of additional areas, thereby influencing hydrogeomorphic feedbacks along these reaches (Shafroth et al. 2010b; see also chapter 23, this volume). Herbicidal control of *Tamarix* along the Rio Puerco in New Mexico permitted substantial lateral bank erosion during a major subsequent flood (Vincent et al. 2009). The sprayed reach experienced an 84% increase in mean width while unsprayed reaches upstream and downstream experienced little erosion. Vincent et al. (2009) proposed that the character and magnitude of this erosional response was likely related to the flashiness of flow on the mostly unregulated Rio Puerco as well as to floodplain attributes, such as the prevalence of sand. Reinforcing the importance of such factors, high flows produced relatively modest channel adjustments following mechanical *Tamarix* removal within Canyon de Chelly, Arizona (Jaeger and Wohl 2011). Despite the complete extraction of individual trees in some treatments, the lower flow competence likely combined with an entrenched channel and more cohesive floodplain materials to limit erosion in this system (Jaeger and Wohl 2011). By contrast, work in Dinosaur National Monument demonstrated increased survival of box elder (*Acer negundo*) under *Tamarix* canopies and increases in desirable vegetation without floodplain sediment disturbance (DeWine and Cooper 2010). More broadly, integrating control efforts with hydrogeomorphic regimes will affect the success of projects. For instance, the timing and magnitude of floodplain inundation following *Tamarix* removal will likely help to determine the species composition of the subsequent plant community. Finally, adaptive river management requires research into the hydrogeomorphic consequences of herbivory by the rapidly spreading *Tamarix* leaf beetle (*Diorhabda* spp.; Lewis et al. 2003; Hudgeons et al. 2007). Large stands of defoliated or dead *Tamarix* now occur along numerous western rivers and streams (O'Meara et al. 2010), but how these areas will respond to flow variation or support native tree establishment remains poorly understood. Key uncertainties for beetle-colonized *Tamarix* stands include the degree to which floods will mobilize sediment by removing standing dead trees or weakening subsurface root reinforcement, the hydraulic properties of vegetation that replaces *Tamarix*, and how such changes might scale up to influence regional patterns of channel change. Taken together, these issues suggest the importance of planning restoration efforts within the context of intrinsic watershed attributes (e.g., drainage area, geology, land cover) and complementary management actions (e.g., dam operation, active revegetation) affecting the target reach (Shafroth et al. 2008).

Conclusion

Flow and sediment regimes shape riparian plant communities and influence the recruitment, survival, and dominance of *Tamarix* along western US rivers. Along many of these rivers, hydrologic alteration resulting from climate shifts, land-use changes, and water-control infrastructure contributed to the spread of *Tamarix* and facilitated its subsequent influence on channel form and function. Nonetheless, as demonstrated by its colonization of both regulated and free-flowing rivers, *Tamarix* possesses a combination of morphological and physiological traits that are well suited to a wide range of conditions found on contemporary western floodplains.

Like native *Populus* and *Salix*, *Tamarix* requires bare, moist surfaces for establishment, such as those created by flood disturbance. However, the timing of flood peaks and subsequent recession may favor *Tamarix* when the seasonal availability of establishment sites matches its pattern of seed release more closely than that of native plants. In addition, reduced flooding and baseflows may favor *Tamarix* relative to native *Populus* and *Salix* because of its greater tolerance of drought, soil salinity, and groundwater fluctuation. The influence of *Tamarix* on local erosional and depositional processes varies according to interactions between its biology (e.g., stem density and size), site characteristics (e.g., valley constraint and prevailing sediment types), and flow regime (e.g., flood frequency and magnitude).

It is likely that *Tamarix* will remain an important component of riparian zones in the American West, though control efforts, coupled with flow management tailored to the needs of native species, may hinder its dominance of new floodplains or those it currently occupies. The introduction and spread of *Tamarix* has provided a unique opportunity to advance understanding of basic relationships among hydrologic, geomorphic, and ecological processes in rivers. Studies of *Tamarix* can continue to provide a focus for research into the reciprocal relationship between riparian trees and hydrogeomorphic dynamics in river ecosystems subject to climate change and water development.

Literature Cited

Alexander, J. S. 2008. The timing and magnitude of channel adjustments in the upper Green River below Flaming Gorge Dam in Browns Park and Lodore Canyon, Colorado: An analysis of the pre- and post-dam river using high-resolution dendrogeomorphology and repeat topographic surveys. Master's thesis. Utah State University, Logan.

Allred, T. W., and J. C. Schmidt. 1999. Channel narrowing by vertical accretion along the Green River near Green River, Utah. Geological Society of America Bulletin 111:1757–1772.

Auerbach, D. A., N. L. Poff, R. R. McShane, D. M. Merritt, M. I. Pyne, and T. Wilding. 2012. Streams past and future: Fluvial responses to rapid environmental change in the context of historical variation. Chapter 16 *in* J. A.Wiens, G. D. Hayward, H.D. Safford, and C. Giffen, editors. *Historical Environmental Variation in Conservation and Natural Resource Management.* Wiley-Blackwell, Hoboken, NJ.

Baker, V. R., and D. F. Ritter. 1975. Competence of rivers to transport coarse bedload material. Geological Society of America Bulletin 86:975–978.

Barnett, T. P., D. W. Pierce, H. G. Hidalgo, C. Bonfils, B. D. Santer, T. Das, G. Bala, A. W. Wood, T. Nozawa, A. A. Mirin, D. R. Cayan, and M. D. Dettinger. 2008. Human-induced changes in the hydrology of the western United States. Science 319:1080–1083.

Beauchamp, V. B., and J. C. Stromberg. 2007. Flow regulation of the Verde River, Arizona encourages *Tamarix* recruitment but has minimal effect on *Populus* and *Salix* stand density. Wetlands 27:381–389.

Birken, A. S., and D. J. Cooper. 2006. Processes of *Tamarix* invasion and floodplain development along the lower Green River, Utah. Ecological Applications 16:1103–1120.

Blackburn, W. H., R. W. Knight, and J. L. Schuster. 1982. Saltcedar influence on sedimentation in the Brazos River. Journal of Soil and Water Conservation 37:298–301.

Brotherson, J. D., and D. Field. 1987. *Tamarix*: Impacts of a successful weed. Rangelands 9:110–112.

Busch, D. 1995. Effects of fire on southwestern riparian plant community structure. Southwestern Naturalist 40:259–267.

Busch, D. E., and S. D. Smith. 1995. Mechanisms associated with decline of woody species in riparian ecosystems of the southwestern U.S. Ecological Monographs 65:347–370.

Campbell, C. J., and W. A. Dick-Peddie. 1964. Comparison of Phreatophyte Communities on the Rio Grande in New Mexico. Ecology 45:492–502.

Cayan, D. R., S. A. Kammerdiener, M. D. Dettinger, J. M. Caprio, and D. H. Peterson. 2001. Changes in the onset of spring in the western United States. Bulletin of the American Meteorological Society 82:399–415.

Cleverly, J. R., S. D. Smith, and A. Sala. 1997. Invasive capacity of *Tamarix ramosissima* in a Mojave Desert floodplain: The role of drought. Oecologia 111:12–18.

Cooper, D. J., D. C. Andersen, and R. A. Chimner. 2003. Multiple pathways for woody plant establishment on floodplains at local to regional scales. Journal of Ecology 91:182–196.

Cooper, D. J., D. M. Merritt, D. C. Andersen, and R. A. Chimner. 1999. Factors controlling the establishment of Fremont cottonwood seedlings on the Upper Green River, USA. Regulated Rivers: Research and Management 15:419–440.

Cooper, D. J., R. A. Chimner, and D. M. Merritt. 2012. Western Mountain Wetlands. Chapter 22 *in* D.P. Batzer and A.H. Baldwin, editors. *Wetland Habitats of North America: Ecology and Conservation Concerns.* University of California Press, Berkeley.

Dean, D. J., and J. C. Schmidt. 2011. The role of feedback mechanisms in historic channel changes of the lower Rio Grande in the Big Bend region. Geomorphology 126:333–349.

Dean, D. J., M. L. Scott, P. B. Shafroth, and J. C. Schmidt. 2011. Stratigraphic, sedimentologic, and dendrogeomorphic analyses of rapid floodplain formation along the Rio Grande in Big Bend National Park, Texas. Geological Society of America Bulletin 123:1908–1925.

Dettinger, M. D., and D. R. Cayan. 1995. Large-scale atmospheric forcing of recent trends toward early snowmelt runoff in California. Journal of Climate 8:606–623.

DeWine, J. M., and D. J. Cooper. 2010. Habitat overlap and facilitation in tamarisk and box elder stands: Implications for tamarisk control using native plant. Restoration Ecology 18:349–358.

Ellis, L. M. 2001. Short-term response of woody plants to fire in a Rio Grande riparian forest, central New Mexico, USA. Biological Conservation 97:159–170.

Ellis, L. M., C. S. Crawford, and M. C. Molles. 1998. Comparison of litter dynamics in native and exotic riparian vegetation along the Middle Rio Grande of central NM, USA. Journal of Arid Environments 38:283–296.

Everitt, B. L. 1998. Chronology of the spread of tamarisk in the central Rio Grande. Wetlands 18:658–668.

Fisher, S. G., J. B. Heffernan, R. A. Sponseller, and J. R. Welter. 2007. Functional ecomorphology: Feedbacks between form and function in fluvial landscape ecosystems. Geomorphology 89:84–96.

Friedman, J. M., G. T. Auble, P. B. Shafroth, M. L. Scott, M. F. Merigliano, M. D. Freehling, and E. R. Griffin. 2005. Dominance of non-native riparian tree in western USA. Biological Invasions 7:747–751.

Friedman, J. M., J. E. Roelle, J. F. Gaskin, A. E. Pepper, and J. R. Manhart. 2008. Latitudinal variation in cold hardiness in introduced *Tamarix* and native *Populus*. Evolutionary Applications 1:598–607.

Friedman, J. M., K. R. Vincent, and P. B. Shafroth. 2005b. Dating floodplain sediments using tree-ring response to burial. Earth Surface Processes and Landforms 30:1077–1091.

Friedman, J. M., and V. J. Lee. 2002. Extreme floods, channel change, and riparian forests along ephemeral streams. Ecology 72:409–425.

Friedman, J. M., W. R. Osterkamp, and W. M. Lewis, Jr. 1996. The role of vegetation and bed-level fluctuations in the process of channel narrowing. Geomorphology 14:341–351.

Gerry, E. 1954. Athel Tamarisk *Tamarix aphylla* (L.) Karst. (= *T. articulata* Vahl.) Family: Tamaricaceae. U.S. Forest Service, Forest Products Laboratory, information leaflet. Foreign Woods Series Report 1986, Madison, WI.

Gladwin, D., and Roelle, J., 1998. Survival of plains cottonwood (*Populus deltoides* susp.) (*monilifera*) and saltcedar (*Tamarix ramosissima*) seedlings in response to flooding. Wetlands 18:669–674.

Glenn, E., R. Tanner, S. Mendez, T. Kehret, D. Moore, J. Garcia, and C. Valdes. 1998. Growth rates, salt tolerance and water use characteristics of native and invasive riparian plants from the delta of the Colorado River, Mexico. Journal of Arid Environments 40:281–294.

Glenn E. P., K. Morino, P. L. Nagler, R. S. Murray, S. Pearlstine, and K. R. Hultine. 2012. Roles of saltcedar (*Tamarix* spp.) and capillary rise in salinizing a nonflooding terrace on a flow-regulated river. Journal of Arid Environments 79:56–65.

Glenn, E. P., and P. L. Nagler. 2005. Comparative ecophysiology of *Tamarix ramosissima* and native trees in western U.S. riparian zones. Journal of Arid Environments 61:419–446.

Graf, W. L. 1999. Dam nation: A geographic census of American dams and their large-scale hydrologic impacts. Water Resources Research 35:1305–1311.

Graf, W. L. 1978. Fluvial adjustments to the spread of tamarisk in the Colorado Plateau region. Geological Society of America Bulletin 89:1491–1501.

Graf, W. L. 1982. Tamarisk and river-channel management. Environmental Management 6:283–296.

Grams, P. E., and J. C. Schmidt. 2005. Equilibrium or indeterminate? Where sediment budgets fail: Sediment mass balance and adjustment of channel form, Green River downstream from Flaming Gorge Dam, Utah and Colorado. Geomorphology 71:156–181.

Grams, P. E., and J. C. Schmidt. 2002. Streamflow regulation and multi-level floodplain formation: Channel narrowing on the aggrading Green River in the eastern Uinta Mountains, Colorado and Utah. Geomorphology 44:337–360.

Griffin, E. R., and J. D. Smith. 2004. Floodplain stabilization by woody riparian vegetation during an extreme flood. Pages 221–236 *in* S. J. Bennett and A. Simon, editors. *Riparian Vegetation and Fluvial Geomorphology*. American Geophysical Union, Washington DC.

Griffin, E. R., J. W. Kean, K. R. Vincent, J. D. Smith, and J. M. Friedman. 2005. Modeling effects of bank friction and woody bank vegetation on channel flow and boundary shear stress in the Rio Puerco, New Mexico. Journal of Geophysical Research 110. F04023. doi:10.1029/2005JF000322.

Hereford, R. 1984. Climate and ephemeral-stream processes: Twentieth-century geomorphology and alluvial stratigraphy of the Little Colorado River, Arizona. Geological Society of America Bulletin 95:654–668.

Horton, J., and J. Clark. 2001. Water table decline alters growth and survival of *Salix gooddingii* and *Tamarix chinensis* seedlings. Forest Ecology and Management 140:239–247.

Hudgeons, J. L., A. E. Knutson, K. M. Heinz, C. J. DeLoach, T. L. Dudley, R. R. Pattison, J. R. Kiniry. 2007. Defoliation by introduced *Diorhabda elongata* leaf beetles (Coleoptera: Chrysomelidae) reduces carbohydrate reserves and regrowth of *Tamarix* (Tamaricaceae). Biological Control 43:213–221.

Jaeger, K. L., and E. Wohl. 2011. Channel response in a semiarid stream to removal of tamarisk and Russian olive. Water Resources Research. 47:W02536. doi:10.1029/2009WR008741.

Jolly, I. D., G. R. Walker, and P. J. Thorburn. 1993. Salt accumulation in semiarid floodplain soils with implications for forest health. Journal of Hydrology 150:589–614.

Karrenberg, S., P. J. Edwards, and J. Kollmann. 2002. The life history of Salicaceae living in the active zone of floodplains. Freshwater Biology 47:733–748.

Katz, G. L., and P. B. Shafroth. 2003. Biology, ecology and management of *Elaeagnus Angustifolia* L. (Russian olive) in Western North America. Wetlands 23:763–777.

Kean J. W., and J. D. Smith. 2004. Flow and boundary shear stress in channels with woody bank vegetation. Pages 237–252 in *Riparian Vegetation and Fluvial Geomorphology*, S. J. Bennett and A. Simon, editors. Water Science and Applications No. 8. American Geophysical Union.

Leopold, L. B., M. G. Wolman, and J. P. Miller. 1964. *Fluvial Processes in Geomorphology*. W.H. Freeman, San Francisco, CA.

Lescia, P., and S. Miles. 2004. Ecological strategies for managing tamarisk on the C. M. Russell National Wildlife Refuge, Montana, USA. Biological Conservation 119:535–543.

Levine, C., and J. Stromberg. 2001. Effects of flooding on native and exotic plant seedlings: Implications for restoring southwestern riparian forests by manipulating water and sediment flows. Journal of Arid Environments 49:111–131.

Lewis, P. A., C. J. DeLoach, A. E. Knutson, J. L. Tracy, and T. O. Robbins. 2003. Biology of *Diorhabda elongata deserticola* (Coleoptera: Chrysomelidae): An Asian leaf beetle for biological control of saltcedars (*Tamarix* spp.) in the United States. Biological Control 27:101–116.

Lite, S. J., and J. C. Stromberg. 2005. Surface water and ground-water thresholds for maintaining *Populus-Salix* forests, San Pedro River, Arizona. Biological Conservation 125:153–167.

Luce, C. H., and Z. A. Holden. 2009. Declining annual stream flow distributions in the Pacific Northwest United States, 1948–2006. Geophysical Research Letters 36:1–6.

Mahoney, J. M., and S. B. Rood. 1998. Streamflow requirements for cottonwood seedling recruitment: An integrative model. Wetlands 18:634–645.

Martin, C. W., and W. C. Johnson. 1987. Historical Channel Narrowing and Riparian Vegetation Expansion in the Medicine Lodge River Basin, Kansas, 1871–1983. Annals of the Association of American Geographers 77:436–449.

Merigliano, M. F. 2005. Cottonwood understory zonation and its relation to floodplain stratigraphy. Wetlands 25:356–374.

Merritt, D. M. 2012. Reciprocal relations between riparian vegetation, fluvial landforms, and channel processes. Chapter 9.14 *in* J. Shroder Jr. and E. Wohl, editors. *Treatise on Geomorphology*. Academic Press, San Diego, CA.

Merritt, D. M., and D. J. Cooper. 2000. Riparian vegetation and channel change in response to river regulation: A comparative study of regulated and unregulated streams in the Green River Basin, USA. Regulated Rivers: Research and Management 16:543–564.

Merritt, D. M., and E. E. Wohl. 2002. Processes governing hydrochory along rivers: Hydraulics, hydrology, and dispersal phenology. Ecological Applications 12:1071–1087.

Merritt, D. M., M. L. Scott, N. L. Poff, G. T. Auble, and D. A. Lytle. 2010. Theory, methods and tools for determining environmental flows for riparian vegetation: Riparian vegetation-flow response guilds. Freshwater Biology 55:206–225.

Merritt, D. M., and N. L. Poff. 2010. Shifting dominance of riparian *Populus* and *Tamarix* along gradients of flow alteration in western North American rivers. Ecological Applications 20:135–152.

Merritt, D. M., and P. B. Shafroth. 2012. Edaphic, salinity, and stand structural trends in chronosequences of native and non-native dominated riparian forests along the Colorado River, USA. Biological Invasions DOI 10.1007/s10530-012-0263-4.

Miller, J. R., and J. M. Friedman. 2009. Influence of flow variability on floodplain formation and destruction, Little Missouri River, North Dakota. Geological Society of America Bulletin 121:752–759.

Molles, M. C., C. S. Crawford, L. M. Ellis, H. M. Valett, and C. N. Dahm. 1998. Managed flooding for riparian ecosystem restoration. BioScience 48:749–756.

Mortenson, S. G., and P. J. Weisberg. 2010. Does river regulation increase the dominance of invasive woody species in riparian landscapes? Global Ecology and Biogeography 19:562–574.

Mote, P. W., A. F. Hamlet, M. P. Clark, and D. P. Lettenmaier. 2005. Declining mountain snowpack in western North America. Bulletin of the American Meteorological Society 86:39–49.

Nagler, P. L., E. P. Glenn, C. S. Jarnevich, and P. B. Shafroth. 2011. Distribution and abundance of saltcedar and Russian olive in the western United States. Critical Reviews in Plant Sciences 30:508–523.

Nagler, P. L., O. Hinojosa-Huerta, E. P. Glenn, J. Garcia-Hernandez, R. Romo, C. Curtis, A. R. Huete, and S. G. Nelson. 2005. Regeneration of native trees in the presence of invasive saltcedar in the Colorado River Delta. Conservation Biology 19:1842–1852.

Naiman, R. J., H. Decamps, and M. E. McClain. 2005. *Riparia: Ecology, Conservation, and Management of Streamside Communities.* Elsevier Academic Press. London

Nilsson, C., and K. Berggren. 2000. Alterations of riparian ecosystems resulting from river regulation. BioScience 50:783–792.

Nilsson, C., and M. Svedmark. 2002. Basic principles and ecological consequences of changing water regimes: Riparian plant communities. Environmental Management 30:468–480.

O'Meara, S., D. Larsen, and C. Owens. 2010. Methods to control Saltcedar and Russian Olive. Chapter 5 *in* P.B. Shafroth, C.A. Brown, and D.M. Merritt, editors. Saltcedar and Russian Olive Control Demonstration Act science assessment. U.S. Geological Survey Scientific Investigations Report 2009–5247.

Ohrtman, M. K. 2009. Quantifying soil and groundwater chemistry in areas invaded by *Tamarix* spp. along the middle Rio Grande, New Mexico. Dissertation. Natural Sciences and Mathematics. University of Denver, CO.

Palmer, M. A., D. P. Lettenmaier, N. L. Poff, S. L. Postel, B. Richter, and R. Warner. 2009. Climate change and river ecosystems: Protection and adaptation options. Environmental Management 44:1053–1068.

Patten, D. T. 1998. Riparian ecosystems of semi-arid North America: Diversity and human impacts. Wetlands 18:498–512.

Perry, L. G., D. C. Andersen, L. V. Reynolds, S. M. Nelson, and P. B. Shafroth. 2012. Vulnerability of riparian ecosystems to elevated CO_2 and climate change in arid and semiarid western North America. Global Change Biology. doi: 10.1111/j.1365-2486. 2011.02588.x.

Pockman, W. T., and J. S. Sperry. 2000. Vulnerability to xylem cavitation and the distribution of Sonoran desert vegetation. American Journal of Botany 87:1287–1299.

Poff, N. L., J. D. Allan, M. B. Bain, J. R. Karr, K. L. Prestegaard, B. D. Richter, R. E. Sparks, and J. C. Stromberg. 1997. The natural flow regime: A paradigm for river conservation and restoration. BioScience 47:769–784.

Poff, N. L., B. P. Bledsoe, and C. O. Cuhaciyan. 2006a. Hydrologic variation with land use across the contiguous United States: Geomorphic and ecological consequences for stream ecosystems. Geomorphology 79:264–285.

Poff, N. L., J. D. Olden, D. M. Merritt, and D. M. Pepin. 2007. Homogenization of regional river dynamics by dams and global biodiversity implications. Proceedings of the National Academy of Sciences of the United States of America 104:5732–5737.

Poff, N. L., J. D. Olden, D. M. Pepin, and B. P. Bledsoe. 2006b. Placing global stream flow variability in geographic and geomorphic contexts. River Research and Applications 22:149–166.

Poff, N. L., B. D. Richter, A. H. Arthington, S. E. Bunn, R. J. Naiman, E. Kendy, M. Acreman, C. Apse, B. P. Bledsoe, M. C. Freeman, J. Henriksen, R. B. Jacobson, J. G. Kennen, D. M. Merritt, J. H. O'Keefe, J. D. Olden, K. Rogers, R. E. Tharme, and A. Warner. 2010. The ecological limits of hydrologic alteration (ELOHA): A new framework for developing regional environmental flow standards. Freshwater Biology 55:147–170.

Pollen-Bankhead, N., A. Simon, K. Jaeger, and E. Wohl. 2009. Destabilization of streambanks by removal of invasive species in Canyon de Chelly National Monument, Arizona. Geomorphology 103:363–374.

Renofalt, B. M., D. M. Merritt, and C. Nilsson. 2007. Connecting variation in vegetation and stream flow: The role of geomorphic context in vegetation response to large floods along boreal rivers. Journal of Applied Ecology 44:147–157.

Reynolds, L. V., and D. J. Cooper. 2010. Environmental tolerance of an invasive riparian tree and its potential for continued spread in the southwestern US. Journal of Vegetation Science 21:733–743.

Ringold, P. L., T. K. Magee, and D. V. Peck. 2008. Twelve invasive plant taxa in US western riparian ecosystems. Journal of the North American Benthological Society 27:949–966.

Robinson, T. W. 1965. Introduction, Spread, and Aerial Extent of Saltcedar (*Tamarix*) in the Western States. U.S. Geological Survey Professional Paper 491-A. Washington, DC.

Rood, S. B., G. M. Samuelson, J. H. Braatne, C. R. Gourley, F. M. R. Hughes, and J. M. Mahoney. 2005. Managing river flows to restore floodplain forests. Frontiers in Ecology and the Environment 3:193–201.

Rood, S. B., and J. M. Mahoney. 1990. Collapse of riparian Poplar forests downstream from dams in western prairies: Probable causes and prospects for mitigation. Environmental Management 14:451–464.

Sabo, J., T. Sinha, L. Bowling, G. Schoups, W. Wallender, M. Campana, K. Cherkauer, P. Fuller, W. Graf, J. Hopmans, J. Kominoski, C. Taylor, S. Trimble, R. Webb, and E. Wohl. 2010. Reclaiming freshwater sustainability in the Cadillac Desert. Proceedings of the National Academy of Sciences of the United States of America 107:21263–21270.

Sandercock, P. J., and J. M. Hooke. 2010. Assessment of vegetation effects on hydraulics and of feedbacks on plant survival and zonation in ephemeral channels. Hydrological Processes 24:695–713.

Sandercock, P. J., J. M. Hooke, and J. M. Mant. 2007. Vegetation in dryland river channels and its interaction with fluvial processes. Progress in Physical Geography 31:107–129.

Schumm, S. A., and R. W. Lichty. 1963. Channel Widening and Flood-Plain Construction along the Cimarron River in Southwestern Kansas: U.S. Geological Survey Professional Paper 352-D. Washington, DC.

Scott, M. L., J. M. Friedman, and G. T. Auble. 1996. Fluvial process and the establishment of bottomland trees. Geomorphology 14:327–339.

Sexton, J. P., J. K. McKay, and A. Sala. 2002. Plasticity and genetic diversity may allow saltcedar to invade cold climates in North America. Ecological Applications 12:1652–1660.

Shafroth, P. B., A. C. Wilcox, D. A. Lytle, J. T. Hickey, D. C. Andersen, V. B. Beauchamp, A. Hautzinger, L. E. McMullen, and A. Warner. 2010a. Ecosystem effects of environmental flows: Modelling and experimental floods in a dryland river. Freshwater Biology 55:68–85.

Shafroth, P. B., C. A. Brown, and D. M. Merritt, editors. 2010b. Saltcedar and Russian olive control demonstration act science assessment. U.S. Geological Survey Scientific Investigations Report 2009–5247.

Shafroth, P. B., G. T. Auble, J. C. Stromberg, and D. T. Patten. 1998. Establishment of woody riparian vegetation in relation to annual patterns of streamflow, Bill Williams River, Arizona. Wetlands 18:577–590.

Shafroth, P. B., J. C. Stromberg, and D. T. Patten. 2002. Riparian vegetation response to altered disturbance and stress regimes. Ecological Applications 12:107–123.

Shafroth, P. B., J. M. Friedman, and L. Ischinger. 1995. Effects of salinity on establishment of *Populus fremontii* (cottonwood) and *Tamarix ramosissima* (saltcedar) in the southwestern United States. Great Basin Naturalist 55:58–65.

Shafroth, P. B., J. R. Cleverly, T. L. Dudley, and J. P. Taylor. 2005. Control of *Tamarix* in the western United States: Implications for water salvage, wildlife use, and riparian restoration. Environmental Management 35:231–246.

Shafroth, P. B., V. B. Beauchamp, M. K. Briggs, K. Lair, M. L. Scott, and A. A. Sher. 2008. Planning riparian restoration in the context of *Tamarix* control in western North America. Restoration Ecology 16:97–112.

Sher, A. A., and D. L. Marshall. 2003. Seedling competition between native *Populus deltoides* (Salicaceae) and exotic *Tamarix ramosissima* (Tamaricaceae) across water regimes and substrate types. American Journal of Botany 90:413–422.

Sher, A. A., D. L. Marshall, and J. P. Taylor. 2002. Establishment patterns of native *Populus* and *Salix* in the presence of invasive nonnative *Tamarix*. Ecological Applications 12:760–772.

Sigafoos, R. S. 1964. Botanical evidence of floods and floodplain deposition. U.S. Geological Survey Professional Paper 485-A. Washington, DC.

Sprenger, M. D., L. M. Smith, and J. P. Taylor. 2001. Testing control of saltcedar seedlings using fall flooding. Wetlands 31:437–441.

Stromberg, J. C. 1998a. Dynamics of Fremont cottonwood (*Populus fremontii*) and saltcedar (*Tamarix chinensis*) populations along the San Pedro River, Arizona. Journal of Arid Environments 40:133–155.

Stromberg, J. C. 1993. Fremont cottonwood–Goodding willow riparian forests: A review of their ecology, threats, and recovery potential. Journal of the Arizona–Nevada Academy of Sciences 26:97–111.

Stromberg, J. C. 1998b. Functional equivalency of saltcedar (*Tamarix chinensis*) and Fremont cottonwood (*Populus fremontii*) along a free-flowing river. Wetlands 18:675–686.

Stromberg, J. C. 1997. Growth and survivorship of Fremont cottonwood, Gooddings willow, and salt cedar seedlings after large floods in central Arizona. Great Basin Naturalist 57:198–208.

Stromberg, J. C., J. A. Boudell, and A. F. Hazelton. 2008. Differences in seed mass between hydric and xeric plants influence seed bank dynamics in a dryland riparian ecosystem. Functional Ecology 22:205–212.

Stromberg, J. C., and M. Chew. 2002. Foreign visitors in riparian corridors of the American Southwest: Is xenophobia justified? Pages 195–219 *in* Tellman, B., editor. *Invasive Exotic Species in the Sonoran Region*. University of Arizona Press, Tucson.

Stromberg, J. C., M. K. Chew, P. L. Nagler, and E. P. Glenn. 2009. Changing perceptions of change: The role of scientists in *Tamarix* and river management. Restoration Ecology 17:177–186.

Stromberg, J. C., S. J. Lite, R. Marler, C. Paradzick, P. B. Shafroth, D. Shorrock, J. M. White, and M. S. White. 2007b. Altered stream-flow regimes and invasive plant species: The Tamarix case. Global Ecology and Biogeography 16:381–393.

Stromberg, J. C., V. B. Beauchamp, M. D. Dixon, S. J. Lite, and C. Paradzick. 2007a. Importance of low-flow and high-flow characteristics to restoration of riparian vegetation along rivers in arid southwestern United States. Freshwater Biology 52:651–679.

Tallent-Halsell, N. G., and L. R. Walker. 2002. Responses of *Salix gooddingii* and *Tamarix ramosissima* to flooding. Wetlands 22:776–785.

Taylor, J. P., D. B. Wester, and L. M. Smith. 1999. Soil disturbance, flood management, and riparian woody plant establishment in the Rio Grande floodplain. Wetlands 19: 372–382.

Tooth, S. 2000. Process, form and change in dryland rivers: A review of recent research. Earth-Science Reviews 51:67–107.

Vandersande, M., E. Glenn, J. Walford. 2001. Comparison of tolerances of riparian plants from the lower Colorado River to salinity, drought and inundation. Journal of Arid Environments 49:147–160.

VanLooy, J. A., and C. W. Martin. 2005. Channel and vegetation change on the Cimarron River, southwestern Kansas,1953–2000. Annals of the Association of American Geographers 95:727–739.

Vincent, K. R., J. M. Friedman, and E. R. Griffin. 2009. Erosional consequence of saltcedar control. Environmental Management 44:218–227.

Warren, D. K., and R. M. Turner. 1975. Saltcedar (*Tamarix chinensis*) Seed production, seedling establishment, and response to inundation. Journal of the Arizona Academy of Science 10:135–144.

Wolman, M. G., and R. Gerson. 1978. Relative scales of time and effectiveness of climate in watershed geomorphology. Earth Surface Processes 3:189–208.

8

Tamarix and Salinity: An Overview

Michelle K. Ohrtman and Ken D. Lair

Plant invasions are often correlated with edaphic environmental changes, such as to soil and water. However, it is not always clear whether the plants themselves are the cause or consequence of such modifications (MacDougal and Turkington 2005). The woody exotic *Tamarix* spp. (*T. chinensis*, *T. ramosissima* or hybrids; salt-cedar, tamarisk), represent an extreme example of this phenomenon; elevated soil and water salinity is generally associated with these species, and is often attributed to them (Glenn and Nagler 2005). As the common name "saltcedar" implies, one of the most often-cited positive feedback mechanisms associated with *Tamarix* is its ability to extract salts from groundwater and subsurface soils, sequester them in its tissues, excrete them through the foliage, and deposit them on the soil surface through leaf senescence. Although this is a known process, all published research to date about the effect of *Tamarix* on environmental salinity has been correlative. Recent publications have called the mechanism into question, pointing out that there are other causes of elevated salinity that could facilitate invasiveness (Stromberg et al. 2009).

There are many sources of salts in riparian systems of the American West, where *Tamarix* is dominant. High levels of environmental salinity have natural origins in ambient geochemistry, fluvial dynamics, and landscape-scale response to naturally occurring drought conditions. Arid regions tend to have higher concentrations of surface salts because evaporation concentrates salts in surface soils through capillary rise from groundwater, and there is insufficient precipitation to remove accumulated compounds by runoff or leaching (Goodall and Perry 1981). More often, however, river systems that have been regulated through anthropogenic impoundments and floodplain alterations (e.g., dams, diversions, drains, and irrigation-system infrastructure) have greatly exacerbated and accelerated the extent and adverse impacts of salinity over the last century (Ohmart et al. 1988; Stromberg 2001). The common impact to existing native plant communities is large-scale replacement of less saline-tolerant riparian species with more salt-tolerant vegetation (see chapter 5, this volume).

Adverse impacts to native riparian vegetation from increases in soil and groundwater salinity are well documented (Richards 1954; Branson et al. 1988; Ohmart et al. 1988; Busch and Smith 1995). Excess salts in the soil reduce the ability of plant roots to withdraw water. This lowers the amount of water available to plants, especially those that lack salt tolerance. Elevated concentrations of soluble sodium (i.e., sodicity) in soils can also cause irreversible damage to soil structure. Sodicity can cause soil dispersion, where clay particles plug soil pores and create an impenetrable barrier to water transport. Soil dispersion often reduces water infiltration, while increasing runoff and erosion; the result is less water available to plants.

Elevated surface soil salinity can inhibit the germination, establishment, and growth of many native species (Busch and Smith 1993; Glenn et al. 1998; Jackson et al. 1990; Rowland et al. 2004; Shafroth et al. 1995; Singh et al. 1999). *Populus* spp. (cottonwood) and *Salix* spp. (willow) performed poorly when exposed to salt levels greater than 1,500 parts per million (ppm) (2.3 mmhos cm^{-1} EC; Jackson et al. 1990). Elevating electrical conductivity (EC) from 0.8 mmhos cm^{-1} to 1.3 mmhos cm^{-1} led to a significant decrease in *Populus* seedling growth for several Rio Grande populations (Rowland et al. 2004). Similarly, *Populus* seed germination decreased significantly when exposed to a growth medium with an EC of 3.5 mmhos cm^{-1} in a controlled outdoor environment (Shafroth et al. 1995).

Given the lower salt tolerance of many species relative to *Tamarix*, riparian salinization can adversely affect our ability to restore native riparian plant communities within large-scale *Tamarix* infestations or subsequent to control measures (Lair and Wynn 2002; Anderson 1995; Pinkney 1992; Anderson and Ohmart 1979, 1982, 1985). Saline/sodic soil and shallow groundwater environments also severely limit the selection and applicability of commonly applied revegetation practices, simultaneously adding complexity to the restoration strategies and technologies that must be considered in these environments.

There is some debate as to whether *Tamarix* is the cause or the opportunistic result of elevated salinity (e.g., Stromberg et al. 2009). This chapter represents a synthesis of the literature devoted to the subject of *Tamarix* and environmental salinity, including the physiological adaptations for salt tolerance by *Tamarix*, how it may enhance environmental salt loading, and a review of the ecological literature attempting to document this linkage.

Tamarix Salt Tolerance

Tamarix is a facultative halophyte, that is, it is not dependent on elevated salt concentrations for survival but can successfully grow and reproduce in the presence of high electrolyte levels (Busch and Smith 1995; Glenn et al. 1998; Jackson et al. 1990; Shafroth et al. 1995). *Tamarix* has been found growing in substrates containing high levels of soluble salts, especially when compared with soils and groundwater supporting native riparian vegetation (Busch and Smith 1995). *Tamarix* was found to survive soil salt concentrations up to 36,000 ppm (56 mmhos cm^{-1} EC) which is equivalent to the salt content of ocean water (Glenn et al. 1998; Jackson et al. 1990). Studies on *Tamarix* spp. in their native range suggest they can tolerate soil salinity levels up to 60,000 ppm (Chen et al. 2008; Cui et al. 2010; He et al. 2011). Increased soil water availability may enhance *Tamarix*'s salt tolerance (Wang et al. 2011). *Tamarix* can also proliferate in areas where the groundwater has a high concentration of salts. Field studies have confirmed *Tamarix* spp. populations in Southern California growing above groundwater with between 2,000 and 10,000 ppm dissolved salts (between 3 and 16 mmhos cm^{-1}; Glenn et al. 2012; Nagler et al. 2008). *Tamarix* has been found growing in Death Valley, California, where the dissolved solid content of the groundwater can be up to 50,000 ppm (Robinson 1965). Although *Tamarix ramosissima* performed poorly at extreme salinity levels (40,000 ppm), photosynthesis and metabolic function increased after about 35 days of exposure (Carter and Nippert 2011). These results and the success of *Tamarix* across a broad range of environmental salt levels (Busch and Smith 1995; Brotherson and Winkel 1986; Carter and Nippert 2012) suggest that *Tamarix* acclimates to elevated salinities in relatively short periods. Though *Tamarix* species can tolerate extreme levels of salinity, plants appear to grow equally well or better where soil and groundwater are little to moderately mineralized (Brotherson and Winkel 1986; Carter and Nippert 2011; Cui et al 2010; Glenn et al. 1998; He et al. 2011; Kleinkopf and Wallace 1974; Li et al. 2010; Sharif and Khan 2009).

TAMARIX SALT UPTAKE

Tamarix is a facultative phreatophyte and can extract water and its solutes both from groundwater and the soil horizons above the water table (Nippert et al. 2010). The tap root of *Tamarix* species in the United States can exceed five meters in length, often showing signs of continuing growth toward a deeper water table (Horton and Campbell 1974; Merkel and Hopkins 1957). *Tamarix* also develops extensive lateral root networks. A 70-year-old *T. taklamakanensis* plant in China had roots stretching a lateral distance of 50 meters (Liu et al. 2008). Lateral roots for a presumably younger *T. ramosissima* in the same environment reached 16 meters from the base of the tree (Arndt et al. 2004b). Three-year-old *T. gallica* plants in Kansas had roots up to 6 meters long and older plants had roots intertwined with neighboring tree root networks 10 meters away (Merkel and Hopkins 1957). *Tamarix* species therefore have the ability to extract water and dissolved salts from large volumes in the soil profile (Waisel 1991b).

Although *Tamarix* can accumulate salts, salt exclusion at the periphery of roots may still control the majority of the salts that would have otherwise entered these plants. Research on other halophytes has shown that up to 80% of salts are prevented from entering plant roots (Waisel et al. 1986). Relationships between elevated groundwater salinity and greater separation from diluting river water in *Tamarix* stands support the occurrence of ion exclusion by *Tamarix* (Glenn et al. 2012; Nagler et al. 2008). Selective water uptake from the aquifer creates a mechanism by which *Tamarix* can increase the salinity of its environment.

SALT TRANSPORT AND COMPARTMENTALIZATION IN *TAMARIX*

Whereas most salt-sensitive species generally rely on total ion exclusion at the root endodermis to survive in saline soils, *Tamarix* can survive in part through ion uptake and compartmentalization in salt glands (Waisel 1991a). Salt glands allow *Tamarix* to avoid harmful effects of salts on foliar mesophyll cells through recretion, the passing of ions through the plant without being metabolized or chemically changed (Bosabalidis and Thomson 1984, 1985; Campbell and Strong 1964; Waisel 1991a). High foliar salt content in *T. ramosissima* leaves relative to other co-occuring species in both its native and introduced range confirms the transport and storage of recreted ions in leaves (Arndt et al. 2004a; Busch and Smith 1995).

The mechanisms of salt transport and sequestration in *Tamarix* salt glands are well understood. However, much of the information available about salt glands, recretion, and secretion (the release of ionic solutions from glands) is derived from work performed on *T. aphylla* (athel), an evergreen species recently identified as invasive in the southwestern United States (Walker et al. 2006). We can presume that the features described for *T. aphylla* also apply to more invasive *Tamarix* species (e.g., *T. chinensis* and *T. ramosissima*) because species within a genus possess similar anatomical structures and mechanisms for compartmentalizing salts (Black 1954; Gucci et al. 1997). Structural and functional similarities have also been observed among *T. aphylla*, *T. pentandra* (Campbell and Strong 1964; Decker 1961; Wilkinson 1966), *T. chinensis* (Dressen and Wangen 1981), and *T. ramosissima* (Kleinkopf and Wallace 1974).

Following uptake by *Tamarix* roots, ions are transported from the vascular tissue to the salt glands by two pathways: (1) between cell walls (*apoplastic*), or (2) from cell to cell via the cytoplasm and plasmodesmata (*symplastic*). The primary pathway for salt movement from the soil to the foliar glands in *T. aphylla* has been demonstrated to be apoplastic transport (Campbell et al. 1974; Campbell and Thomson 1975). However, the ability to use bimodal transport could speed the rate of salt transport and thereby increase the recretion efficiency of *Tamarix*.

Tamarix salt glands are eight-celled complexes sunken slightly in the foliar epidermis. The six outer cells are primarily cytoplasmic, secretory cells, and the two inner cells are highly vacuolated collecting cells (see figure 8.1; Bosabalidis and Thomson 1984, 1985). High concentrations of salts first accumulate in collecting

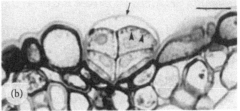

FIGURE 8.1 *Tamarix aphylla* salt gland. (a) SEM micrograph illustrating a salt gland (*asterisk*) on the leaf surface. (b) LM micrograph of a secreting salt gland. The arrow shows the cuticle detached from the apical cell walls, which bear a system of internal protuberances (*arrowheads*).

Source: Bosabalidis (2010).

cells prior to secretion (Bosabalidis and Thomson 1984, 1985). Solutes travel between gland cells by means of wall protuberances (Bosabalidis 2010). Microvacuoles containing salty solution release their content into the space between the cuticle and the larger, outer secretory cells (Bosabalidis 2010; Thomson et al. 1969). This collecting chamber is where secretory fluid accumulates. After the development of hydrostatic pressure with an osmotic flow of water, the chamber expands and salt exudates are released through the cuticular pores (Bosabalidis and Thomson 1985). Secretion is an efficient and active process, evidenced by the large number of intercellular connections and mitochondria in the middle and outer secretory cells (Bosabalidis and Thomson 1985; Bosabalidis 2010). These structures provide transport pathways and energy within the glands to continuously drive the removal of ionic contaminants, a process that occurs even in conditions of low salinity (Berry 1970; Waisel 1961).

Salt glands may have other functions that further enhance the ability of *Tamarix* to tolerate conditions of elevated salinity and drought. Exudate-laden leaf surfaces have been hypothesized to trap carbon dioxide and reduce the amount of water lost through transpiration (Waisel 1991a). Salt accumulations in salt glands may also perform osmotic work (Osonubi and Davies 1978; Abrams 1988), acting as a significant pulling force that further enhances continued soil water extraction (Sala et al. 1996; Wiesenborn 1996). This adaptation allows *Tamarix* to extract water under conditions of low soil moisture or elevated salinity, permitting establishment and growth in environments where a depressed water table and elevated salinity would otherwise constrain a plant's ability to extract water from the soil.

Osmotic adjustment used by *Tamarix* spp. to tolerate elevated salinity is also made possible by the production free intracellular-compatible solutes. *Tamarix jordanis* has been observed to synthesize proline in the presence of high NaCl content (Solomon et al. 1994). Proline concentrations increase in *Tamarix ramosissima* with greater salt exposure (Carter and Nippert 2011; Ruan et al. 2009); leading to the hypothesis that proline promotes internal osmotic balance for *Tamarix* spp. during salinity stress.

TAMARIX RELATIVE TO OTHER HALOPHYTES

Tamarix leaf anatomy provides these plants with a greater capacity to take up and sequester salts than in co-occuring halophytes with similar mechanisms for salt tolerance. *Atriplex* spp. (e.g., saltbush, quailbush) are the most common halophytic shrubs found coexisting with *Tamarix* that compartmentalize and remove salts via leaf structures. Salt-sequestering structures in *Atriplex* are found only on leaf surfaces. In contrast, *Tamarix* salt glands are known to be abundant on both the upper and lower leaf surfaces as well as on young stems (Campbell and Strong 1964; Sookbirsingh et al. 2010; Wilkinson 1966). *Tamarix* plants are also able to replace aging salt glands with more-productive new glands throughout leaf maturation (Bosabalidis 1992). Most halophytic species do not produce new glands to replace those that have senesced. These differences in gland production allow *Tamarix* spp. to process more salts, which provides these plants with an advantage in high salinity conditions (Bosabalidis 1992).

Tamarix stands can grow to be extremely dense and can have a high leaf area index (LAI), that is, leaf surface per unit ground area (DiTomaso 1998; see also chapter 9, this volume). *Tamarix* LAI varies by site but has been observed to be between 3.5 m² m^{-2} and 7.4 m² m^{-2} (Dahm et al. 2002; Sala et al. 1996; Nagler et al. 2003; Nagler et al. 2008). *Atriplex* spp. are more commonly associated with a lower LAI (< 3 m² m^{-2}; Groenveld 1997, Steinwand et al. 2001), although measurements between 4 and 5 m² m^{-2} have been reported (Groenveld 1997; Jordan et al. 2009). A higher LAI, in combination with more continuous coverage of salt glands on leaf and stem surfaces, suggests that *Tamarix* may have a larger number of salt-removing structures per unit area than other halophytic shrubs that inhabit the same environments.

Tamarix is known to tolerate elevated environmental salinity in two ways: (1) via salt uptake, compartmentalization in salt glands, and secretion, and (2) through selective exclusion of salts by its roots. These mechanisms used to regulate internal salt content in *Tamarix* may in turn disturb the external environment by increasing soil and groundwater salinity at the local scale. These positive feedback mechanisms enhance further proliferation of *Tamarix* over native salt-sensitive riparian vegetation.

Tamarix Salt Secretion

Salt glands have been observed to secrete numerous salts and minerals, both macro and micronutrients, for *T. aphylla* (Berry 1970; Bosabalidis and Thomson 1984; Storey and Thomson 1994; Thomson et al. 1969), *T. chinensis* (Dressen and Wangen 1981), and *T. ramosissima* (Kleinkopf and Wallace 1974). Sodium (Na), potassium (K), nitrate (NO$_3$), calcium (Ca), magnesium (Mg), sulfur

(S), sulfate (SO_4), bicarbonate (HCO_3), and chloride (Cl) have been observed to account for more than 99% of the total ions secreted by *T. aphylla* (Berry 1970). Waisel (1961) observed a sequence of ion secretion from *T. aphylla* salt glands to occur in decreasing order of concentration: Na > Ca > K. Similarly, Na has been reported to be the dominant ion in secretions and leaf litter of *T. ramosissima* (Arndt et al. 2004a; Glenn et al. 2012; Kleinkopf and Wallace 1974; Ma et al. 2011). The broad range of secreted ions clearly indicates that the glands of *Tamarix* have a low level of selectivity for salts (Berry 1970; Kleinkopf and Wallace 1974).

Environmental salt exposure appears to dictate the rate and content of ionic secretions in *Tamarix*. The composition of secreted salts depends on the salt composition of the rooting medium for *T. aphylla* in Southern California (Berry 1970; Berry and Thomson 1967; Thomson et al. 1969) and for populations of this species in its native range (Hagemeyer and Waisel 1988; Waisel 1961). Exudates and soils associated with *T. ramosissima* growing in New Mexico and Texas also had similar salt constituents (Sookbirsingh et al. 2010). *Tamarix* secretions contain substantially higher salt concentrations than the soil (Berry 1970; Kleinkopf and Wallace 1974; Thomson et al. 1969) and secretion rates increase with increasing salt content in the environment (Kleinkopf and Wallace 1974; Ma et al. 2011; Thomson et al. 1969). *Tamarix* spp. ability to recrete a variety of ions and increase the rate of salt transport and secretion based on environmental levels has undoubtedly contributed to their survival in many different saline and/or sodic soil types.

The rate of secretion by *Tamarix* is likely dependent on other environmental factors besides soil and water salt concentrations. Secretion rates in *T. ramosissima* are highest in the morning and decline with increasing solar radiation and temperature exposure (Chen et al. 2010; Wiesenborn 1996). Secretion is not related to photosynthesis, however, because it has been observed to occur at night (Sookbirsingh et al. 2010; Chen et al. 2010). High atmospheric humidity has been observed to increase ionic secretion by *Tamarix* during the day (Chen et al. 2010), and this moisture combines with crystallized secretions to form brine solutions on plant surfaces (Decker 1961; Waisel 1991a).

Salts are initially secreted from *Tamarix* salt glands as liquid exudates but under dry atmospheric conditions the water evaporates and the salts crystallize (Decker 1961; Sookbirsingh et al. 2010; see figure 8.2). Salt exudates have been reported to contain up to 41,000 ppm dissolved solids (64 mmhos cm^{-1} EC) (Gatewood et al. 1950). These salt accumulations are easily removed by rain and wind, or fall to the soil surface with senesced leaves. Because invasive *Tamarix* species are deciduous, leaf fall deposits a major pulse of salts to the soil surface at the end of each growing season (see figure 8.3). The ability of *Tamarix* to uptake, compartmentalize, and deposit salts may enhance its salt tolerance, and has even been referred to as a form of elemental allelopathy (Brock 1994; Morris et al. 2009).

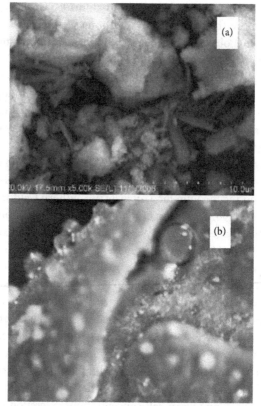

FIGURE 8.2 Exudates from cultivated *Tamarix ramosissima* plants. (a) SEM of salt crystals removed from a plant. (b) Spherical briny drops are secreted; these ionic concentrates later harden into salt crystal aggregates.

Source: Sookbirsingh et al. (2010).

FIGURE 8.3 *Tamarix* leaf litter covers the ground surface following fall leaf senescence.

Source: Photo courtesy of Tim Carlson, Tamarisk Coalition.

Tamarix **and Environmental Salinity**

Of great interest has been the relationship between *Tamarix* and environmental salinity; many authors have speculated that these plants have a strong impact on soil chemistry (Brotherson and Field 1987; Busch and Smith 1993; DiTomaso 1998; Sala et al. 1996; Shafroth et al. 1995). Other investigators argue that any observed positive relationships between *Tamarix* and elevated soil salinity result from correlations confirming the known higher salinity tolerance of this plant (Stromberg et al. 2009). Numerous studies have quantified soil salinity in *Tamarix*-invaded areas, but the literature appears highly discordant with regard to the contribution of *Tamarix* to soil salinity. Here we will explore the contexts and possible explanations for these differences.

It is likely that *Tamarix* contributes to salt loading in groundwater and soils through continuous groundwater extraction and salt redistribution to the soil surface. However, it is unclear whether these salt contributions are significant relative to other environmental salinity predictors, such as flood suppression and surface evaporation, in arid environments. This section summarizes the available literature that relates *Tamarix* invasion to elevated soil and groundwater salinity and attempts to separate correlation from causation.

Among the most convincing evidence to date supporting a causal role of *Tamarix* in floodplain salinization documents soil and groundwater salinity in *Tamarix* spp. stands along the lower Colorado River in Southern California. Similar ratios of cations and anions were observed in groundwater, *Tamarix* leaves, and surface soils, but these salt signatures were different (i.e., more concentrated) than soils outside the plant canopy (Glenn et al. 2012). The presence of comparable salt concentrations along the pathway from groundwater to plant tissues to surface soils illustrates salt redistribution by *Tamarix* is an active process in the field. Higher groundwater salinity was also observed at greater distances from the river, a factor that was attributed to progressive extraction of water from the aquifer by *Tamarix* roots in the absence of diluting river flows (Glenn et al. 2012; Nagler et al. 2008). It is still unclear, however, whether environmental factors, such as river disconnect, are more important for understanding groundwater salinity.

Several studies are frequently cited as those that demonstrate a direct role *Tamarix* plays in soil and groundwater salinization (by cycling salt to the soil surface or excluding it in the groundwater), however most of these simply correlate high salinity with the presence of *Tamarix*. Carman and Brotherson (1982) observed that sites inhabited by *T. pentandra* in Utah were more saline than those containing Russian olive (*Elaeagnus angustifolia*), another invasive riparian tree in the western United States. Another study in Wyoming found similar results; soil salinity was highest underneath *Tamarix* spp. canopies in upland areas and lowest within native stands on lower floodplain terraces (Ladenburger et al. 2006). In Spain, *T. africana* and *T. canariensis* were associated with higher groundwater salinity than associated *Salix* populations (Salinas et al. 2000). Similarly, research performed near Moab, Utah, observed higher groundwater salinity within a *T. ramosissima*–invaded site compared with a native-dominated site (Pataki et al. 2005). This research suggests that *Tamarix* may be responsible for salt loading in soils and groundwater because adjacent or nearby sites would appear to have similar ambient soils and groundwater conditions. However, these studies are observational and confounded by site; without an experimental approach it is difficult to confirm whether elevated salinity is the result of *Tamarix* invasion or some other environmental factor.

Tamarix may also indirectly alter the solute dynamics of its edaphic environment by trapping sediment and increasing fire potential. *Tamarix* stands can be considerably denser than naturally occurring riparian vegetation (Cleverly et al. 1997; Sala et al. 1996). Increased *Tamarix* density can enhance deposition of fine sediments during floods (e.g., clays, which have a greater capacity to adhere to mobilized salts; Everitt 1980; Stromberg 1998). *Tamarix* plants have many stems with high rates of twig mortality, creating continuous accumulations of fuels that can increase the frequency and severity of fires (Busch and Smith 1993; Ellis 2001; Ellis et al. 1998; see also chapter 14, this volume). Ash deposits following fire are highly saline (> 200 mmhos cm^{-1}; Lair and O'Meara 2009). Data collected along the Bill Williams River in Arizona show that fire can increase soil EC from 1.5 to 4.5 mmhos cm^{-1} (Busch and Smith 1993). Fires are becoming more frequent in riparian areas, largely because of *Tamarix* invasion and flood deprivation (Busch and Smith 1995; Ellis 2001). This disturbance represents another source of environmental salts that should be recognized when attempting to determine the drivers of floodplain salinity.

There are observational studies that found no difference when comparing soil salinity under *Tamarix* with other vegetation (Bagstad et al. 2006; Campbell and Dick-Peddie 1964; Gary 1965). However, in these studies, fewer than 100 soil samples were used for salinity comparisons and in some cases sampling spanned over 200 kilometers. Because riparian areas are highly heterogenous ecosystems, intense sampling is necessary to adequately quantify salinity. It is conceivable that no relationship was observed because small sample sizes may be inadequate to uncover any correlation. There is one well-replicated study that found no difference in soil salinity; Stromberg (1998) found no statistical difference between soil salinity underneath *T. chinensis* versus *Populus*. Without considering other environmental variables that may influence salinity (e.g., flooding, plant age), these results suggested that *Tamarix* contributions to soil salinity are similar to that of *Populus*.

Another way to determine *Tamarix* impact on environmental salinity is to compare soil salinity under *Tamarix* versus open areas (i.e., without woody vegetation cover); however, such studies show mixed results. Soils underneath *Tamarix* spp. canopies have been observed to have salinity values averaging more than twice those of adjacent open areas in Montana and Wyoming, but it should be noted that salt concentrations were within the range of native riparian vegetation tolerance (i.e., ≤ 4 mmhos cm⁻¹; Ladenburger et al. 2006; Lesica and DeLuca 2004). Another study recorded higher salinity under *Tamarix* growing in its native range; Yin et al. (2010) observed salinity within *Tamarix* spp. mounds (i.e., soil aggregations around mature shrubs from wind deposition) in China had significantly higher salinity than adjacent open areas. Along the Middle Rio Grande in central New Mexico a contradictory result was found. Plots experimentally cleared of *T. ramosissima* had higher surface salinity than soils underneath *Tamarix* canopies (Taylor et al. 1999). The same result was found when pairwise comparisons were made between soils under *Tamarix* canopies versus adjacent open areas on the same river reach (Ohrtman et al. 2012).

The remaining studies that quantify soil salinity under *Tamarix* generally lack appropriate reference sites and so cannot be used to determine the role of *Tamarix* in contributing to salinity. Two studies simply documented soil conditions within *Tamarix*-invaded areas to address other research objectives (Brotherson and Winkel 1986; Van Hylckama 1980). In a third, salt levels underneath *Tamarix* spp. along the lower Colorado River were compared to farmlands that potentially receive regular irrigation with relatively high solute concentrations (Nagler et al. 2008). Without sampling appropriate sites for comparison, it is impossible to use these data to either support or refute a causal relationship between *Tamarix* and elevated soil salinity.

UNDERSTANDING THE DISCREPANCIES IN THE DATA

We have emphasized the importance of location effects in understanding the relationship between *Tamarix* invasion and soil salinity. We now address environmental parameters the research has found important for understanding salinity in riparian areas and discuss the details of a research study that tested the importance of these variables for predicting soil salinity in an arid *Tamarix*-invaded floodplain.

Surface water flows are expected to reduce soil salinity by horizontal salt transport (e.g., downstream) or vertical leaching to greater soil depths. Because flooding can remove surface salts, research performed within active floodplains may be influenced by this flushing mechanism. Two studies observed no difference between surface soil salinity underneath *Tamarix* and *Populus* canopies within a floodplain that is still exposed to flooding (Bagstad et al. 2006; Stromberg 1998). It is possible that no difference was detected because flooding removed salt deposits from the soils in these areas (see figure 8.4).

Just as flooding could be responsible for a lack of salt-level variation among vegetation types within the active floodplain, this environmental factor could explain significant differences in soil salinity between flooded and nonflooded sites. Lower soil salinity levels underneath native vegetation in a Wyoming drainage compared with soils associated with upland *Tamarix* vegetation was likely the result of a more active hydrology removing soil salts within the lower terrace

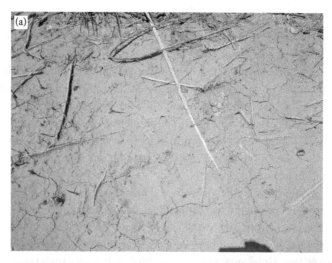

FIGURE 8.4 *Tamarix* spp. litter composition for (a) flooded, and (b) nonflooded site along the Middle Rio Grande, New Mexico. Flooding removes salt-rich *Tamarix* leaf litter and other surface salt deposits thereby reducing surface soil salinity.

Source: Photos courtesy of Michelle Ohrtman.

(Ladenburger et al. 2006). Similarly, lower elevation areas adjacent to *Tamarix* mounds in China showed clear signs of water erosion (Yin et al. 2010), suggesting that water may have removed surface salt accumulations in open areas.

One study that does support salt-loading by *Tamarix* in the context of flooding was performed along the middle Rio Grande. Significantly higher surface soil salinity was observed in areas dominated by *Tamarix* when compared with sites previously invaded but restored with native vegetation between five and 17 years prior to sampling (Ohrtman et al. 2012). However, looking at the interaction, it is clear that this is only true where overbank flooding no longer occurs (Ohrtman et al. 2012;

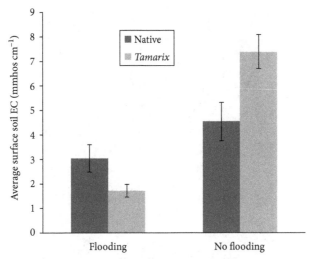

FIGURE 8.5 Surface soil (0- to 15-cm depth) electrical conductivity (EC) under the canopy of native riparian woody vegetation (native) or *Tamarix* spp. averaged across trees sampled in flooded (n = 138) and nonflooded (n = 125) areas along the Middle Rio Grande, New Mexico. Mean ± 1 SE.

see figure 8.5). Salinity was much more variable under native trees and even sometimes higher than under *Tamarix* with flooding. Flooded *Tamarix*-invaded sites had significantly lower salt levels than adjacent stands that are no longer exposed to flooding. These results support claims that overbank flows flush the soils of accumulated salts and lower salinity.

Evaporation from the soil surface draws salt upward in the soil profile via capillary action, increasing surface salinity. In areas where evaporation exceeds precipitation, woody vegetation can slow the capillary rise of salts in the soil profile by (1) reducing the amount of solar radiation reaching the soil surface (Yang et al. 1999), and (2) roots intercepting rising water before it reaches surface soils (Tamea et al. 2009). Greater canopy cover may explain why soils underneath *Tamarix* had significantly lower EC when compared with adjacent open-canopy soils for research performed in New Mexico (Ohrtman et al. 2012; Taylor et al. 1999; see figure 8.6). Increased soil exposure to solar radiation may not have increased soil salinity in the Montana (Lesica and DeLuca 2004) and Wyoming studies (Ladenburger et al. 2006) because these areas have lower temperatures and higher precipitation than the New Mexico study locations. In addition, *Tamarix* trees are known to have high rates of transpiration (Cleverly et al. 1997; Sala et al. 1996); trees actively extract soil moisture and groundwater to replenish lost water. Along the lower Colorado River, soils under dense *Tamarix* stands, with higher LAI and greater rates of transpiration, had lower surface soil salinity relative to less-dense growth (Glenn et al. 2012). The authors attribute this difference to greater soil water uptake

FIGURE 8.6 Canopy cover in (a) dense, and (b) more sparse *Tamarix* spp. stands along the Middle Rio Grande, New Mexico. Decreased solar radiation means less surface evaporation, which may explain why soils under mature *Tamarix* are sometimes found to have lower salinity than adjacent open areas.

Source: Photos courtesy of Michelle Ohrtman.

by plant roots in dense stands that reduced the upward movement of salts in the soil profile (Glenn et al. 2012).

If *Tamarix* simply establishes where soils are more saline but does not subsequently contribute in any way to that salinity, we would not expect any correlations between stand features (e.g., age) and salinity, except perhaps a negative one, indicating that *Tamarix* may do best with lower salinity. Stromberg (1998) observed that soil salinity increases with greater *T. chinensis* age (i.e., from 4 to 43 years) in the presence of flooding. Although this is consistent with the hypothesis that *Tamarix* contributes to soil salinity, this study was confounded by relationships between age and soil texture. *Tamarix* is known to trap finer sediments when exposed to flooding, and clay content was observed to increase with increasing stand age (Stromberg 1998). If older *Tamarix* stands are associated with higher clay content, and clay soils contain more salts, then it is possible that the increase in salinity with increasing *Tamarix* age was the result of greater accumulations of fine sediments. This explanation applies to a similar observed relationship between larger *Tamarix* size and higher salinity in response to wind-deposited sediments in China (Yin et al. 2010). On the contrary, native *T. chinensis* plants in China were observed to have lower soil salinity with increasing size (Cui et al. 2010). Larger individuals (6- to 10-cm diameter) were located closer to the old and active river channels (Cui et al. 2010) suggesting that surface flows may have influenced salinity in these areas. Sexton et al. (2006) in Montana found no relationship between *Tamarix* spp. age and soil salinity. These results may represent the lower importance

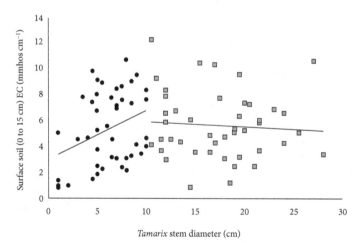

FIGURE 8.7 Surface soil (0- to 15-cm depth) electrical conductivity (EC) by stem size (cm) for soils sampled under the canopy of *Tamarix* spp. trees within nonflooded areas along the Middle Rio Grande, New Mexico. Separate linear regressions were performed for *Tamarix* plants with primary stems less than 10-cm diameter, i.e., less than 22 years of aboveground age ([*black circles*]; n = 47, R^2 = 0.10, p < 0.02), and with stems greater than 10-cm diameter ([*grey squares*]; n = 44, R^2 = -0.02, NS).

of vegetation cover in mediating salt content in this climate. Ohrtman et al. (2012) observed surface soil salinity to increase with greater *Tamarix* spp. stem diameter (a predictor of age) in an arid region deprived of flooding. In this study, however, soil salinity leveled off with increasing stem size past a certain threshold, with some of the lowest soil salinities found among larger trees (>10-cm diameter or 22 years of age; see figure 8.7). This relationship is unexpected if *Tamarix* simply establishes where conditions are more saline, especially in areas that are no longer flooded. The pattern of surface soil salt loading relative to *Tamarix* age may be related to flooding, temperature, and sediment dynamics but also can be the result of the physical changes of *Tamarix* with age (e.g., stem size and configuration, foliage density and content), although no research has tested this hypothesis.

Conclusion

Elevated salt concentration is a commonly cited cause of ecosystem degradation associated with *Tamarix* invasion. However, scientists who have quantified salinity in *Tamarix*-invaded areas are divided on whether *Tamarix* is a significant contributor to elevated salinity. While it appears that *Tamarix* does increase groundwater salinity through root exclusion, and that this effect increases with distance from the river, the research on soils is more complex. Unusually high soil salinity under *Tamarix* has been well documented; however, in most cases these results could be explained by high salinity exclusion of all other vegetation than *Tamarix*. Whereas much of the research that fails to connect *Tamarix* with elevated salinity is based on small sample sizes, there are two well-replicated studies that found either no difference between soil salinity under *Tamarix* and natives (Stromberg 1998) or higher soil salinity outside *Tamarix* canopies (Taylor et al. 1999), which appear to refute the claim that *Tamarix* is responsible for surface soil salt loading. These findings could be due to either the lack of influence of *Tamarix* or that there were environmental factors such as flooding and surface evaporation that were more commanding drivers of soil salinity. A study designed to investigate these hypotheses found that flooding and surface evaporation are significant contributors to elevated soil salinity but that *Tamarix* plants may also increase salinity based on unexplained relationships with age (Ohrtman et al. 2012). Thus, it seems likely that *Tamarix* can increase soil and groundwater salinity but that other environmental factors such as surface evaporation in arid climates and overbank flooding can mitigate or mask these effects to a large degree.

Regardless of cause, elevated salinity associated with *Tamarix* stands can decrease our ability to restore native riparian plant communities following *Tamarix* removal. If *Tamarix* is increasing environmental salt levels, then the removal of this vegetation can halt further concentration of salts in surface soils and groundwater by this plant. Previous salt contributions by *Tamarix* and those from other environmental sources, however, cannot be reversed or negated without expensive and/or logistically complex methods (Sher et al. 2010; see also chapter 20, this volume). Land managers must weigh the costs of removing *Tamarix* in areas where flood deprivation and surface evaporation may elevate surface salinity above the tolerance of desirable replacement vegetation. Healthy *Tamarix* are undeniably better habitat than botched removal and restoration projects, some of which result in rapid *Tamarix* recolonization (see chapter 6, this volume). In addition, although *Tamarix* may redistribute salts, from the salinity standpoint *Tamarix* cover may be better than no cover. Management efforts should focus on prioritizing *Tamarix* stands for treatment based on current and projected environmental factors (Shafroth et al. 2008). A key consideration should be to establish realistic goals that include selecting appropriate revegetation species that are able to tolerate elevated salt concentrations characteristic of *Tamarix*-invaded habitats.

Literature Cited

Abrams, M. D. 1988. Comparative water relations of three successional hardwood species in Central Wisconsin. Tree Physiology 4:263–273.

Anderson, B. W. 1995. Salt cedar, revegetation and riparian ecosystems in the Southwest. Proceedings of the California Exotic Plant Pest Council Symposium, vol. 1. [Online]. http://www.cal-ipc.org/symposia/archive/pdf/1995_symposium_proceedings1797.pdf.

Anderson, B. W., and R. D. Ohmart. 1982. Revegetation for wildlife enhancement along the lower Colorado River. U.S. Bureau of Reclamation, Lower Colorado Region. Boulder City, NV.

Anderson, B. W., and R. D. Ohmart. 1979. Riparian revegetation: An approach to mitigating for a disappearing habitat in the Southwest. Pages 481–487 *in* The mitigation symposium: A national workshop on mitigating losses of fish and wildlife habitats. USDA Forest Service, Rocky Mountain Region, Former Rocky Mountain and Range Exp. Station Publications. General Technical Report RM-65, Fort Collins, CO.

Anderson, B. W., and R. D. Ohmart. 1985. Riparian revegetation as a mitigating process in stream and river restoration. Pages 41–80 *in* J. A. Gore, editor. *The Restoration of Rivers and Streams*. Butterworth, Boston, MA.

Arndt, S. K., C. Arampatsis, A. Foetzki, X. Li, F. Zeng, and X. Zhang. 2004a. Contrasting patterns of leaf solute accumulation and salt adaptation in four phreatophytic desert plants in a hyperarid desert with saline groundwater. Journal of Arid Environments 59:259–270.

Arndt, S. K., A. Kahmen, C. Arampatsis, M. Popp, and M. Adams. 2004b. Nitrogen fixation and metabolism by groundwater-dependent perennial plants in a hyperarid desert. Oecologia 141:385–394.

Bagstad, K. J., S. J. Lite, and J. C. Stromberg. 2006. Vegetation, soils, and hydrogeomorphology of riparian patch types of a dryland river. Western North American Naturalist 66(1):23–44.

Berry, W. L. 1970. Characteristics of salts secreted by *Tamarix aphylla*. American Journal of Botany 57:1226–1230.

Berry, W. L., and W. W. Thomson. 1967. Composition of salt secreted by salt glands of *Tamarix aphylla*. Canadian Journal of Botany 45:1774–1775.

Black, R. F. 1954. The leaf anatomy of Australian members of the genus *Atriplex*. I. *Airiplex vesicaria* Heward and *A. nummularia* Lindl. Australian Journal of Botany 2:269–286.

Bosabalidis, A. M. 1992. A morphological approach to the question of salt gland lifetime in leaves of *Tamarix aphylla* L. Israel Journal of Botany 41:115–121.

Bosabalidis, A. M. 2010. Wall protuberance formation and function in secreting salt glands of *Tamarix aphylla* L. Acta Botanica Croatica 69:229–235.

Bosabalidis, A. M., and W. W. Thomson. 1984. Light microscopical studies on salt gland development in *Tamarix aphylla* L. Annals of Botany 54:169–174.

Bosabalidis, A. M., and W. W. Thomson. 1985. Ultrastructural development and secretion in the salt glands of *Tamarix aphylla*. Journal of Ultrastructure Research 92:55–62.

Branson, F. A., R. F. Miller, and S. K. Sorenson. 1988. Tolerances of plants to drought and salinity in the western United States. Report 88-4070. U.S. Geological Survey Water Resources Investigations.

Brock, J. H. 1994. *Tamarix* spp. (salt cedar), an invasive exotic woody plant in arid and semi-arid riparian habitats of western USA. Pages 27–44 *in* L. C. De Waal, L. E. Child, P. M. Wade, and J. H. Brock, editors. *Ecology and Management of Invasive Riverside Plants*. John Wiley & Sons, New York.

Brotherson, J. D., and D. Field. 1987. *Tamarix*: Impacts of a successful weed. Rangelands 9:110–112.

Brotherson, J. D., and V. Winkel. 1986. Habitat relationships of saltcedar (*Tamarix ramosissima*) in central Utah. Great Basin Naturalist 46:535–541.

Busch, D. E., and S. D. Smith. 1993. Effects of fire on water and salinity relations of riparian woody taxa. Oecologia 94:186–194.

Busch, D. E., and S. D. Smith. 1995. Mechanisms associated with decline of woody species in riparian ecosystems of the southwestern U.S. Ecological Monographs 65:347–370.

Campbell, C. J., and W. A. Dick-Peddie. 1964. Comparison of phreatophyte communities on the Rio Grande in New Mexico. Ecology 45:492–502.

Campbell, C. J., and J. E. Strong. 1964. Salt gland anatomy in *Tamarix pentandra* (Tamaricaceae). Southwestern Naturalist 9:232–238.

Campbell, N., and W. W. Thomson. 1975. Chloride localization in the leaf of *Tamarix*. Protoplasma 83:1–14.

Campbell, N., W. W. Thomson, and K. Platt. 1974. The apoplastic pathway of transport to salt glands. Journal of Experimental Botany 25:61–69.

Carman, J. G., and J. D. Brotherson. 1982. Comparisons of sites infested and not infested with saltcedar (*Tamarix pentandra*) and Russian olive (*Elaeagnus angustifolia*). Weed Science 30:360–364.

Carter, J. M., and J. B. Nippert. 2012. Leaf-level physiological responses of *Tamarix ramosissima* to increasing salinity. Journal of Arid Environments 77:17–24.

Carter, J. M., and J. B. Nippert. 2011. Physiological responses of *Tamarix ramosissima* to extreme NaCl concentrations. American Journal of Plant Sciences 2:808–815.

Chen, J. X., L. Q. Qiao, S. H. Gou, Y. X. Wang, and H. B. He. 2008. Experiment of salinity tolerance of *Tamarix*. Shandong Forest Science and Technology 1:18–19.

Chen, Y., H. Wang, F. Zhang, J. Xi, and Y. He. 2010. The characteristic of salt excretion and its affected factors on *Tamarix ramosissima* Ledeb under desert saline-alkali habitat in Xinjiang Province. Acta Ecologica Sinica 30:511–518.

Cleverly, J. R., S. D. Smith, A. Sala, and D. A. Devitt. 1997. Invasive capacity of *Tamarix ramosissima* in a Mojave Desert floodplain: The role of drought. Oecologia 111:12–18.

Cui, B., Q. Yang, K. Zhang, X. Zhao, and Z. You. 2010. Responses of saltcedar (*Tamarix chinensis*) to water table depth and soil salinity in the Yellow River delta, China. Plant Ecology 209:279–290.

Dahm, C. N., J. R. Cleverly, J. E. A. Coonrod, J. R. Thibault, D. E. McDonnell, and D. J. Gilroy. 2002. Evapotranspiration at the land/water interface in a semi-arid drainage basin. Freshwater Biology 47:831–843.

Decker, J. P. 1961. Salt secretion by *Tamarix pentandra* Pall. Forest Science 7:214–217.

DiTomaso, J. M. 1998. Impact, biology, and ecology of saltcedar (*Tamarix* spp.) in the southwestern United States. Weed Technology 12:326–336.

Dressen, D. R., and L. E. Wangen. 1981. Elemental composition of saltcedar (*Tamarix chinensis*) impacted by effluents from a coal-fired power plant. Journal of Environmental Quality 10:410–416.

Ellis, L. M. 2001 Short-term response of woody plants to fire in a Rio Grande riparian forest, Central New Mexico, USA. Biological Conservation 97:159–170.

Ellis, L., C. Crawford, and M. Molles. 1998. Comparison of litter dynamics in native and exotic riparian vegetation along the Middle Rio Grande of central NM, USA. Journal of Arid Environments 38:283–296.

Everitt, B. L. 1980. Ecology of saltcedar: A plea for research. Environmental Geology 3:77–84.

Gary, H. L. 1965. Some site relations in three floodplain communities in Central Arizona. Journal of the Arizona Academy of Science 3:209–212.

Gatewood, J. S., T. W. Robinson, B. R. Colby, J. D. Hem, and L. C. Halpenny. 1950. Use of water by bottom-land vegetation in lower Safford Valley, Arizona. United States Geological Survey Water-Supply Paper 1103. Los Angeles, CA.

Glenn, E., R. Tanner, S. Mendez, T. Kehret, D. Moore, J. Garcia, and C. Valdes. 1998. Growth rates, salt tolerance and water use characteristics of native and invasive riparian plants from the delta of the Colorado River, Mexico. Journal of Arid Environments 40:281–294.

Glenn, E. P., K. Morino, P. L. Nagler, R. S. Murray, S. Pearlstein, and K. Hultine. 2012. Role of saltcedar (*Tamarix* spp.) and capillary rise in salinizing a non-flooding terrace on a flow-regulated desert river. Journal of Arid Environments 79:56–65.

Glenn, E. P., and P. L. Nagler. 2005. Comparative ecophysiology of *Tamarix ramosissima* and native trees in western U.S. riparian zones. Journal of Arid Environments 61:419–446.

Goodall D. W., and R. A. Perry. 1981. *Arid Land Ecosystems: Structure, Functioning, and Management.* Cambridge University Press, Cambridge, UK.

Groenveld, D. P. 1997. Vertical point quadrat sampling and an extinction factor to calculate leaf area index. Journal of Arid Environments 36:475–485.

Gucci, R., G. Aronne, L. Lombardini, M. Tattini. 1997. Salinity Tolerance in *Phillyrea*. New Phytologist 135:227–234.

Hagemeyer, J., and Y. Waisel. 1988. Excretion of ions (cadmium, lithium, sodium, and chlorine) by *Tamarix aphylla*. Physiologia Plantarum 73:541–546.

He, Q., B. Cui, and Y. An. 2011. The importance of facilitation in the zonation of shrubs along a coastal salinity gradient. Journal of Vegetation Science 22:828–836.

Horton, J., and C. Campbell. 1974. Management of phreatophyte and riparian vegetation for maximum multiple use values. U.S. Forest Service Research Paper RM-117, Fort Collins, CO.

Jackson, J., J. T. Ball, and M. Rose. 1990. Assessment of the salinity tolerance of eight Sonoran desert riparian trees and shrubs. U.S. Bureau of Reclamation, Yuma, AZ.

Jordan, F. L., M. Yoklic, K. Morino, P. Brown, R. Seaman, and E. P. Glenn. 2009. Consumptive water use and stomatal conductance of *Atriplex lentiformis* irrigated with industrial brine in a desert irrigation district. Agricultural and Forest Meteorology 149(5):899–912.

Kleinkopf, G. E., and A. Wallace. 1974. Physiological basis for salt tolerance of *Tamarix ramosissima*. Plant Science Letters 3:157–163.

Ladenburger, C. G., A. L. Hild, D. J. Kazmer, and L. C. Munn. 2006. Soil salinity patterns in *Tamarix* invasions in the Bighorn Basin, Wyoming, USA. Journal of Arid Environments 65:111–128.

Lair, K. D., and S. O'Meara. 2009. Interim report: Revegetation strategies and plant materials selection for restoration of xeric *Tamarix* infestation sites following fire. Tech. Memo. No. 86-68220-09-05. Environmental Applications and Ecological Research Group, Technical Service Center, Bureau of Reclamation, Denver, CO.

Lair, K. D., and S. L. Wynn. 2002. Revegetation strategies and technology development for restoration of xeric *Tamarix* infestation sites. Tech. Memo. No. 8220-02-04. Bureau of Reclamation, Technical Service Center, Denver, CO.

Lesica, P., and T. H. DeLuca. 2004. Is tamarisk allelopathic? Plant and Soil 267:357–365.

Li, W., M. A. Khan, X. Zhang, and X. Liu. 2010. Rooting and shoot growth of stem cuttings of saltcedar (*Tamarix chinensis* Lour) under salt stress. Pakistan Journal of Botany 42: 4133–4142.

Liu, G. J., X. M. Zhang, X. R. Li, J. Wei, and L. S. Shan. 2008. Adaptive growth of *Tamarix taklamakanensis* root systems in response to wind action. Chinese Science Bulletin 53(S2):164–168.

Ma, H. Y., C. Y. Tian, G. Feng, and J. F. Yuan. 2011. Ability of multicellular salt glands in *Tamarix* species to secrete Na+ and K+ selectively. Science China Life Sciences 54:282–289.

MacDougal, A. S., and R. Turkington. 2005. Are invasive species the drivers or passengers of change in degraded ecosystems? Ecology 86:42–55.

Merkel, D. L., and H. H. Hopkins. 1957. Life history of salt cedar (*Tamarix gallica*). Transactions of the Kansas Academy of Science 60:360–369.

Morris, C., P. R. Grossl, and C. A. Call. 2009. Elemental allelopathy: Processes, progress, and pitfalls. Plant Ecology 202:1–11.

Nagler, P. L., E. P. Glenn, K. Didan, J. Osterberg, F. Jordan, and J. Cunningham. 2008. Wide-area estimates of stand structure and water use of *Tamarix* spp. on the Lower Colorado River: Implications for restoration and water management projects. Restoration Ecology 16(1):136–145.

Nagler, P. L., E. P. Glenn, and T. L. Thompson. 2003. Comparison of transpiration rates among saltcedar, cottonwood and willow trees by sap flow and canopy temperature methods. Agricultural and Forest Meteorology 116: 73–89.

Nippert, J. B., J. J. Butler Jr., G. J. Kluitenberg, D. O. Whittemore, D. Arnold, S. E. Spal, and J. K. Ward. 2010. Patterns of *Tamarix* water use during a record drought. Oecologia 162:283–292.

Ohrtman, M. K., A. A. Sher, and K. D. Lair. 2012. Quantifying soil salinity in areas invaded by *Tamarix* spp. Journal of Arid Environments 85:114–121.

Ohmart, R. D., B. W. Anderson, and W. C. Hunter. 1988. Ecology of the Lower Colorado River from Davis Dam to Mexico–United States boundary: A community profile. U.S. Fish and Wildlife Service Biological Report 85(7.19):296.

Osonubi, O., and W. J. Davies. 1978. Solute accumulation in leaves and roots of woody plants subjected to water stress. Oecologia 32:323–332.

Pataki, D. E., S. E. Bush, P. Gardner, D. K. Solomon, and J. R. Ehleringer. 2005. Ecohydrology in a Colorado River riparian forest: Implications for the decline of *Populus fremontii*. Ecological Applications 15:1009–1018.

Pinkney, F. C. 1992. Revegetation and enhancement of riparian communities along the lower Colorado River. USDI Bureau of Reclamation, Ecological Resources Division. Denver, CO.

Richards, L. A. Editor. 1954. Diagnosis and improvement of saline and alkaline soils. USDA Handbook 60. U.S. Government Printing Office, Washington DC.

Robinson, T. W. 1965. Introduction, spread and aerial extend of saltcedar (*Tamarix*) in the western states. Professional Paper 491-A. U.S. Geological Survey. Washington, DC.

Rowland, D. L., A. A. Sher, and D. L. Marshall. 2004. Inter- and Intra-population variation in seedling performance of Rio Grande cottonwood under low and high salinity. Canadian Journal of Forest Research 34:1458–1466.

Ruan, X., Q. Wang, C. Pan, Y. Chen, and H. Jian, 2009. Physiological acclimation strategies of riparian plants to environment change in the delta of the Tarim River, China. Environmental Geology 57:1761–1773.

Sala, A., S. D. Smith, and D. A. Devitt. 1996. Water use by *Tamarix ramosissima* and associated phreatophytes in a Mojave Desert floodplain. Ecological Applications 6:888–898.

Salinas, M. J., G. Blanca, and A. T. Romero. 2000. Riparian vegetation and water chemistry in a basin under semiarid Mediterranean climate, Andarax River, Spain. Environmental Management 26(5):539–552.

Sexton, J. P., A. Sala, and K. Murray. 2006. Occurrence, persistence, and expansion of saltcedar (*Tamarix* spp.) populations in the Great Plains of Montana. Western North American Naturalist 66(1):23–44.

Shafroth, P. B., J. M. Friedman, and L. S. Ischinger. 1995. Effects of salinity on establishment of *Populus fremontii* (cottonwood) and *Tamarix ramosissima* (saltcedar) in southwestern United States. Great Basin Naturalist 55:58–65.

Shafroth, P. B., V. B. Beauchamp, M. K. Briggs, K. Lair, M. L. Scott, and A. A. Sher. 2008. Planning riparian restoration in the context of *Tamarix* control in western North America. Restoration Ecology 16:97–112.

Sharif, F., and A. U. Khan. 2009. Alleviation of salinity tolerance by fertilization in four thorn forest species for the reclamation of salt-affected sites. Pakistan Journal of Botany 41: 2901–2915.

Sher, A., K. Lair, M. DePrenger-Levin, and K. Dohrenwend. 2010. Best Management Practices for Revegetation in the Upper Colorado River Basin. Denver Botanic Gardens. Denver, CO.

Singh, M., M. Jain, and R. Pant. 1999. Clonal variability in photosynthetic and growth characteristics of *Populus deltoides* under saline irrigation. Photosynthetica 36:605–609.

Solomon, A., S. Beer, Y. Waisel, G. P. Jones, and L. G. Paleg. 1994. Effects of NaCl on the carboxylating activity of Rubisco from *Tamarix jordanis* in the presence and absence of proline-related compatible solutes. Physiologia Plantarum 90:198–204.

Sookbirsingh, R., K. Castillo, T. E. Gill, and R. R. Chianelli. 2010. Salt separation processes in the saltcedar *Tamarix ramosissima* (Ledeb.). Communications in Soil Science and Plant Analysis 41:1271–1281.

Steinwand, A. L., R. F. Harrington, and D. P. Groeneveld. 2001. Transpiration coefficients for three Great Basin shrubs. Journal of Arid Environments 49:555–567.

Storey, R., and W. W. Thomson. 1994. An X-ray microanalysis study of the salt glands of intracellular calcium crystals of *Tamarix*. Annals of Botany 73:307–313.

Stromberg, J. 1998. Functional equivalency of saltcedar (*Tamarix chinensis*) and Fremont cottonwood (*Populus fremontii*) along a free-flowing river. Wetlands 18:675–686.

Stromberg, J. C. 2001. Restoration of riparian vegetation in the southwestern United States: Importance of flow regimes and fluvial dynamism. Journal of Arid Environments 49:17–34.

Stromberg, J. C., M. K. Chew, P. L. Nagler, and E. P. Glenn. 2009. Changing perceptions of change: The role of scientists and river management. Restoration Ecology 17:177–186.

Tamea, S., F. Laio, L. Ridolfi, and P. D. Odorico. 2009. Ecohydrology of groundwater-dependent ecosystems: 2. Stochastic soil moisture dynamics. Water Resources Research 45:1–13.

Taylor, J. P., D. B. Wester, and L. M. Smith. 1999. Soil disturbance, flood management, and riparian woody plant establishment in the Rio Grande floodplain. Wetlands 19:372–382.

Thomson, W. W., W. L. Berry, and L. L. Liu. 1969. Localization and secretion of salt by the salt glands of *Tamarix aphylla*. Proceedings of the National Academy of Science USA 63:310–317.

Van Hylckama, T. E. A. 1980. Weather and evapotranspiration studies in a saltcedar thicket, Arizona. Geological Survey Professional Paper 491-F, U.S. Government Printing Office, Washington DC.

Waisel, Y. 1961. Ecological studies on *Tamarix aphylla* (L.) Karst. II. The salt economy. Plant Soil 13:356–364.

Waisel, Y. 1991a. The glands of *Tamarix aphylla*: A system for salt secretion or for carbon concentration? Physiologia Plantarum 83:506–510.

Waisel, Y. 1991b. Adaptation to salinity. Pages 357–381 *in* A. S. Raghavendra, editor. *Physiology of Trees*. John Wiley & Sons, New York.

Waisel, Y., A. Eshel, and M. Agami. 1986. Salt balance of leaves of the mangrove *Avicennia marina*. Physiologia Plantarum 67:67–72.

Walker, L. R., P. L. Barnes, and E. A. Powell. 2006. *Tamarix aphylla*: A newly invasive tree in southern Nevada. Western North American Naturalist 66:191–201.

Wang, W., R. Wang, Y. Yuan, N. Du, and W. Guo. 2011. Effects of salt and water stress on plant biomass and photosynthetic characteristics of Tamarisk (*Tamarix chinensis* Lour.) seedlings. African Journal of Biotechnology 10:17981–17989.

Wiesenborn, W. E. 1996. Saltcedar impacts on salinity, water, fire frequency, and flooding. Proceedings of the Saltcedar Management Workshop. Exotic Pest Plant Council, P 9, Rancho Mirage, CA.

Wilkinson, R. E. 1966. Seasonal development of anatomical structures of saltcedar foliage. Botanical Gazette 127:231–234.

Yang, Z. L., Y. Dai, R. E. Dickinson, and W. J. Shuttleworth. 1999. Sensitivity of ground heat flux to vegetation cover fraction and leaf area index. Journal of Geophysical Research 104:19,505–19,514.

Yin, C. H., G. Feng, F. Zhang, C. Y. Tian, and C. Tang. 2010. Enrichment of soil fertility and salinity by tamarisk in saline soils on the northern edge of the Taklamakan Desert. Agricultural Water Management 97:1978–1986.

PART II

Ecology

9

Tamarix from Organism to Landscape

Kevin Hultine and Tom Dudley

The ecological role that a plant performs in an ecosystem is a product of its distinctive life history traits and responses to local environmental factors, and this is no less true with a nonindigenous taxon than with native species. The establishment and dominance of *Tamarix* spp. across a variety of riparian and wetland ecosystems in North America is particularly representative of how phenotypic plasticity and wide tolerance of environmental conditions have facilitated the success of this group of species and their hybrids in many environments in its adventive range. In this chapter, we consider characteristics of *Tamarix* growth and reproduction that can influence interactions with associated vegetation and fauna under different environmental conditions. We also address how human impacts on these environments may determine the invasion success of this taxon, and whether its impacts can be considered benign or detrimental to native plant biodiversity and the functions of riparian ecosystems.

In the American Southwest, *Tamarix* is a pioneer species in physically disturbed riparian settings that can prosper in relatively stressful environments while simultaneously displaying highly competitive traits over much of its life history (see chapter 1, this volume). As with other dominant woody riparian trees, such as cottonwoods (*Populus* spp.) and willows (*Salix* spp.), *Tamarix* can rapidly colonize recently flooded and scoured floodplain habitats. This classic "ruderal" type strategy (Grime 1977) allows *Tamarix* to quickly occupy bare substrates as floodwaters recede. *Tamarix* can also persist in areas where river regulation (damming) and groundwater extraction have transformed productive riparian systems into physiologically stressful environments. On the other hand, *Tamarix* populations display traits that confer high survival and competitive ability in hot, dry and saline conditions, such as fast growth rates, dense canopies, relatively deep rooting systems, and high litter production (but may be at a competitive disadvantage to cottonwood and willow at the seedling stage; Sher and Marshall 2003; Stromberg et al. 2007). The combination of rapid establishment after disturbance, high stress tolerance and competitive life strategies has led to the successful colonization of *Tamarix* spp. in riparian areas and wetlands throughout the western United States and northern Mexico, reflecting its broad geographical distribution across its home ranges in Eurasia and North Africa.

Here, we examine many of the characteristics that underpin the complex interspecific interactions that *Tamarix* forms with other plant species. Our review starts with the basic phenology and growth of *Tamarix* and looks at how these traits relate to patterns of establishment in comparison with other species. We then focus on the ecophysiological mechanisms that promote the establishment of *Tamarix* in floodplains, its patterns of resource use and its competitive interactions with other species. We review these processes in the context of scale, from the individual organism to plant communities to landscape as a whole. The chapter addresses how large-scale change (including climate change, alterations in fluvial hydrology, and episodic defoliation by the *Tamarix* leaf beetle) affect the interspecific interactions between *Tamarix* and other dominant western North American plant species. Our goal is to shed light on key traits related to the success of *Tamarix* across a complex web of resource availability gradients, competition and plant community structure.

Reproductive Phenology and Establishment

The reproductive strategy of *Tamarix* spp. mirrors that of many pioneer species in that it quickly reaches reproductive maturity and produces many, easily dispersible seeds. *Tamarix* plants can reproduce within one year of germination, and mature plants produce a half a million or more tiny seeds, less than 0.5 mm in length, that are easily dispersed by wind and water, and that have a high germination rate (Young et al. 2004; Neil 1985). The seeds of most *Tamarix* species in North America can be produced over much of the growing season, conferring an ecological advantage over other woody riparian species that only produce seed for a few weeks in the spring or early summer (Stromberg 1997; Shafroth et al. 1998; Cooper et al. 1999). *Tamarix* seeds are highly sensitive to desiccation and lose viability within weeks (Young et al. 2004), so there is no seed bank produced for later germination when suitable conditions might occur. However, the long seed production period improves the probability for germination along river reaches where the seasonal timing of overbank flooding is highly variable. For example, flood events throughout much of the western United States reflect a largely bimodal precipitation pattern where overbank flooding can occur in the spring from rapid snowmelt or in summer following episodic monsoon rainfall. By distributing seeds over most of the growing season, *Tamarix* spp. opportunistically takes advantage of both spring runoff and/or monsoon derived flash flooding (see chapter 7, this volume). *T. parviflora*, from the Mediterranean region, flowers for a relatively short period in spring prior to full leaf flush, and appears to be less invasive than *T. ramossisima* and other longer-flowering forms with which it often co-occurs (Dudley and Bean 2011). In this regard *T. parviflora* is similar to the native taxa with their relatively short "window of opportunity" for matching hydrology and germination success.

Despite their much shorter seed-dispersal period, *Populus* and *Salix* both hold an important competitive advantage over *Tamarix* in that they disperse seeds earlier in the growing season, a trait that can preempt resource use and suppress *Tamarix* recruitment (Sher et al. 2000; Lesica and Miles, 2001). This advantage is realized on river reaches where the timing and intensity of overbank flooding is maintained by unaltered flow regimes (Stromberg et al. 2007). However, shifts in spring runoff caused by earlier snowmelt, now predicted for the southwestern United States, may result in runoff/flood patterns that are out of phase with adaptive seed dispersal phenology. Climate change, therefore, may offset this important phenological advantage that native riparian tree species hold over *Tamarix* (see chapter 25, this volume).

Once riparian plants have successfully germinated, they must grow roots quickly to stay in contact with moist soil as surface layers dry after floods. At the same time, they must produce fast-growing shoots to compete for sunlight with other recently germinated plants. At this stage of its life cycle, *Tamarix* is an inferior competitor to other dominant riparian tree species such as *Populus* and *Salix* (Sher et al. 2000). Competition experiments conducted by Sher et al. (2000, 2002), and Sher and Marshall (2003) along the Rio Grande in New Mexico showed that *Populus* and *Salix* seedlings grew faster than *Tamarix* seedlings in both static and receding water tables during spring and summer. *Tamarix* growth was suppressed in the presence of *Populus* seedlings while native seedlings were mostly unaffected by the presence of *Tamarix* seedlings. Similar results were reported along the Hassayampa River in Arizona where *Populus* and *Salix* seedlings grew taller than *Tamarix* during the first growing season after germination (Stromberg 1997). The faster early growth in *Populus* and *Salix* is likely related to their larger seeds that supply more energy to support rapid shoot and root growth.

Fast growth rate can also protect plants from displacement caused by scouring or burial during post-germination floods. Sediment deposited by spring floods creates seedbeds for germination. However, new sediment deposited on seedlings in summer (July–September) can easily kill the recently germinated plants (see chapter 7, this volume). Likewise, a pulse flood return interval of less than three years can result seedling death, regardless of the seasonal timing. Therefore, rapid stem growth is an important factor that could reduce the likelihood of scour displacement or complete burial of seedlings by floods. The fast growth and seedling height in native riparian plants compared to *Tamarix* suggests that natives are better adapted to survive summer flooding (Lovell et al. 2009). For example, in Anza-Borrego State Park in the western Mojave Desert, a flood representing an approximate 10-year return interval reduced *Tamarix* sapling abundance far more than that of native *Populus* and *Salix*, presumably related to faster biomass production by the native taxa (D'Antonio and Dudley 1997; Dudley et al. 2000). On the other hand, *Tamarix* often produces many shoots during establishment. If a single shoot can emerge from a deposition the plant has a high chance of survival (Levine and Stromberg 2001). Likewise, summer flooding may also create new seedbeds, especially for *Tamarix* species and hybrids that distribute seeds throughout the year. Thus, the community structure of riparian tree species is largely driven by the complex interactions between the seasonal timing of pulse flooding, seed dispersal phenology, and growth.

Survival after germination is also linked to rapid root growth. As flood waters recede, the freshly deposited sediments begin to dry. Riparian plant species generally have poor drought tolerance, and therefore must maintain contact with moist soils by rapid root growth into progressively deeper soil layers. In general *Tamarix* seedlings and saplings are more tolerant of declining water tables than *Populus* and *Salix* (Shafroth et al. 2000), in part due to faster root elongation rates than native riparian tree species (Horton and Clark 2001). Therefore, episodic flooding followed by intense drought may yield high rates of *Tamarix* establishment relative to native species, and contribute to expansion of *Tamarix* onto upper floodplain terraces. On the other hand, native plants may be more tolerant of floodplain scouring after germination. For example, *Salix goodingii* seedlings maintain much larger root-to-shoot ratios than do *Tamarix* spp. (Hayes et al. 2009). These high root-to-shoot ratios in *Salix* may prevent uprooting as soils are scoured during flood events, whereas *Tamarix* may be relative susceptible to these events.

Of course the balance between establishment and survival described above is dependent on maintaining the natural flow regime and the fluvial hydrology of a given reach. A recent investigation along 13 perennial rivers in the arid and semi-arid western United States shows that successful *Populus* recruitment is dependent on maintaining natural flow conditions to the extent that even modest flow regulation will inhibit *Populus* recruitment (Merritt and Poff 2010). *Tamarix* recruitment is also highest in free-flowing river reaches, although, unlike *Populus*, some recruitment was detected in the most highly regulated (dammed river reaches that saw little or no overbank flooding) stream reaches (Merritt and Poff 2010). Certainly *Tamarix* spp. do not *require* altered hydrologic regimes to establish aggressively, as dominant stands are found in the Virgin River (Mortenson and Weisberg 2010), the unregulated upper portion of the Gila River (Whiteman 2006), Coyote Creek/Anza-Borrego State Park, and other systems with relatively natural flow regimes but where natural flood-scouring may be insufficiently frequent to favor native over non-native plants.

Groundwater reduction due to human use can also dramatically affect recruitment and survival of riparian tree species. A natural decline in groundwater depth from −0.86 m to −1.97 m along the Bill Williams River in Arizona resulted in the death of nearly 100% of *Populus* and *Salix* saplings whereas only 0% to 13% of *Tamarix* saplings died (Shafroth et al. 2000). *Populus* and *Salix* are dominant along reaches of the San Pedro River in southern Arizona where groundwater fluctuations are less than 0.5 meters, and average depth to groundwater is above 2.6 meters (Lite and Stromberg 2005). Sites with deeper water tables tended to favor *Tamarix* shrublands over *Populus/Salix* woodlands (Lite and Stromberg 2005).

This combination of altered stream flows and a large temporal seed dispersal/recruitment window are just two of the characteristics that have facilitated the replacement of native riparian vegetation with *Tamarix* shrublands across the western United States. A full understanding *Tamarix* impacts on the structure and function of riparian plant communities in the southwestern United States also requires a broad recognition of the complex interaction between ecohydrological processes and *Tamarix* biology within discrete river reaches.

Foliage Phenology and Resource Uptake

Of the eight or more Eurasian species in the genus *Tamarix* introduced into North America (see chapters 2 and 19, this volume), only *Tamarix aphylla* is evergreen. The remaining species and their hybrids are winter deciduous with canopies that flush in the spring and senesce in the fall. As with all widely distributed plant species, the timing of leaf flush and leaf drop varies across its range. Near its northern limits, *Tamarix* leaves emerge in late April to mid-May and usually drop by early October. However, at lower latitudes, leaves emerge in early to late March (see figure 9.1) and senesce in mid to late October. A recent common garden study of *Tamarix* genotypes collected across a broad latitudinal gradient showed that spring phenology was largely a plastic response to temperature while fall senescence was a function of genetically controlled responses to photoperiod (Friedman et al. 2011). These results suggest that warmer spring temperatures that are predicted for the region under climate change scenarios will lengthen the period of *Tamarix* resource uptake and use. Most likely, preemptive resource use by *Tamarix* will be most significant along upper riparian terraces where *Tamarix*, once established, directly competes with xeric shrubs or native annuals that are late spring or summer active (Hultine and Bush 2011). *Tamarix* populations can develop closed canopies within the first few years of establishment, even along drier upper floodplain terraces. Under these conditions, the availability of sunlight, nutrients and water can be dramatically reduced to the extent that competing plants have few opportunities to complete their life cycle, even if germination is successful.

Exotic plants such as *Tamarix* often disproportionately influence carbon, water and nutrient cycles, in part through increased biomass and net primary production (Ehrenfeld 2003). Although data are limited, it appears that *Tamarix*

(a)

(b)

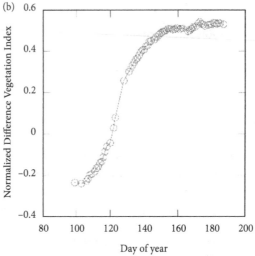

Day of year

FIGURE 9.1 *Tamarix* leaf phenology measured on the Dolores River in southeastern Utah during 2009 with web-based "pheno-cams." (a) Instrumentation elevated above the tamarisk canopy used to measure (b) normalized difference vegetation index (NDVI) from April 9 (day 99) to July 6 (day 187). The NDVI data (an analog for canopy greenness) begins to increase around mid-April before reaching a maximum in early June.

follows these general patterns. As with most pioneering species, a recently established *Tamarix* stand displays a high leaf area index ([LAI]; the total one-sided leaf area per unit ground area) that support high rates of above-ground net primary productivity ([ANPP]; Sala et al. 1996). In productive stands, LAI can reach 4 m² m⁻² (Cleverly et al. 2002; Nagler et al. 2009). However, these plants are relatively short-lived (most *Tamarix* species rarely live more than 80 years; Horton 1977). Therefore, high ANPP rates are usually only maintained in the first few decades or after recruitment. Established *Tamarix* stands along river reaches that have had little or no recent overbank flooding have low LAIs and comparatively low water-use rates (Hultine et al. 2010b), reducing ANPP.

Net primary productivity (NPP) is a function of photosynthetic rate (A), which in turn is influenced by nitrogen concentrations of the leaves. Because N is a large component of chlorophyll, plants with higher leaf nitrogen are expected to have higher A, which is linked to high stomatal conductance (g_s), and therefore high water loss (Katul et al. 2003). Nitrogen concentrations of fresh leaves and leaf litter are higher in *Tamarix* than in co-occurring *Populus* and *Salix* species. For example, fresh leaves and leaf litter have been measured with N concentrations ranging from 23% to 100% higher in *Tamarix* than in co-occurring *Populus* (Pataki et al. 2005; Follstad Shah and Dahm 2008; Moline and Poff 2008). Nitrogen concentrations in *Tamarix* leaf litter along two Southern California streams were four times higher than *Salix* leaf litter (Going and Dudley 2008). Likewise, on the Dolores River near Moab, Utah, *Tamarix* maintained higher leaf N concentrations than co-occurring *Populus fremontii* by

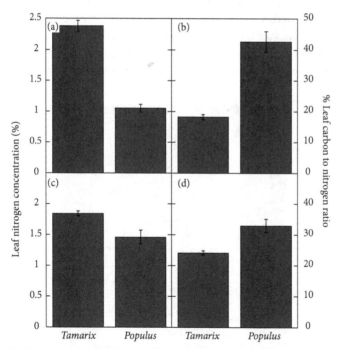

FIGURE 9.2 Leaf nitrogen concentrations and leaf carbon to nitrogen ratios of co-occurring mature *Tamarix* trees (n = 37) and *Populus* trees (n = 4) on the Dolores River in southeastern Utah. (a) Nitrogen concentrations of leaves collected in September 2008. (b) Carbon to nitrogen ratios of leaves collected in September 2008. (c) Nitrogen concentrations of leaves collected in June 2009. (d) Carbon to nitrogen ratios of leaves collected in June 2009. Error bars represent ± 1 standard error of the means. All *P*-value comparisons between *Tamarix* and *Populus* were significant at the 0.05 level.

130% in September and by 30% in June (Hultine and Bush 2011; see figure 9.2). Thus we would expect that *Tamarix* has higher photosynthetic rates and therefore expresses higher leaf-level gas exchange rates than co-occurring *Populus* and *Salix*. Under high soil-moisture conditions on the Bill Williams River in central Arizona, g_s and net photosynthesis in mature *Tamarix* trees were about two and three times higher, respectively, than rates expressed by co-occurring *Populus* and *Salix* trees (Horton et al. 2001). Likewise, *Tamarix* has photosynthetic twigs and small stems that facilitate re–assimilation of respired CO_2 that might otherwise be lost (see figure 9.3), a trait that not only improves plant water use efficiency (defined as the molar ratio of carbon assimilated per water loss) but also increases the gross amount of carbon uptake. We currently lack data for *Tamarix*, but studies of other tree species show that up to 70% of a stem's annual respiration is converted back into photosynthetic materials, and 11% of a stem's biomass is directly supported by the re-assimilation of respired CO_2 (Cernusak and Hutley 2011). These data suggest that under some conditions *Tamarisk* displays higher stand-level ANPP rates than native species, although NPP is not always linked to leaf-level gas exchange. Other resource allocations, such as defense against herbivory, maintenance respiration, reproduction, and belowground allocation, can have critical impacts on ANPP. Thus, more research is needed to better understand patterns of riparian species ANPP and the cascading impact these carbon fluxes have on plant community structure.

FIGURE 9.3 Photo showing green photosynthetic *Tamarix* twigs that have the capacity to re-assimilate a large percentage of respired CO_2.

Resource Allocation and Nutrient Cycling

Plants in terrestrial environments face the difficult choice of maximizing growth and fitness while simultaneously defending against the onset of disease, pathogens, and herbivory. This trade-off requires plants to balance the allocation of resources, such as photosynthates (i.e., carbohydrate and starch reserves) and mineral nutrients, to growth, reproduction, defense (i.e., secondary metabolites) or internal storage (Bloom et al. 1985; Chapin et al. 1990; Ayres 1993). If all else is equal, plants that allocate a larger proportion of resources to growth must do so at the expense of allocating fewer resources to other areas such as internal storage. Riparian tree species such as *Tamarix* have evolved with fast rates of growth following recruitment, because of intense competition for sunlight and space that occurs for early successional species in systems with frequent disturbance (Grime 1977). However, for perennial plants such as *Tamarix*, maintaining relatively large storage pools is also necessary, not only to facilitate the seasonal growth of new tissues, but also to provide critical resources to maintain function and/or replace tissues after a catastrophic event such as fire (Chapin et al. 1990). Unlike many native plants, *Tamarix* is very well adapted to even very intense fires (see chapter 14, this volume). Plants that are well adapted to episodic disturbance, like *Tamarix*, often "bank" internal resources within complex storage organs or recycle materials from expendable tissues with high turnover rates (i.e., leaves and fine roots).

To understand this seeming paradox, we must also look at the role of herbivory. Many *Tamarix* species have evolved under intense pressure from herbivory. In its home range, *Tamarix* is consumed by 325 specialist insects and mites from 88 genera (Kovalev 1995). This plant-insect coevolution has likely resulted in a diverse set of defensive strategies across its range, including maintenance of relatively high storage pools as soluble or total nonstructural carbohydrates (TNC) in stems or roots (Bloom et al. 1985). This stored energy allows the plant to recover after tissues have been eaten. TNC concentrations in the root crowns of mature *Tamarix* trees can reach 9% (Hudgeons et al. 2007), and similar values have been measured in small *Tamarix stems* (< 1 cm) during dormancy in southeastern Utah (Hultine, unpublished). Total non-structural carbohydrate concentrations in the fine roots may be much higher than values reported in the root crown and stems. In fact, TNC can constitute 10% to 40% of total root dry mass in some plants (Kobe 1997; Canham et al. 1999). However, the total TNC storage pool is difficult to quantify in mature *Tamarix* because of its extensive root systems. Nevertheless, it seems likely that the selective pressure of both herbivory and fire disturbance has created a plant that is able to allocate resources to rapid growth at some life-history stages and significant storage at others.

Allocation of photosynthates to TNC storage is highly dependent on availability of other resources including water, sunlight, and mineral nutrients, such as N (Kobe et al. 2010). *Tamarix* invasion increases plant-available and total soil N in riparian areas, through sediment and litter accumulation (Adair et al. 2004). Once established, *Tamarix* roots trap high volumes of sediment that often have relatively high concentrations of N and other nutrients. In fact, freshly deposited sediment can account for the majority of N that accumulates in floodplain soils dominated by *Tamarix* (Adair et al. 2004). Likewise, *Tamarix* stands produce high leaf areas that result in high volumes of leaf litter. *Tamarix* leaf litter has high concentrations of N (Hultine and Bush 2011; figure 9.2a, c), relatively low C:N ratios (Moline and Poff 2008; Follstad Shah and Dahm 2008; figure 9.2b, d), and high rates of leaf decomposition relative to those of some native plants (Bailey et al. 2001). Taken together, these data suggest that the presence of *Tamarix* increases rates of N cycling and the availability of mineralized N. Given that western riparian plants and animals are often more limited by N availability than water, such inputs by *Tamarix* might be expected to have significant effects on the ecosystem, although nutrient enrichment is more likely to benefit *Tamarix* itself than natives.

Rapid growth response to nutrient enrichment is a life-history trait that contributes to the success of many invasive plants (Grime 1977). A greenhouse study of seedling responses to N and P concentrations revealed that *Tamarix ramosissima* seedlings produce significantly more stems, a higher shoot biomass, and higher total biomass at progressively higher nutrient concentrations, whereas aboveground productivity for both *Populus fremontii* and *Salix goodingii* leveled off at lower concentrations (Marler et al. 2001). The authors concluded that high plant-available nutrient concentrations likely favor the establishment of *Tamarix*, while lower concentrations may increase the potential establishment of *Populus/Salix* riparian stands (Marler et al. 2001), a common mechanism of invasion by many nonnative species (Vitousek and Walker 1989, Brooks 2003). In another study, lower soil N was associated with dominance of *Populus/Salix* over *Tamarix* during the first three years of seedling establishment in the field, although it was not clear whether this was a cause or an effect (Sher et al. 2002). Moreover, there are several reasons enhanced nutrient concentrations that accumulate under mature *Tamarix* stands might not facilitate the growth and establishment of native plant species. These include limited sunlight penetration to the soil surface, enhanced surface salinity (see chapter 8, this volume), and negative associations that *Tamarix* roots have with native mycorrhizal communities (see chapter 13, this volume).

Tamarix Water Relations

Tamarix has several physiological advantages over *Populus*, *Salix*, and other riparian phreatophytes for tolerating unfavorable soil water conditions, including greater rooting depths and extremely fast root growth compared to *Populus* and *Salix* (Shafroth et al. 2000; Horton and Clark 2001). Both traits favor access to groundwater where the water table may be too deep and/or too variable to support most native phreatophytic plant species (Horton and Campbell, 1974; Shafroth et al. 2000; Horton et al. 2003; Lite and Stromberg, 2005).

Another key advantage *Tamarix* has over native species is its xylem structure and function, which allow it to maintain a higher hydraulic conductance relative to many native riparian species. Under conditions where soil water is always available (such as in many riparian areas), and/or evaporative demand is low to moderate, plants build xylem conduits with large diameters that can transport water at high rates to the canopy. These conditions allow plants to construct large canopies that produce high transpiration rates per unit ground area. However, larger diameter conduits are also more susceptible to drought-induced xylem cavitation, the condition in which the water column within a xylem conduit is broken, thus preventing further water transport to the canopy (Zimmerman 1983; Tyree and Sperry 1989). This is similar to the risk of an air bubble breaking the siphon in a large-diameter hose. If xylem cavitation occurs throughout the plant, the canopy becomes decoupled from soil water, causing significant hydraulic failure, carbon starvation, canopy dieback, and in many cases plant death (Sperry et al. 1993; McDowell et al. 2008). Data from southern Arizona show that *Tamarix* shoots are considerably more resistant to xylem cavitation than co-occurring *P. fremontii* and *S. goodingii* (Pockman and Sperry 2000). In fact, xylem pressures of 2 megapascals (MPa) would cause complete cavitation in both *Populus* and *Salix*, while *Tamarix* shoots can retain conductivity at water potentials of −4 MPa or below (Pockman and Sperry 2000) (see figure 9.4). Not surprisingly, *Populus* and *Salix* in northern Arizona had significant canopy dieback when midday water potentials dropped below −2 MPa, while co-occurring *Tamarix* did not (Horton et al. 2001).

Data from several other studies also suggest greater drought tolerance in *Tamarix* compared to native riparian tree species (Sala et al. 1996; Smith et al. 1998; Horton et al. 2003). A greater overall drought tolerance allows *Tamarix* to persist in habitats that are not suitable for most other obligate riparian species, in many cases resulting in new riparian tree canopy cover along river reaches that historically supported little or riparian vegetation (Hultine and Bush 2011; see also chapters 5 and 6, this volume).

Despite its greater cavitation resistance, *Tamarix* can take up and transport water through its stems at surprisingly high rates. The mean specific conductivity of *Tamarix* shoots (defined as the maximum stem hydraulic conductivity divided by the total stem area that is conducting water) measured in southern Arizona was 23% and 46% higher than those of *P. fremontii* and *S. goodingii* shoots, respectively (Pockman and Sperry 2000). Likewise, the mean conduit diameters of *Tamarix* shoots was about 30% larger than either *Populus* or *Salix* (Pockman and Sperry 2000). Consequently, *Tamarix* shoots likely operate with a higher stem hydraulic conductance than either *Populus* or *Salix* regardless of shoot water potential down to about −6 MPa (figure 9.4). These observations may have important implications for several ecophysiological processes. *Tamarix* can maintain a consistently high leaf area to unit shoot area and/or high stomatal conductance over a wide range of soil or groundwater conditions. This allows *Tamarix* to maintain high transpiration rates, high nutrient uptake rates, and maintain a favorable carbon balance

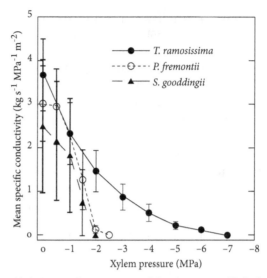

FIGURE 9.4 Relationship between xylem water potential and mean specific hydraulic conductivity in co-occurring *Tamarix*, *Populus*, and *Salix* trees in southern Arizona. Error bars represent ± 1 standard deviation from the means.

Source: Adapted from Pockman and Sperry (2000).

across a wide range of environmental gradients. This is not to say that water use by *Tamarix* does not vary in response to water potential gradients, as several studies have shown that it does (Sala et al. 1996; Cleverly et al. 1997; Devitt et al.1997a, b; Nagler et al. 2009). Nevertheless, the dynamic structure and function of *Tamarix* xylem is an apparent mechanism underpinning its distribution across broad environmental gradients relative to other riparian tree species.

Less understood are the below-ground components of the plant water transport system (i.e., the soil-plant-atmosphere continuum) with respect to dominant riparian tree species. We can only speculate how below-ground parameters vary among species and over environmental gradients. An important observation is that *Tamarix* often takes up a higher proportion of water from shallow unsaturated soils than do either *Populus* or *Salix* (Busch et al. 1992; Smith et al. 1998). Its status as a facultative rather than an obligate phreatophyte, such as *Salix*, or semi-obligate phreatophyte such as *Populus* (Snyder and Williams 2000) suggests that *Tamarix* is able to maintain a higher root hydraulic conductance in unsaturated soils. If so, in areas where access to groundwater is highly variable, *Tamarix* would be able to maintain high rates of water uptake and subsequently avoid canopy dieback and mortality.

Tamarix Effects on Plant Biodiversity and Riparian Ecosystem Function

A fundamental question is whether *Tamarix* negatively affects plant biodiversity and riparian ecosystem structure and function as a whole. In many cases, *Tamarix* preempts resource use, increases soil salinity (see chapter 8, this volume) disrupts soil ecological processes (see chapter 13, this volume), increases fire frequency and intensity (see chapter 14, this volume), and alters nutrient cycles (see previous section, "Resource Allocation and Nutrient Cycling"). To varying degrees, all of these factors can impair riparian ecosystem structure and function at multiple scales. Nevertheless, evidence that supports the traditional view that *Tamarix* invasion directly reduces plant biodiversity is at best mixed. Recent evidence suggests that *Tamarix* is often the symptom of reduced biodiversity and not the direct cause (Stromberg and Chew 2002; Stromberg et al. 2009). Differentiating one from the other is critical for defining riparian management objectives and designing restoration projects.

Although *Tamarix* seedlings do not compete well with native riparian tree seedlings (Sher et al. 2000; Sher and Marshall 2003), *Tamarix* plants that do successfully germinate in recently disturbed riparian areas produce thickets that are often considerably denser than those produced by recently established native tree species. These dense thickets effectively crowd and shade understory vegetation, often resulting in dramatic reductions in herbaceous plant composition and abundance (DiTomaso 1998). Land managers may view this preemptive resource use favorably, given the potential for *Tamarix* to exclude, if not displace invasive herbaceous perennials such as Russian knapweed (*Acroptilon repens*) and perennial pepperweed (*Lepidium latifolium*) among other species. Data from the Verde River in northern Arizona indirectly support this view: sites that were clear of *Tamarix* had a greater abundance of invasive plants and more xeric adapted vegetation compared to sites that were dominated by *Tamarix* (Johnson et al. 2010). On the other hand, sites with *Tamarix* displayed a higher overall abundance of native understory species, native trees, and native perennial wetland species (Johnson et al. 2010). Again, these results do not establish whether *Tamarix* was the cause or effect of native/invasive vegetation assemblages, but do highlight the difficulty of broadly characterizing *Tamarix* impacts on plant community structure and riparian function. It is plausible that the presence of *Tamarix* in ecosystems altered by other global processes (i.e., drought, increased temperature, increased land use by humans) can synergistically reduce available resources for co-occurring plant species, depending on life history, life form, physiology, and phenology. Future research should address these complex interactions in order to better understand the effects of *Tamarix* on riparian plant biodiversity.

Cascading Impacts of Biological Control on Plant Community Structure

Recent releases of the *Tamarix* leaf beetle (*Diorhabda* spp.), a biological control agent from Eurasia, has brought widespread herbivory to *Tamarix* throughout the western United States (Tracy and Robbins 2009; Hultine et al. 2010a; see also chapter 22, this volume). The beetle feeds exclusively on *Tamarix* foliage, causing episodic defoliation that lasts several weeks or longer. Specifically, the beetles scrape the leaf cuticle before attacking the mesophyll and the vascular system, resulting in leaf dessication and subsequent leaf drop (Dudley 2005). Repeated defoliation (either within one season or over several consecutive growing seasons) results in carbon starvation that, in turn, reduces foliage production and growth, in some cases killing the tree (Dudley and Bean 2011).

The impacts of these disturbance events on *Tamarix* productivity, reproduction, resource use and competitive interactions vary across the landscape as the beetle spreads across the region. Defoliation events over the first few years generally do not produce significant rates of tree mortality. However, they reduce carbohydrate storage and subsequent leaf production after each event (Hudgeons et al. 2007). With fewer leaves, canopy LAI is lower and competition for sunlight is reduced. Likewise, plants exposed to herbivory are forced to produce more leaves during the growing season. Partially eaten leaves fall as litter that has higher N concentrations and lower C:N ratios compared to leaves that normally drop in the fall (Snyder et al. 2010). Therefore, rapid defoliation may have the short-term effect of increasing N availability because of increased litter amounts with faster decomposition rates (Hultine et al. 2010a).

Episodic *Tamarix* defoliation may have large-scale impacts on plant community structure, particularly where *Tamarix* populations have extensive dieback or high mortality rates. Assuming that competitive interactions with other taxa are due to aggressive *Tamarix* growth and resource monopolization, the stress of herbivory should ameliorate these interactions, and potentially release associated plants from competitive dominance, as is being documented in Texas and other locations where *Diorhabda* is well established (DeLoach, unpublished data). In some cases, large areas may remain bare if there is substantial mortality, particularly where stream-flow regulation prevents flooding and subsequent plant establishment. Conversely, the accumulation of sediment following *Tamarix* colonization could create plant habitats high in N (Adair et al. 2004). No allelopathic effects of *Tamarix* have been shown, and *Tamarix* litter can actually facilitate native plant growth if the soil salinity is not too high (Lesica and DeLuca 2004). High inputs of high-quality litter caused by repeated defoliation could facilitate high rates of NPP in riparian zones where N is more limiting to growth than available water.

Conclusions

The successful colonization of *Tamarix* in riparian areas and wetlands throughout the western United States can be attributed to a unique biology that allows this plant to express a combination of ruderal-type strategies, competitive abilities, and stress tolerance during different phases of its life history. *Tamarix* can rapidly colonize recently disturbed river reaches by taking advantage of its long and prolific seed production and rapid germination. Although *Tamarix* seedlings do not compete well with co-occurring *Populus* and *Salix* seedlings for resources, they often avoid direct competition by germinating outside the narrow phenological windows of native species. Once established, however, *Tamarix* becomes an effective competitor for space and resources because of its rapid growth, dense canopy, high resource-uptake efficiency, and preemptive resource monopolization. *Tamarix* can also tolerate resource limitation to the extent that it often persists and thrives along upper floodplain terraces that have no riparian vegetation. Moreover, the higher stress tolerance allows *Tamarix* to outperform *Populus* and *Salix* along flow-regulated river reaches, or where groundwater pumping has adversely affected plant water availability. This unique combination of traits has turned *Tamarix* into one of the most successful nonnative plant species in western North America. One outcome of this success has been to spur aggressive management activities to control and/or remove *Tamarix* along highly valued river reaches and watersheds. The widespread distribution of the *Tamarix* leaf beetle will likely reduce the competitive ability of *Tamarix* populations by lowering their resource use and shifting water, nutrient, and carbon cycling processes. How these changes will affect riparian plant community structure and function remains to be seen. Future research and management activities should focus on the long-term impact of *Tamarix* defoliation on riparian community structure, ecohydrology, and ecosystem services as a whole.

Literature Cited

Adair, E. C., D. Binkley, D. C. Andersen. 2004. Patterns of nitrogen accumulation and cycling in riparian floodplain ecosystems along the Green and Yampa rivers. Oecologia 139:108–116.

Ayres, M. P. 1993. Plant defense, herbivory and climate change. Pages 75–94 *in* P. M. Kareiva, J. G. Kingsolver, and R. B. Huey, editors. *BioticInteractions and Global Change*. Sinauer Associates, Sunderland, MA.

Bailey, J., J. Schweitzer, and T. Whitham. 2001. Saltcedar negatively affects biodiversity of aquatic macroinvertebrates. Wetlands 21:442–447.

Bloom A. J., F. S. Chapin III, and H. A. Mooney. 1985. Resource limitation in plants: An economic analogy. Annual Review of Ecology and Systematics. 16:363–392.

Brooks, M. L. 2003. Effects of increased soil nitrogen on the dominance of alien annual plants in the Mojave Desert. Journal of Applied Ecology 40:344–353.

Busch, D. E., N. L. Ingraham, and S. D. Smith. 1992. Water uptake in woody phreatophytes of the southwestern United States: A stable isotope study. Ecological Applications 2:450–459.

Busch, D. E., and S. D. Smith. 1995. Mechanisms associated with decline of woody species in riparian ecosystems of the southwestern U.S. Ecological Monographs 65:347–370.

Canham, C. D., R. K. Kobe, E. F. Latty, and R. L. Chazdon. 1999. Interspecific and Intraspecific variation in tree seedling survival: Effects of allocation to roots versus carbohydrate reserves. Oecologia 121:1–11.

Cernusak, L. A., and Hutley, L. B. 2011. Stable isotopes reveal the contribution of cortic-ular photosynthesis to growth in branches of *Eacalyptus miniata*. Plant Physiology 155:515–523.

Chapin III F. S., E. D. Shulze, and H. A. Mooney. 1990. The ecology and economics of storage in plants. Annual Review of Ecology and Systematics 21:423–447.

Cleverly, J. R., C. N. Dahm, J. R. Thibault, D. J. Gilroy, and J. E. Allred Coonrad. 2002. Seasonal estimates of actual evapotranspiration from *Tamarix ramosissima* stands using three-dimensional eddy covariance. Journal of Arid Environments 52:181–197.

Cleverly, J. R., S. D. Smith, A. Sala, and D. A. Devitt. 1997. Invasive capacity of *Tamarix ramosissima* in a Mojave floodplain: The role of drought. Oecologia 111:12–18.

Cooper, D. J., D. M. Merritt, D. C. Anderson, and R. A. Chimner. 1999. Factors controlling the establishment of Fremont cottonwood seedlings on the upper Green River, USA. Regulated Rivers: Research and Management 15:419–440.

D'Antonio, C. and T. Dudley. 1997. Saltcedar as an invasive component of the riparian vegetation of Coyote Creek, Anza-Borrego State Park. Final Report to California Department of Parks and Recreation, San Diego, CA. 56 pp.

Devitt, D. A., A. Sala, K. A. Mace, and S. D. Smith. 1997a. The effect of applied water on the water use of saltcedar in a desert riparian environment. Journal of Hydrology 192:233–246.

Devitt, D. A., J. M. Piorkowski, S. D. Smith, J. R. Cleverly, and A. Sala. 1997b. Plant water relations in *Tamarix ramosissima* response to the imposition and alleviation of soil moisture stress. Journal of Arid Environments 36:527–540.

DiTomaso, J. M. 1998. Impact, biology, and ecology of saltcedar (*Tamarix* spp.) in the southwestern United States. Weed Technology 12:326–336.

Dudley, T. L. 2005. Progress and Pitfalls in the Biological Control of Saltcedar (*Tamarix* spp.) in North America. Proceedings of the 16th U.S. Department of Agriculture interagency research forum on gypsy moth and other invasive species; 18–21 Jan. 2005. Annapolis, MD. USDA Forest Service General Technical Report NE-337, Morgantown, WV.

Dudley, T. L., C. J. DeLoach, J. Lovich, and R. I. Carruthers. 2000. Saltcedar invasion of western riparian areas: Impacts and new prospects for control. Pages 345–381, Transactions of the 65th North American Wildlife and Natural Resources Conference, March 2000, Chicago, IL. Wildlife Management Insitute Publications.

Dudley, T L., and D W. Bean. 2011. Tamarisk biocontrol, endangered species effects and resolution of conflict through riparian restoration. Biological Control 57:331–347.

Ehrenfeld, J. G. 2003. Effects of exotic plant invasions on soil nutrient cycling. Ecosystems 6:503–523.

Follstad Shah, J. J., and C. N. Dahm. 2008. Flood regime and leaf fall determine soil inorganic nitrogen dynamics in semiarid riparian forests. Ecological Applications 18:771–788.

Friedman, J. M., J. E. Roelle, and B. S. Cade. 2011. Genetic and environmental influence on leaf phenology and cold hardiness of native and introduced riparian trees. International Journal of Biometeorology. doi:10.1007/s00484-011-0494-6.

Going, B. M., and T. L. Dudley. 2008. Invasive riparian plant litter alters aquatic insect growth. Biological Invasions 10:1041–1051.

Grime, J. P. 1977. Evidence for the existence of three primary strategies in plants and its relevance to ecological and evolutionary theory. American Naturalist 11:1169–1194.

Hayes, W. E., L. R. Walker, and E. A. Powell. 2009. Competitive abilities of *Tamarix aphylla* in southern Nevada. Plant Ecology 202:159–167.

Horton, J. 1977. The Development of Perpetuation of the Permanent Tamarisk Type in the Phreatophyte Zone of the Southwest. USDA. U.S/Rocky Mountain Forest Range Experimental Station. General Technical Report 43. Fort Collins, CO.

Horton, J., and C. Campbell. 1974. Management of Phreatophyte and Vegetation for Multiple Use Values. U.S. Forest Service Research Paper RM-117, Fort Collins, CO.

Horton, J. L., and J. L. Clark. 2001. Water table decline alters growth and survival of *Salix goodingii*, and *Tamarix chinensis* seedlings. Forest Ecology and Management 140:239–247.

Horton, J. L., S. C. Hart, and T. E. Kolb. 2003. Physiological condition and water source use of Sonoran Desert riparian trees at the Bill Williams River, AZ, USA. Isotopes in Environmental Health Studies 39:69–82.

Horton, J. L., T. E. Kolb, and S. C. Hart. 2001. Responses of riparian trees to interannual variation in ground water depth in a semi-arid river basin. Plant, Cell and Environment 24:293–304.

Hudgeons, J. L., A. E. Knutson, K. M. Heinz, C. J. DeLoach, T. L. Dudley, R. R. Pattison, and J. R. Kiniry. 2007. Defoliation by introduced *Diorhabda elongata* leaf beetles (Coleoptera Chrysoelidae) reduces carbohydrate reserves and regrowth of *Tamarix* (Tamaricaceae). Biological Control 43:213–221.

Hultine, K. R., J. Belnap, C. van Riper III, J. R. Ehleringer, P. E. Dennison, M. E. Lee, P. L Nagler, K. A. Snyder, S. E. Uselman, and J. B. West. 2010a. Tamarisk biocontrol in the western United States: Ecological and societal implications. Frontiers of Ecology and the Environment 8:467–474.

Hultine, K. R., P. L Nagler, K. Morino, S. E. Bush, K. G. Burtch, P. E. Dennison, E. P. Glenn, and J. R. Ehleringer. 2010b. Sap flux-scaled transpiration by tamarisk (*Tamarix* spp.) before, during and after episodic defoliation by the saltcedar leaf beetle (*Diorhabda carinulata*). Agricultural and Forest Meteorology 150:1467–1475.

Hultine, K. R., and S. E. Bush. 2011. Ecohydrological consequences of non-native riparian vegetation in the southwestern U.S: A review from an ecophysiological perspective. Water Resources Research 47 (7) [Online] http://dx.doi.org/10.1029/2010WR010317.

Johnson, T. D., T. E. Kolb, and A. L. Medina. 2010. Do riparian plant community characteristics differ between *Tamarix* (L.) invaded and non-invaded sites on the upper Verde River, Arizona? Biological Invasions 12:2487–2497.

Katul, G., R. Luening, and R. Oren. 2003. Relationships between plant hydraulic and biochemical properties derived from a steady-state coupled water and carbon transport model. Plant, Cell and Environment 26:339–350.

Kobe, R. K. 1997. Carbohydrate allocation to storage as a basis of interspecific variation in sapling survivorship and growth. Oikos 80:226–233.

Kobe, R. K., M. Iyer, and M. B. Walters. 2010. Optimal partitioning theory revisited: Nonstructural carbohydrates dominate root mass responses to nitrogen. Ecology 91:166–179

Kovalev, O. V. 1995. Co-evolution of tamarisk (Tamaricaceae) and pest arthroods (Insecta: Arachnida: Acarina), with special reference to biological control prospects. Pensoft, Sofia. 110 pp.

Lesica, P., and S. Miles. 2001. Tamarisk growth at the northern margin of its naturalized range in Montana, USA. Wetlands 21:240–246.

Lesica, P., and T. H. DeLuca. 2004. Is tamarisk allelopathic? Plant and Soil 267:357–365

Levine, C. M., and J. C. Stromberg. 2001. Effects of flooding on native and exotic plant species: Implications for restoring southwestern riparian forests and manipulating water and sediment flows. Journal of Arid Environments 49:111–131.

Lite, S. J., and J. C. Stromberg. 2005. Surface water and groundwater thresholds for maintaining *Populus-Salix* forests, San Pedro River, Arizona. Biological Conservation 125:153–167.

Lovell, J. T., J. Gibson, and M. S. Heschel. 2009. Disturbance regime mediates riparian forest dynamics and physiological performance, Arkansas River, CO. The American Midland Naturalist 162:289–304.

Marler, R. J., J. C. Stromberg, and D. T. Patten. 2001. Growth response of *Populus fremontii*, *Salix goodingii*, and *Tamarix ramosissima* seedlings under different nitrogen and phosphorous concentrations. Journal of Arid Environments 49:133–146.

McDowell, N., W. T. Pockman, C. D. Allen, D. D. Breshears, N. Cobb, T. Kolb, J. Plaut, J. Sperry, A. West, D. G. Williams, and E. A. Yepez. 2008. Mechanisms of plant survival and mortality during drought: Why do some plants survive while others succumb to drought? New Phytologist 178:719–739.

Merritt, D. M., and N. L. Poff. 2010. Shifting dominance of riparian *Populus* and *Tamarix* along gradients of flow alteration in western North American rivers. Ecological Applications 20:135–152.

Moline, A. B., and N. L. Poff. 2008. Growth of an invertebrate shredder on native (*Populus*) and non-native (*Tamarix, Eleagnus*) leaf litter. Freshwater Biology 53:1012–1020.

Mortenson, S. G., and P. J. Weisberg. 2010. Does river regulation increase the dominance of invasive woody species in riparian landscapes? Global Ecology and Biogeography 19:562–574.

Nagler, P. L., K. Morino, K. Didan, J. Erker, J. Osterberg, K. R. Hultine, and E. P. Glenn. 2009. Wide-area estimates of saltcedar (*Tamarix* spp.) evapotranspiration on the lower Colorado River measured by heat balance and remote sensing methods. *Ecohydrology* 2:18–33.

Neil, W. 1985. Tamarisk. Fremontia 12:22–23.

Pataki, D. E., S. E. Bush, P. Gardener, D. K. Solomon, and J. R. Ehleringer. 2005. Ecohydrology in a Colorado River riparian forest: Implications for the decline of *Populus fremontii*. Ecological Applications 15:1009–1018

Pockman, W. T., and J. S. Sperry. 2000. Vulnerability to xylem cavitation and the distribution of Sonoran Desert vegetation. American Journal of Botany 87:1287–1299.

Sala, A., S. D. Smith, and D. A. Devitt. 1996. Water use by *Tamarix ramosissima* and associated phreatophytes in a Mojave Desert floodplain. Ecological Applications 6:888–898.

Shafroth, P. B., G. T. Auble, J. C. Stromberg, and D. T. Patten 1998. Establishment of woody vegetation in relation to annual patterns of streamflow, Bill Williams River, AZ. Wetlands 18:557–590.

Shafroth, P. B., J. C. Stromberg, and D. T. Patten. 2000. Woody riparian vegetation response to different alluvial water table regimes. Western North American Naturalist 60:66–76.

Sher, A. A., and D. L. Marshall. 2003. Seedling competition between native *Populus deltoides* (Salicaceae) and exotic *Tamarix ramosissima* (Tamaricaceae) across water regimes and substrate types. American Journal of Botany 90:413–422.

Sher, A. A., D. L. Marshall, and J. P. Taylor. 2002. Establishment patterns of native *Populus* and *Salix* in the presence of invasive, non-native *Tamarix*. Ecological Applications 12:760–772.

Sher, A. A., D. L. Marshall, and S. A. Gilbert. 2000. Competition between native *Populus deltoides* and invasive *Tamarix ramosissima* and the implication for reestablishing flooding disturbance. Conservation Biology 14:1744–1754.

Sher, A. A., K. Wiegand, and D. Ward. 2010. Do *Acacia* and *Tamarix* trees compete for water in the Negev desert? Journal of Arid Environments 74:338–343.

Smith, S. D., D. A. Devitt, A. Sala, J. R. Cleverly, and D. Busch. 1998. Water relations of riparian plants from warm desert regions. Wetlands 18:687–696.

Snyder, K. A., and D. G. Williams. 2000. Water sources used by riparian trees varies among stream types on the San Pedro River. Agricultural and Forest Meteorology 105:227–240.

Snyder, K. A., S. M. Uselman, T. J. Jones, and S. Duke. 2010. Ecophysiological responses of salt cedar (*Tamarix* spp. L.) to the northern tamarisk beetle (*Diorhabda carinulata* Desbrochers) in a controlled environment. Biological Invasions 12:3795–3808.

Sperry, J. S., N. N. Alder, and S. E. Eastlick. 1993. The effect of reduced hydraulic conductance on stomatal conductance and xylem cavitation. Journal of Experimental Botany 44:1075–1082.

Stromberg, J. C. 1997. Growth and survivorship of Fremont cottonwood, Gooding willow, and salt cedar seedlings after large floods in central Arizona. Great Basin Naturalist 57:198–208.

Stromberg, J. C., S. J. Lite, R. Marler, C. Paradzick, P. B. Shafroth, D. Shorrock, J. M. White, and M. S. White. 2007. Altered stream-flow regimes and invasive plant species: The *Tamarix* case. Global Ecology and Biogeography 16:381–393.

Stromberg, J. C., and M. K. Chew. 2002. Foreign visitors in riparian corridors of the American Southwest: is xenophytophobia justified? Pages 195–219 *in* B. Tellman, editor. Invasive exotic species in the Sonoran region. University of Arizona Press, Tucson.

Stromberg, J. C., M. K. Chew, P. L. Nagler, and E. P. Glenn. 2009. Changing perceptions of change: The role of scientists in *Tamarix* and river management. Restoration Ecology 17:177–186.

Tracy, J. L., and T. O. Robbins. 2009. Taxonomic revision and biogeography of the Tamarix-feeding *Diorhabda elongata* (Brullé, 1832) species group (Coleoptera: Chrysomelidae: Galerucinae: Galerucini) and analysis of their potential in biological control of tamarisk. Zootaxa 2101:1–152.

Tyree, M. T., and J. S. Sperry. 1989. Vulnerability of xylem to cavitation and embolism. Annual Review of Plant Physiology and Molecular Biology 40:19–38.

Vitousek, P. M., and L. R. Walker. 1989. Biological invasion by *Myrica faya* in Hawai'i: Plant demography, nitrogen fixation, ecosystem effects. *Ecological Monographs* 59:247–265.

Whiteman, K. E. 2006. Distribution of salt cedar (*Tamarix* spp. L) along an unregulated river in south-western New Mexico, USA. Journal of Arid Environments 64:364–368.

Young, J. A., C. D. Clements, and D. Harmon. 2004. Germination of seeds of *Tamarix ramosissima*. Rangeland Ecology & Management 57:475–481.

Zimmerman, M. H. 1983. *Xylem Structure and the Ascent of Sap*. Springer-Verlag, New York.

10

Tamarix as Wildlife Habitat

Heather L. Bateman, Eben H. Paxton, and William S. Longland

Riparian areas of floodplains provide a mosaic of productive habitats (Stanford et al. 2005; Latterell et al. 2006) capable of supporting a rich community of vertebrate species. In the semiarid southwestern United States, more than 50% of land birds depend upon riparian habitats for breeding, and most other avian species use riparian environments at some point in their life cycle (Anderson et al. 1977; Knopf et al. 1988). Small mammals in the arid and semiarid Southwest, represented by insectivorous and granivorous species, are often more numerous in riparian habitats than in adjacent uplands (Stamp and Ohmart 1979; Doyle 1990; Falck et al. 2003). Amphibians and reptiles (collectively referred to as "herpetofauna") are often overlooked in these systems; however, they provide an important link from arthropod to vertebrate predators. Together, birds, mammals, and herpetofauna are secondary consumers within a riparian food web, and these wildlife species can provide a tool to evaluate how the establishment and proliferation of exotic species such as *Tamarix* can effect ecosystem-level changes (Bateman and Paxton 2010).

Comparison of *Tamarix* versus Native or Mixed Habitats

Understanding wildlife use or avoidance of habitats dominated by exotic species is important for natural resource managers, who often have to balance nonnative species control with protection of wildlife and their habitats (Bateman et al. 2010). Comparisons of wildlife abundance and diversity in *Tamarix* habitats and native habitats or in mixed *Tamarix* and native habitats have mostly focused on birds, although research investigating small mammal and herpetofauna communities is accumulating. We define mixed habitats as stands composed of nonnative *Tamarix* and native tree species, mainly cottonwood (*Populus* spp.) and willow (*Salix* spp.), and in some studies, mesquite (*Prosopis* spp.).

BIRDS

Across the arid western United States, and in particular, the desert Southwest, riparian woodlands are critical habitat for birds. A majority of land birds that breed in the Southwest are directly dependent on riparian habitats, and most other land bird species utilize this habitat at some point in their annual cycle (Anderson et al. 1977; Knopf et al. 1988). Although a number of authors have assumed a priori that *Tamarix* would negatively affect birds (DeLoach et al. 2000; Dudley and DeLoach 2004), the evidence to date suggests a mixed effect that varies by species and geographic region (Sogge et al. 2008; van Riper et al. 2008). Research focused on bird abundance and breeding behavior in southwestern riparian systems supports the idea that *Tamarix* can be suitable for generalist species; but *Tamarix* can be unsuitable for some specialist species if key habitat requirements are missing.

Many studies have documented that *Tamarix* can provide habitat for breeding-bird communities in some parts of the Southwest (Brown et al. 1987; Hunter et al. 1988; Fleishman et al. 2003; Holmes et al. 2005; Sogge et al. 2005; Hinojosa-Huerta 2006). Corman and Wise-Gervais (2005) found that 76% of low- to midelevation breeding riparian bird species nested in *Tamarix*, and Sogge et al. (2008) documented 49 species throughout the western United States for which there are records of nesting in *Tamarix*. On the other hand, some studies have shown that in some areas bird abundance and diversity can be lower in *Tamarix* than in nearby native-dominated riparian vegetation. On the lower Colorado River in Arizona and Mexico, avifaunal diversity is lower in *Tamarix*-dominated areas compared with native-plant-dominated areas, and some riparian bird species are absent (Hunter et al. 1988; Hinojosa-Huerta et al. 2004; Hinojosa-Huerta 2006). Similarly, along the upper San Pedro River in Arizona, highest avian diversity is associated with native woodlands (Brand et al. 2010).

The primary avian users of riparian woodlands are the passerines and other land birds (e.g., cuckoos, doves, and hummingbirds). Certain land birds, such as doves, may use *Tamarix* for nesting substrate, even those that primarily forage away from riparian woodlands (Cunningham et al. 1997). Herons, egrets, and other water birds (e.g., green herons and great-crested cormorants) nest in riparian trees, occasionally including *Tamarix* (Corman and Wise-Gervais 2005; Sogge et al. 2008). *Tamarix* use is most common among riparian generalists (i.e., birds that breed in a variety of different native riparian habitat types), but *Tamarix* is unlikely to be suitable habitat for all native riparian birds. Some that have very specific habitat requirements—such as woodpeckers, secondary cavity nesters, or raptors that require large branches to support their nests—may be less numerous or absent in

Tamarix stands owing to a lack of essential habitat components (Anderson et al. 1977; Hunter et al. 1988; Ellis 1995; Walker 2006).

The value of *Tamarix* as habitat for birds varies regionally and from site to site, and it may be poor habitat for birds with specific habitat needs; however, *Tamarix* appears to be suitable for a majority of generalist avian species (see chapter 11, this volume). For example, 76% of Arizona low- and mid-elevation riparian breeding bird species nest in *Tamarix* (Corman and Wise-Gervais 2005). A small increase in native vegetation in tamarisk stands can greatly increase avian diversity and abundance, even over native habitats (van Riper et al. 2008), suggesting diversity of structure and prey base may be key to understanding the value of *Tamarix* to birds. More research is needed to understand what ecological services riparian habitats are providing, and what factors drive variation among sites and even patches within sites.

MAMMALS

Studies of the use of *Tamarix* habitats by mammals are relatively scarce, particularly in the case of medium- and large-sized mammals for which knowledge of *Tamarix* habitat use is primarily anecdotal. In a large-scale survey of middle Rio Grande riparian environments in New Mexico, Hink and Ohmart (1984) recorded 18 species of "large" mammals, including larger rodent species. Only blacktail jackrabbits (*Lepus californicus*) were noted as occurring commonly in *Tamarix*; of the remaining 17 species recorded, only a single individual of one species, a porcupine (*Erethizon dorsatum*), was noted in *Tamarix* as opposed to the other riparian vegetation associations that were surveyed. We have made casual observations of other large mammal species in *Tamarix*-dominated habitats, including striped skunks (*Mephitus mephitus*), javelina (*Pecari tajacu*), mule deer (*Odocoileus hemionus*), black bears (*Ursus americanus*), cougars (*Puma concolor*), gray foxes (*Urocyon cinereoargenteus*), and coyotes (*Canis latrans*; E. H. Paxton, W. S. Longland, pers. obs.). However, it is likely that at least some of these were simply traveling through *Tamarix*.

Beavers (*Castor canadensis*) can occur in river stretches where *Tamarix* dominates the bank vegetation (W. S. Longland, pers. obs.). Although beavers have been noted to include *Tamarix* in their diets, they exhibit strong dietary preferences for other tree species, particularly cottonwood and willow (Kimball and Perry 2008). The selective removal of native trees resulting from beaver foraging may accelerate or exacerbate *Tamarix* invasion of native riparian stands through competitive release (Lesica and Miles 2004; Longcore et al. 2007; Mortenson et al. 2008).

Other than very limited feeding by beavers, we are unaware of any records of medium or large mammals native to North America feeding on *Tamarix*. Although we have observed mule deer in *Tamarix* stands—perhaps because of the cover they provide (if *Tamarix* density is not excessive) when passing through to open water in riparian environments—*Tamarix* is probably nutritionally inferior to native plants. In a detailed study of forage use by mule deer in the Colorado River region of southeastern California, *Tamarix* was not included among 34 taxa of plants consumed by deer, despite the fact that the plant occurred commonly along riparian corridors in the study area (Marshal et al. 2004). Domestic goats can be conditioned to consume *Tamarix* (Richards and Whitesides 2006), and various *Tamarix* spp. have been reported in the diets of domestic cattle (Badri and Hamed 2000). However, the only published report of North American mammals consuming introduced *Tamarix* documents a small mammal species, the valley pocket gopher (*Thomomys bottae*), which caused fatal or severe root damage to more than 12% of *Tamarix* plants at an eastern California site (Manning et al. 1996). We have observed *Tamarix* branches with gnaw marks associated with desert woodrat (*Neotoma lepida*) middens (H. L. Bateman, W. S. Longland, pers. obs.), but woodrats also incorporate many items into the structure of middens that are not necessarily included in their diets.

Two studies in the Southwest employed ultrasonic detectors to document use of *Tamarix* habitats by bats. Chung-MacCoubrey and Bateman (2006) recorded bats foraging along the middle Rio Grande above riparian forest consisting of mixed stands of Rio Grande cottonwood (*P. deltoides wislizenii*), *Tamarix*, and Russian olive (*Elaeagnus angustifolia*). In an Arizona study along the Gila and San Pedro Rivers, Buecher and Sidner (2006) also documented bats foraging above *Tamarix*, although their activity levels were significantly greater in native cottonwood forest.

On the middle Rio Grande, both Ellis et al. (1997) and Hink and Ohmart (1984) found that species richness of small mammals was greater in *Tamarix* than in native cottonwoods due to the presence of species adapted to dry, sparse shrub or grass habitats, such as desert-dwelling rodent species in the family Heteromyidae. Hink and Ohmart (1984) recorded 4 to 12 small mammal species along the *Tamarix*-dominated river stretches they sampled, but they also found lower overall numbers of small mammals in *Tamarix* habitats compared to native cottonwood and willow habitats. Abundance, reproductive activity, and sex ratios of the most common rodent species (*Peromyscus leucopus*) were similar in native habitats and *Tamarix* habitats in the Ellis et al. (1997) study, although three native rodent species (western harvest mouse, *Reithrodontomys megalotis*; deer mouse, *P. maniculatus* [see figure 10.1a]; and silky pocket mouse, *Perognathus flavus*) were found either exclusively or nearly so in *Tamarix*. The white-throated woodrat (*N. albigula*) was the only species that consistently showed a strong affinity for native habitat and avoidance of *Tamarix* in studies on both the middle Rio Grande (Ellis et al. 1997) and the lower Colorado River (Anderson and Ohmart 1984; Andersen and Nelson 1999). However, a few additional native rodent species that were captured in low frequencies, such as northern grasshopper mice (*Onychomys leucogaster*; Ellis et al. 1997) and desert woodrats (Anderson and Ohmart 1984) were captured exclusively in native habitats. An insectivore, Crawford's gray shrew (*Notiosorex crawfordi*), was captured frequently in willow stands, but was seldom found in *Tamarix* by Hink and Ohmart (1984), and Chung-MacCoubrey et al. (2009) captured more

FIGURE 10.1 Examples of small mammals typically associated with *Tamarix* habitats include both generalist species such as (a) deer mouse (*Peromyscus maniculatus*), which also occur frequently in native riparian habitats; (b) heteromyid rodent species such as Merriam's kangaroo rat (*Dipodomys merriami*); and (c) Ord's kangaroo rat (*Dipodomys ordii*), which tend to favor *Tamarix* over native riparian habitats.

Source: Photographs courtesy of William S. Longland.

than 2,000 shrews in mixed cottonwood, willow, and *Tamarix* habitats along the middle Rio Grande.

Two ongoing live-trapping assessments of small mammal use of *Tamarix* habitats are expanding the geographic range of such studies to the northern Mojave and Great Basin Deserts. Preliminary results from a study on the Virgin River documented similar small mammal abundance in both monotypic *Tamarix* and mixed habitats (i.e., stands containing more than 50% *Tamarix* in addition to native Fremont cottonwood, *P. fremontii*; Gooding's willow, *S. gooddingii*; and mesquite, *Prosopis* spp.), largely driven by deer mice, which are ubiquitous in both habitats and represent more than 50% of total small mammal abundance (Bateman and Ostoja 2012). However, in the same study, diversity was greater in mixed sites, and no species of small mammals occurred in greater numbers in *Tamarix* stands. In the Great Basin study, small mammal use of paired native riparian and *Tamarix* sites was monitored for up to 11 consecutive years at five locations in western Nevada and eastern California (Longland 2012). Although species richness was generally similar between habitats, the overall number of trap captures combined among species was greater in *Tamarix* than in native habitats, in direct contrast to the results of Hink and Ohmart (1984) along the middle Rio Grande. Heteromyid rodent species (see figure 10.1b, c) occurred more commonly in *Tamarix* habitats at the Great Basin sites, as has been found in previous studies (Hink and Ohmart 1984; Ellis et al. 1997), while montane voles (*Microtus montanus*) and western harvest mice (*Reithrodontomys megalotis*) occurred in significantly greater frequencies in native habitat (Longland 2012). Both of the ongoing studies in the Mojave and Great Basin Deserts have found that desert woodrats (*N. lepida*) appear to be negatively impacted by *Tamarix* invasion; they were captured more frequently in native habitats at two of the three paired-habitat sites where they occurred in the Great Basin study, and were captured nearly exclusively in native riparian habitat in the Virgin River Valley (Bateman and Ostoja 2012), as found by Anderson and Ohmart (1984).

Current literature on *Tamarix* habitat use indicates that small mammal abundance and diversity do not differ consistently in *Tamarix*-invaded habitat relative to native riparian vegetation. In some respects, the literature implies that native riparian vegetation may offer superior habitat for various mammalian taxa. *Tamarix* appears to provide a poor and generally unused food resource for herbivorous mammals occurring in riparian habitats. Bats forage above *Tamarix* habitats, but forage more under native cottonwood canopy. A few species of small mammals appear to occur more frequently or even nearly exclusively in native riparian forests and not in *Tamarix*. Certainly, there are also mammal species that benefit from the habitat modifications associated with *Tamarix* invasion. The most consistent finding of the various studies that have compared small mammal communities between *Tamarix* and native habitats is that species occurring more often in *Tamarix* are generally desert-adapted taxa that benefit from habitat desertification that often accompanies invasion, such as heteromyid rodents. Other common rodents in *Tamarix* habitats may be generalist species, such as deer mice, which can be common in a variety of habitats (O'Farrell 1980). When its density is not excessive, *Tamarix* may provide a suitable proxy for native vegetation for mammalian fauna that occur in desert environments typified by sparse shrub cover.

HERPETOFAUNA

Amphibians and reptiles may be common wildlife, but they are often overlooked in surveys of riparian systems. General riparian occurrence of herpetofauna information is described in field guides (i.e., Degenhardt et al. 1996; Brennan and Holycross 2006) and for site-specific locations (e.g., Rosen 2005); however, beyond research along the middle Rio Grande and Virgin River systems little information is available. Because reptiles are strongly dependent on habitat structure for their survival (Pianka 1966), and because most amphibian species use wetland or temporary water sources during part of their life cycle, herpetofauna can be excellent model organisms to examine wildlife responses to disturbances such as nonnative plant invasion (Valentine et al. 2007).

Research along the middle Rio Grande in New Mexico has documented several species of amphibians, lizards, and snakes occurring in mixed stands with native Rio Grande cottonwood overstory and nonnative *Tamarix* and Russian olive dominant in the understory. These studies found 11 species of lizards (Bateman et al. 2008a), 9 species of amphibians (Bateman et al. 2008c), and 13 species of snakes (Bateman et al. 2009) in mixed sites. The Rio Grande studies did not consider monotypic stands of *Tamarix*.

Preliminary results from a live-trapping study in the Virgin River Valley on the Arizona-Nevada border found differences in lizard communities between monotypic stands of *Tamarix* and mixed stands (i.e., stands of Fremont cottonwood, Gooding's willow, and mesquite; Bateman and Ostoja 2012). *Tamarix* and mixed stands had similar measures of diversity, but total lizard abundance was greater in mixed sites. Tiger whiptails (*Aspidoscelis tigris*; see figure 10.2a) represented more than 65% of total lizard captures and were more numerous in mixed stands. Tiger whiptails are a generalist species and can occupy a variety of habitats. In the same study, desert spiny lizards (*Sceloporus magister*, see figure 10.2b) occurred almost exclusively in mixed stands, and habitat models predicted their occurrences in habitats with cottonwood and willow trees and woody debris ground cover. Spiny lizards are often observed on logs and overhead on large diameter cottonwoods (H. Bateman pers. obs.). The somewhat arboreal nature of these lizards may preclude them from *Tamarix* stands dominated by small-diameter stems. No species of lizard had greater abundances in *Tamarix* habitats.

As part of the same work along the Virgin River, Nielsen (2012) found that common side-blotched lizards (*Uta stansburiana*; see figure 10.2c) had similar abundances in monotypic and mixed stands, but lizards preferred microhabitats with more open understories and avoided dense, shady habitats. Juvenile common side-blotched lizards had lower recapture rates and body condition indices (snout-to-vent length/body mass) in sites with high canopy cover (> 75% cover). Interestingly, compared to mixed stands, monotypic *Tamarix* stands had among the highest canopy cover values.

In contrast to studies focused on birds and mammals, there are few comparisons of reptile and amphibian communities in *Tamarix* and native or mixed

FIGURE 10.2 Examples of lizard species found in *Tamarix* and native riparian habitats: (a) Tiger whiptails (*Aspidoscelis tigris*) are habitat generalists found in both *Tamarix* and native riparian habitats; (b) desert spiny lizards (*Sceloporus magister*) are semiarboreal species tied to native riparian habitats with large woody debris; and (c) common side-blotched lizards (*Uta stansburiana*) are short-lived species that prefer habitats with open understories.

Source: Photographs courtesy of Heather L. Bateman.

riparian habitats. There is evidence that *Tamarix* can support high numbers of habitat generalist species. However, species that are dependent upon large woody debris or open understories may avoid dense *Tamarix* stands because of their density and architecture (Bateman and Ostoja 2012).

The Impact of *Tamarix* Control on Wildlife

Tamarix control programs use a variety of control techniques, including chemical (Duncan and McDaniel 1998); mechanical, including burning; and biological (see figure 10.3; see also chapter 20, this volume). These control measures can alter riparian areas and can also affect a variety of habitat types used by wildlife (Bateman et al. 2008b; Bateman and Paxton 2010). Few studies have evaluated the quality of habitat that *Tamarix* provides to birds, mammals, and herpetofauna, which certainly varies regionally and among species. Therefore, general conclusions regarding (both positive and negative) effects of *Tamarix* control are difficult to make. The degree to which wildlife use exotic habitats such as *Tamarix*, and the availability of suitable native habitat, should be a consideration when control measures are planned and executed (Paxton et al. 2011).

BIRDS

Studies indicate that *Tamarix* seldom supports the same avian-species richness, guilds, and population sizes as native habitat; however, *Tamarix* can fulfill an

FIGURE 10.3 Defoliated *Tamarix* along the Virgin River in Arizona and Nevada with native (not defoliated) mesquite (*Prosopis* spp.).

Source: Photograph courtesy of Heather L. Bateman.

important habitat role for some species (U.S. Fish and Wildlife Service 2002; Walker 2006; Walker 2008; Paxton et al. 2011), especially in areas where degraded riparian systems preclude the reestablishment of native vegetation (Shafroth et al. 2005). If a *Tamarix*-dominated area that currently supports riparian breeding birds is replaced by nonriparian vegetation, or by a much smaller amount of native riparian habitat, there may be a net loss of riparian habitat value (Shafroth et al. 2005; Paxton et al. 2011) and possible local/regional loss of some or all riparian birds due to changes in the vegetation structure (Fleishman et al. 2003; Walker 2006). For example, restoration efforts that involved clearing exotic vegetation under cottonwood gallery forests in New Mexico decreased lower- and midstory avian species, presumably due to the loss of tall vegetation (Bateman et al. 2008a). Yellow-billed cuckoos have all but disappeared in the lower Pecos River valley from Six-Mile Dam near Carlsbad, New Mexico, to the Texas border, following a large-scale *Tamarix* removal project from 1999 through 2006 (Hart et al. 2003; Travis 2005). The southwestern willow flycatcher (*Empidonax traillii extimus*) recovery plan (U.S. Fish and Wildlife Service 2002) expressed concerns about large-scale *Tamarix* control or removal at occupied flycatcher sites because flycatcher breeding sites require very dense vegetation (see chapter 11, this volume).

Whether particular avian species would be negatively affected by *Tamarix* eradication efforts depends in large part on the value of the particular *Tamarix* stands as habitat and the extent and pace of both *Tamarix* loss and the development of replacement habitat. Geographic factors (e.g., climate and elevation), stand characteristics, and the type and structure of adjacent and interspersed habitats are key factors in determining the habitat value of *Tamarix* (Hunter et al. 1988; Livingston and Schemnitz 1996; Walker 2008). Likewise, the return of native riparian woodlands following *Tamarix* control is far from certain (Harms and Hiebert 2006), and the degree to which recovery occurs is influenced by many physical, ecological, and technical factors (Shafroth et al. 2008). Therefore, careful restoration planning, execution, and follow-up are needed to ensure that *Tamarix* is replaced by native vegetation, and not by species that have even lower habitat value or greater negative effects, such as other exotic vegetation (D'Antonio and Meyersen 2002; Harms and Hiebert 2006; Shafroth et al. 2008).

Two species of conservation concern that use *Tamarix* in some places highlight the challenges to its eradication. The southwestern willow flycatcher is a federally endangered species, having declined markedly over the last 100 to 200 years from loss of riparian breeding habitat (U.S. Fish and Wildlife Service 2002). Although nearly half (43%) of southwestern willow flycatcher territories are found in riparian patches consisting primarily (greater than 90%) of native trees such as willow, 6% of known breeding territories are in monotypic (greater than 90%) *Tamarix*; 22% are in habitats dominated by *Tamarix* (50%–100%), and another 28% are in native habitats where *Tamarix* and other exotics provide 10% to 50% of the habitat structure (Durst et al. 2007). Flycatchers likely select their breeding sites based more on the structural characteristics of vegetation than on species compostion (U.S. Fish and Wildlife Service 2002). Because the flycatcher breeds in both native and exotic habitat types, often in the same drainage, it is possible to evaluate whether flycatchers breeding in *Tamarix* habitats are affected negatively by a poor food base, reduced survivorship, and low productivity, or whether *Tamarix* is functionally of a similar quality to native habitat. Recent research on flycatchers breeding in *Tamarix* has found no evidence of a limited diet (DeLay et al. 1999; Drost et al. 2001; Durst 2004), and Owen et al. (2005) concluded that the physiological condition of birds breeding in *Tamarix* did not differ from that of birds nesting in native habitats. Similarly, no evidence was found of reduced survivorship or productivity among flycatchers breeding in *Tamarix* habitats compared to those breeding in native vegetation in central Arizona (Sogge et al. 2006; Paxton et al. 2007). Thus, *Tamarix* appears to provide habitat quality similar to that provided by native vegetation for flycatchers in at least some locations and is considered an important habitat for recovery of this species (U.S. Fish and Wildlife Service 2002).

However, much of the *Tamarix* along riparian systems is not used by flycatchers and is presumably unsuitable; for example, flycatchers are absent today from some areas where they historically bred and where *Tamarix* is now dominant and widespread (e.g., the lower Colorado River near Yuma, Arizona; Unitt 1987; U.S. Fish and Wildlife Service 2002). Furthermore, fire is considered one of the greatest threats to flycatcher breeding sites (U.S. Fish and Wildlife Service 2002), and the presence of *Tamarix* may increase the likelihood of large fires because of its flammability (see chapter 14, this volume). Additional research is needed to evaluate whether *Tamarix* in these unoccupied areas fails to provide the necessary ecological functions and environmental conditions for flycatchers, or whether southwestern willow flycatchers simply do not have the population numbers necessary to occupy all suitable habitats present in the Southwest.

The yellow-billed cuckoo (*Coccyzus americanus*) has been extirpated from much of its western range; currently the western population is a candidate for Federal Endangered Species listing (U.S. Fish and Wildlife Service 2001). Cuckoos generally prefer mature riparian habitats and are most commonly associated with Fremont cottonwood or other native forests (Hughes 1999). However, yellow-billed cuckoos breed extensively in the dense *Tamarix* stands along parts of the Pecos River in New Mexico (Hunter et al. 1988; Livingston and Schemnitz 1996). Although cuckoos in this region are not considered to be part of the western population, Howe (1986) described how a large cuckoo breeding population developed along the Pecos River by the mid-1980s concurrent with the establishment of large stands of *Tamarix* that created new riparian woodlands. Although there are no specific studies on the relative breeding success of cuckoos in *Tamarix*, the notable population expansion along the Pecos River (Howe 1986) suggests that successful breeding did occur (Livingston and Schemnitz 1996). The frequency of *Tamarix* use varies geographically in cuckoos, with *Tamarix* use common on the Pecos River, more limited on the Rio Grande (and usually associated with a native component), and absent on the Gila River (Howe 1986; Hunter et al. 1988; Woodward et al. 2003) within New Mexico. Outside New Mexico, cuckoos have not been found breeding in *Tamarix*-dominated habitats (Johnson et al. 2006; Johnson et al. 2007), though *Tamarix* can be a component of the habitat patch, as in mixed stands. This suggests that the suitability of *Tamarix* as breeding habitat for cuckoos, as with other bird species, varies across the landscape, with local environmental factors determining its relative habitat value.

Overall, there are likely to be winners and losers with *Tamarix* control, and management of bird communities across the landscape should take into consideration the changing distribution of riparian habitat, both native and exotic.

MAMMALS

Other than studies addressing how riparian habitat conversion to *Tamarix* affects mammals, limited data document *Tamarix* control and riparian restoration effects on mammals. At a Colorado River site in Arizona that had undergone mechanical *Tamarix* control and revegetation with native woody vegetation five years earlier, Andersen (1994) found nine small mammal species present out of 15 potentially occurring in the local species pool of small mammals. Although the associations of native plant species in the revegetated area did not resemble natural riparian habitats, the area supported stable or increasing populations of most of the common rodent species present. On the Pecos River in southeastern New Mexico, Andersen et al. (2000) trapped small mammals in *Tamarix* habitats that had been treated with herbicides and in untreated control plots. Herbicide treatment increased capture rates overall, although these higher rates may simply have been due to greater numbers of recaptured unmarked individuals on treated plots.

Small mammal monitoring in the Great Basin occurred at three sites where the *Tamarix* has been defoliated using the biological control agent *Diorhabda carinulata* (Longland 2012). Most species have not shown any trends in abundance that might be attributable to increasing *Tamarix* defoliation, but capture frequencies of western harvest mice (*R. megalotus*) have increased significantly over the ten years, since the release of *D. carinulata* at the site where the beetles were established first and where they have caused the greatest death of *Tamarix*. The trend, however, has not occurred at other release sites.

Three studies have focused on the effects of *Tamarix* control on mammals other than rodents. Chung-MacCoubrey et al. (2009) captured numerous Crawford's grey shrews along the middle Rio Grande and found that the removal of nonnative plants, including *Tamarix*, did not affect shrew capture rates. Chung-MacCoubrey and Bateman (2006) documented increased bat activity at sites where exotic *Tamarix* and Russian olive had been removed, relative to nonremoval sites. It is possible that clearing vegetation altered the structure of the understory, permitting

FIGURE 10.4 Juvenile pocket mouse in pitfall trap with biocontrol beetle (*Diorhabda carinulata*) larvae. The *Diorhabda* provides its prey for such omnivorous rodents.
Source: Photo courtesy of Heather L. Bateman.

bats of various foraging styles and maneuverabilities to access additional foraging areas.

In a third experimental study along the San Pedro River, Hagen (2010) compared the abundance of aquatic insects, terrestrial insects, and bat activity along stream reaches with a mixed stand (Freemont cottonwood, Gooding's willow, and *Tamarix*) to reaches with *Tamarix* stands having less structural complexity (or clutter) and no cottonwood or willow. Aquatic insect abundance and bat activity were higher along the mixed vegetation reach compared to the *Tamarix* reach.

We need further studies of how *Tamarix* control affects mammal populations (see figure 10.4). For example, small mammals (particularly deer mice) will consume *D. carinulata* (W. Longland pers. obs.); however, the benefits of this food resource to rodent populations have not been investigated. Future study should include a comparison of the effects of biological control relative to alternate means, such as mechanical control.

HERPETOFAUNA

Methods used to control *Tamarix* can alter the structural and thermal environment of a habitat and may affect some reptiles. For example, a study along the middle Rio Grande in New Mexico found that treatments to remove *Tamarix*, Russian olive, and woody fuels appeared beneficial or at least nondamaging to species of lizards (Bateman et al. 2008b). Compared to nonremoval sites, plateau fence lizards (*Sceloporus tristichus*, formerly *S. undulatus*) and New Mexico whiptails (*A. neomexicana*) increased in abundance after *Tamarix* removal. During the period of study, no negative effects were detected for any species of lizard or amphibian. Chihuahuan spotted whiptails (*Aspidoscelis exsanguis*), desert grassland whiptails (*A. uniparens*), and common side-blotched lizards were either positively associated with habitat in removal sites or negatively associated with habitat in nonremoval sites. The open understory found in removal sites may have provided more basking opportunities for reptiles by allowing solar radiation to reach ground level (Bateman et al. 2008b). Also along the middle Rio Grande, no negative effects were detected for toads (*Anaxyrus woodhousii* and *A. cognatus*, formerly *Bufo* spp.), which instead responded to hydrologic variables such as spring flooding and summer precipitation (Bateman et al. 2008c).

The use of insect herbivores as biocontrol agents for *Tamarix* may affect insectivorous herpetofauna. Several species of common insectivorous lizards (*Aspidoscelis, Sceloporus, Uta*, and *Urosaurus* spp.) have been shown to consume *D. carinulata* (Bateman 2010). In the short term, *Tamarix* defoliation by the beetle may raise ambient temperatures, benefiting generalist ectothermic wildlife, and insectivorous lizards may capitalize on beetle food resources (see figure 10.5). However, the long-term trend for wildlife populations in habitats affected by biocontrol or other control measures remains uncertain.

Data Gaps and Future Focus

Given the vast extent of *Tamarix* in riparian habitats across North America (Friedman et al. 2005), it is important to understand as fully as possible the benefits and costs (both financial and ecological) of *Tamarix* control and riparian restoration to wildlife (Hultine et al. 2009). Alterations to riparian areas resulting from nonnative plant control can change habitat quality, such as the surface and thermal environment for reptiles, structural breeding habitat for birds (Paxton et al. 2011),

FIGURE 10.5 *Diorhabda* sp. elytra (wings) in lizard scat, evidence that confirms the biocontrol beetle acts as prey.

Source: Photograph courtesy of Heather L. Bateman.

cover for terrestrial mammals and herpetofauna, and aerial foraging habitat for bats (Bateman et al. 2008a).

In terms of avifauna, habitats provide food, shelter from predators and weather, and structure for nesting, and *Tamarix* appears to provide these services for many species in at least some areas. However, more research is needed to match up which ecological services *Tamarix* provides to which species, and to learn how and why the value of such services varies across the landscape. Such studies include comparisons of productivity, physiological stress, site fidelity, and survivorship. Additionally, the response of bird populations to *Tamarix* control and to the inevitable lag time between control efforts and eventual replacement of vegetation (whether native vegetation or exotic) remains largely unstudied (Paxton et al. 2011).

Clearly, there is need for further research on the effects of *Tamarix* invasion and *Tamarix* control on the use of riparian habitats by mammals. There are few comparative studies of mammals in *Tamarix*-invaded and native riparian habitats, and those that exist focus exclusively on bats or small terrestrial mammals, mostly rodents. Investigations of large mammals in *Tamarix* habitats are lacking; radiotelemetry studies, for example, detailing use of *Tamarix* versus native riparian vegetation could be illuminating in situations where both of these vegetation types are available. Little is known about effects of *Tamarix* control on mammals, suggesting a need for increased monitoring of mammal populations in areas undergoing *Tamarix* control. Because effects of *Tamarix* invasion on small mammals are often particular to species and site, such monitoring should be conducted throughout the range of *Tamarix* invasion in the western United States.

Amphibians and reptiles are common but often-overlooked inhabitants of riparian areas. Since reptiles respond to structural changes to their habitat (Pianka 1967), their presence and abundance can be good indicators of healthy riparian ecosystem structure and function. Most of the few studies that have focused on herpetofauna relate the abundance of common reptiles, such as diurnal lizards, to the habitat characteristics of exotic and native habitats. Much less is known of aquatic or semiaquatic amphibians and reptiles, such as turtles. For example, Western pond turtles (*Clemmys marmorata*) occur in habitats where *Tamarix* has invaded, but there exist no comparisons of their occurrences in native habitats (Lovich and Meyer 2002).

Unfortunately, we have a limited understanding of how most species of wildlife are affected by the proliferation of *Tamarix*, which hinders management. From studies focused mostly on passerine birds (particularly threatened and endangered species), rodents, and lizards, we have seen a trend for common, generalist species to be the most successful in *Tamarix* habitats. However, much less is understood about specialist or uncommon species in these exotic habitats, particularly if species are not monitored because of their endangered status.

We need research that focuses on multiple taxa, including uncommon species. These should be studies with controls and replications, tracking the response over ecologically meaningful periods, and experiments to help distinguish between changes in plant species composition from changes in vegetation structure. We encourage experimental projects comparing *Tamarix*-invaded habitats to native habitats and *Tamarix* removal sites to both native and nonremoval sites. In addition, monitoring sites after control treatments will be important to understanding the short- to long-term effects of control efforts on wildlife. Moreover, investigating effects of *Tamarix* invasion and control beyond simple monitoring of abundances would be illuminating. For example, detailed studies of effects of *Tamarix* invasion and control on reproductive parameters, such as those that have focused on bird species of conservation concern (e.g., Sogge et al. 2006), would add greatly to our current understanding of the quality of *Tamarix* as wildlife habitat.

Literature Cited

Anderson, B. W., A. Higgins, and R. D. Ohmart. 1977. Avian use of *Tamarix* communities in the Lower Colorado River valley, *in* Johnson, and D. A. Jones, technical coordinators. Importance, preservation, and management of riparian habitat: A symposium. General Technical Report RM-43. USDA Forest Service, Fort Collins, CO.

Anderson, B. W., and R. D. Ohmart. 1984. A vegetation management study for wildlife enhancement along the lower Colorado River. U.S. Bureau of Reclamation, Lower Colorado Region. Boulder City, NV.

Andersen, D. C. 1994. Demographics of small mammals using anthropogenic desert riparian habitat in Arizona. Journal of Wildlife Management 58:445–454.

Andersen, D. C., and S. M. Nelson. 1999. Rodent use of anthropogenic and "natural" desert riparian habitat, lower Colorado River, Arizona. Regulated Rivers: Research and Management 15:377–393.

Andersen, M. C., J. Hollenbeck, K. Kintigh, T. Barlow, R. Konkle, M. Livingston, and S. Schemnitz. 2000. Effects on wildlife of saltcedar (*Tamarix chinensis*) control by aerial application of herbicides along the Pecos River, southeastern New Mexico. Project summary for New Mexico Department of Game and Fish and Pecos River Native Restoration Organization. Department of Fisheries and Wildlife Sciences, New Mexico State University. Las Cruces.

Badri, M. A., and A. I. Hamed. 2000. Nutrient value of plants in an extremely arid environment (Wadi Allaqi Biosphere Reserve, Egypt). Journal of Arid Environments 44:347–356.

Bateman, H. L. 2010. Saltcedar, beetles, and lizards: Evaluating the impacts of biocontrol on herpetofauna. 95th Annual Meeting of the Ecological Society of America. 1–6 August. Pittsburgh, PA.

Bateman, H. L., A. Chung-MacCoubrey, D. M. Finch, H. L. Snell, and D. L. Hawksworth. 2008a. Impacts of native plant removal on vertebrates along the Middle Rio Grande (New Mexico): Ecological Restoration 26:193–195.

Bateman, H. L., A. Chung-MacCoubrey, and H. L. Snell. 2008b. Impact of non-native plant removal on lizards in riparian habitats in the southwestern United States: Restoration Ecology 16:180–190.

Bateman, H. L., A. Chung-MacCoubrey, H. L. Snell, and D. M. Finch. 2009. Abundance and species richness of snakes along the Middle Rio Grande riparian forest in New Mexico. Herpetological Conservation and Biology 4:1–8.

Bateman, H. L., and E. H. Paxton. 2010. Saltcedar and Russian olive interactions with wildlife. Pages 49–63 *in* P.B. Shafroth, C.A. Brown, and D.M. Merritt, editors. Saltcedar and Russian olive control demonstration act science assessment. U.S. Geological Survey Scientific Investigations Report 2009–5247.

Bateman, H. L., M. J. Harner, and A. Chung-MacCoubrey. 2008c. Abundance and reproduction of toads (*Bufo*) along a regulated river in the southwestern United States: Importance of flooding in riparian ecosystems. Journal of Arid Environments 72:1613–1619.

Bateman, H. L., and S. M. Ostoja. 2012. Invasive woody plants affect the composition of native lizard and small mammal communities in riparian woodlands. Animal Conservation 15:294–304.

Bateman, H L., T. L. Dudley, D. W. Bean, S. M. Ostoja, K. R. Hultine, and M. J. Kuehn. 2010. A river system to watch: Documenting the effects of saltcedar (*Tamarix* spp.) biocontrol in the Virgin River Valley. Ecological Restoration 28:405–410.

Brand, L. A., J. C. Stromberg, and B. R. Noon. 2010. Avian density and nest survival on the San Pedro River: Importance of vegetation type and hydrological regime. Journal of Wildlife Management 74:739–754.

Brennan, T. C., and A. T. Holycross. 2006. Amphibians and reptiles in Arizona. Arizona Game and Fish Department. Phoenix, AZ.

Brown, B. T., S. W. Carothers, and R. R. Johnson. 1987. *Grand Canyon Birds.* University of Arizona Press, Tucson.

Buecher, D. C., and R. Sidner. 2006. A comparison of bat-use between native cottonwood galleries and non-native *Tamarix* groves near Winkelman, Arizona. Bat Research News 47:91–92.

Chung-MacCoubrey, A., and H. L. Bateman. 2006. Bosque restoration effects on bats and herpetofauna. Chapter 4 *in* D. M. Finch, A. Chung-MacCoubrey, R. Jemison, D. Merritt, B. Johnson, and M. Campana, editors. Effects of fuel reduction and exotic plant removal on vertebrates, vegetation and water resources in the Middle Rio Grande, New Mexico. Final report for Joint Fire Sciences Program. USDA Forest Service Rocky Mountain Research Station. Fort Collins, CO.

Chung-MacCoubrey, A., H. L. Bateman, and D. M. Finch. 2009. Captures of Crawford's gray shrews (*Notiosorex crawfordi*) along the Rio Grande in central New Mexico. Western North American Naturalist 69:260–262.

Corman, T., and C. Wise-Gervais, editors. 2005. *Arizona Breeding Bird Atlas.* University of New Mexico Press. Albuquerque.

Cunningham, S. C., R. W. Engel-Wilson, P. M. Smith, and W. B. Ballard. 1997. Food habits and nesting characteristics of sympatric mourning and white-winged doves in south-central Arizona. Journal of Wildlife Research 2:242–253.

D'Antonio, C. D., and L. A. Meyerson. 2002. Exotic plant species as problems and solutions in ecological restoration: A synthesis. Restoration Ecology 10:703–713.

Degenhardt, W. G., C. W. Painter, and A. H. Price. 1996. Amphibians and reptiles of New Mexico. The University of New Mexico Press. Albuquerque, NM.

DeLay, L., D. M. Finch, S. Brantley, R. Fagerlund, M. D. Mearns, and J. F. Kelly. 1999. Arthropods of native and exotic vegetation and their associations with Willow Flycatchers and Wilson's Warblers. Pages 216–221 *in* D. M. Finch, J. C. Whitney, J. F.

Kelly, and S. R. Loftin, technical coordinators. Rio Grande ecosystems: Linking land, water and people. Proceedings RMRS-P-7. Department of Agriculture Forest Service. Rocky Mountain Research Station. Ogden, UT.

DeLoach, C. J., R. I. Carruthers, J. E. Lovich, T. L. Dudley, and S. D. Smith. 2000. Ecological interactions in the biological control of *Tamarix* (*Tamarix* spp.) in the United States: Toward a new understanding. Pages 819–873 *in* N. R. Spencer, editor. Proceedings of the 10th international symposium on the biological control of weeds. Montana State University, Bozeman.

Doyle, A.T. 1990. Use of riparian and upland habitats by small mammals. Journal of Mammalogy 71:14–23.

Drost, C. A., E. H. Paxton, M. K. Sogge, and M. J. Whitfield. 2001. Food Habits of the endangered Southwestern Willow Flycatcher. USGS report to U.S. Bureau of Reclamation. Salt Lake City, UT.

Dudley, T. L., and C. J. DeLoach. 2004. *Tamarix* (*Tamarix* spp.), endangered species, and biological weed control: Can they mix? Weed Technology 18:1542–1551.

Duncan, K.W., and K.C. McDaniel. 1998. Saltcedar (*Tamarix* spp.) management with imazapyr. Weed Technology 12:337–344.

Durst, S. L. 2004. Southwestern Willow Flycatcher potential prey base and diet in native and exotic habitats. M.S. thesis. Northern Arizona University, Flagstaff.

Durst, S. L., M. K. Sogge, S. D. Shay, S. O. Williams, B. E. Kus, and S. J. Sferra. 2007. Southwestern Willow Flycatcher breeding site and territory summary—2006. Report to U.S. Bureau of Reclamation. U.S. Geological Survey. Southwest Biological Science Center. Flagstaff, AZ.

Ellis, L. M. 1995. Bird use of *Tamarix* and cottonwood vegetation in the middle Rio Grande Valley of New Mexico, USA. Journal of Arid Environments 30:339–349.

Ellis, L. M., C. S. Crawford, and M. C. Molles. 1997. Rodent communities in native and exotic riparian vegetation in the Middle Rio Grande Valley of central New Mexico. Southwestern Naturalist 42:13–19.

Falck, M. J., K. R. Wilson, and D. C. Andersen. 2003. Small mammals within riparian habitats of a regulated and unregulated aridland river. Western North American Naturalist 63:35–42.

Fleishman, E., N. McDonal, R. MacNally, D. D. Murphy, J. Walters, and T. Floyd. 2003. Effects of floristics, physiognomy and non-native vegetation on riparian bird communities in a Mojave Desert watershed. Journal of Animal Ecology 72:484–490.

Friedman, J. M., G. T. Auble, and P. B. Shafroth. 2005. Dominance of non-native riparian trees in western USA. Biological Invasions 7:747–751.

Hagen, E. M. 2010. Spatial and temporal patterns in insectivorous bat activity in river-riparian landscapes. PhD dissertation. Arizona State University, Tempe.

Harms, R. S., and R. D. Hiebert. 2006. Vegetation response following invasive tamarisk (*Tamarix* spp.) removal and implications for riparian restoration. Restoration Ecology 14:461–472.

Hart, C. R., L. D. White, A. McDonald, and Z. Sheng. 2003. Saltcedar control and water salvage on the Pecos River, Texas, 1999–2003. Journal of Environmental Management 75:399–409.

Hink, V. C., and R. D. Ohmart. 1984. Middle Rio Grande biological survey. Final report to the US Army Corps of Engineers No. DACW47-81-C-0015. Center for Environmental Studies, Arizona State University, Tempe.

Hinojosa-Huerta, O. M. 2006. Birds, water, and *Tamarix*: Strategies for riparian restoration in the Colorado River Delta. Doctoral dissertation, University of Arizona, Tucson.

Hinojosa-Huerta, O. M., H. Iturribarria-Rojas, Y. Carrillo-Guerrero, M. de la Garza-Trevino, and E. Zamora-Hernandez. 2004. Bird conservation plan for the Colorado River Delta. Pronatura Noroeste, Direccion de Conservacion Sonora San Luis Rio Colorado. Sonora, Mexico.

Holmes, J. J., J. Spence, and M. K. Sogge. 2005. Birds of the Colorado River in Grand Canyon: A synthesis of status, trends, and dam operation effects. Pages 123–138 *in* S. P. Gloss, J. E. Lovich, and T. E. Melis, editors. The state of the Colorado River ecosystem. USGS Circular 128 U.S. Geological Survey. Reston, VA.

Howe, W. H. 1986. Status of the Yellow-billed Cuckoo (*Coccyzus americanus*) in New Mexico: 1986. Final report, contract 516.6-75-09. New Mexico Department of Game and Fish. Sante Fe, NM.

Hughes, J. M. 1999. Yellow-billed Cuckoo (*Coccyzus americanus*). Pages 1–28 *in* A. Poole, and F. Gill, editors. *The Birds of North America*, no. 418. The Birds of North America, Inc. Philadelphia, PA.

Hultine, K. R., J. Belnap, C. van Riper, J. R. Ehleringer, P. E. Dennison, M. E. Lee, P. L. Nagler, K. A. Snyder, S. M. Uselman, and J.B. West. 2009. Tamarisk biocontrol in the western United States: Ecological and societal implic ations. Frontiers in Ecology and the Environment 8:467–474.

Hunter, W., R. Ohmart, and B. Anderson. 1988. Use of exotic *Tamarix* (*Tamarix chinensis*) by birds in arid riparian systems. Condor 90:113–123.

Johnson, M. J., J. A. Holmes, C. Calvo, I. Samuels, S. Krantz, and M. K. Sogge. 2007. Yellow-billed Cuckoo distribution, abundance, and habitat use along the Lower Colorado and tributaries: 2006 annual report. USGS Open File Report 2007–1097. U.S. Geological Survey. Reston, VA.

Johnson, M. J., J. A. Holmes, and R. Weber. 2006. Yellow-billed Cuckoo distribution and abundance, habitat use, and breeding ecology in select habitats of the Roosevelt Habitat Conservation Plan, 2003–2006. Final report submitted to the Salt River Project. Northern Arizona University, Colorado Plateau Research Station. Flagstaff, AZ.

Kimball, B. A., and K. R. Perry. 2008. Manipulating beaver (*Castor canadensis*) feeding responses to invasive tamarisk (*Tamarix* spp.). Journal of Chemical Ecology 8:1050–1056.

Knopf, F. L., R. R. Johnson, T. Rich, F. B. Samson, and R. C. Szaro. 1988. Conservation of riparian ecosystems in the United States. Wilson Bulletin 100:272–284.

Latterell, J. J., J. S. Bechtold, T. C. O'Keefe, R. Van Pelt, and R. J. Naiman. 2006. Dynamic patch mosaics and channel movement in an unconfined river valley of the Olympic Mountains. Freshwater Biology 51:523–544.

Lesica, P., and S. Miles. 2004. Beavers indirectly enhance the growth of *Elaeagnus* and tamarisk along eastern Montana rivers. Western North American Naturalist 64:93–100.

Lewis, P. A., C. J. DeLoach, A. E. Knutson, J. L. Tracy, and T. O. Robbins. 2003. Biology of *Diorhabda elongata deserticola* (Coleoptera: Chrysomelidae), an Asian leaf beetle for biological control of saltcedars (*Tamarix* spp.) in the United States. Biological Control 27:101–116.

Livingston, M. F., and S. D. Schemnitz. 1996. Summer bird/vegetation associations in tamarisk and native habitat along the Pecos River, southeastern New Mexico. Pages 171–180 *in* D. W. Shaw and D. M. Finch, technical coordinators, Desired future conditions for southwestern riparian ecosystems: Bringing interests and concerns together. USDA General Technical Report RM-GTR-272. Albuquerque, NM.

Longcore, T., C. Rich, and D. Muller-Schwarze. 2007. Management by assertion: Beavers and songbirds at Lake Skinner (Riverside County, California). Environmental Management 39: 460–471.

Longland, W. S. 2012. Small mammals in saltcedar (*Tamarix ramosissima*)-invaded and native riparian habitats of the western Great Basin. Invasive Plant Science and Management 5:230–237.

Lovich, J., and K. Meyer. 2002. The western pond turtle (*Clemmys marmorata*) in the Mojave River, California: Highly adapted survivor or tenuous relict? Journal of Zoological Society of London 256:537–545.

Manning, S. J., B. L. Cashore, and J. M. Szewezak. 1996 Pocket gophers damage *Tamarix* (*Tamarix ramosissima*) roots. The Great Basin Naturalist 56:183–185.

Marshal, J. P., V. C. Bleich, N. G. Andrew, and P. R. Krausman. 2004. Seasonal forage use by desert mule deer in southeastern California. Southwestern Naturalist 49:501–505.

Mortenson S. G., P. J. Weisberg, and B. E. Ralston. 2008. Do beavers promote the invasion of non-native *Tamarix* in the Grand Canyon riparian zone? Wetlands 3:666–675.

Nielsen, D. P., and H. L. Bateman. In press. Population metrics and habitat utilization of common side-blotched lizards (Uta stansburiana) in saltcedar *Tamarix* habitat. Southwestern Naturalist.

O'Farrell, M. J. 1980. Spatial relationships of rodents in a sagebrush community. Journal of Mammalogy 61:589–605.

Owen, J. C., M. K. Sogge, and M. D. Kern. 2005. Habitat and gender differences in the physiological condition of breeding Southwestern Willow Flycatchers. Auk 122:1261–1270.

Paxton, E. H., M. K. Sogge, S. L. Durst, T. C. Theimer, and J. R. Hatten. 2007. The Ecology of the Southwestern Willow Flycatcher in central Arizona: A 10-year Synthesis Report. USGS Open File Report 2007–1381.

Paxton, E. H., M. K. Sogge, and T. C. Theimer. 2011. Biocontrol of exotic tamarisk thorough introduced beetle defoliation: Potential demographic consequences for riparian passerine birds in the southwestern United States. Condor 113:255–265.

Pianka, E. R. 1966. Convexity, desert lizards, and spatial heterogeneity. Ecology 47:1055–1059.

Pianka, E. R. 1967. On lizard species diversity: North American flatland deserts. Ecology 48:332–351.

Richards, R., and R. E. Whitesides. 2006. Tamarisk control by grazing with goats compared to herbicides. Abstract. *Tamarix* Research Conference. Current Status and Future Directions, Fort Collins, CO.

Rosen, P. C. 2005. Lowland riparian herpetofaunas: The San Pedro River in Southeastern Arizona. USDA Forest Service Proceedings, RMRS-P-36.

Shafroth, P. B., J. R. Cleverly, T. L. Dudley, J. P. Taylor, C. van Riper, E. P. Weeks, and J. N. Stuart. 2005. Control of *Tamarix* spp. in the western U.S.: Implications for water salvage, wildlife use, and riparian restoration. Environmental Management 35:231–246.

Shafroth, P. B., V. B. Beauchamp, M. K. Briggs, K. Lair, M. L. Scott, and A. A. Sher. 2008. Planning Riparian Restoration in the Context of *Tamarix* Control in Western North America. Restoration Ecology 16:97–112.

Sogge, M. K., D. L. Felley, and M. Wotawa. 2005. A quantitative model of avian community and habitat relationships along the Colorado River in the Grand Canyon. Pages 161–192 *in* C. van Riper and D. Mattson, editors. *The Colorado Plateau II: Biophysical, Socioeconomic, and Cultural Research.* University of Arizona Press, Tuscon.

Sogge, M. K., E. H. Paxton, and A. A. Tudor. 2006. *Tamarix* and Southwestern Willow Flycatchers: Lessons from long-term studies in central Arizona. Pages 238–241 *in* C. Aguirre-Bravo, P. J. Pellicane, D. P. Burns, and S. Draggan, editors. Monitoring Science and Technology symposium: Unifying knowledge for sustainability in the western hemisphere. Proceedings RMRS-P-42CD. USDA Forest Service, Rocky Mountain Research Station. Fort Collins, CO.

Sogge, M. K., S. J. Sferra, and E. H. Paxton. 2008. *Tamarix* as habitat for birds: Implications for riparian restoration in the southwestern United States. Restoration Ecology 16:146–154.

Stamp, N. E., and R. D. Ohmart. 1979. Rodents of desert shrub and riparian woodland habitats in the Sonoran desert. Southwestern Naturalist 24:279–289.

Stanford, J. A., M. S. Lorang, and R. F. Hauer. 2005. The shifting habitat mosaic of river Ecosystems. Verhandlungen Internationale Vereinigung für Theoretische und Angewandte Limnologie 29:123–136.

Travis, J. R. 2005. Statewide Yellow-billed Cuckoo survey: 2004. New Mexico Department of Game and Fish. Sante Fe, NM.

U.S. Fish and Wildlife Service. 2001. 12-month finding for a petition to list the yellow-billed cuckoo (*Coccyzus americanus*) in the western continental United States. [Online]. http://ecos.fws.gov/docs/federal_register/fr3780.pdf. Accessed July 29, 2012.

U.S. Fish and Wildlife Service. 2002. Southwestern Willow Flycatcher (*Empidonax traillii extimus*) final recovery plan. U.S. Fish and Wildlife Service. Albuquerque, NM.

Unitt, P. 1987. *Empidonax traillii extimus*: An endangered subspecies. Western Birds 18:137–162.

Valentine, L. E., B. Roberts, and L. Schwarzkopf. 2007. Mechanisms driving avoidance of non-native plants by lizards. Journal of Applied Ecology 44:228–237.

van Riper, C., K. L. Paxton, C. O'Brien, P. B. Shafroth, and L. J. McGrath. 2008. Rethinking avian response to *Tamarix* on the lower Colorado River: A threshold hypothesis. Restoration Ecology 16:155–167.

Walker, H. A. 2008. Floristics and physiognomy determine migrant landbird response to tamarisk (*Tamarix ramosissima*) invasion in riparian areas. Auk 125:520–531.

Walker, H. A. 2006. Southwestern avian community organization in exotic tamarisk: Current patterns and future needs. Pages 274–286 *in* C. Aguirre-Bravo, P. J. Pellicane, D. P. Burns, and S. Draggan, editors. Monitoring science and technology symposium: Unifying knowledge for sustainability in the western hemisphere. Proceedings RMRS-P-42CD. U.S. Department of Agriculture Forest Service, Rocky Mountain Research Station. Fort Collins, CO.

Woodward, H. D., S. H. Stoleson, and D. M. Finch. 2003. Yellow-billed Cuckoo on the Gila National Forest: Presence-absence, abundance and habitat. Final report for the 2002 field season. U.S. Department of Agriculture Forest Service, Rocky Mountain Research Station. Albuquerque, NM.

11

Tamarisk in Riparian Woodlands: A Bird's Eye View

Mark K. Sogge, Eben H. Paxton, and Charles van Riper III

Although nonnative plant and animal species are common in many ecosystems, the ecological ramifications of their presence are generally not well studied or understood. Yet in the conservation, resource management, and policy communities, there has for decades been a strong bias against nonnative species (Westman 1990; Davis et al. 2011). One reason for this sentiment is that plants and animals introduced into new ecosystems have sometimes had negative impacts on native species and their communities. Nonetheless, the realities and complexities of how nonnative species interact with native flora, fauna, and ecosystems are often oversimplified, and issues become framed as a "native good, alien bad" argument (Goodenough 2010). Nonnative species are frequently vilified and portrayed as "enemies of nature and man" (Davis et al. 2011; see also chapter 16, this volume), and become the targets of aggressive control or eradication programs. Policy and management actions—although generally based on good intent—can result in unintended ecological outcomes and consequences (Westman 1990; D'Antonio and Meyerson 2002; Goodenough 2010).

In this chapter we present a "bird's eye" view of tamarisk (*Tamarix* spp., salt-cedar) and some issues surrounding management of this introduced plant in riparian habitats. We focus on birds not because they are the only wildlife group of concern or necessarily the most important aspect of the issue, but rather because they are a relatively well-studied group that can provide important insights into the role of tamarisk in riparian ecosystems, and because impacts on wildlife—especially birds—have been raised by all sides in the tamarisk management debate

(DeLoach et al. 2000; Dudley and DeLoach 2004; Sogge et al. 2008; van Riper et al. 2008; Paxton et al. 2011; see also chapter 10, this volume). We begin with an overview of the perceptions and realities of bird use of tamarisk, and then discuss tamarisk control and its effects on birds. We close by describing some of the changing perspectives about the management of tamarisk and riparian habitats in western North America.

Early Views of Tamarisk and Birds

Riparian habitat—shrub and tree vegetation bordering streams, rivers, lakes and reservoirs—occupies less than 10% of the landscape yet is disproportionately important to wildlife, including many bird species (Knopf et al. 1988). Birds use riparian areas for breeding, migration stopovers, and wintering habitat in which they can find food resources, nest sites, shelter from inclement weather, protection from predators. and other ecological services. Native riparian habitats comprising primarily cottonwood (*Populus* spp.), willow (*Salix* spp.), and mesquite (*Prosopis* spp.) have declined greatly over the last 100 years with land conversion to agriculture, urbanization, overgrazing, and disruption of hydrologic processes via damming, water withdrawals, and diversions (Knopf et al. 1988; Graf 1992; but see also chapter 6, this volume). With this decline in native trees has come reduction of many riparian breeding bird populations (Johnson et al. 2010) and negative impacts on migrating species (McGrath and van Riper 2005; McGrath et al. 2008).

Tamarisk is a tree/shrub that was intentionally introduced into the United States from southern Europe, Asia, and North Africa in the 1800s, originally as an ornamental plant and subsequently to provide shade and erosion control in the arid west (see chapters 16 and 19, this volume). Since then, especially during the dam construction and water-diversion era of the 1940s to 1960s, tamarisk spread widely, and is now found in all western states (see chapter 1, this volume). Although there is currently no credible estimate of overall tamarisk abundance (Nagler et al. 2010; see also chapter 3, this volume), this introduced plant is among the most common riparian trees and/or shrubs along many river systems and reservoirs in Arizona, New Mexico, Utah, Southern California and Nevada, and in western Oklahoma and Texas. In some locations tamarisk can form large monocultures that cover thousands of hectares (e.g., the Pecos River and lower Colorado River). In areas along highly regulated and altered rivers, tamarisk can grow and persist where native riparian trees cannot (Shafroth et al. 2008; Stromberg et al. 2009; Nagler et al. 2010; see also chapters 5, 6, 9, and 24, this volume).

Because the decline of native riparian habitat occurred concurrently with the spread of tamarisk, this introduced plant has been portrayed as a key factor in the reduction of riparian breeding bird numbers (Kunzman et al. 1989; DeLoach et al. 2000). Through the 1980s and 1990s, there was a general view that tamarisk was an undesirable habitat, despite data showing extensive use of tamarisk by birds. For example, Brown et al. (1987), Hunter et al. (1988), and Rosenberg et al. (1991) described riparian birds breeding in tamarisk riparian habitats along major river systems in New Mexico, Arizona, and Colorado. Although their studies showed that some birds were less abundant or absent in tamarisk when compared to native riparian vegetation, it was also clear that tamarisk supported larger local and regional bird populations than would have occurred if local native vegetation was absent. However, even after these results were published, the more positive aspects of bird use of tamarisk were largely overlooked, and managers and conservationists continued to focus on the shortcomings of tamarisk rather than on its habitat value for birds.

More evidence of the positive value of tamarisk to birds accumulated in the 1990s. Livingston and Schemnitz (1996) reported that tamarisk along the Pecos River in New Mexico supported diverse and abundant bird communities, adding to Howe's (1986) earlier report of yellow-billed cuckoo (*Coccyzus americanus*) populations in tamarisk habitats along that river. In addition, researchers began to document that the federally endangered southwestern willow flycatcher (*Empidonax traillii extimus*) was breeding in tamarisk and likely using it as migration stopover habitat (Sogge et al. 1997). Over the next decade, additional studies were published showing diverse riparian bird communities in tamarisk at a variety of river systems in the arid Southwest (see review in Sogge et al. 2008).

Despite peer-reviewed studies suggesting a more complex ecological relationship between birds and tamarisk, however, there continued to be a widespread view that this introduced plant was unsuitable and provided poor-quality habitat for birds, that only the southwestern willow flycatcher bred in tamarisk, and that tamarisk eradication would have overwhelmingly positive effects for birds and other wildlife (Kunzmann et al. 1989; DeLoach et al. 2000). That perception of tamarisk as a habitat "sink" continued to drive many aspects of resource management policy, practice, and legislation regarding tamarisk control (Shafroth and Briggs 2008).

Birds' Use of Tamarisk: Patterns, Mechanisms, and Considerations

Sogge et al. (2008), Paxton et al. (2011) and Bateman et al. (see chapter 10, this volume) detail the extent and nature of tamarisk use by breeding birds. We summarize here some major points and general considerations. First, tamarisk is an important component of avian habitat throughout many western riparian areas. However, the quality of tamarisk habitat varies among sites; geographic, structural, and hydrologic factors are key determinants of its habitat suitability in any given location (Hunter et al. 1988; Yard et al. 2004; Hinojosa-Huerta 2006; Walker 2006; see also chapter 10, this volume). While it is now widely known that the endangered southwestern willow flycatcher uses tamarisk as breeding habitat (USFWS 2002), this is not a "single-species" issue; approximately 50 different species of birds are documented as breeding in tamarisk (Sogge et al. 2008). These species use tamarisk differently, depending on their specific ecological needs, just as is true for native habitats (Paxton et al. 2008; Sogge et al. 2008; Paxton et al. 2011). Birds that use a range of native riparian habitats ("riparian generalists"; see figure 11.1) can be frequently found in tamarisk, while those with specialized riparian-habitat needs generally are not. For example, larger cavity nesters and raptors are generally not found breeding in tamarisk (Sogge et al. 2008) because tamarisk does not have stems that are big enough for cavities, or branches strong enough to support large nests, although the larger athel (*Tamarix aphylla*) may be an exception.

Why and how do birds use tamarisk? They seek food, including arthropods; shelter from predators and weather; song and foraging perches; and nest sites. They obtain these where available from tamarisk in much the same way as from any native riparian habitat. But because the ecological needs of birds vary depending on whether they are resident breeders, migrants, or are wintering, their use of habitats—native or tamarisk—will vary during these different annual stages. A bird's use of tamarisk is also influenced by whether it can use several habitat types or depends solely on riparian vegetation. This complexity in behavior and habitat use is not always fully considered, and complicates the task of fully understanding how and where tamarisk provides resources for birds. Just as with the overall issue of whether tamarisk is good or bad generally (see chapters 1 and 18, this volume), the answer as to whether tamarisk is good habitat for birds is likely to be "it depends."

Unfortunately, detailed studies comparing productivity, survivorship, and other parameters for birds that breed in tamarisk and native habitats are scarce (Sogge et al. 2008). However, we can look at these questions from the perspective of the southwestern willow flycatcher, one of the most extensively studied birds

FIGURE 11.1 The nest of an Abert's towhee (a riparian generalist) in tamarisk near the Verde River in Arizona.

Source: Photo by Greg Clark; used with permission.

that uses both native and tamarisk vegetation in the Southwest. A medium-sized neotropical migrant that builds open-cup nests (see figure 11.2) and consumes a variety of arthropods, the willow flycatcher shares key life-history traits with many other riparian breeding birds; therefore, insights drawn from flycatcher studies can help us understand tamarisk use by other riparian avifauna. The flycatcher spends its breeding season within and adjacent to riparian woodlands, where relatively tall and dense stands provide shelter from predators and weather (USFWS 2002). Flycatchers breed extensively in dense tamarisk, where they nest in areas of high leaf cover and favor the high twig density and branching structure of tamarisk (see figure 11.3), even when native willows are present nearby (Allison et al. 2003; Ellis et al. 2008; Sogge et al. 2010). If tamarisk were ineffective as shelter or as a nest substrate, we would expect to see increased nest predation, failure, abandonment, and brood parasitism, or possibly increased adult mortality through predation. But this is not the case: long-term studies in central Arizona (Paxton et al. 2007; Ellis et al. 2008) have shown no difference in success of nests placed in native versus tamarisk habitat, nor in survivorship of adults that bred in tamarisk or young that are fledged within tamarisk habitat.

Tamarisk is sometimes perceived and portrayed as providing inadequate food resources for insectivorous birds (e.g., DeLoach et al. 2000), in large part because very few arthropods in the western United States develop on or directly consume tamarisk. However, the food resources it provides are dependent on how many insects and other arthropods can be found in tamarisk stands (which often include native plant components), not on the more restrictive question of how many develop in it. It is clear that tamarisk habitats can have abundant and diverse arthropod communities (Durst et al. 2008a; see also chapter 12, this volume). Arthropod species and abundance may differ in tamarisk compared to native or mixed habitat, but such differences are to be expected and do not necessarily

FIGURE 11.2　Southwestern willow flycatcher nest in tamarisk tree in Arizona.

Source: Photo courtesy of the U.S. Geological Survey.

FIGURE 11.3 Southwestern willow flycatcher nest in tamarisk tree, surrounded by native willows, along the Rio Grande in New Mexico.
Source: Photo courtesy of the U.S. Geological Survey.

mean that one is "inferior" for birds. What matters to the birds' diet is whether the arthropods that are present in a habitat are sufficient to meet energetic and nutritional needs. The southwestern willow flycatcher, and likely other generalist insectivore birds, will compensate for differing arthropod conditions by varying their diet in different habitats (Durst et al. 2008b), and there is no evidence that these arthropod differences result in any meaningful difference in flycatcher body condition or productivity (Owen et al. 2005; Paxton et al. 2011).

DeLoach et al. (2000) proposed that because the willow flycatcher and other southwestern riparian birds did not evolve with tamarisk as a habitat component, tamarisk is inherently inferior or unsuitable habitat. However, the flycatcher evolved with and historically used a variety of early-stage successional riparian habitats—not just willows (Sogge et al. 2010). Thus, the flycatcher's habitat selection behavior would have been shaped by the ability to take advantage of different riparian plants and habitat structures, as long as these provided the necessary features and resources. This flexibility in habitat use would be selectively advantageous for any bird that favors dynamic habitat such as riparian systems in the Southwest (USFWS 2002). Even a tight evolutionary linkage between a bird and a specific habitat would not necessarily mean that more recent alternatives are inherently inferior. For example, cavity nesting bluebirds (*Sialia* spp.) readily and very successfully use artificial nest boxes, which bear little resemblance to a real tree. Like nest boxes, tamarisk is a new habitat feature and does not in all ways mirror the native habitat the birds used before, but it can attract and provide habitat value for southwestern willow flycatchers and other birds.

It is interesting to note that southwestern willow flycatchers nesting in box elder (*Acer negundo*) along the Gila River in New Mexico (Stoleson and Finch 2003), or in live oak (*Quercus agrifolia*) along the upper San Luis Rey River in California (Haas 2003) did not raise concerns about habitat suitability, even though flycatchers' use of these tree species is very rare, much more so than their use of tamarisk. This difference between how atypical native habitats are perceived compared to tamarisk may be another manifestation of the native good, alien bad paradigm (Goodenough 2010; Davis et al. 2011; see also chapter 16, this volume).

Much less is known about the use of tamarisk by migrating birds. However, at the most basic level, we know that birds often use a broader array of habitats during migration than they do for breeding (Walker 2008; Carlisle et al. 2009). So it is not surprising that tamarisk habitats are used by migrant birds, including many species that do not breed in them. These migrants are obtaining other resources such as shelter, food, roosting perches, and predator protection. But given how the habitat characteristics of tamarisk patches vary over the landscape, we should expect its use by migrants to vary spatially and among bird species, and to negate simple "tamarisk bad, native good" assumptions. From what is known to date, this appears to be the case. For example, migrating birds along the lower Colorado River prefer mixed native and tamarisk riparian habitats over either pure native or pure tamarisk riparian habitats (van Riper et al. 2008). This is largely because of the increased vertical plant diversity provided by the understory tamarisk, as is the case on the Rio Grande in New Mexico (Yong et al. 1998). In early spring migration, McGrath et al. (2008) found that Wilson's warblers (*Wilsonia pusilla*) along the lower Colorado River forage preferentially on honey mesquite (*Prosopis glandulosa*), then switch to foraging almost exclusively in tamarisk later in the spring (Paxton et al. 2008). Contrary to what would be expected if tamarisk offered no benefits to birds, Cerasale and Guglielmo (2010) found that compared to pure native patches, mixed tamarisk/native riparian habitat on the San Pedro River, Arizona, offered superior refueling habitat for migrating Wilson's warblers. Although these studies looked only at Wilson's warblers, other insectivorous migrant birds may use and respond to tamarisk similarly (Walker 2008). Clearly, tamarisk provides habitat value to a wide range of migrating species, although as with breeding birds the degree and nature of use will vary by bird species, time of the migratory season, and specific characteristics of the tamarisk habitat. Given the high value and relative rarity of native riparian habitats as migratory stopovers, and the strong conservation interest in preserving overall migration habitat and connectivity (Knopf et al.

1988; Anderson et al. 2004; Carlisle et al. 2009), the importance of tamarisk as migratory bird habitat should not be summarily dismissed. This is an area in which new research or synthesis of existing information would greatly expand our understanding.

Tamarisk Control and Birds: A Brief History

Although tamarisk introduction was originally considered an environmental success, it later became the target of decades of control and eradication efforts (Shafroth et al. 2005; see also chapters 16 and 20, this volume), including the Salt Cedar and Russian Olive Control Demonstration Act, national legislation promoting elimination of tamarisk (summarized in chapter 17, this volume). Birds have been at the center of such tamarisk control efforts, both as a justification for tamarisk control (to improve habitat for birds and other wildlife) and as a brake on some control efforts because of concerns about habitat loss. Until recently, most tamarisk control efforts were small to moderate scale using mechanical removal, chemical spraying, or other localized techniques and were successful in removing or killing tamarisk trees (see chapter 20, this volume). But such measures are labor intensive, costly, and generally not considered practical on the large scale needed to control or eradicate the plant at landscape levels. Biological control was believed to promise a better option (DeLoach et al. 2006).

The history of saltcedar biocontrol is recounted in Dudley and DeLoach (2004), DeLoach et al. (2006), Dark (2009), and Bean et al. (see chapter 16, this volume), and here we summarize key points. Saltcedar biocontrol investigations began in California in the 1960s, and in 1986, the U.S. Department of Agriculture (USDA) Animal and Plant Health Inspection Service (APHIS) began exploring for biocontrol agents in the native Eurasian range of tamarisk. In 1994, the biocontrol researchers asked APHIS for permission to release the tamarisk leaf beetle (*Diorhabda oblongata*) at multiple sites throughout the western United States. This timing coincided with the discovery that the southwestern willow flycatcher—which had been at the time proposed for endangered species status—was using tamarisk during migration and as breeding habitat at many sites in the Southwest. The U.S. Fish and Wildlife Service (USFWS) expressed concerns about negative impacts of tamarisk biocontrol on the flycatcher's habitats and on other riparian wildlife species. APHIS consulted the USFWS, and in 1997 submitted a Biological Assessment that projected only beneficial or benign effects on the 50 plant and animal species that they assessed (including the southwestern willow flycatcher). APHIS then proposed releasing beetles at 10 sites at least 200 miles from any sites where southwestern willow flycatchers were known to nest in tamarisk, and a planned three-year research program and subsequent review before full implementation of the beetle biocontrol project. The USFWS concurred in 1998, and APHIS was granted permits to start the field cage trials in July 1999. That same summer, and again in 2000, *Diorhabda* beetles were released into field cages at sites in Texas, Colorado, Wyoming, Montana, Nevada, California, and Utah. After the two-year field cage trials, APHIS conducted open field releases in 2001.

From 2001 to 2010, beetle propagation and release expanded and populations became well established at numerous sites across the southwestern United States. The spread of the beetle was facilitated by official releases by state and federal agencies, and unofficial relocations by local governments and private persons. Although it was projected that the *Diorhabda* beetles would expand their range only slowly, in some areas the natural and human-facilitated rate of spread surpassed expectations (DeLoach et al. 2006). The rapid spread in Utah, Nevada, and other release sites in the Southwest raised concerns about the program and its effects on the southwestern willow flycatcher and other birds and wildlife (e.g., Sogge et al. 2008; Dark 2009; Stromberg et al. 2009; Hultine et al. 2010), and prompted a 2009 lawsuit charging that agencies responsible for the *Diorhabda* release program did not adequately mitigate for impacts to the southwestern willow flycatcher (Center for Biological Diversity 2009; see also chapter 1, this volume). In addition, scientific research and other articles published after the beetle's release called into question some of the predictions and assumptions underlying tamarisk control plans, including the high use of water by saltcedar, the potential for widespread regeneration of native riparian vegetation following tamarisk removal, the poor quality of tamarisk as bird habitat, and the view of only benign or positive effects of biocontrol on birds (e.g., Harms and Hiebert 2006; Walker 2006; Bay and Sher 2008; Nagler et al. 2009; Shafroth et al. 2008; Sogge et al. 2008; Stromberg et al. 2009; van Riper et al. 2008). Some newspapers, academics, and environmental organizations began to describe a more complex view of tamarisk and its management (e.g., Gelt 2008; Dark 2009; Glenn et al. 2009).

In 2010, a decade after first introducing *Diorhabda* beetles into the United States, APHIS terminated the beetle release program, citing concerns over impacts to southwestern willow flycatcher habitat (USDA 2010). Despite cessation of the formal APHIS biocontrol program, the *Diorhabda* beetle continues to thrive in the wild and to expand its range. As of 2011, the beetle was well established throughout much of the arid southwest, creating large-scale defoliation of tamarisk riparian areas along major rivers, including the Rio Grande near Albuquerque and the Colorado River in the Grand Canyon (see chapter 22, this volume).

Tamarisk Control and Birds: Potential Impacts

Although tamarisk control in general, and biological control in particular, was promoted as having many benefits and no expected risk (DeLoach et al. 2000), the actual results are much less certain (USFWS 2002; Stromberg et al. 2009; Paxton et al. 2011; see also chapter 22, this volume). Tamarisk is currently a major structural and ecological component of many western riparian areas, and provides some key ecological services. Its elimination or reduction will certainly alter riparian systems and the ways that birds and other wildlife use them (see chapter 10, this volume). Is it possible to predict how and to what degree tamarisk control may affect birds and other wildlife?

Clearly, the types and levels of impacts (positive and negative) will be strongly influenced by the scale and other aspects of tamarisk control (Sogge et al. 2008; Paxton et al. 2011). Locally targeted tamarisk control projects such as those that occur on many federal and state parks, wildlife refuges, and other lands primarily have local effects that are easier to assess and make it easier to ensure that the desired outcomes for bird and wildlife habitat occur. But in the past, many such projects focused on removal of tamarisk without including provisions to ensure restoration of native riparian vegetation (e.g., see Harms and Hiebert 2006). Therefore, to avoid unintended consequences to birds and other wildlife, even local projects would benefit from objective analysis of local conditions, knowledge of birds and tamarisk over the larger landscape, and use of the tamarisk control and riparian restoration guidelines and best practices that are now available (see chapters 23 and 24, this volume).

The effects of biocontrol are harder to predict but likely to be substantial because biocontrol is occurring over large areas of the landscape, is relatively uncontrolled once the beetles reach an area, and beetles will persist in areas with tamarisk. As noted by Bean et al. (chapter 22, this volume), the *Diorhabda* beetle could change western riparian areas as much as did the invasion of tamarisk itself. Therefore, it is important to recognize that the release of the *Diorhabda* beetle is a large-scale ecological experiment with unknown but potentially widespread consequences (Paxton et al. 2011; see also chapter 9, this volume).

In considering the effects of biocontrol on birds, Paxton et al. (2011) described a conceptual model that includes both short-term and long-term effects. For example, short-term defoliation by beetles will cause loss of foliage cover and cascading effects such as changes to the arthropod community (see chapter 12, this volume). The increased *Diorhabda* beetle populations will provide a food source to those insectivorous birds that can prey upon it (Longland and Dudley 2008), including nonriparian birds from adjacent habitats. But many chrysomelid beetles produce secondary compounds disfavored by birds (Hilker and Köpf 1994) so one cannot assume that all insectivorous species will eat it. As importantly, riparian birds need more than just a food source—for successful breeding they also need suitable vegetation structure and leaf cover, the precise things that the beetles eliminate as they defoliate the tamarisk (see figure 11.4). Short-term loss of tamarisk vegetation can discourage some riparian birds from nesting. It is also possible that the timing of tamarisk beetle defoliation can be an ecological trap for birds that build nests in leafy tamarisk early in the season, that fail when beetle defoliation occurs (Paxton et al. 2011; see figure 11.5).

The long-term effects of tamarisk control, particularly biocontrol, depend on what, and how quickly vegetation establishes after the tamarisk is defoliated,

FIGURE 11.4 Tamarisk patch along Colorado River near Moab Utah (a) before, and (b) after *Diorhabda* beetle defoliation.

Source: Photos courtesy of Tim Higgs (a) and Wright Robinson (b), Grand County Weed Department, Utah.

removed, or killed (Shafroth et al. 2005; Sogge et al. 2008; Paxton et al. 2011). If tamarisk woodlands are quickly replaced by native riparian woodlands that provide the various ecological services needed by riparian birds, then the long-term impacts of tamarisk control will be either negligible or positive. On the other hand, if tamarisk-dominated woodlands are replaced by nonriparian vegetation (e.g., introduced herbaceous weeds or upland vegetation) or other vegetation without the same ecological services, there will be a net loss of habitat available to riparian birds (Sogge et al. 2008; Paxton et al. 2011). Research and monitoring of the outcome of tamarisk control will help us better understand the outcomes, and improve our approach to similarly complex challenges in the future. However, the

FIGURE 11.5 Adult southwestern willow flycatcher on nest in tamarisk near St. George, Utah, following defoliation of canopy cover from *Diorhabda* beetle.
Source: Photo courtesy of P. Wheeler, Utah Division of Wildlife Resources.

current scope of monitoring, research, restoration, and revegetation activities is limited compared with the anticipated geographic scope and pace of *Diorhabda* beetle spread and tamarisk defoliation.

Conclusions

Our acceptance of tamarisk as habitat for birds does not mean that all tamarisk habitats are equal in value to native habitats or that tamarisk serves as suitable habitat in all cases. Similarly, our discussion of the challenges of tamarisk control efforts does not mean that tamarisk management and native riparian restoration are not valid concepts or actions. In general, the stated objectives and desired outcomes for tamarisk removal are positive from a conservation and natural resource management perspective. However, it is important to recognize that control of nonnative species can be a technically and ecologically complex issue, especially in the context of native habitat restoration or involving biological control (Westman 1990; D'Antonio and Meyerson 2002; Harms and Hiebert 2006; Shafroth et al. 2008). Such is the case for management of tamarisk in the western United States (Shafroth et al. 2005; Stromberg et al. 2009; Davis et al. 2011; see also chapter 25, this volume). Eliminating tamarisk is itself straightforward compared to understanding and shaping what will take its place following removal. This is part of the challenging equation facing those who struggle with the conservation and management of inherently complex ecosystems.

The issues and controversies surrounding nonnative species can be as much about perception as they are about ecological reality (D'Antonio and Meyerson 2002; see also chapters 18 and 25, this volume). But resource managers seldom have the luxury to defer decisions and actions until all uncertainties are resolved; decisions must sometimes be made on the best available data, even though there may be different perspectives and risks. However, as additional research and field activities are conducted, our understanding of natural systems and their components improves. It is important to regularly assess the state of knowledge, and to reevaluate objectives and outcomes within an adaptive management framework. We must ask ourselves: what have we learned? In the case of tamarisk ecology, habitat use by birds, and riparian system restoration, substantial new science has been conducted in recent decades and information synthesized in ways that allow us to better understand the complexities and challenges (e.g., see Bay and Sher 2008; Shafroth et al. 2008; Sogge et al. 2008; Stromberg et al. 2009; Shafroth et al. 2010; Paxton et al. 2011).

Many researchers, including Westman (1990), Shafroth et al. (2005), van Riper et al. (2008), Nagler et al. (2009), Stromberg et al. (2009), Davis et al. (2011), and others describe the need to approach the management of tamarisk and other nonnative species from an ecological, not just a weed control, perspective. This means moving beyond the simple question of how to efficiently eliminate an undesirable "weedy" invasive plant. It requires considering the full suite of roles— positive and negative—that tamarisk plays in the environment, and recognizing its current value to birds and other wildlife. Such an approach is consistent with the increasing calls for a more objective and nuanced view of nonnative species, their role in ecosystems, and their management (Walker 2006; Stromberg et al. 2009; Goodenough 2010; Shafroth et al. 2010; Davis et al. 2011). It is time to recognize that for many bird species, and likely other wildlife, there are many locations where tamarisk has habitat value and contributes ecological services. Indeed, in some areas tamarisk may be the only option for a functioning riparian forest.

The general perception of the value of tamarisk as habitat for birds in the arid western United States has changed from being almost universally negative to a much more nuanced view that recognizes some benefits as well. This change of attitude of tamarisk as bird habitat is mirrored in other aspects of the tamarisk issue. As a result, riparian habitat management alternatives are being proposed that go beyond the simple elimination of tamarisk (Shafroth et al. 2005; Sogge et al. 2008; Stromberg et al. 2009; Shafroth et al. 2010; Paxton et al. 2011; see also chapters 23 and 24, this volume) including:

- Basing decisions about tamarisk control on a comprehensive and objective understanding of its ecological roles, using current scientific data—not its ancestry or provenance, perceived risks, and outdated information
- Recognizing that tamarisk control will benefit some birds and other wildlife species but negatively affect others, and that those effects will vary based on site-specific, species-specific, and project-specific characteristics

- In conserving and managing riparian systems via nonnative species control, focusing on the final outcome of riparian restoration, not tamarisk control as the first step (see chapter 23, this volume)
- Developing explicit, measurable goals and objectives, site-specific plans, and postimplementation monitoring and maintenance for all riparian restoration projects
- Staging and balancing tamarisk removal and native habitat restoration, geographically and over time, to avoid rapid loss of tamarisk habitats for birds and other wildlife until native habitats can be developed.
- Managing some riparian areas for mixed native-tamarisk habitat, rather than total elimination of tamarisk
- Proactively protecting existing stands of native riparian vegetation in areas where tamarisk may be subject to defoliation by *Diorhabda* beetles, to provide refugia for birds and wildlife affected by large-scale tamarisk loss
- Conducting riparian restoration activities within an adaptive management framework, in which outcomes are objectively evaluated and modifications made based on monitoring results.

As we think about riparian habitat management, it is helpful to recognize that birds are not burdened with our human value system, which tends to favor native versus introduced species (Davis et al. 2011). Instead, birds—and most other wildlife—will use whatever types of habitat that can provide for their basic requirements. In the case of many riparian bird species, and the southwestern willow flycatcher in particular, the ability to use tamarisk as riparian habitat has allowed them to persist in their historic range and in some cases to spread in areas where they would not occur otherwise. It is hard to argue that tamarisk has not provided some real ecological or conservation benefits.

Literature Cited

Allison, L. J., C. E. Paradzick, J. W. Rourke, and T. C. McCarthey. 2003. A characterization of vegetation in nesting and non-nesting plots for Southwestern Willow Flycatchers in central Arizona. Studies in Avian Biology 26:81–90.

Anderson, B. W., P. E. Russell, and R. D. Ohmart. 2004. Riparian revegetation: An account of two decades of experience in the arid southwest. Avvar Books. Blythe, CA.

Bay, R. F., and A. A. Sher. 2008. Success of active revegetation after *Tamarix* removal in riparian ecosystems of the southwestern United States: A quantitative assessment of past restoration projects. Restoration Ecology 16:113–128.

Brown, B. T., S. W. Carothers, and R. R. Johnson. 1987. *Grand Canyon Birds*. University of Arizona Press, Tucson.

Carlisle, J. D., S. K. Skagen, B. E. Kus, C. van Riper III, K. L. Paxton, and J. F. Kelly. 2009. Landbird migration in the American West: Recent progress and future research directions. Condor 111:211–225.

Center for Biological Diversity. 2009. Agriculture department forced to re-examine tamarisk leaf-eating beetle program that hurts endangered songbird. [Online]. http://www.biologicaldiversity.org/news/press_releases/2009/southwestern-willow-flycatcher-06-17-2009.html.

Cerasale, D. J., and C. G. Guglielmo. 2010. An integrative assessment of the effects of tamarisk on stopover ecology of a long-distance migrant along the San Pedro River, Arizona. Auk 127: 636–646.

D'Antonio, C. D., and L. A. Meyerson. 2002. Exotic plant species as problems and solutions in ecological restoration: A synthesis. Restoration Ecology 10:703–713.

Dark, S. 2009. Beetles attack! An imported leaf-eating bug is chewing up scenery from Moab to Salt Lake City. City Weekly. [Online]. http://www.cityweekly.net/utah/print-article-7512-print.html.

Davis, M. A., M. K. Chew, R. J. Hobbs, A. E. Lugo, J. J. Ewel, G. J. Vermeij, J. H. Brown, M. L. Rosenzweig, M. R. Gardener, S. P. Carroll, K. Thompson, S. T. A. Pickett, J. C. Stromberg, P. Del Tredici, K. N. Suding, J. G. Ehrenfeld, J. P. Grime, J. Mascaro, and J. C. Briggs. 2011. Don't judge species on their origins. Nature 474:153–154.

DeLoach, C. J., R. Milbrath, R. Carruthers, A. E. Knutson, F. Nibling, D. Eberts, et al. 2006. Overview of saltcedar biological control. Pages 92–99 *in* C. Aguirre-Bravo, P. J. Pellicane, D. P. Burns, and S. Draggan, editors. Monitoring science and technology symposium: Unifying Knowledge for Sustainability in the Western Hemisphere. 20–24 September 2004, Denver, Colorado. Proceedings RMRS-P-42CD. U.S. Department of Agriculture, Forest Service, Rocky Mountain Research Station, Fort Collins, CO.

DeLoach, C. J., R. I. Carruthers, J. Lovich, T. L. Dudley, and S. D. Smith. 2000. Ecological Interactions in the Biological Control of Saltcedar (*Tamarix* spp.) in the U.S.: Toward a new understanding. Pages 819–874 *in* N. R. Spencer, editor. Proceedings of X International Symposium on Biological Control, July 1999. Montana State University, Bozeman.

Dudley, T. L., and C. J. DeLoach. 2004. Saltcedar (*Tamarix* spp.), endangered species, and biological weed control: Can they mix? Weed Technology 18:1542–1551.

Durst, S. L., T. C. Theimer, E. H. Paxton, and M. K. Sogge. 2008b. Age, habitat, and yearly variation in the diet of a generalist insectivore, the Southwestern Willow Flycatcher. Condor 110:514–525.

Durst, S. L., T. C. Theimer, E. H. Paxton, and M. K. Sogge. 2008a. Temporal variation in the arthropod community of desert riparian habitats with varying amounts of saltcedar (*Tamarix ramosissima*). Journal of Arid Environments 72:1644–1653.

Ellis, L. A., D. M. Weddle, S. D. Stump, H. C. English, and A. E. Graber. 2008. Southwestern willow flycatcher final survey and monitoring report. Arizona Game and Fish Department, Research Technical Guidance Bulletin #10, Phoenix, AZ.

Gelt, J. 2008. Saltcedar found to be friend, not foe of western waterways. Arizona Water Resources 17(2):1, 3.

Glenn, E. P., P. L. Nagler, and J. E. Lovich. 2009. The surprising value of saltcedar. Southwest Hydrology 8(3):10–11.

Goodenough, A. E. 2010. Are the ecological impacts of alien species misrepresented? A review of the "native good, alien bad" philosophy. Community Ecology 11:13–21.

Graf, W. L. 1992. Science, public-policy, and western American rivers. Transactions of the Institute of British Geographers 17:5–19.

Haas, W. E. 2003. Southwestern willow flycatcher field season 2002 data summary. Varanus Biological Services, San Diego, CA.

Harms, R. S., and R. D. Hiebert. 2006. Vegetation response following invasive tamarisk (*Tamarix* spp.) removal and implications for riparian restoration. Restoration Ecology 14:461–472.

Hilker, M., and A. Köpf. 1994. Evaluation of the palatability of chrysomelid larvae containing anthraquinones to birds. Oecologia 100:421–429.

Hinojosa-Huerta, O. M. 2006. Birds, water, and saltcedar: Strategies for riparian restoration in the Colorado River Delta. PhD dissertation. University of Arizona, Tucson.

Howe, W.H. 1986. Status of the Yellow-Billed Cuckoo (*Coccyzus americanus*) in New Mexico: 1986. Final Report (contract 516.6-75-09) to the New Mexico Department of Game and Fish. Santa Fe, NM.

Hultine, K. R., J. Belnap, C. van Riper III, J. R. Ehleringer, P. E. Dennison, M. E. Lee, P. L Nagler, K. A. Snyder, S. E. Uselman, and J. B. West. 2010. Tamarisk biocontrol in the western United States: Ecological and societal implications. Frontiers of Ecology and the Environment 8:467–474.

Hunter, W., R. Ohmart, and B. Anderson. 1988. Use of exotic saltcedar (*Tamarix chinensis*) by birds in arid riparian systems. Condor 90:113–123.

Johnson, M. J., R. T. Magill, and C. van Riper III. 2010. Yellow-billed cuckoo distribution and habitat associations in Arizona, 1998–1999. Pages 197–212 *in* C. van Riper III, B. F. Wakeling, and T. D. Sisk, editors. *The Colorado Plateau IV: Integrating Research and Resources Management for Effective Conservation.* University of Arizona Press, Tucson.

Knopf, F. L., R. R. Johnson, T. Rich, F. B. Samson, and R. C. Szaro. 1988. Conservation of riparian ecosystems in the United States. Wilson Bulletin 100:272–284.

Kunzmann, M. R., R. R. Johnson, and P. S. Bennett. 1989. Tamarisk control in southwestern United States. Special Report No. 9. U.S. Department of the Interior, Cooperative National Park Resources Studies Unit, National Park Service. Tucson, AZ.

Livingston, M. F., and S. D. Schemnitz. 1996. Summer bird/vegetation associations in tamarisk and native habitat along the Pecos River, southeastern New Mexico. Pages 171–180 *in* D. W. Shaw and D. M. Finch, technical coordinators. Desired future conditions for southwestern riparian ecosystems: bringing interests and concerns together. USDA Forest Service General Technical Report RM-GTR-272.

Longland, W. S., and T. Dudley. 2008. Effects of a biological control agent on the use of saltcedar habitat by passerine birds. Great Basin Birds 10:21–26.

McGrath, L. J., and C. van Riper III. 2005. Influences of riparian tree phenology on Lower Colorado River spring-migrating birds: Implications of flower cueing. USGS OFR 2005–1140. U.S. Geological Survey Open-file Report 2005-1140 [Online]. http://pubs.usgs.gov/of/2005/1140.

McGrath, L. J., C. van Riper III, and J. J. Fontaine. 2008. Flower Power: Tree flowering phenology as a settlement cue for migrating birds. Journal of Animal Ecology 78:22–30. [Online]. doi:10.1111/j.1365-2656.2008.01464.x.

Nagler, P. L, E. P. Glenn, C. S. Jarnevich and P. B. Shafroth. 2010. Distribution and Abundance of Saltcedar and Russian Olive in the Western United States. *In* P.B. Shafroth, C. A. Brown, and D. M. Merritt, editors. Saltcedar and Russian Olive Control Demonstration Act Science Assessment: U.S. Geological Survey Scientific Investigations Report 2009-5247. http://pubs.usgs.gov/sir/2009/5247.

Nagler, P. L., K. Morino, K. Didan, J. Erker, J. Osterberg, K. R. Hultine, and E. P. Glenn. 2009. Wide-area estimates of saltcedar (*Tamarix* spp.) evapotranspiration on the lower Colorado River measured by heat balance and remote sensing methods. Ecohydrology 2:18–33.

Owen, J. C., M. K. Sogge, and M. D. Kern. 2005. Habitat and gender differences in the physiological condition of breeding Southwestern Willow Flycatchers. Auk 122:1261–1270.

Paxton, E. H., M. K. Sogge, S. L. Durst, T. C. Theimer, and J. Hatten. 2007. The ecology of the Southwestern Willow Flycatcher in Central Arizona: A 10-year Synthesis Report. USGS Open-File Report 2007–1381. [Online]. http://pubs.usgs.gov/of/2007/1381.

Paxton, E. H., T. C. Theimer, and M. K. Sogge. 2011. Biocontrol of exotic tamarisk: Potential demographic consequences for riparian birds in the southwestern United States. Condor 113:255–265.

Paxton, K. L., C. van Riper III, and C. O'Brien. 2008. Movement patterns and stopover ecology of Wilson's Warblers during spring migration on the lower Colorado River in southwestern Arizona. Condor 110: 672–681.

Rosenberg, K. V., R. D. Ohmart, W. C. Hunter, and B. W. Anderson. 1991. *Birds of the Lower Colorado River Valley*. The University of Arizona Press, Tucson.

Shafroth, P. B., C. A. Brown, and D. M. Merritt, editors. 2010. Saltcedar and Russian Olive Control Demonstration Act Science Assessment: U.S. Geological Survey Scientific Investigations Report 2009–5247. [Online]. http://pubs.usgs.gov/sir/209/5247.

Shafroth, P. B., J. R. Cleverly, T. L. Dudley, J. P. Taylor, C. van Riper III, E. P. Weeks, and J. N. Stuart. 2005. Control of *Tamarix* in the western United States: Implications for water salvage, wildlife use, and riparian restoration. Environmental Management 35:231–246.

Shafroth, P. B., and M. K. Briggs. 2008. Restoration ecology and invasive riparian plants: An introduction to the special section on *Tamarix* spp. in western North America. Restoration Ecology 16:94–96.

Shafroth, P. B., V. B. Beauchamp, M. K. Briggs, K. Lair, M. L. Scott, and A. A. Sher. 2008. Planning riparian restoration in the context of *Tamarix* control in western North America. Restoration Ecology 16:97–112.

Sogge, M. K., D. Ahlers, and S. J. Sferra. 2010. A natural history summary and survey protocol for the Southwestern Willow Flycatcher: U.S. Geological Survey Techniques and Methods 2A-10. [Online]. http://pubs.usgs.gov/tm/tm2a10/.

Sogge, M. K., R. M. Marshall, T. J. Tibbitts, and S. J. Sferra. 1997. A Southwestern Willow Flycatcher Natural History Summary and Survey Protocol. National Park Service Technical Report NPS/NAUCPRS/NRTR-97/12.

Sogge, M. K., S. J. Sferra, and E. H. Paxton. 2008. Saltcedar as habitat for birds: Implications to riparian restoration in the Southwest. Restoration Ecology 16:146–154.

Stromberg, J. C., M. K. Chew, P. L. Nagler and E. P. Glenn. 2009. Changing perceptions of change: The role of scientists in *Tamarix* and river management. Restoration Ecology 17:177–186.

Stoleson, S. H., and D. M. Finch. 2003. Microhabitat use by breeding Southwestern Willow Flycatchers on the Gila River, NM. Studies in Avian Biology 26:91–95.

USDA (U.S. Department of Agriculture). 2010. USDA APHIS PPQ Moratorium for Biological Control of Saltcedar (*Tamarix* species) using the biological control agent Diorhabda species (Coleoptera: Chrysomelidae). Memo from Alan Dowdy to PPQ State Plant Health Directors. July 15, 2010.

USFWS (U.S. Fish and Wildlife Service). 2002. Southwestern Willow Flycatcher (*Empidonax traillii extimus*) final recovery plan. U.S. Fish and Wildlife Service. Albuquerque, NM.

van Riper, C. III, K. Paxton, C. O'Brien, P. Shafroth, and L. McGrath. 2008. Rethinking avian response to tamarisk on the lower Colorado River: A threshold hypothesis. Restoration Ecology 16:155–167.

Walker, H. A. 2008. Floristics and physiognomy determine migrant landbird response to tamarisk (*Tamarix ramosissima*) invasion in riparian areas. Auk 125:520–531.

Walker, H. A. 2006. Southwestern avian community organization in exotic tamarisk: Current patterns and future needs. Pages 274-286 *in* C. Aguirre-Bravo, P. J. Pellicane, D. P. Burns, and S. Draggan, editors. Monitoring science and technology symposium: unifying knowledge for sustainability in the Western hemisphere. 20–24 September 2004, Denver, Colorado. Proceedings RMRS-P-42CD. U.S. Department of Agriculture. Forest Service. Rocky Mountain Research Station. Fort Collins, CO.

Westman, W. E. 1990. Park management of exotic plan species: Problems and issues. Conservation Biology 4:251–260.

Yard H. K., C. van Riper III, B.T. Brown, and M. J. Kearsley. 2004. Diets of insectivorous birds along the Colorado River in Grand Canyon, Arizona. Condor 106:106–115.

Yong, W., D. M. Finch, F. R. Moore, and J. F. Kelly. 1998. Stopover ecology and habitat use of migratory Wilson's Warblers. Auk 115:829–842.

12

Tamarix as Invertebrate Habitat

Stephanie Strudley and Peter Dalin

Invertebrates are a major component of the riparian ecosystem in areas where *Tamarix* is densest. Understanding invertebrate diversity and abundance in *Tamarix* stands is therefore critical to comprehending its ecology. We present evidence here that the impact from tamarisk on habitat characteristics such as detritus quality and quantity, predator-prey relationships, food sources, and vegetation structure is likely promoting changes to the arthropod assemblage.

Riparian corridors provide many resources for invertebrates, including vegetation used as food and structural substrate (Strong et al. 1984). Invertebrate assemblages are dominated in most riparian systems by arthropods, mainly of the classes Insecta and Arachnida, and occupy all trophic positions from detritivores to primary (herbivores) and secondary (predators) consumers. These organisms provide many important ecosystem services, such as decomposition, pollination, parasitism, nutrient cycling, seed dispersal, and soil generation and aeration. Their relatively high reproductive capacity and short generation times enable rapid recovery from disturbances and high abundance in these ecosystems. In turn, invertebrates are consumed by a diverse group of terrestrial and aquatic organisms, including fish, herpetofauna (amphibians and reptiles), small mammals, bats, and birds, as well as by other invertebrates.

From a theoretical perspective, high spatial heterogeneity and biodiversity (as in native riparian ecosystems) enhance the stability and efficiency of ecosystem services as compared with simpler systems (Polis 1999). Dominance by a single taxon reverses that relationship, disrupting trophic connectivity and complexity. Structural and trophic simplification subsequently reduces function and resilience, further diminishing the diversity of other organisms that rely on habitat or resource heterogeneity.

The establishment and dominance of nonnative plants in native ecosystems can disrupt functional relationships that support invertebrates and the ecosystem services they provide, particularly when plants alter the physical structure of the vegetation. The modified substrate changes the nature of refugia and habitat traits for invertebrates and predators, while nonnative plant phenology (flowering pattern, seed production, foliage production and senescence) and food resource quality will greatly alter herbivore and pollinator associations (Schaffers et al. 2008). If invertebrate species cannot adapt to living on or with a nonnative plant, then reductions in trophic connectivity will diminish ecosystem function and biodiversity.

How each nonnative plant affects the invertebrate assemblage may be unique to that species, but there are broad similarities in the responses. For example, Herrera and Dudley (2003) found that riparian zones in Northern California dominated by native vegetation had nearly twice the abundance, biomass, and diversity of aerial invertebrates and, in spring, nearly twice the abundance and higher diversity of ground-dwelling invertebrates compared with invertebrate assemblages in areas dominated by the invasive giant reed (*Arundo donax*). Similarly, invertebrates in native vegetation were twice that of areas dominated by an invasive knotweed (*Fallopia* spp.); furthermore, invertebrates in knotweed exhibited lower abundance and morphospecies richness (Gerber et al. 2008). Simao et al. (2010) found that invasion of Japanese stiltgrass (*Microstegium vimineum*) lowered native vegetation richness which corresponded to decreased abundance and richness of arthropods and changed the composition. In this case, carnivore abundance was reduced more than herbivore abundance with the decrease in native plant richness. Other considerations of nonnative plant-invertebrate interactions reflects similar trends (e.g., Slobodchikoff and Doyen 1977; Able et al. 2003; Osgood et al. 2003; Levin et al. 2006; Mgobozi et al. 2008; Wu et al. 2009): ecosystems dominated by nonnative plants appear to lose invertebrate biodiversity and abundance. The effect this has on the ecosystem function is largely unknown.

Value of Tamarisk as Invertebrate Habitat

In the western United States and northern Mexico, the invasive species tamarisk (*Tamarix* spp.) has monopolized vast expanses of riverine ecosystem (1–1.6 million acres; Shafroth et al. 2005). Tamarisk has established in, and sometimes displaced, many ecosystem types: riparian (*Populus* spp. and *Salix* spp.); upland, mesquite (*Prosopis* spp.); desert-shrub; salt marsh; and wetland (Zouhar 2003). Each of these vegetation types supports specific assemblages of invertebrates, so the risk of community disruption extends beyond just riparian corridors to many diverse landscapes. The invasion of tamarisk may even change the invertebrate assemblages by creating habitat that is more hospitable and/or preferable to exotic species than to native species, for example, the exotic isopods and tamarisk leafhopper, thereby altering the ecosystem even further from its natural state. Factors associated with tamarisk that can alter invertebrate associations include

1) altered plant assemblage and composition (see chapter 9, this volume);
2) change in the leaf chemistry, litter detritus quality, and quantity of the dominant primary producer (see chapter 8, this volume, and references therein);
3) an increase in surface soil salinity and decrease in soil moisture (see chapters 6, 8, and 14, this volume);
4) altered wildlife composition and, in some cases, richness and/or abundance (see chapter 10, this volume).

In many riparian areas, tamarisk has supplanted native plants and compromised their abundance and diversity, therefore altering forest structure and spatial heterogeneity (see chapter 9, this volume). This changes the physical substrate that many invertebrates depend upon for feeding, refuge, and habitat. Tamarisk is able to tolerate drought as well as saline, low-nutrient soils, and it excretes soluble salt ions through specialized foliar glands (Arndt et al. 2004; Sorensen et al. 2009; see also chapter 8, this volume), accumulating salt crystals on the foliage. These adaptations are different nutritive chemistry from that of native plants (see chapter 8, this volume). Altogether, tamarisk chemistry, palatability, and physiology may be unsuitable for most native herbivorous invertebrates.

Tamarisk produces a different quality and quantity of litter than native detritus (Ellis et al. 1998; Bailey et al. 2001; Whitcraft et al. 2008). Litter is critical to invertebrates because it serves as (1) substrate and refuge for several life stages; (2) a food source for detritivores; (3) material for invertebrates to perform ecological processes (nutrient cycling, soil generation, and aeration; and (4) a source of energy that is recycled back into the trophic network. Riparian ecosystems dominated by tamarisk are reported to have lower litter production, and more uniformly distributed detritus than cottonwood-dominated vegetation (Ellis et al. 1998; Ellis et al. 2000). Habitat attributes such as litter development, structure, and depth all affect the species composition and abundance of terrestrial invertebrates (Bultman and Uetz 1984; McIver et al. 1992). For some arthropod predators, such habitat attributes may actually be more valuable than food availability because they can reduce intraguild predation and cannibalism and provide a preferable microclimate (Bultman and Uetz 1984; Gratton and Denno 2003).

The quality of tamarisk detritus is chemically distinct from that of native vegetation. Tamarisk increases litter and thus soil salinity (see chapter 8, this volume). Salinity decreases terrestrial arthropod abundance and richness in agricultural settings (Liu and Stiling 2006). Tannin levels also are often higher in tamarisk than in native litter, which reduces the digestibility of the litter for some detritivorous invertebrates (Motomori et al. 2001; Kennedy and Hobbie 2004). Finally, nitrogen levels and carbon to nitrogen (C:N) ratios are different in tamarisk litter compared to native litter (Pomeroy et al. 2000; Bailey et al. 2001; Whitcraft et al. 2008; Going and Dudley 2008). Because some detritivores show preferences (Hassell and Rushton 1984), these substrate differences influence their composition. Cumulatively, we expect a detrimental response, or at least compositional shifts, in the invertebrate assemblage due to litter characteristics in places where tamarisk has become dominant.

In aquatic ecosystems, many invertebrates use riparian litter as their main source of nutritional input. The rate at which tamarisk litter decomposes and its effects on invertebrate assemblage richness and abundance in aquatic ecosystems are unclear (Pomeroy et al. 2000; Bailey et al. 2001; Whitcraft et al. 2008; Kennedy and Hobbie 2004). Invertebrates that colonized tamarisk litter in one system did not use it as a primary source of food, suggesting that it is mainly used for substrate and refuge (Pomeroy et al. 2000). In other systems, however, tamarisk litter provided an adequate food source for crane fly (*Tipula* spp.) and caddis (*Lepidostoma unicolor*) larvae, supporting greater growth than did the litter of native plants (Moline and Poff 2008; Going and Dudley 2008). The improved insect performance was attributed to the higher tamarisk nitrogen content and leaf morphology (small, scaly leaf blades, easier to consume and handle compared to native plants' leaves). Although nutritional quality was suitable for consumption, tamarisk litter is quickly broken down and does not provide a sustained resource for many aquatic invertebrates (Bailey et al. 2001; Kennedy and Hobbie 2004; Moline 2007; Going and Dudley 2008).

Invertebrate Assemblages Observed in Tamarisk

Given the ways in which tamarisk differs from cottonwood and other native riparian species, it is not surprising that surveys of invertebrate communities consistently find differences between these types of vegetation. Most frequently, invertebrate assemblages are found to be less abundant and less diverse in tamarisk-dominated vegetation (Liesner 1971; Stevens 1985; Miner 1989; DeLay et al. 1999; Pendleton et al. 2011). However, in other cases, similar abundance was found, but with different community composition, when compared to native vegetation (Mund-Meyerson 1998; Ellis et al. 2000; Anderson et al. 2004). There are two foci in studies regarding invertebrate abundance, richness, and composition in tamarisk stands. One type considers ground-dwelling invertebrates (e.g., ground-collection via pitfall traps), and the other considers arboreal and aerial invertebrates (e.g., collected by sweep-nets or sticky traps). Although the assemblages are not mutually exclusive, the different types of studies provide insight about functional and behavioral components of the trophic spectrum.

Studies of trophic composition and abundance under tamarisk and native vegetation in riparian ecosystems are scarce. A survey by Ellis et al. (2000) in New Mexico compares the trophic spectrum of ground-dwelling invertebrates in tamarisk to that found in native vegetation. In this case, invertebrates under tamarisk had similar abundance but different composition compared to populations found in native riparian vegetation. Predators, primarily spiders, were a more diverse and abundant functional group under tamarisk-dominated vegetation. Insect abundance and composition fluctuated more under tamarisk-dominated vegetation. With the exception of ants, insects had lower abundance at tamarisk-dominated sites than in native vegetation. Exotic isopod (*Armadillidium vulgare* and *Porcellio laevis*) abundance was high in both vegetation types but more so in native riparian vegetation. Michels et al. (2010) compared abundance and diversity of predatory ground beetles (Carabidae) between tamarisk and native vegetation in Texas. This study also found similar abundance between the vegetation types, but composition and richness was lower in tamarisk than in native vegetation. Both studies suggest that predatory invertebrates are affected by tamarisk-dominated vegetation.

Strudley (2009) also quantified the abundance and richness of ground-dwelling insects and the abundance of spiders under tamarisk-dominated vegetation (see figure 12.1). In this study the insect assemblage was influenced by the biological control agent, *Diorhabda carinulata*. Ants (Formicidae) were the most abundant insects. It is important to note that ants can be predators but may also behave like herbivores or omnivores. Their strength as predators may help determine the efficacy of tamarisk biological control (see figure 12.2). Ellis et al. (2000) found that predators had the greatest trophic diversity in tamarisk stands,

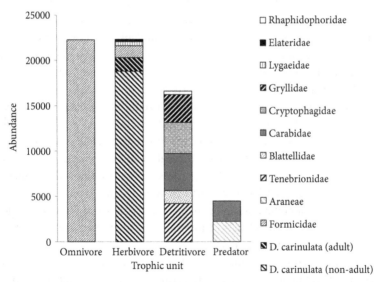

FIGURE 12.1 Ground-dwelling insects and spiders in tamarisk-dominated riparian ecosystems under the influence of an herbivorous biological control (*Diorhabda carinulata*). The data presented here represents approximately 99% of the total abundance of insects from this study. All members from the order Araneae (spiders) are combined into a single category. *D. carinulata* (biological control) is in a separate category and counted as an herbivore. Some families contain species belonging to more than one trophic unit. Insect family categorizes all other data.

FIGURE 12.2 Ant (Formicidae) carrying a *Diorhabda* larva in Nevada.
Source: Photograph courtesy of Tom Dudley.

followed by detritivores. However, in that study predators were dominated by wolf spiders (Lycosidae) and ground spiders (Gnaphosidae, Order Araneida), followed by predatory beetles (Carabidae), while herbivores were largely absent from the riparian floor. In contrast, Strudley (2009) found that after ants, the herbivorous biocontrol beetle, *D. carinulata*, was the most abundant insect. Of the 11 families represented, members from the Lygaeidae and Elateridae family were the only other frequent herbivores. Detritivores were abundant, represented by 14 families. Spiders and Carabid beetles were the only abundant predators out of the eight insect families and spiders represented. Predator and ant abundances may have been influenced by the availability of *Diorhabda* spp. as a food source.

Studies of aerial and arboreal invertebrates in tamarisk also consistently report different assemblages in tamarisk, with reduced abundance compared to invertebrates in willows or cottonwood midstory (DeLay et al. 1999; Pendelton et al. 2011). Pendelton et al. (2011) reported insect assemblages in tamarisk vegetation represented by 10 orders and 46 families and dominated by Hemiptera, whereas willows were represented by 14 orders and 72 families and dominated by Homoptera and Diptera. Willows also had an overall higher insect abundance, and a greater number of pollinators and ants than tamarisk. DeLay et al. (1999) found dipterans were less abundant in midstory tamarisk-dominant vegetation than in cottonwood-dominated vegetation.

Even though few native herbivores are known to consume tamarisk, some of the species found on tamarisk are primarily herbivores. Apache cicada (*Diceroprocta apache*) is one of the only native herbivores reported in high abundance in tamarisk; this was attributed to the preferable plant structure rather than food supply (Glinski and Ohmart 1984; Ellingson and Anderson 2002). Many pollinators visit tamarisk during its flowering period but require native vegetation as larval habitat. Insect populations reportedly fluctuate more on tamarisk than on native vegetation

(Sudbrock 1993; DeLay et al. 1999). The pollinators and other visiting invertebrates may cause this fluctuation, instead of the resident insects on tamarisk. Common predators on tamarisk included spiders, ants, heteropterans, coccinellids, and carabids (P. Dalin; T. Dudley; and D. Bean unpublished data). Fifty-five spider species from 15 orders were identified in tamarisk vegetation (Knutson et al. 2010); however, there was no available comparison in native vegetation.

The scant surveys suggest that several functional groups of invertebrates are being affected by the dominating tamarisk vegetation in riparian ecosystems, particularly predators, ants, and herbivores. We have limited knowledge of the life histories, diet, behavior, and habitat preferences of many invertebrates, making it difficult to determine their relationship with tamarisk. Additionally, natural ecosystems (unlike controlled experiments) present unique challenges in developing replicable, unbiased, controlled experiments. Despite these challenges, more research is essential to confirm how these invertebrates are being impacted. Clarifying the mechanisms that affect the trophic spectrum of invertebrates also deserves attention. As invertebrates play such a valuable role in ecosystem processes, determining their relationship with tamarisk is critical to preserving riparian ecosystems.

Tamarisk-Invertebrate Herbivore Interactions

More than 300 herbivorous invertebrate species are known to use tamarisk as host plants in the Old World (Kovalev 1995). About one-third of these are specialists that only feed on plants in the family Tamaricaceae (Tracy and Robbins 2009). In the United States, however, tamarisk has very few specialist herbivores. This lack of natural herbivores may partly explain why the plants have become invasive in the introduced range. Although some generalists may occasionally be found feeding on tamarisk in North America, the impact of native invertebrates has thus far been negligible (DeLoach et al. 2004). Why so few native herbivores have colonized tamarisk is most likely explained by the taxonomic isolation of *Tamarix* in the United States (see chapter 2, this volume). The flora of the United States contains no native plant species in Tamaricaceae. The only native plants in the same order (Tamaricales) are a few species in the genus *Frankenia* (Frankeniaceae), which grow in salt marsh or xeric shrubland habitats.

Exotic tree species are usually colonized by herbivorous insects from closely related plant species in the new ecosystem (Dalin and Björkman 2006; Roques et al. 2006). These insects are often preadapted to feed on the new plant species because of physiological and phytochemical similarities between the introduced plant and their native hosts. While introduced *Diorhabda* spp. from the same original range can freely feed and develop on *Tamarix* and *Frankenia* plants under laboratory conditions (Herr et al. 2009), there are few (if any) native herbivores in the United States that can be considered preadapted to use *Tamarix*.

In addition to *Diorhabda* spp. intentionally introduced for the biological control of tamarisk, a few naturalized specialist invertebrate species are associated with tamarisk in the United States. These invertebrates were probably introduced accidentally from the Old World along with their plant hosts. The leafhopper *Opsius stactogalus* (Homoptera: Cicadellidae) is a common foliage-feeding herbivore on *Tamarix* in North America (Wiesenborn 2005). The hoppers feed on the phloem and cause chlorosis of plant tissue. Density of *O. stactogalus* may vary considerably over time and space (Wiesenborn 2005; Louden 2010). During times of local high-population densities, whole plants can become discolored, and chlorosis reduces the photosynthetic capacity of the trees. Two scale insects, *Chinoaspis etrusca* and *C. gilli* (Homoptera: Diaspididae), are also common on tamarisk in the United States (Wiesenborn 2005), but their impact on plant growth is unknown. Moreover, DeLoach et al. (2003) found that two mite species from Asia are associated with tamarisk in the United States. Recently, a specialist herbivorous weevil, the splendid tamarisk weevil (*Coniatus splendidulus*, Coleoptera: Curculionidae), was reported from Arizona and subsequently detected in Nevada and other southwestern states. (Eckberg and Foster 2011; Dudley and Bean 2012). Although the leafhopper *O. stactogalus* may affect the growth of individual trees during high population densities (DeLoach et al. 2004; Virla et al. 2010), the impact of these invertebrates in reducing the spread and invasiveness of tamarisk is probably insignificant.

Tamarisk also varies in food quality for herbivorous invertebrates depending on plant genotypes and growth conditions, influencing the abundance and species composition of herbivores. Invasive tamarisk in the United States comprise at least six species and novel hybrids between species (Gaskin and Schaal 2002; Gaskin and Schaal 2003). Most of our knowledge about the quality of tamarisk for herbivorous insects in the United States is derived from studies with the biological control agent *Diorhabda* spp. There are several studies investigating the preference and performance of *Diorhabda* spp. on tamarisk genotypes (DeLoach et al. 2003; Lewis et al. 2003; Dalin et al. 2009; Moran et al. 2009; Herr et al. 2009; Tracy and Robbins 2009; Thomas et al. 2010; Dudley et al. in press) that suggest *Diorhabda* prefer some *Tamarix* types to others and that the beetles' performance may be affected by genotypic variation in host-plant quality.

Plant quality is also affected by growing conditions, such as soil nutrient availability. Sucking insects (e.g., the leafhopper *O. stactogalus*) are often limited by nitrogen content in the plant vascular tissue and decline if their host-plant grows on nutrient-poor soils (Strong et al. 1994). Only a few studies have investigated how growth conditions affect the performance of *Diorhabda carinulata* (Sorensen et al. 2009; Guenther et al. 2011). Tamarisk plants sometimes grow on extremely nutrient-poor soils; larval survival was near zero when the beetles were reared on foliage with nitrogen levels below 2% (Guenther et al. 2011). The level of potassium in the foliage was also shown to affect egg production by adult beetles (Guenther et al. 2011), suggesting that plant growth conditions can influence the quality of plants for *D. carinulata* and other herbivores. High concentrations of salts also affect feeding by herbivorous insects. Sorensen et al. (2009) showed that larval performance of *D. carinulata* was reduced on plants growing in soils containing selenium. Tamarisk sometimes grows in areas affected by anthropogenic pollutants that accumulate in the plant tissue or as minerals on the foliage. Thus, considering the great variety of plant genotypes and the sometimes extreme growth conditions, tamarisk plants can vary considerably in quality for herbivorous invertebrates and affect their abundances and species compositions.

Effects of *Diorhabda* spp. (Biological Control) on Invertebrate Assemblage of Tamarisk Stands

The release of the beetle *Diorhabda* spp. as a biological control agent has increased herbivore biomass load on tamarisk plants in parts of the western United States. Beetle populations have expanded rapidly from several release sites in Texas, Utah, Colorado, and Nevada (DeLoach et al. 2003; Lewis et al. 2003; DeLoach et al. 2004; Hudgeons et al. 2007; Moran et al. 2009; Thomas et al. 2010; see also chapter 22, this volume). This increase in herbivorous insects affects not just target plant populations, but also organisms at other trophic levels in the ecosystem (Dudley and DeLoach 2004; Paxton et al. 2011; Dudley and Bean 2012). These effects are important yet understudied components of classical biological control. The effects can be both direct and indirect (e.g., via defoliation). Direct effects include the responses that may occur if organisms at higher trophic levels (predators) are able to use the beetles as a food resource. *Diorhabda* eggs, larvae, pupae, and adults may be eaten by a variety of native species, such as birds (Longland and Dudley 2008), spiders (Knutson et al. 2010), and ants (DeLoach et al. 2004), which therefore benefit from the presence of *Diorhabda* spp.

Defoliation by the beetles may also affect organisms indirectly through altered habitat conditions (Dudley and DeLoach 2004; Paxton et al. 2011). Dense populations of beetles can cause complete defoliation of tamarisk plants, altering the habitat for other organisms living in the plants' canopies. For example, Louden (2010) showed that abundance of the nonnative leafhopper *O. stactogalus* was reduced on trees heavily damaged by *D. carinulata* beetles. Foliage chlorosis caused by the leafhopper also reduced the growth of *D. carinulata* larvae (Louden 2010). Thus, there may be indirect interactions between these two herbivores on tamarisk. Invertebrate predators that hunt for prey on tamarisk foliage may also be negatively affected by defoliation. For example, many web-building spiders construct nets for flying insects in the canopy of tamarisk trees (Knutson et al. 2010). Foliage removal due to *Diorhabda* beetles may therefore reduce the amount of hunting substrate for spiders and other invertebrate predators in the foliage. Thus, there are likely negative short-term effects of tamarisk defoliation. In the long term, however, the biological control of tamarisk can benefit arboreal invertebrates living on riparian vegetation.

Tamarisk dieback from extensive and continuous beetle defoliation will increase the structural heterogeneity of the vegetation if native plant species reestablish. Many invertebrate predators and herbivores will benefit from increased availability of alternative prey/hosts and a greater variety of microhabitats.

To understand how native invertebrate predators respond to the presence of *D. carinulata* on tamarisk, Dalin et al. (unpublished data) conducted a field and laboratory study on *D. carinulata* in northern Nevada during spring and summer 2006. *D. carinulata* released in 2001 (DeLoach et al. 2004) spread rapidly, and thousands of hectares of tamarisk were defoliated in subsequent years (Pattison et al. 2010). The study compared the abundance of *D. carinulata* and predatory invertebrates on tamarisk between release sites and nonrelease sites (which were selected at least 30 kilometers away from release sites). Biocontrol beetles were captured at all sites, but release sites had significantly higher densities than nonrelease sites. Predatory invertebrates were grouped into spiders, ants, Heteropterans, Coccinellidae, and Carabidae. Predator densities did not differ between release and nonrelease sites, though this may be attributed to the low number of replicates. In May 2006, there was a positive relationship between total predator density (densities of spiders, Heteropterans, Coccinellidae, and Carabidae added together) and *D. carinulata*. Spiders may have accounted for most of this since they were the most abundant predatory invertebrates found on tamarisk, and there was a positive relationship between spiders and *D. carinulata* in May.

That *D. carinulata* was found at all six study sites, even at relatively remote and isolated tamarisk stands about 60 kilometers from the original release site, confirms that the beetles have dispersed over a wide area in Nevada. The density of *D. carinulata* fluctuated substantially over the growing season. At one site, high population densities of adults in May and larvae in June almost completely defoliated tamarisk by July. The plants re-sprouted by September, but the lack of summer foliage reduced invertebrate numbers on the trees. At a different site, also containing high larval densities in June, much less extensive defoliation was observed. This site contained the highest density of spiders by the end of the season. Thus, the positive relationship between spiders and *D. carinulata* at the beginning of the season may reveal between-year responses of spiders to the presence of *D. carinulata*. However, the fact that *D. carinulata* can severely defoliate plants during high population densities, such between-year responses may be overridden by foliage removal during the summer. In a similar study done in Utah, Strudley (2009) found higher richness and abundance of predator species in tamarisk stands where *D. carinulata* had been present the previous year, but there were no relationships within the same year. Thus, the relationship between predatory invertebrates and *D. carinulata* is not fully clear.

In Utah during 2007, Strudley (2009) quantified common insect families and spiders by trophic level when *D. carinulata* was present in tamarisk-dominated vegetation. Insects and spiders were collected in pitfall traps four times throughout one season, and their abundance and diversity were calculated (see figure 12.1). Predator, omnivore and herbivore diversity all increased with biocontrol abundance. Histeridae, Lygeaidae, Ptinidae, and Formicidae abundance, as well as omnivores in general, also increased with *Diorhabda* abundance. However, Carabidae and spiders did not respond to *Diorhabda* abundance. This suggests that in general, invertebrates are positively responding to the biocontrol treatment, at least in the short-term. A positive relationship was also found between *Formica* spp. and *D. carinulata* abundance; predation by these relatively large ants on both larvae and adult biocontrol beetles is commonly observed in the field, so this relationship was not unexpected. The manner and speed by which the adult beetles feed, reproduce, and move explains some of these fluctuations in short-term responses to some trophic levels and insect families.

Dalin's laboratory feeding trials revealed that many of the invertebrate predators found on tamarisk in Nevada could feed on *D. carinulata* larvae (see table 12.1). With only a few replicates for each predator type, there were no statistical analyses of feeding rates among predators. Although jumping spiders were observed to attack *D. carinulata* in the field, they did not consume any larvae in the feeding trials. Knutson et al. (2010) observed orb-weaving spiders (Araneidae) feeding on *D. carinulata* in the field in Texas. Predation by spiders may be difficult to study under lab conditions, suggesting caution in extrapolating the results from our laboratory experiment to field conditions. It is possible that some of the predators that consumed larvae in the feeding trials rarely attack *D. carinulata* under field conditions.

Thus, many native predatory invertebrates in the United States may be able to use *D. carinulata* as a food resource and benefit from the biological control program. However, arboreal invertebrates are likely to also be adversely affected by *D. carinulata* indirectly via defoliation. Community-wide responses may be difficult to predict because the direction and strength of these effects can vary. However, differing between-year responses of predators to *D. carinulata* in both Nevada and Utah (Dalin, unpublished data; Strudley 2009) warrant further study on long-term responses of invertebrates to *D. carinulata* in the United States.

Impacts on Vertebrates

Higher trophic groups, including fish, herpetofauna (amphibians and reptiles), small mammals, bats, and birds consume invertebrates from many functional groups and life stages in riparian areas (see chapter 10, this volume). Changes in invertebrate abundances can therefore be detrimental to these consumers. In one study, 91% of insectivorous bird diets were composed of terrestrial arthropods (Yard et al. 2004), illustrating the important trophic link between terrestrial insects and avian predators. The loss of native riparian vegetation and replacement by tamarisk, in some cases, has affected bird species composition, abundance, and habitat use by avifauna (see chapter 10, this volume) which may be driven in part by changes in invertebrate

TABLE 12.1

Feeding rates in the laboratory by potential invertebrate predators on third instar *Diorhabda carinulata* larvae

Predator taxon	n	Feeding rate (larvae consumed/day ± S.E.)
INSECTA		
Carabid sp. (Col.: Carabidae)	3	1.00 ± 0.29
Coccinella septempunctata (Col.: Coccinellidae)	5	0.10 ± 0.10
Coccinella transversoguttata (Col.: Coccinellidae)	9	1.50 ± 0.25
Hippodamia convergens (Col.: Coccinellidae)	2	0.30 ± 0.30
Mirid sp. (Het.: Miridae)	7	0.29 ± 0.15
Nabid sp. (Het.: Nabidae)	19	0.94 ± 0.21
Pentatomid sp. (Het.: Pentatomidae)	4	0.38 ± 0.24
Assasin bug sp. (Het.: Reduvidae)	4	2.00 ± 0.29
Lacewing spp. (Neuroptera: Chrysopidae)	2	0.50 ± 0.00
Mantis spp. (Mantodea)	2	2.00 ± 0.29
ARACHNIDA		
Misumenops sp. (Aranae: Thomisidae)	10	0.05 ± 0.05
Jumping spiders spp. (Aranae: Salticidae)	8	0
Spiders spp. (Aranae)	20	0.13 ± 0.06

Abbreviations: Col. = Coleoptera; Het. = Heteroptera.

communities (but see chapter 11, this volume). Native vegetation provides preferable foraging structure and higher insect abundance and diversity (van Riper et al. 2004). According to some, birds that feed in tamarisk vegetation mostly consume the tamarisk leafhopper *O. stactogalus* (Drost et al. 2001; Yard et al. 2004). With the availablity of *Diorhabda* spp., food resource for some birds increases (Longland and Dudley 2008; Dudley and Bean 2012). Herpetofauna and small mammals also consume invertebrates and contribute to litter structure/dynamics (Pauley et al. 1999; Dickerson 2001; see also chapter 10, this volume). While small mammals appear to be adaptable to the impacts from tamarisk (Ellis et al. 1997), herptofauna diversity and density have declined compared to populations in native vegetation (Jakle and Gatz 1985; see also chapter 10, this volume). Reduction in invertebrate food sources, along with other changes to habitat, could cause such change. However, the impact of invertebrate habitat modifications on predator composition and density in tamarisk-dominated ecosystems is still largely unknown.

Conclusion

Invertebrates are responsible for many ecological functions that are necessary to sustain riparian ecosystems. They are also the food sources for members of many higher trophic guilds. The dominance of tamarisk in riparian corridors compromises habitat architectural structure, detritus, nutritional chemistry, food sources, and predator-prey relations, which influence invertebrate assemblages. Trophic studies provide evidence that predatory and herbivorous invertebrates have responded to altered composition and abundance in tamarisk-dominated vegetation. A shift in the invertebrate assemblage due to tamarisk dominance could have a cascading effect through the dependent ecosystems. The broader impacts this will have on the riparian food web remain unclear, but pose important and intriguing questions for riparian research.

Literature Cited

Able, K.W., S. M. Hagan, and S. A. Brown. 2003. Mechanisms of marsh habitat alteration due to *Phragmites*: Response of young-of-the-year mummichog (*Fundulus heteroclitus*) to treatment for *Phragmites* removal. Estuaries 26:484–494.

Anderson, B. W., P. E. Russell, and R. D. Ohmart, 2004. *Riparian Revegetation: An Account of Two Decades of Experience in the Arid Southwest.* Avvar Books. Blythe, CA.

Arndt, S. K., C. Arampatsis, A. Foetzki, X. Y. Li, F. J. Zeng, and X. M. Zhang. 2004. Contrasting patterns of leaf solute accumulation and salt adaptation in four phreatophytic desert plants in a hyperarid desert with saline groundwater. Journal of Arid Environments 59:259–270.

Bailey J. K., J. A. Schweitzer, and T. G. Whitham. 2001. Salt cedar negatively affects biodiversity of aquatic macroinvertebrates. Wetlands 21:442–447.

Bultman, T. L., and G. W. Uetz. 1984. Effects of structure and nutritional quality of litter on abundance of litter dwelling arthropods. American Midland Naturalist. 111:165–172.

Dalin, P., and C. Björkman. 2006. Native insects colonizing introduced tree species: Patterns and potential risks. Pages 63–77 *in* T. D. Paine, editor. *Invasive Forest Insects, Introduced Forest Trees, and Altered Ecosystems: Ecological Pest Management in Global Forests of a Changing World.* Springer. Dordrecht, the Netherlands.

Dalin, P., M. J. O'Neal, T. Dudley, and D. W. Bean. 2009. Host plant quality of *Tamarix ramosissima* and *T. parviflora* for three sibling species of the biocontrol insect *Diorhabda elongata* (Coleoptera: Chrysomelidae). Environmental Entomology 38:1373–1378.

DeLay, L., D. M. Finch, S. Brantley, R. Fagerlund, M. D. Mearns, and J. F. Kelly. 1999. Arthropods of native and exotic vegetation and their associations with willow flycatchers and Wilson's warblers. Pages 216–221 *in* D. M. Finch, J. C. Whitney, J. F. Kelly, and S. R. Loftin, technical coordinators. Rio Grande ecosystems: linking land, water and people. Proceedings RMRS-P-7. USDA Forest Service. Rocky Mountain Research Station, Ogden, UT.

DeLoach, C. J., P. A. Lewis, J. C. Herr, R. I. Carruthers, J. L. Tracy, and J. Johnson. 2003. Host specificity of the leaf beetle, *Diorhabda elongata deserticola* (Coleoptera: Chrysomelidae) from Asia, a biological control agent for saltcedars (*Tamarix*: Tamaricaceae) in the western United States. Biological Control 27:117–147.

DeLoach, C. J., R. I. Carruthers, T. L. Dudley, D. Eberts, D. J. Kazmer, A. E. Knutson, D. W. Bean, J. Knight, P. A. Lewis, L. R. Milbrath, J. L. Tracy, N. Tomic-Carruthers, J. C. Herr, G.

Abbott, S. Prestwich, G. Harruff, J. H. Everitt, D. C. Thompson, I. Mityaev, R. Jashenko, B. Li, R. Sobhian, A. Kirk, T. O. Robbins, and E. S. Delfosse. 2004. First results for control of saltcedar (*Tamarix* spp.) in the open field in the western United States. *In* R. Cullen, editors. XI International Symposium on Biological Control of Weeds, Canberra, Australia.

Dickerson, D. D. 2001. Riparian habitat management for reptiles and amphibians on Corps of Engineers projects. ERDC TN-EMRRP-S1-22, Technical Notes collection. U.S. Army Engineer Research and Development Center. Major Shared Resource Center. Vicksburg, MA.

Drost, C. A., E. H. Paxton, M. K. Sogge, and M. J. Whitfield. 2001. Food habits of the endangered Southwestern Willow Flycatcher. USGS Report to U.S. Bureau of Reclamation, Salt Lake City, Utah. USGS Southwest Biological Science Center. Flagstaff, AZ.

Dudley, T. L., and C. J. DeLoach. 2004. Saltcedar (*Tamarix* spp.), endangered species, and biological weed control: Can they mix? Weed Technology 18:1542–1551.

Dudley, T. L., and D. W. Bean. 2012. Tamarisk biocontrol, endangered species risk and resolution of conflict through riparian restoration. Biological Control 57: 331–347.

Eckberg, J. R., and M. E. Foster. 2011. First account of the splendid tamarisk weevil, *Conatius splendidulus* Fabricius, 1781 (Coleoptera: Curculionidae) in Nevada. Pan-Pacific Entomologist 87:51–53.

Ellingson, A. R., and D. C. Andersen. 2002. Spatial correlations of *Diceroprocta apache* and its host plants: Evidence for a negative impact from *Tamarix* invasion. Ecological Entomology 27:16–24.

Ellis, L. M., C. S. Crawford, and M. C. Molles, Jr. 1997. Rodent communities in native and exotic riparian vegetation in the Middle Rio Grande Valley of Central New Mexico. Southwestern Naturalist. 42:13–19.

Ellis, L. M., C. S. Crawford, and M. C. Molles, Jr. 1998. Comparison of litter dynamics in native and exotic riparian vegetation along the Middle Rio Grande of central New Mexico, U.S.A. Journal of Arid Environments. 38: 283–296.

Ellis, L. M., M. C. Molles Jr., C. S. Crawford, and F. Heinzelmann. 2000. Surface-active arthropod communities in native and exotic riparian vegetation in the Middle Rio Grande Valley, New Mexico. Southwestern Naturalist 45:456–471.

Gaskin, J. F., and B. A. Schaal. 2002. Hybrid *Tamarix* widespread in US invasion and undetected in native Asian range. Proceedings of the National Academy of Sciences of the United States of America 99:11256–11259.

Gaskin, J. F., and B. A. Schaal. 2003. Molecular phylogenetic investigation of US invasive *Tamarix*. Systematic Botany 28:86–95.

Gerber, E., C. Krebs, C. Murrell, M. Moretti, R. Rocklin, and U. Schaffner. 2008. Exotic invasive knotweeds (*Fallopia* spp.) negatively affect native plant and invertebrate assemblages in European riparian habitats. Biological Conservation 141:646–654

Glinski, R. L., and R. D. Ohmart. 1984. Factors of reproduction and population densities in the Apache cicada (*Diceroprocta apache*). Southwestern Naturalist 29:73–79.

Going, B. M., and T. L. Dudley. 2008. Invasive riparian plant litter alters aquatic insect growth. Biological Invasions 10:1041–1051.

Gratton C., and R. F. Denno. 2003. Seasonal shift from bottom-up to top-down impact in phytophagous insect populations. Oecologia 134:487–495.

Guenther, D. A., K. T. Gardner, and D. C. Thompson. 2011. Influence of nutrient levels in *Tamarix* on *Diorhabda sublineata* (Coleoptera: Chrysomelidae) survival and fitness with implications for biological control. Environmental Entomology 40:66–72.

Hassall, M., and S. P. Rushton. 1984. Feeding behaviour of terrestrial isopods in relation to plant defenses and microbial activity. Pages 487–505 *in* S. L. Sutton, and D. L. Holdich, editors. *The Biology of Terrestrial Isopods*. Symposia of the Zoological Society of London No 53. Clarendon, Oxford, UK.

Herr, J. C., R. I. Carruthers, D. W. Bean, C. J. DeLoach, and J. Kashefi. 2009. Host preference between saltcedar (*Tamarix* spp.) and native non-target *Frankenia* spp. within the *Diorhabda elongata* species complex (Coleoptera: Chrysomelidae). Biological Control 51:337–345.

Herrera, A. M., and T. L. Dudley. 2003. Reduction of riparian arthropod abundance and diversity as a consequence of giant reed (*Arundo donax*) invasion. Biology Invasions 5:167–177.

Hudgeons, J. L., A. E. Knutson, C. J. DeLoach, K. M. Heinz, W. A. McGinty, and J. L. Tracy. 2007. Establishment and biological success of *Diorhabda elongata elongata* on invasive *Tamarix* in Texas. Southwestern Entomologist 32:157–168.

Jakle, M. D., and T. A. Gatz. 1985. Herpetofaunal use of four habitats of the middle Gila River drainage, Arizona. Pages 355–358 *in* R. R. Johnson, C. D. Ziebell, D. R. Patton, P. F. Ffolliott, and R. H. Hamre, technical coordinators. Riparian ecosystems and their management: Reconciling conflicting uses. USDA Forest Service General Technical Report RM-120. USDA Forest Service. Rocky Mountain Forest and Range Experiment Station. Fort Collins, CO.

Kennedy, T. A., and S. E. Hobbie. 2004. Saltcedar (*Tamarix ramosissima*) invasion alters organic matter dynamics in a desert stream. Freshwater Biology 49:65–76

Knutson, E. M. 2010. A modified sampling method to capture arboreal arthropods. Southwestern Entomologist 35:289–294.

Knutson, E. M., D. B. Richman, and C. Doetkott. 2010. Arboreal spider ecology on saltcedar (*Tamarix*) at Big Spring, Howard County, Texas. Southwestern Entomologist 35:513–523.

Kovalev, O.V., 1995. *Co-evolution of the Tamarisks (Tamaricaceae) and Pest Arthropods (Insecta; Arachnida: Acarina), with Special Reference to Biological Control Prospects*. Pensoft Publishers. Sofia, Bulgaria.

Levin, L. A., C. Neira, and E. D. Grosholz. 2006. Invasive cordgrass modifies wetland trophic function. Ecology 87:419–432

Lewis, P. A., C. J. DeLoach, A. E. Knutson, J. L. Tracy, and T. O. Robbins. 2003. Biology of *Diorhabda elongata deserticola* (Coleoptera: Chrysomelidae), an Asian leaf beetle for biological control of saltcedars (*Tamarix* spp.) in the United States. Biological Control 27:101–116.

Liesner, D. R. 1971. Phytophagous insects of *Tamarix* spp. in New Mexico. Master of Science thesis, New Mexico State University. Las Cruces.

Liu, Y., Z. Yu, W. Gu, J. C. Axmacher. 2006. Diversity of carabids (Coleoptera, Carabidae) in the desalinized agricultural landscape of Quzhou county, China. Agriculture, Ecosystems and Environment 113: 45–50.

Longland, W. S., and T.D. Dudley. 2008. Effects of a biological control agent on the use of saltcedar habitat by passerine birds. Great Basin Birds 10:21–26.

Louden, N. P. 2010. Asymmetric interspecific competition between specialist herbivores that feed on tamarisk in western Colorado. Masters of Science thesis. Utah State University, Logan.

McIver, J. D., G. L. Parsons, and A. R. Moldenke. 1992. Litter spider succession after clear-cutting in a western coniferous forest. Canadian Journal of Forest Research 22:984–992.

Mgobozi, M. P., M. J. Somers, and A. S. Dippenaar-Schoeman. 2008. Spider responses to alien plant invasion: The effect of short- and long-term *Chromolaena odorata* invasion and management. Journal of Applied Ecology 45:1189–1197.

Michels Jr., G. J., V. A. Carney, E. N. Jones, and D. A. Pollock. 2010. Species diversity and qualitative assessment of ground beetles (Coleopterea: Carabidae) in three riparian habiats. Environmental Entomology 39:738–752.

Miner, K. L. 1989. Foraging ecology of the Least Bell's Vireo, *Vireo bellii pusillus*. Master of Science thesis. San Diego State University, San Diego, CA.

Moline, A. B. 2007. A Survey of Colorado Plateau Stream Insect Communities: The roles of riparian leaf litter and hydrologic variation on species growth and community structure. PhD dissertation, Colorado State University. Fort Collins.

Moline, A. B., and N. L. Poff. 2008. Growth of an invertebrate shredder on native (*Populus*) and non-native (*Tamarix, Elaeagnus*) leaf litter. Freshwater Biology 53:1012–1020.

Moran, P. J., C. J. DeLoach, T. L. Dudley, and J. Sanabria. 2009. Open field host selection and behavior by tamarisk beetles (*Diorhabda* spp.) (Coleoptera: Chrysomelidae) in biological control of exotic saltcedars (*Tamarix* spp.) and risks to non-target athel (*T. aphylla*) and native *Frankenia* spp. Biological Control 50:243–261.

Motomori, K., H. Mitsuhashi, and S. Kakano. 2001. Influence of leaf litter quality on the colonization and consumption of stream invertebrate shredders. Ecological Resources 16:173–183.

Mund-Meyerson, M.J. 1998. Arthropod abundance and composition on native vs. exotic vegetation in the Middle Rio Grande riparian forest as related to avian foraging. Masters of Science thesis. University of New Mexico, Albuquerque.

Osgood, D. T., D. J. Yozzo, R. M. Chambers, D. Jacobson, T. Hoffman, and J. Wnek. 2003. Tidal hydrology and habitat utilization by resident nekton in Phragmites and non-Phragmites marshes. Estuaries 26:522–533.

Pattison, R. R., C. M. D'Antonio, T. L. Dudley, K. K. Allander, and B. Rice. 2010. Early impacts of biological control on canopy cover and water use of the invasive saltcedar tree (*Tamarix* spp.) in western Nevada, USA. Oecologia. 165 (3). doi:10.1007/s00442-010-1859-y 12 p.

Pauley, T. K., J. C. Mitchell, R. R. Buech, and J. J. Moriarty. 1999. Ecology and management of riparian habitats for amphibians and reptiles. Pages 169–192 *in* E. S. Verry, J. W. Hornbeck, and C. A. Dolloff, editors. *Riparian Management in Forests of the Continental Eastern United States*. Lewis Publishers. Boca Raton, FL.

Paxton, E. H., T. C. Theimer, and M. K. Sogge. 2011. *Tamarix* biocontrol using tamarisk beetles: Potential consequences for riparian birds in the southwestern United States. Condor 113:255–265.

Pendleton, R. L., B. K. Pendleton, and D. Finch. 2011. Displacement of native riparian shrubs by woody exotics: Effects on arthropod and pollinator community composition. Natural Resources and Environmental Issues 16(25):1–12.

Polis, G. A. 1999. Why are parts of the world green? Multiple factors control productivity and the distribution of biomass. Oikos 86:3–15.

Pomeroy, K. E., J. P. Shannon, and D. W. Blinn. 2000. Leaf breakdown in a regulated desert river: Colorado River, Arizona, USA. Hydrobiologia 434:193–199.

Roques, A., M. A. Auger-Rozenberg, and S. Boivin. 2006. A lack of native congeners may limit colonization of introduced conifers by indigenous insects in Europe.

Canadian Journal of Forest Research / Revue Canadienne de Recherche Forestiere 36:299–313.

Schaffers, A. P., I. P. Raemakers, K. V. Sýkora, and C. J. F. Braak. 2008. Arthropod assemblages are best predicted by plant species composition. Ecology 89:782–794.

Shafroth, P. B., J. R. Cleverly, T. L. Dudley, J. P. Taylor, C. Van Riper III, E. P. Weeks, and J. N. Stuart. 2005. Control of *Tamarix* in the western United States: Implications for water salvage, wildlife use and riparian restoration. Environmental Management 35:231–246.

Simao, M. C. M., S. L. Flory, and J. A. Rudgers. 2010. Experimental plant invasion reduces arthropod abundance and richness across multiple trophic levels. Oikos 119:1553–1562.

Slobodchikoff, C. N., and J. T. Doyen. 1977. Effects of *Ammophila arenaria* on sand dune arthropod communities. Ecology 58:1171–1175.

Sorensen, M. A., D. R. Parker, and J. T. Trumble. 2009. Effects of pollutant accumulation by the invasive weed saltcedar (*Tamarix ramosissima*) on the biological control agent *Diorhabda elongata* (Coleoptera: Chrysomelidae). Environmental Pollution 157:384–391.

Stevens, L. E., 1985. Invertebrate herbivore community dynamics on *Tamarix chinensis* Loueiro and *Salix exigua* Nuttal in the Grand Canyon, Arizona. Master of Science thesis. Northern Arizona University, Flagstaff.

Strong D. R. J., J. H. Lawton, and T. R. E. Southwood. 1984. *Insects on Plants: Community Patterns and Mechanisms*. Blackwell Scientific. Oxford, UK.

Strudley, S. 2009. Impacts of tamarisk biocontrol (*Diorhabda elongata*) on the trophic dynamics of terrestrial insects in monotypic tamarisk stands. Masters of Science thesis. University of Denver, CO.

Sudbrock, A. 1993. Tamarisk control I, fighting back: An overview of the invasion, and a low-impact way of fighting it. Restoration Management Notes 11:31–34.

Thomas, H. Q., F. G. Zalom, and R. T. Roush. 2010. Laboratory and field evidence of post-release changes to the ecological host range of *Diorhabda elongata*: Has this improved biological control efficacy? Biological Control 53:353–359.

Tracy, J. L., and T. O. Robbins. 2009. Taxonomic revision and biogeography of the *Tamarix*-feeding *Diorhabda elongata* (Brulle, 1832) species group (Coleoptera: Chrysomelidae: Galerucinae: Galerucini) and analysis of their potential in biological control of tamarisk. Zootaxa 2101:1–152.

van Riper, C. III, K. L. Ecton, C. O. Brien, and L. J. McGrath. 2004. Avian response to tamarisk invasion on the Lower Colorado River: A threshold hypothesis. US Geological Survey Open-File Report SBSC/SDRS/No. 2004, 1003.

Virla, E. G., G. A. Logarzo, and S. L. Paradell. 2010. Occurrence of the tamarix leafhopper, *Opsius stactogalus* Fieber (Hemiptera: Cicadellidae), in Argentina. Journal of Insect Science 10:23.

Whitcraft, C. R., L. A. Levin, D. Talley, and J. A. Crooks. 2008. Utilization of invasive tamarisk by salt marsh consumers. Oecologia 158:259–272.

Wiesenborn, W. D. 2005. Biomass of arthropod trophic levels on *Tamarix ramosissima* (Tamaricaceae) branches. Environmental Entomology 34:656–663.

Wu, Y.-T., C.-H. Wang, X.-D. Zhang, B. Zhao, L.-F. Jiang, J.-K. Chen, and B. Li. 2009. Effects of saltmarsh invasion by *Spartina alterniflora* on arthropod community structure and diets. Biological Invasions 11:635–649.

Yard, H. K., C. van Riper III, B. T. Brown, and M. J. Kearsley. 2004. Diets of insectivorous birds along the Colorado River in Grand Canyon, Arizona. Condor 106:106–115.

Zouhar, K. 2003. *Tamarix* spp. Fire Effects Information System. [Online]. U.S. Department of Agriculture. Forest Service. Rocky Mountain Research Station. Fire Sciences Laboratory (Producer). http://www.fs.fed.us/database/feis/. Accessed November 13, 2011.

13

Tamarix and Soil Ecology

Kelley A. Meinhardt and Catherine A. Gehring

An important characteristic of *Tamarix* is that it can translocate chemical compounds from deep soil layers and deposit them on the soil surface. This deposition can drastically affect soil chemistry, changing the soil microbial community and, ultimately, native-plant species composition. In this chapter, we delve into these "from the ground up" changes. We begin by briefly discussing the compounds excreted and produced by *Tamarix*, the resulting abiotic soil changes, and the consequences for below-ground communities. We provide a short synopsis of how these compounding changes affect native plant communities, with reference to broader ecological theory on native/invasive plant interactions, and we conclude by outlining implications for restoring riparian habitats.

It is important to note that our review includes not only research on the North American *Tamarix* species but on other species as well. This is primarily because our review of the literature suggests similarities among all species in the chemical compounds in their tissues, as well as in their ability to take up chemicals from the soil and excrete them from their leaves. These characteristics are likely inherent to the *Tamarix* genus and generalizable across species. Further, species identifications in the literature are not reliable, as *Tamarix* species can be difficult to distinguish from one another and are often misidentified (see chapters 2 and 19, this volume). In North America, this is likely due to hybridization among the species (Gaskin and Schaal 2003). Traits that are unique to one species may be passed on to hybrids; an estimated 83% of the *Tamarix* population in North America are novel hybrids (Gaskin and Kazmer 2009). Thus, we believe that the traits we describe below are common to the genus.

Chemical Compounds Transported and Produced by *Tamarix*

Tamarix can take up chemical compounds from the soil, transport them upward, and deposit them on the soil surface by excreting them through salt glands (vesiculated trichomes) found in their leaves and branches (Sookbirsingh et al. 2010) or via leaf fall. Several plant families (e.g., Frankeniaceae, Poaceae, and Tamaricaceae) have evolved glands that regulate salt balance by excreting excess salts, thus allowing the plant to osmoregulate in saline environments (Liphschitz and Waisel 1974). What makes *Tamarix* unique, however, is that, in addition to the sodium chloride (NaCl) predominantly secreted by most halophytes, *Tamarix* can take up, transport, and sequester or secrete many other chemical compounds in its soil environment (see chapter 8, this volume). Ions (e.g., Waisel 1961; Berry 1970; Sookbirsingh et al. 2010); heavy metals (e.g., Consea et al. 2006; Kadukova and Kalogerakis 2007); and even pollutants (e.g., Dreesen and Wangen 1981; Urbansky et al. 2000; Sorensen et al. 2009) can all be taken up by *Tamarix*.

In addition to the compounds stored and secreted by *Tamarix*, compounds produced by the plant itself, such as polyphenols, can be at high levels in *Tamarix* tissues (Drabu et al. 2012). The plant is thought to produce these compounds as a chemical defense against herbivores and pathogens (Brock 1994; Ksouri et al. 2009). Some of them have been shown to have antioxidant, antibacterial, and antifungal activity (Sultanova et al. 2001; Saïdana et al. 2008; Ksouri et al. 2009). For example, when several human pathogenic bacteria, including *Escherichia coli*, *Micrococcus luteus*, and *Staphylococcus aureus*, were exposed to extracts of one *Tamarix* species, their growth was reduced, probably because of the polyphenols' adsorption to the bacterial cell membranes or interaction with bacterial enzymes and substrates. These compounds also showed weak antifungal activity against five *Candida* (yeast) species at high concentrations (Ksouri et al. 2009). Similarly, extracts of another *Tamarix* species showed a 62.5% inhibition of the growth of the human pathogenic fungus *Aspergillus niger* (Sultanova et al. 2001). Although these secondary compounds are not secreted through foliar glands, they likely end up in the soil through root secretions or in fallen branches, leaves, and flowers.

Abiotic Soil Changes

Deposition of salts, metals, pollutants, and natural plant compounds may alter the chemistry of the soil below *Tamarix*. Most well noted have been changes in soil salinity or electrical conductivity (see chapter 8, this volume). Under *Tamarix*, surface soils are generally more saline than the deeper soil layers (e.g., Yin et al. 2009). Other abiotic soil changes under *Tamarix* have been recognized, including alterations in pH, nutrient concentrations, and organic matter content (see table 13.1). Reports of changes in pH in the presence of *Tamarix* have been conflicting; Lesica and DeLuca (2004) and Ladenburger et al. (2006) reported a decrease

TABLE 13.1

Studies of chemical constituents of *Tamarix* spp.

Characteristic investigated	Tamarix species	General conclusions	References
Salt secretions	T. aphylla	- The root environment determines the composition and concentration of excretions. - Glands are nonselective and rare and/or toxic ions in the growth medium can be taken up, transported, and excreted; some are stored in tissues.	Waisel (1961), Thomson et al. (1969), Berry (1970), Hagemeyer and Waisel (1988), Storey and Thomson (1994)
Foliar nutrient concentrations	T. aphylla T. chinensis T. nilotica T. pentandra	- High Na+ concentrations (Na:K ratio = 1.87) - Tissues enriched in various micronutrients - Ions found at high concentrations in tissues also found at high concentrations in soil and groundwater.	Berry (1970), Dreesen and Wangen (1981), Busch and Smith (1995)
Phenolic compounds, volatile oils, acids	T. africana T. aphylla T. boveana T. canariensis T. gallica T. hispida T. nilotica T. pentandra T. ramosissima T. smyrnensis	- Leaves, stems, flowers, and roots contain polyphenols or other compounds that may exhibit antioxidant, antimicrobial, and/or antifungal activity.	Harborne (1975), Nawwar et al. (1982), Sultanova et al. (2001), Saidana et al. (2008), Ksouri et al. (2009), Orabi et al. (2009)
Metals	T. nilotica T. smyrnensis	- Metals can be found in leaves, stems, and roots, sometimes at very high concentrations. They are also excreted in salt crystals. - Heavy metal content was higher in surface soil than subsoil.	Soltan et al. (2004), Fawzy et al. (2006), Manousaki et al. (2009)
Pollutants	T. ramosissima	- Perchlorate can be taken up and stored in tissues.	Urbansky et al. (2000)

in pH in soils associated with *Tamarix*, whereas Yin et al. (2009) reported an increase. The addition of organic acids from decaying leaf litter may explain lower pH (Ladenburger et al. 2006), while increased pH may be due to the breakdown of organic matter within the soil and the subsequent release of CO_2 (Yan et al. 1996). There is also evidence that *Tamarix* promotes soil fertility. Researchers have often observed increases in calcium ([Ca]; Brotherson and Winkel 1986), nitrogen ([N]; Lesica and DeLuca 2004; Landenburger et al. 2006; Zhaoyong et al. 2006), phosphorus ([P]; Ladenburger et al. 2006; Zhaoyong et al. 2006; Yin et al. 2009), and potassium ([K]; Yin et al. 2009) up to 100 centimeters deep in soils beneath *Tamarix*. This fertilization effect of *Tamarix* has been hypothesized to be a result of leaf-litter fall (Ladenburger et al. 2006), leaf secretions (Lesica and DeLuca 2004), or increased rhizodeposition after the widespread root system has scavenged for nutrients in the surrounding open spaces (Yin et al. 2009). Large leaf-litter inputs and the lower leaf-litter C:N ratio of *Tamarix* compared to some species of native vegetation (Going and Dudley 2008; Whitcraft et al. 2008; see also chapter 9, this volume) can result in rapid decomposition rates (e.g., Bailey et al. 2001; Whitcraft et al. 2008) and increased soil organic matter (SOM) content (Ladenburger et al. 2006; Yin et al. 2009). Lastly, it is important to mention that while there is evidence for the production of polyphenolic compounds and the sequestration and excretion of metals and pollutants, there have been no studies on the effects of these constituents on the soil below *Tamarix*.

Responses of Soil Organisms and the Processes They Mediate

Given that *Tamarix* tissues contain higher salt, nutrient, and metal concentrations and different defensive compounds than native vegetation, it is not surprising that *Tamarix* alters soil microbial communities (bacteria, archaea, and fungi) and the processes they mediate. These processes include decomposition, the transformation of N into other chemical forms (e.g., N fixation, nitrification, etc.), and facilitation of nutrient uptake by plants. Only one study focused on N fixation (the conversion of atmospheric nitrogen into organic compounds); it showed that there was no difference in fixation rates for soils under *Tamarix* relative to those under native salt marsh species in southern California (Moseman et al. 2009).

Greater differences have been found in the decomposition of *Tamarix* litter. The C:N ratio of tamarisk litter is less than 20:1 (Going and Dudley 2008; Whitcraft et al. 2008), while many native riparian or salt marsh plants, such as members of the genera *Juncus*, *Acer*, and *Quercus* and the family Salicaceae, have C:N ratios that exceed 35:1 (Motomori et al. 2001; Going and Dudley 2008; Whitcraft et al. 2008). Members of the genus *Alnus* are an exception to this pattern: their litter has a low C:N ratio because of their association with N-fixing bacteria (Motomori et al. 2001; Going and Dudley 2008). These chemical differences, along with altered leaf architecture, were hypothesized to contribute to the more rapid decomposition of

Tamarix litter than of Fremont cottonwood (*Populus fremontii*) litter after three weeks in a stream (Bailey et al. 2001). The low C:N ratio of *Tamarix* in a salt marsh ecosystem contributed to its rapid decomposition and a leaf residence time within this habitat of only 29 days (Whitcraft et al. 2008). These differences are likely responsible for observations that *Tamarix* litter is associated with lower macro-invertebrate species richness and different species composition relative to native species litter (see chapter 12, this volume).

The most recent studies of *Tamarix* leaf-litter decomposition have been conducted in the context of its biological control by the northern tamarisk beetle (*Diorhabda carinulata*), which has been released in several western states (DeLoach et al. 2003; see also chapter 22, this volume). Feeding by both the larval and adult stages of the beetle on *Tamarix* leaves causes the remaining foliage to fall from the tree during the summer growing season, rather than in the late fall. This change reduces the retranslocation of nutrients that usually occurs as leaves senesce, resulting in litter with higher N and P concentrations and more favorable C:N ratios and lignin to nitrogen ratios than *Tamarix* leaves that senesce normally (Uselman et al. 2010). These changes in leaf chemistry promote increased rates of decomposition and greater release of N and P. Because *Tamarix* shrubs often survive initial bouts of defoliation and produce new leaves in the same growing season, the quantity of leaf litter increases substantially under beetle-affected canopies (Uselman et al. 2010). However, a more open canopy resulting from beetle defoliation does not accelerate leaf decomposition by ultraviolet B (UVB) radiation (Uselman et al. 2011). Taken together, these studies provide an explanation for the higher soil nutrient concentrations frequently observed in riparian ecosystems occupied by *Tamarix* and suggest that these changes could be intensified by the introduction of the northern tamarisk beetle. Defoliation resulting in widespread *Tamarix* death could also promote erosion and loss of N from riparian ecosystems (Hultine et al. 2010).

Tamarix, Native Plants, and Mycorrhizal Fungi

The relationships between *Tamarix* and soil biota have been most extensively studied in plant-fungal symbioses called mycorrhizae. Mycorrhizae are predominantly mutualistic associations between fungi and the roots of many plant species in which the fungi facilitate the plants' soil nutrient uptake in exchange for photosynthetic carbon (Smith and Read 2008). Associations with mycorrhizal fungi can improve the performance of individual plants and can also have higher-order effects on plant community structure, plant productivity, soil stability, and nutrient cycling (Smith and Read 2008; Wilson et al. 2009). Studies of interactions between mycorrhizal fungi and *Tamarix* have focused on the two most widespread types of mycorrhizal fungal associations: the arbuscular mycorrhizae (AM) and the ectomycorrhizae ([EM]; see figure 13.1). Arbuscular mycorrhizae are formed by a wide range of plants including mosses, ferns, gymnosperms, and angiosperms in a variety of

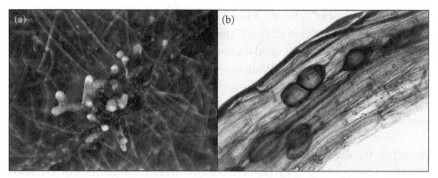

FIGURE 13.1. (a) Ectomycorrhizal (EM) fungi colonize the inside and outside of this woody pine root to create swollen, branched structures with white tips. (b) Arbuscular mycorrhizal (AM) fungi form long, filamentous, tubelike structures called hyphae for resource exchange, and hollow oblong sacs for storage called vesicles inside their host plant's roots.

habitats, and their fungal associates belong to the division Glomeromycota, an ancient lineage of only 140 to 160 described species (Johnson and Gehring 2007). Ectomycorrhizae occur primarily in woody gymnosperms and angiosperms, and the fungi forming these associations are diverse (more than 6000 species) and belong to the divisions Ascomycota and Basidiomycota (Johnson and Gehring 2007; Tedersoo et al. 2010). Significant functional variation among species has been demonstrated for both AM and EM fungi in terms of tolerance of environmental extremes and resource use (e.g., Taylor and Bruns 1999; Hart and Reader 2002). Both genera in the family Salicaceae, the cottonwoods and willows that dominate many native riparian areas in the United States, are among the few plant groups that form both EM and AM associations (Gehring et al. 2006).

Like many nonnative plant species that have successfully colonized habitats outside their native range (Vogelsang and Bever 2009), *Tamarix* appears to be less dependent on associations with mycorrhizal fungi than are many native riparian plants. *Tamarix* roots are generally not colonized by EM fungi. Root colonization by AM fungi (Titus et al. 2002) is low in the field (Beauchamp et al. 2005) and in the greenhouse (Beauchamp et al. 2005; Meinhardt and Gehring 2012), or is completely absent. Furthermore, when *Tamarix* are grown singly in pots, inoculation with AM fungi does not promote increased height or biomass growth (Beauchamp et al. 2005). Spores are reproductive cells produced by fungi for dispersal; their presence can be a useful indicator of the existence and diversity of mycorrhizae in the soil. In arid shrublands in China, the abundance and species richness of AM fungal spores was lower beneath the canopies of *Tamarix* shrubs than beyond them (Zhaoyong et al. 2006), suggesting that the genus also lacks mycorrhizal associations in its native habitat.

There is evidence that, instead of benefiting from associations with mycorrhizal fungi, the presence of mycorrhizal fungi in neighboring native vegetation can adversely affect *Tamarix* growth. When grown together with established *P.*

fremontii seedlings, *Tamarix* seedlings had lower aboveground growth (height and biomass) in pots inoculated with AM and EM fungi than in pots that were not inoculated or were treated with a fungicide (Beauchamp et al. 2005). Meinhardt and Gehring (2012) also examined *Tamarix* performance when grown with cottonwoods, but they inoculated the controls with sterilized AM and EM fungi and added AM-only, EM-only, or AM-plus-EM treatments. Contrary to the results of the previous study, *Tamarix* shoot growth was unaffected by any of the mycorrhizal fungal inocula, but *Tamarix* root growth was reduced by inoculation with AM fungi, either alone or with EM fungi. These results are consistent with studies on other invasive plant species, such as Russian thistle (*Salsola kali*), which responds negatively to mycorrhizal fungi (e.g., Johnson 1998). Clearly, more work is needed in this area, but these preliminary studies suggest that AM fungi may negatively affect *Tamarix*, could shift the competitive balance toward cottonwoods.

The negative relationship between mycorrhizae and *Tamarix* appears to be reciprocal. *Tamarix* can adversely affect both AM and EM, disrupting their mutualisms with native plant species. Meinhardt and Gehring (2012) conducted a field survey and a greenhouse experiment to investigate the effect of a close *Tamarix* neighbor on cottonwoods and their mycorrhizae. In the field, the mycorrhizal colonization of 22 cottonwoods was observed: 11 had a close *Tamarix* neighbor while the other 11 had no *Tamarix* within their root zones. Cottonwoods without a *Tamarix* neighbor had twofold greater AM and EM colonization than cottonwoods with a *Tamarix* neighbor. When looking for EM, the researchers recorded many living, uncolonized root tips in cottonwoods with a *Tamarix* neighbor, suggesting that *Tamarix* may have prevented the colonization of available fine roots by the mycorrhizae (see figure 13.2). Additionally, cottonwoods with a *Tamarix* neighbor had a different community composition, that is, number and presence of EM species, than those without *Tamarix* nearby. The community composition of AM was unaffected, however. Colonization by other types of fungi, including pathogens, was also higher in cottonwood roots when *Tamarix* was present. Results from the greenhouse study were similar for EM, but not for AM. When cottonwoods were grown in the same pot with a *Tamarix* neighbor, EM fungal colonization was reduced by over 50% compared to cottonwoods grown with a cottonwood neighbor. Arbuscular mycorrhizae colonization of cottonwoods was not affected by *Tamarix*, strengthening the idea that this type of mycorrhizal fungus can withstand the presence of *Tamarix* and may negatively affect its growth. Ectomycorrhizae were consistently more negatively affected by *Tamarix* than AM in the field and the greenhouse, which is especially important because EM provided the greatest shoot biomass benefit to their cottonwood hosts (Meinhardt and Gehring 2012).

Tamarix can adversely affect AM and EM, but AM may inhibit *Tamarix* growth—so which of these interactions matters most? It is likely that the negative effects of *Tamarix* on mycorrhizae will be most important, at least in situations where *Tamarix* and cottonwoods are roughly the same size. In the greenhouse,

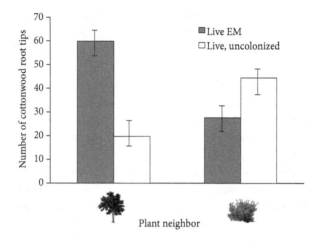

FIGURE 13.2. Cottonwoods with a conspecific or willow neighbor (🌳) had twice the number of live EM-colonized tips than did cottonwoods with a *Tamarix* neighbor (🌿). Conversely, cottonwoods with a *Tamarix* neighbor had more than twice the number of live, uncolonized root tips compared with cottonwoods without a *Tamarix* nearby. This suggests that *Tamarix* inhibits root colonization by mycorrhizal fungi. Bars represent means ±1 SE.

Source: Aadapted from Meinhardt and Gehring (2012).

cottonwoods grown with a *Tamarix* neighbor and inoculated with AM-only, EM-only, or AM-plus-EM, grew no better than cottonwoods with sterile inoculum (i.e., no mycorrhizae; see figure 13.3). This suggests that *Tamarix* affected the mycorrhizae to the point that they provided no measurable benefit to their cottonwood hosts. In contrast, although *Tamarix* root biomass was significantly reduced by AM (Meinhardt and Gehring 2012), the reduction in overall *Tamarix* biomass by AM was minimal (5%; figure 13.3). Together, these results suggest that the negative effects of *Tamarix* on the mycorrhizae are likely to be more damaging to natives than any negative effects of AM on *Tamarix*.

The effects of *Tamarix* on mycorrhizae may not only affect the growth of a few neighboring species but can also degrade the mutualistic community, making the survival and establishment of native, mutualist-dependent species more difficult. The continuous disruption of mycorrhizal mutualisms by an invasive species such as *Tamarix* reduces the colonization of native plant roots by mycorrhizae. Without a host, mycorrhizal fungi cannot reproduce, and so their abundance in the soil declines over time. Meinhardt and Gehring (2012) found that EM propagules (spores) were far fewer in soils beneath *Tamarix* than in soils beneath native species. Invasive species do not need to disrupt native plant–mycorrhizae relationships to degrade the mutualistic community. A high abundance of nonmycorrhizal invasive species can cause the same changes. Native California cudweed (*Gnaphalium californicum*) was inhibited by soils that had previously supported a mix of invasive species, but it grew well in soils previously supporting a mix

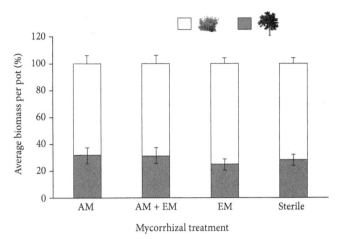

FIGURE 13.3 Cottonwoods inoculated with mycorrhizal fungi and growing with a *Tamarix* neighbor grew no better than cottonwoods with sterile inoculum after four months in the greenhouse. Although they may have been somewhat impaired by AM fungi, *Tamarix* plants made up nearly 70% of the biomass per pot in all treatment groups. Bars represent means ±1 SE.
Source: Data from Meinhardt and Gehring (2012).

of native species. The opposite results were observed with invasive Italian thistle (*Carduus pycnocephalus*), which grew very well in soil without AM that had previously supported a mix of invasive species; its growth was inhibited in soils that promoted native species growth. This positive feedback cycle—where invasive species grow better in soils previously inhabited and conditioned by invasives—can further promote the dominance of invasive species (Vogelsang and Bever 2009).

No mechanism has yet been identified, but one or more of the changes to soil chemistry by *Tamarix* may contribute to the disruption of mycorrhizal mutualisms important to native plants. First, increased salt concentrations in soils near *Tamarix* may reduce both the abundance of mycorrhizae and their ability to colonize native plant roots. Dixon et al. (1993) found that the biomass of six EM fungal species decreased when they were exposed to various salts, and the colonization of seedling host-plant roots also decreased. Likewise, high soil salinity has been shown to reduce AM colonization (Asghari et al. 2008) and spore germination (Juniper and Abbot 2006). Second, soil fertilization can reduce the need for natives to form mycorrhizal associations. *Tamarix* excretions and leaf litter can increase soil N and P (Ladenburger at al. 2006; Yin et al. 2009). When nutrients are abundant, there is less need for plants to form associations with mycorrhizae because they can procure nutrients without them (Johnson 2003). However, without mycorrhizae, environmental pressures (e.g., salt stress or pathogen stress) will not be as well mediated for a plant. Third, the natural compounds produced by *Tamarix* may negatively affect mycorrhizae. The polyphenols contained in the tissues of some *Tamarix* species have been shown to have antioxidant, antibacterial, and antifungal acivity (e.g., Ksouri et al. 2009). Colonization of native plants by AM decreased with the presence of similar chemicals extracted from *Artemisia princeps* (Yun and Choi 2002).

Implications for Riparian Restoration

Tamarix is capable of causing changes in soil chemistry and biota, and these changes may persist long after *Tamarix* removal. These long-term changes can be called "legacy effects," that is, competitive influences that persist after the causal plant species is removed (Grman and Suding 2010). After *Tamarix* is removed, how can legacy effects be diminished so that soils can return to a more native state? Over time, native bacteria and mycorrhizal fungi will arrive at a site, and, as their abundance increases, they will be able to assist native species in recolonization. However, this may take many years to happen naturally. To speed up the process of native plant establishment at sites previously invaded by *Tamarix*, soil restoration may be required (see chapter 23, this volume). Because of the impact *Tamarix* can have on soil chemistry, it would be beneficial for land managers to test soil chemical properties and, if needed and feasible, use controlled flooding to leach contaminants (e.g., salts) from the soil (Sher et al. 2010). The presence of mycorrhizae is crucial for the establishment and growth of many native species. Ectomycorrhizae seem to provide the greatest growth benefits to native cottonwoods, while AM may be able to inhibit the growth of *Tamarix* if it begins to reestablish (Beauchamp et al. 2005; Meinhardt and Gehring 2012). Land managers could assess the fungal propagule (e.g. spores, fragments of fungal tissue) abundance of both types of mycorrhizae at restoration sites, and if it is low, propagules can be reintroduced via live soil from a nearby site dominated by native vegetation or generated through trap cultures. Trap cultures are used to produce mycorrhizal inoculum and consist of a host plant grown in a medium, such as sand or perlite, and fungal propagules. The fungi colonize the host plant, grow, and reproduce, yielding highly colonized roots or spore-filled soil that can be used as inoculum.

It is likely that after large-scale *Tamarix* removal or death, little woody vegetation will remain, and more active methods of riparian restoration may be needed. Active restoration, such as the planting of native vegetation and the soil restoration approaches described in this chapter, could be particularly important in sites affected by the tamarisk biocontrol beetle. The premature leaf drop caused by the beetle increases soil N concentrations due to altered litter chemistry. Higher N concentrations are known to favor the establishment of nonnative plants over native plants (e.g., Alpert and Maron 2000), as do highly saline soils (Hultine et al. 2010) and soils with low levels of mycorrhizae (Vogelsang and Bever 2009). Sites dominated by dead *Tamarix* are likely to have all these soil properties, creating very difficult environments for native plant reestablishment, even following the restoration planting of seedlings or cuttings of native vegetation. Other nonnatives that are tolerant of high salt and N and not dependent on mycorrhizae, such as Russian olive (*Elaeagnus angustifolia*), may rapidly establish (Hultine et al. 2010) if the soil is not amended in ways that favor native plants. Where soil resources, such as N and water, are more readily available, nonnative understory plants considered noxious weeds, such as Russian knapweed *(Acroptilon repens)*, have been observed to colonize *Tamarix* stands defoliated by the tamarisk beetle (Hultine et al. 2010).

Conclusions and Directions for Future Research

The studies we have discussed show that *Tamarix* alters both the chemistry and biology of the soil and that these changes contribute to its ability to establish and compete in nonnative habitats. There is also evidence that the effects of *Tamarix* on mycorrhizae may impose a "degraded mutualism" for native plant species, dramatically slowing reestablishment of these plants following *Tamarix* removal and necessitating soil restoration. However, many questions remain to be answered. First, there are few solid mechanistic links between the chemical and biological effects of *Tamarix* on the soil, limiting our ability to mitigate these effects. Second, the research on soil microbes is still very limited; only a few types of microbes have been studied, and these only in a few locations. An important group of microbes to focus on in the future are the dark septate endophytes (DSE). These fungi colonize *Tamarix* roots extensively (Beauchamp et al. 2005; Meinhardt and Gehring 2012), and a review of studies from several ecosystems suggests they might function similarly to mycorrhizae (Newsham 2011). These findings raise the question of whether DSE could help *Tamarix* compete against native plants. Third, the legacy effects of *Tamarix* on the soil appear to be substantial, but they remain poorly understood, precluding prediction of when soils will return to their original state following *Tamarix* removal. Finally, the large-scale reduction of *Tamarix* following the introduction of *D. carinulata* presents an opportunity to address many of these questions and to improve our understanding of the importance of soil ecology to the success of *Tamarix* and other nonnative invasive species.

Some of the most important, yet least obvious, consequences of *Tamarix* invasion are changes to the soil and its biota. Invasive species can change components of the N, C, and water cycles, productivity and nutrient mineralization rates, and nutrient distribution in an ecosystem (Ehrenfeld 2003). Invasive species, such as *Tamarix*, can catalyze these "from the ground up" ecosystem changes by either directly (e.g., chemical compound deposition, direct root interactions) or indirectly (e.g., replacement of native host plants) affecting the microbes that play a role in such processes. A widespread shift in soil dynamics and the loss of or change in the communities of soil microbes may negatively affect native species. These changes may promote further *Tamarix* invasion or facilitate secondary invasion by other nonnative species following tamarisk removal.

Literature Cited

Alpert, P., and Maron, J. L. 2000. Carbon addition as a countermeasure against biological invasion by plants. Biological Invasions 2:33–40.

Asghari, H. R., M. R. Amerian, and H. Gorbani. 2008. Soil salinity affects arbuscular mycorrhizal colonization of halophytes. Pakistan Journal of Biological Sciences 11:1909–1915.

Bailey, J. K., J. A. Schweitzer, and T. G. Whitham. 2001. Salt cedar negatively affects biodiversity of aquatic macroinvertebrates. Wetlands 21:442–447.

Beauchamp, V., J. Stromberg, and J. Stutz. 2005. Interactions between *Tamarix ramosissima* (saltcedar), *Populus fremontii* (cottonwood), and mycorrhizal fungi: Effects on seedling growth and plant species coexistence. Plant and Soil 275:221–231.

Berry, W. L. 1970. Characteristics of salts secreted by *Tamarix aphylla*. American Journal of Botany 57:1226–1230.

Brock, J. H. 1994. *Tamarix* spp. (salt cedar): An invasive exotic woody plant in arid and semi-arid riparian habitats of western USA. Pages 27–44 *in* L. C. de Waal, L. E. Child, P. M. Wade, and J. H. Brook, editors. *Ecology and Management of Invasive Riverside Plants*. John Wiley and Sons, Chichester, UK.

Brotherson, J. D., and V. Winkel. 1986. Habitat relationships of saltcedar (*Tamarix* ramosissima) in central Utah. Great Basin Naturalist 46:535–541.

Busch, D. E., and S. D. Smith. 1995. Mechanisms associated with decline of woody species in riparian ecosystems of the Southwestern U.S. Ecological Monographs 65:347–370.

Conesa, H. M., A. Faz, and R. Arnaldos. 2006. Heavy metal accumulation and tolerance in plants from mine tailings of the semiarid Cartagena-La Unión mining district (SE Spain). Science of the Total Environment 366:1–11.

DeLoach, C. J., P. A. Lewis, J. C. Herr, R. I. Carruthers, J. L. Tracy, and J. Johnson. 2003. Host specificity of the leaf beetle, *Diorhabda elongata deserticola* (Coleoptera: Chrysomelidae) from Asia, a biological control agent for saltcedars (*Tamarix*: Tamaricaceae) in the western United States. Biological Control 27:117–147.

Dixon, R. K., M. V. Rao, and V. K. Garg. 1993. Salt stress affects in-vitro growth and in-situ symbioses of ectomycorrhizal fungi. Mycorrhiza 3:63–68.

Drabu, S., S. Chaturvedi, and M. Sharma. 2012. *Tamarix gallica:* An overview. Asian Journal of Pharmaceutical and Clinical Research 5:17–19.

Dreesen, D. R., and L. E. Wangen. 1981. Elemental composition of saltcedar (*Tamarix chinensis*) impacted by effluents from a coal-fired power plant. Journal of Environmental Quality 10:410–416.

Ehrenfeld, J. G. 2003. Effects of exotic plant invasion on soil nutrient cycling processes. Ecosystems 6:503–523.

Fawzy, E. M., M. E. Soltan, and S. M. Sirry. 2006. Mobilization of different metals between *Tamarix* parts and their crystal salts: Soil system at the banks of river Nile, Aswan, Egypt. Toxicological and Environmental Chemistry 88:603–618.

Gaskin, J. F., and B. A. Schaal. 2003. Molecular phylogenetic investigation of U.S. invasive *Tamarix*. Systematic Botany 28:86–95.

Gaskin, J. F., and D. J. Kazmer. 2009. Introgression between invasive saltcedars (*Tamarix chinensis* and *T. ramosissima*) in the USA. Biological Invasions 11:1121–1130.

Gehring, C. A., R. C. Mueller, and T. G. Whitham. 2006. Environmental and genetic effects on the formation of ectomycorrhizal and arbuscular mycorrhizal associations in cottonwoods. Oecologia 149:158–164.

Going, B. M., and T. L. Dudley. 2008. Invasive riparian plant litter alters aquatic insect growth. Biological Invasions 10:1041–1051.

Grman, E., and K. N. Suding. 2010. Within-year soil legacies contribute to strong priority effects of exotics on native California grassland communities. Restoration Ecology 18:664–670.

Hagemeyer, J., and Y. Waisel. 1988. Excretion of ions (Cd2+, Li+, Na+, and Cl-) by *Tamarix aphylla*. Physiologia Plantarum 73:541–546.

Harborne, J. B. 1975. Flavonoid bisulphates and their co-occurances with ellagic acid in the Bixaceae, Frankeniaceae and related families. Phytochemistry 14:1331–1337.

Hart, M. M., and Reader, R. J. 2002. Taxonomic basis for variation in the colonization strategy of arbuscular mycorrhizal fungi. New Phytologist 153: 335–344.

Hultine, K. R., J. Belnap, C. van Riper III, J. R. Ehleringer, P. E. Dennison, M. E. Lee, P. L. Nagler, K. A. Snyder, S. M. Uselman, and J. B. West. 2010. Tamarisk biocontrol in the western United States: Ecological and societal implications. Frontiers in Ecology and the Environment 8: 467–474

Johnson, N. C. 2003. Can fertilization of soil select less mutualistic mycorrhizae? Ecological Applications 3:749–757.

Johnson, N. C. 1998. Responses of *Salsola kali* and *Panicum virgatum* to mycorrhizal fungi, phosphorus and soil organic matter: Implications for reclamation. Journal of Applied Ecology 35:86–94.

Johnson, N. C., and C. A. Gehring. 2007. Mycorrhizas: Symbiotic Mediators of Rhizosphere and Ecosystem Processes. Pages 73–100 *in* Z. Cardon and J. Whitbeck, editors. *The Rhizosphere: An Ecological Perspective*, Academic Press, New York.

Juniper, S., and L. Abbott. 2006. Soil salinity delays germination and limits growth of hyphae from propagules of arbuscular mycorrhizal fungi. Mycorrhiza 16:371–379.

Kadukova, J., and N. Kalogerakis. 2007. Lead accumulation from non-saline and saline environment by *Tamarix smyrnensis* Bunge. European Journal of Soil Biology 43:216–223.

Ksouri, R., H. Falleh, W. Megdiche, N. Trabelsi, B. Mhamdi, K. Chaieb, A. Bakrouf, C. Magné, and C. Abdelly. 2009. Antioxidant and antimicrobial activities of the edible medicinal halophyte *Tamarix gallica* L. and related polyphenolic constituents. Food and Chemical Toxicology 47:2083–2091.

Ladenburger, C. G., A. L. Hild, D. J. Kazmer, and L. C. Munn. 2006. Soil salinity patterns in *Tamarix* invasions in the Bighorn Basin, Wyoming, USA. Journal of Arid Environments 65:111–128.

Lesica, P., and T. H. Deluca. 2004. Is tamarisk allelopathic? Plant and Soil 267:357–365.

Liphschitz, N., and Y. Waisel. 1974. Existence of salt glands in various genera of the Gramineae. New Phytologist 73:507–513.

Manousaki, E., F. Kokkali, and N. Kalogerakis. 2009. Influence of salinity on lead and cadmium accumulation by the salt cedar (*Tamarix smyrnensis* Bunge). Journal of Chemical Technology and Biotechnology 84:877–883.

Meinhardt, K. A., and C. A. Gehring. 2012. Disrupting mycorrhizal mutualisms: A potential mechanism by which exotic tamarisk outcompetes native cottonwoods. Ecological Applications 22:532–549.

Moseman, S. M., R. Zhang, P. Y. Qian, and L.A. Levin. 2009. Diversity and functional responses of nitrogen-fixing microbes to three wetland invasions. Invasion Biology 11:225–239.

Motomori, K., H. Mitsuhashi, and S. Kakano. 2001. Influence of leaf litter quality on the colonization and consumption of stream invertebrate shredders. Ecological Resources 16:173–183.

Nawwar, M. A. M., J. Buddrus, and H. Bauer. 1982. Dimeric phenolic constituents from the roots of *Tamarix nilotica*. Phytochemistry 21:1755–1758.

Newsham, K. K. 2011. A meta-analysis of plant responses to dark septate root endophytes. New Phytologist 190:783–793.

Orabi, M. A. A., S. Taniguchi, T. Hatano. 2009. Monomeric and dimeric hydrolysable tannins of *Tamarix nilotica*. Phytochemistry 70:1286–1293.

Saïdana, D., M. A. Mahjoub, O. Boussaada, J. Chriaa, I. Chéraif, M. Daami, Z. Mighri, and A. N. Helal. 2008. Chemical composition and antimicrobial activity of volatile compounds of *Tamarix boveana* (Tamaricaceae). Microbiological Research 163:445–455.

Sher, A. A., K. Lair, M. DePrenger-Levin, and K. Dohrenwend. 2010. *Best Management Practices for Revegetation in the Upper Colorado River Basin*. Denver Botanic Gardens, Denver, CO.

Smith, S. E., and D. J. Read. 2008. *Mycorrhizal Symbiosis*, 3rd edition. Academic Press, New York, NY.

Soltan, M. E., S. M. N. Moalla, M. N. Rashed, and E. M. Fawzy. 2004. Assessment of metals in soil extracts and their uptake and movement within *Tamarix nilotica* at Lake Nasser banks, Egypt. Chemistry and Ecology 20:137–154.

Sookbirsingh, R., K. Castillo, T. E. Gill, and R. R. Chianelli. 2010. Salt separation process in the saltcedar *Tamarix* ramosissima (Ledeb.). Communications in Soil Science and Plant Analysis 41:1271–1281.

Sorensen, M. A., D. R. Parker, and J. T. Trumble. 2009. Effects of pollutant accumulation by the invasive weed saltcedar (*Tamarix ramosissima*) on the biological control agent *Diorhabda elongata* (Coleoptera: Chrysomelidae). Environmental Pollution 157:384–391.

Storey, R., and W. W. Thomson. 1994. An x-ray microanalysis study of the salt glands and intracellular calcium crystals of *Tamarix*. Annals of Botany 73:307–313.

Sultanova, N., T. Makhmoor, Z. A. Abilov, Z. Parwee, V. B. Omurkamzinova, Atta-ur-Rahman, and M. Iqbal Choudhary. 2001. Antioxidant and antimicrobial activities of *Tamarix ramosissima*. Journal of Ethnopharmacology 78:201–205.

Taylor D. L., and T. D. Bruns. 1999. Community structure of ectomycorrhizal fungi in a *Pinus muricata* forest: Minimal overlap between the mature forest and resistant propagule communities. Molecular Ecology 8:1837–1850

Tedersoo, L., T. W. May, and M. E. Smith. 2010. Ectomycorrhizal lifestyle in fungi: Global diversity, distribution, and evolution of phylogenetic lineages. Mycorrhiza 20: 217–263.

Thomson, W. W., W. L. Berry, and L. L. Liu. 1969. Localization and secretion of salt by the salt glands of *Tamarix aphylla*. Proceedings of the National Academy of Sciences 63:310–317.

Titus, J. H., P. J. Titus, R. S. Nowak, and S. D. Smith. 2002. Arbuscular mycorrhizae of Mojave Desert Plants. Western North American Naturalist 62:327–334.

Urbansky, E. T., M. L. Magnuson, C. A. Kelty, and S. K. Brown. 2000. Perchlorate uptake by salt cedar (*Tamarix ramosissima*) in the Las Vegas Wash riparian ecosystem. Science of the Total Environment 256:227–232.

Uselman, S. M., K. A. Snyder, and R. R. Blank. 2010. Insect biological control accelerates leaf litter decomposition and alters short-term nutrient dynamics in a *Tamarix*-invaded riparian ecosystem. Oikos 120:409–417.

Uselman, S.M., K. A. Snyder, R. R. Blank, and T. J. Jones. 2011. UVB exposure does not accelerate rates of litter decomposition in a semiarid riparian ecosystem. Soil Biology and Biochemistry 43:1254–1265.

Vogelsang, K. M., and J. D. Bever. 2009. Mycorrhizal densities decline in association with nonnative plants and contribute to plant invasion. Ecology 90:399–407.

Waisel, Y. 1961. Ecological studies on *Tamarix aphylla* (L.) Karst., III. The salt economy. Plant and Soil 13:356–364.

Whitcraft, C.R., L. A. Levin, D. Talley, and J. A. Crooks. 2008. Utilization of invasive tamarisk by salt marsh consumers. Oecologia 158:259–272.

Wilson, G. W. T., C. W. Rice, M. C. Rillig, A. Springer, and D. C. Hartnett. 2009. Arbuscular mycorrhizal fungi control soil aggregation and carbon sequestration. Ecology Letters 12:452–461.

Yan, F., S. Schubert, and K. Mengel. 1996. Soil pH increase due to biological decarboxylation of organic anions. Soil Biology and Biochemistry 28:617–624.

Yin, C. H., G. Feng, F. Zhang, C. Y. Tian, and C. Tang. 2009. Enrichment of soil fertility and salinity by tamarisk in saline soils on the northern edge of the Taklamakan Desert. Agricultural Water Management 97:1978–1986.

Yun, K. W., and S. K. Choi. 2002. Mycorrhizal colonization and plant growth affected by aqueous extract of *Artemisia princeps* var. *orientalis* and two phenolic compounds. Journal of Chemical Ecology 28:353–361.

Zhaoyong, S., L. Zhang, G. Feng, C. Peter, T. Changyan, and L. Xiaolin. 2006. Diversity of arbuscular mycorrhizal fungi associated with desert ephemerals growing under and beyond the canopies of tamarisk shrubs. Chinese Science Bulletin 51:132–139.

14

Fire Ecology of *Tamarix*

Gail M. Drus

Invasive nonnative plants are often associated with increased wildfire risk, primarily as a consequence of their growth form and structural characterisics that differ from the plants that have been replaced (Brooks et al. 2004). All plants are flammable and can support wildfire when dry enough, but many invasive plants are particularly fire-prone because they develop high density and/or monocultural stands. More contiguous plant material means fewer gaps that might have disrupted the spread of flames in heterogeneous native vegetation, so fires tend to carry over a greater extent in homogenous stands (Brooks et al. 2004). When natural or anthropogenic fires start, the likelihood of wildfire spread increases under these conditions. This relationship is illustrated throughout western North America where various invasive grasses have promoted greater frequency and extent of severe wildfires in recent decades and led to further alteration of plant community composition (D'Antonio and Vitousek 1992). Increased incidence of wildfire costs billions of dollars a year in damage to structures, natural resources, and in the costs of suppressing fires (Zybach et al. 2009).

Tamarisk (*Tamarix* spp.) invasion is similarly associated with increased wildfire frequency and intensity in riparian ecosystems of the arid West (Busch and Smith 1995), and with negative socioeconomic and ecological consequences (Shafroth et al. 2005). When riparian terraces adjacent to agricultural fields, housing, and other structures become densely invaded by tamarisk and other nonnative plants, such as giant reed (*Arundo* spp.), the associated increase in fire risk threatens people and property alike (Coffman et al. 2010). Riparian zone fires also increase erosion, sedimentation, and algal blooms in desert rivers that provide limited water resources supporting agriculture, hydroelectric power, and other infrastructure vital to the cities in the arid southwest (Dwire and Kauffman 2003;

Subhadra 2010). Riparian biodiversity is also threatened by increased flammability and fire frequency in riparian zones, as native species may be unable to adapt to increased fire frequency. For example, cottonwood (*Populus* spp.) and willow (*Salix* spp.) have declined dramatically in recent decades along regulated (i.e., dammed) rivers throughout western North America concurrent with tamarisk invasion (Friedman et al. 2005). Although cottonwoods and willows have been shown to resprout following fire (Ellis 2001), their ability to recover declines as fire intensity increases, whereas tamarisk is more tolerant to more-frequent and higher-intensity fires (Heywood 1989; Ellis 2001). Cottonwoods are especially sensitive to fire (Pattison et al. 2011); they provide not only valuable habitat structure for wildlife, but are a desirable species because their size and shade are amenities for public land use (Knight 1997). The loss of native species and associated biodiversity may have lasting negative impacts on desert riparian ecosystems that formerly provided stable water sources and sustained desert wildlife populations.

As dense tamarisk monocultures replace diverse, open, native gallery forests dominated by cottonwood and willow, desert riparian systems are being converted from barriers to pathways for wildfire spread (Busch and Smith 1992; Dudley et al. 2000). Fires in native gallery forests are relatively rare and occur only after severe drought when vegetation becomes desiccated (Gregory et al. 2003). In contrast, a regular fire interval of 10 to 20 years has been documented in tamarisk-dominated areas in the lower Colorado River (Ohmart and Anderson 1982; Lovich et al. 1994), and is becoming the norm in other riparian areas of the American Southwest (Dudley et al. 2011). While this suggests that tamarisk is altering fire regimes in riparian zones, the role of fire in riparian ecosystems in general is poorly known because riparian fire is historically infrequent (Pettit and Naiman 2007).

Fire Characteristics in *Tamarix* Stands

FUEL COMPOSITION

Tamarisk is considered a hazardous fuel throughout the southwestern states because its branch, twig, and leaf structure is very fine and dense, and it is highly flammable even when green and hydrated (Racher 2003; Dwire et al. 2010). Fine fuels are defined as leaves and stems with diameters less than 0.625 centimeters (Pyne et al. 1996). They are more easily ignited than coarser, woody fuels because their high surface-to-volume ratio accelerates moisture release from leaf tissues and speeds the rate of ignition and combustion (White and Zipperer 2010; Sala and Smith 1996). The surface-to-volume ratio measured in native species such as Gooding's willow (*Salix goodingii*) and Fremont cottonwood (*Populus fremontii*) is lower than in tamarisk because their leaves are larger (see figure 14.1). This means that tamarisk ignites more easily than native species; in timed ignition trials tamarisk burned faster than foliage from both *Populus fremontii* and *Salix goodingii* (Dudley et al. 2011).

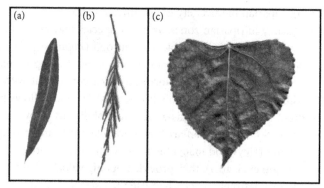

FIGURE 14.1 Leaf structure of (a) Gooding's willow (*Salix goodingii*), (b) tamarisk (*Tamarix* spp.) and (c) Fremont cottonwood (*Populus fremontii*). Tamarisk foliage is more finely divided than either willow or cottonwood.

Source: Photos courtesy of Gail Drus.

The amount of fine fuels, including foliage, stems, and litter, and the way they are distributed within the tree architecture also influence flammability of vegetation (van Wagtendonk 2006). Destructive sampling of tamarisk trees has shown that of total dry biomass, one third is composed of fine fuels, and that the fuels are distributed evenly throughout the vertical profile of the tree (Dudley et al. 2011). Dead stems and branches are often interspersed among living parts of the tree, as tamarisk often has partial die-back in response to prolonged drought or inundation (Horton 1977; Everitt 1980). As tamarisk drops its foliage from mid-autumn to early winter in preparation for dormancy, litter is often caught in dense branches (Wilkinson 1966). Litter also accumulates in the understory, creating a layer of duff that is flammable year-round (Weisenborn 1996), particularly in the absence of overbank flooding that removes litter from the understory (Ellis et al. 1998; see also chapter 8, this volume). Collectively, the distribution of foliage, litter, and woody debris creates a laddered fuel structure (see figure 14.2) that allows fire to climb up from the understory into the tree canopy. Under large tamarisk trees, deep layers of duff accumulate and can smolder for days following fire. Prolonged smoldering has been shown to damage root cambium in many woody species (Sugihara et al. 2006), but tamarisk can resprout vigorously from scorched root crowns (see figure 14.3).

Foliar moisture, in addition to fuel-particle size and packing density, is an important factor contributing to the flammability of vegetation. Moisture content declines as leaves senesce, or during prolonged periods of drought. Desiccated foliage is more easily burned than hydrated foliage because less energy is needed to raise the material to the temperature of ignition. However, the effect of foliar desiccation on flammability is moderate when compared to the high overall flammability of tamarisk vegetation (Dudley et al. 2011). Tamarisk leaves contain volatile compounds (Saïdana et al. 2008), which likely contribute to its high flammability

FIGURE 14.2 Accumulation of litter and debris in tamarisk vegetation.
Source: Photo courtesy of Gail Drus.

FIGURE 14.3 Tamarisk vigorously resprouts from the root crown's meristematic tissue protected under the soil after a fire kills aboveground growth.
Source: Photo courtesy of Tim Carlson, Tamarisk Coalition.

even when hydrated. Such organic compounds are hydrocarbons with low boiling points, which elevate flammability (Owens et al. 1998).

VEGETATION STRUCTURE AND FIRE BEHAVIOR

Basic measures of fire behavior include flame lengths, rate of spread and fire intensity, which are influenced by weather and topography, and are positively related to density and fuel loads that tend to increase with nonnative plant invasions (Sugihara et al. 2006). Dense canopy cover, high fuel loads (biomass of flammable materials), and flammable foliage promote fire behavior that can threaten human life and property. Flame lengths and rate of spread are enhanced by dry windy conditions and by closed canopy cover in dense vegetation (Racher et al. 2001). Flame lengths greater than 40 meters have been reported in tamarisk stands with closed canopies in southern Nevada and New Mexico (Racher et al. 2001; Dudley et al. 2011). Flame lengths greater than 30 meters are considered extreme, and have resulted in the loss of human life and property in many fire-prone systems, such as chaparral (Riggan et al. 1994).

Accurate predictions of rate of spread and flame lengths in various fuel structures and environmental conditions are vital to the ability of managers to contain fires, and to safely and effectively conduct prescribed burns. "Fire brands," small, air-borne pieces of burning fuel, facilitate the spread of fires in tamarisk vegetation by igniting fires as far away as 162 meters during prescribed burns, and are likely to travel much farther during wildfires (Racher 2003). During prescribed burns, fire intensity can be increased by adding fuels or piling slash, or by slowing the fire front (e.g., using backing fires instead of head fires; DiTomaso et al. 2006).

TAMARISK RESPONSE TO FIRE

Tamarisk trees have many adaptations that allow them to tolerate fire. Its deep roots are not affected by surface fire, because its root crown meristems are below the surface of the soil. Root crowns are the part of the root system from which stems arise, and provide carbohydrate storage in tamarisk and in other woody plants (Sosebee 2004, 85–95). When fire kills stems, new ones can grow from the root crown. This resprouting following fire requires substantial carbohydrate reserves in woody plants (DeLoach et al. 2000). When aboveground tissues are damaged or consumed by fire, the root to shoot ratio is substantially increased (i.e., there is more root biomass than shoot biomass), which enhances nitrogen and phosphorus consumption and improves plant water status (tissue water content), thereby further accelerating the initial growth from meristems (Carreira and Niell 1992). As aboveground tissues regrow, carbohydrate availability becomes a more important determinant of recovery than even access to water or nutrients (DeSouza et al. 1986).

In addition to being able to regrow from the root crown, tamarisk has other characteristics that allow it to tolerate fire. One of these is salt tolerance (chapter 8, this volume). During combustion, basic cations are liberated from organic compounds and deposited in the ash, which increases soil salinity and pH (van Wagtendonk 2006). Ash deposits from tamarisk-fueled fires are particularly high in sodic salt and boron, compounds that are not well tolerated by other species (Busch and Smith 1993). In contrast, tamarisk is not only salt tolerant, but also boron resistant, facilitating its more rapid post-fire recovery relative to other species (Zhang et al. 2002). Further, tamarisk's drought tolerance (see chapters 5 and 6, this volume), also contributes to its postfire recovery. Wildfires in tamarisk vegetation most commonly occur in the summer (Dudley et al. 2011) when the ambient temperature is high and the relative humidity is low; surviving plants are further stressed by high solar radiation and elevated rates of evapotranspiration in burned areas. The deep root system and drought-tolerance of tamarisk enable it to overcome these stressful conditions.

Community Level Dynamics

A TRAJECTORY TOWARD DOMINANCE

Riparian vegetation is considered fire resistant owing to the humidity associated with high foliar moisture (Dwire and Kauffman 2003). These characteristics may inhibit the movement of wildfire from adjacent upland areas into the stream environment. Because tamarisk is more flammable than native species, it has fueled fires in many desert river systems, especially the Colorado River and its major tributaries across the lower Colorado basin (Arizona and New Mexico), in the Mojave Desert (the Virgin River in Nevada and Utah), and in the Rio Grande and the Pecos River in the Chihuahuan Desert (New Mexico and Texas; Ohmart et al. 1988; Taylor and McDaniel 1998). Along the lower Colorado and Bill Williams Rivers, 37% of tamarisk-dominated riparian areas burned during a 12-year period, and an increase in fire number and area was observed during this period (Busch 1995).

As tamarisk increases fire frequency, native riparian species have been displaced, because although forests dominated by native trees are less likely to burn, native trees do not have the same suite of adaptations as tamarisk to tolerate fire. To document this pattern, we conducted a survey of riparian areas across the American Southwest where fires occurred and observed that more native trees were killed in areas of high tamarisk density than where there were few or no tamarisk (see figure 14.4; Dudley et al. 2011). Many riparian species resprout following fire, but the ability to resprout has been shown to decrease with increased fire intensity (Ellis 2001), and species such as Fremont cottonwood are especially sensitive (Rood et al. 2007). Other investigators have observed less recovery by tamarisk than by natives when tamarisk is sparse in the understory (Stromberg and Rychener 2010), which is likely due to much lower fire intensity. As tamarisk density increases, high fuel loads likely promote higher fire intensity, making recovery by native species especially difficult.

A positive feedback cycle develops in which tamarisk dominance is established and maintained by fire. Such a relationship has been documented in other invasive species (Brooks et al. 2004). In systems where tamarisk has expanded to monocultures, high intensity fires have become very common. Each fire promotes greater abundance of tamarisk in relation to the native plants, a trajectory resulting in a dense monoculture with a 10- to 20-year fire cycle (Ohmart and Anderson 1982, 433–479). Greater fire intensity has been measured where the stem density and standing biomass of litter and woody materials (fuel loads) are greater (Dudley et al. 2011). The greater flammability of tamarisk relative to native vegetation,

FIGURE 14.4 A 2008 fire along the Rio Grande burned a vegetation assemblage consisting of a Fremont cottonwood (*Populus fremontii*) overstory and an understory consisting of tamarisk (*Tamarisk* spp.) and coyote willow (*Salix exigua*). The willows and tamarisk were completely consumed by the fire. In 2010 (*photo*), the remaining Fremont cottonwoods were dead, and tamarisk resprouts dominated the understory.
Source: Photo courtesy of Gail Drus.

higher survival rates following fire, and increased fire intensity as tamarisk density increases are contributing to a fire cycle that results in decline and eventual loss of native species and promotes tamarisk dominance (see figure 14.5).

BIOLOGICAL CONTROL MAY ALTER THE TRAJECTORY

Biocontrol by the tamarisk leaf beetle (*Diorhabda carinulata*) has caused considerable tamarisk death in many areas (Pattison et al. 2010), but there have been concerns that desiccation associated with herbivory increases tamarisk flammability (USDA 2010). Higher tamarisk flammability, even over the short term, could have large-scale impacts since more than 600,000 hectares of riparian systems in the Southwestern United States are invaded by tamarisk, and the tamarisk leaf beetle has rapidly spread across at least nine states (Dudley and DeLoach 2004; Dalin et al. 2009; see also chapter 22, this volume). Thus it is critical to understand how this impact to tamarisk stands may alter fire frequency and intensity, and ultimately plant community composition.

PHYSICAL EFFECTS OF BIOLOGICAL CONTROL ON TAMARISK FIRE BEHAVIOR

Leaf beetle herbivory moderately enhances fire risk and behavior in tamarisk-dominated areas by altering fine fuel properties. When the beetles

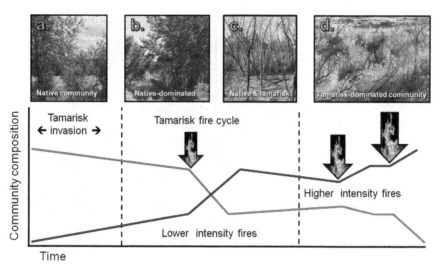

FIGURE 14.5 (a) Native riparian community composed of willows and cottonwoods. (b) Tamarisk colonizes openings in the native-dominated community increasing the proportion of flammable fine fuels. (c) Fires in native-dominated communities allow both natives and tamarisk to resprout. (d) Tamarisk is a more efficient resprouter than natives and comprises a greater proportion of the assemblage following each fire, which contributes to greater flammable biomass and greater fire intensity with successive fires. A tamarisk monoculture results, with a fire interval of 10 to 20 years. *Source:* Photos courtesy of Gail Drus.

consume tamarisk foliage, they scrape the cuticle of the leaves and stems, which damages the vascular system causing the foliage to turn brown and die (see chapter 22, this volume). Much of the dead, dry foliage remains intact and distributed throughout all levels of the tamarisk vegetation structure for a time before falling to the forest floor. Individual trees and entire tamarisk stands can become desiccated quickly, resulting in large areas of dry vegetation (Pattison et al. 2011). However, after the foliage has dropped, tamarisk trees can regrow leaves within the same season, decreasing fire risk but also creating a new food source for the beetle. Foliar desiccation can therefore fluctuate over a single growing season as beetle populations can undergo several life cycles defoliating the same trees several times in a year (Bean et al. 2007).

We conducted prescribed burn experiments at sites in the Great Basin and Mojave Desert to examine the influence of herbivory on the fire behavior and recovery of tamarisk (see figure 14.6). The continuity of fuels was an important factor influencing fire behavior. At the Great Basin site, beetle herbivory produced patchy distributions of dead foliage, thus creating variability in desiccation, as some trees were able to regrow more foliage than others after prolonged herbivory. This variability in fuels added an element of discontinuity to the vegetation structure, and the effect of herbivory on fire behavior at this site was relatively minor (Brooks et al. 2008). In contrast, when herbivory was simulated with herbicide

FIGURE 14.6 Application of prescribed burn treatments at the (a) Mojave and (b) Great Basin sites.

Source: Photos courtesy of Gail Drus.

at the Mojave site, the effect of foliar desiccation on fire behavior (longer flame lengths and faster rates of spread) and fire intensity was more obvious (Dudley et al. 2011). This is probably because the herbicide produced uniform dead foliage throughout the trees at the stand, and fine fuels were more evenly desiccated and continuously distributed throughout the trees. Similar effects of foliar condition on fire behavior have been observed in coniferous forests when bark-boring beetles damage the vascular system, resulting in dead needles that remain intact and evenly distributed throughout the tree canopy (Kulakowski et al. 2003; Knight 1987). However, the effect of herbivory desiccation on flame lengths and other fire behavior measurements was modest in comparison to effects of weather and fuel structure (Dudley et al. 2011).

PHYSIOLOGICAL EFFECTS OF BIOLOGICAL CONTROL ON POST-FIRE TAMARISK RECOVERY

Herbivory has been shown to enhance the ability of fire to kill plants in other systems, such as slash pine forests, where bark boring beetle infestations make the trees disproportionately more susceptible to fire damage (Lichtenthaler 1998; McCullough et al. 1998; Menges and Deyrup 2001). This is also the case in tamarisk exposed to repeated leaf beetle herbivory. At our Great Basin site, variability in herbivory impact and recovery among tamarisk individuals was related to the plants' starch content. Decreased root-crown carbohydrates, starch in particular, are associated with tamarisk exposed to greater beetle herbivory damage (Hudgeons et al. 2007). Greater post-fire mortality occurred in tamarisk with greater herbivory damage and lower carbohydrate stores, as fewer resources were available for resprouting (Brooks et al. 2008). Carbohydrate availability has been shown to be a more important determinant of post-fire recovery than access to water or nutrients (DeSouza et al. 1986).

We observed complex relationships between fire timing, plant physiological status and tamarisk mortality at the Great Basin site (Brooks et al. 2008). Mortality caused by integrated fire and herbivory treatments was greater than mortality due to fire alone (0%–10%; Drus unpublished data; C. Deuser pers. comm.) or biocontrol alone (60%; Pattison et al. 2010), and summer burns caused greater mortality than did the fall burn (Brooks et al. 2008). Greater tamarisk mortality during summer months is attributed to higher fire intensity, with greater damage to root crown tissues (Howard et al. 1983). Higher fire intensity is promoted by high temperatures and low humidity during summer months (van Wagtendonk 2006). However, greater mortality observed in summer relative to fall burn treatments are probably due to seasonal plant physiology, carbohydrates in particular, and may play a larger role in post-fire tamarisk mortality than fire intensity alone (Drus, unpublished data).

Carbohydrate stores in tamarisk and other deciduous species fluctuate throughout the seasons; stores are low during periods of growth when energy is needed to produce leaves, and high during periods of dormancy to sustain the plant until the next growing season (Howard et al. 1983). Because carbohydrate stores fluctuate over the year, the ability of plants to recover from fire will also vary with timing. Accordingly, we observed greater tamarisk mortality at the Great Basin site as a result of the summer burn when carbohydrates were supporting active growth, and lower mortality in the fall burn when the plants were beginning to store carbohydrates in preparation for dormancy (Drus unpublished data). In the same area, almost no tamarisk mortality occurred in a fire when the plants were dormant, indicating that carbohydrate stores were abundant (Drus unpublished data), although it is also possible that less root damage occurred.

At the Great Basin site, herbivory and fire stress interacted synergistically to produce tamarisk mortality greater than the sum of their individual effects. One stressor (herbivory) causes physiological strain to the organism that puts it at greater risk to additional stressors, such as fire (Sih et al. 1998; Alexieva et al. 2003). Synergisms have received considerable attention from ecologists concerned that anthropogenic stressors such as pollutants may alter established species interactions, and because synergistic interactions have important applications in pest management (Schindler 1987; Hatcher 1995; Hay 1996). For example, in the biological control program for water hyacinth (*Eichhornia crassipes*), weevil (*Neochetina* spp.) herbivory increased the plant's susceptibility to secondary microbial infection (Charudattan 1986). At the Great Basin site, we compared observed tamarisk mortality to hypothetical mortality generated by the sum of individual fire and herbivory effects. The difference between the observed versus predicted mortality was a measure of the strength of the interaction (synergism) between fire and herbivory. The strongest interaction between fire and herbivory occurred at lower herbivory levels during the summer burn than the fall burn, indicating that greater herbivory stress is necessary to cause tamarisk mortality during the fall burn to compensate for lower fire intensity (Drus unpublished data).

Management Implications

Prescribed fire is sometimes used to manage tamarisk because the vegetation structure promotes rapid fire spread, high fire intensity, and good fuel consumption. Large amounts of aboveground biomass can be removed quickly by prescribed fire, and fire costs less than other control techniques. However, burning is only effective as a control measure in combination with other approaches because of tamarisk's adaptations to tolerate fire. Burning biomass after herbicide application, biological control, or root-raking are all common methods for removal (see chapters 22 and 21, this volume). In general, the ability of tamarisk to regrow following fire precludes the use of fire as an actual control method.

Although fire itself is rarely employed to kill tamarisk, other forms of tamarisk control can increase wildfire risk if they result in the accumulation of dry woody or leaf material; this is generally of most concern in the context of biocontrol. While tamarisk flammability is moderately increased over the short-term by beetle herbivory because of increasing the availability of dry fuels, flammability will decline over the long term because herbivory reduces the production of fine fuels (Pattison et al. 2011). Managers can reduce the fire risk associated with tamarisk vegetation more effectively and frugally with the knowledge that fire risk is moderately enhanced when the beetle establishes and defoliates tamarisk; fire suppression efforts can be focused in areas with the greatest amount of desiccated fine fuels where wildfires are most likely to occur, and land managers can apply prescribed fires when maximum tamarisk mortality is likely to occur. Management plans should also consider the vegetation structure in each tamarisk-invaded area, because denser infestations may prove a greater fire risk than less-dense infestations, especially those desiccated by the leaf beetle. When used in combination with biocontrol, prescribed burning can be a valuable management tool that can quickly reduce fuel loads in high-risk areas, thus reducing the risk of wildfire.

Conclusion

In the American Southwest, the shift in riparian community composition from cottonwood-willow gallery forests to tamarisk monocultures is proving a difficult trajectory to alter, given the complexity of the physical and physiological factors promoting the establishment and dominance of tamarisk. The reduced flooding,

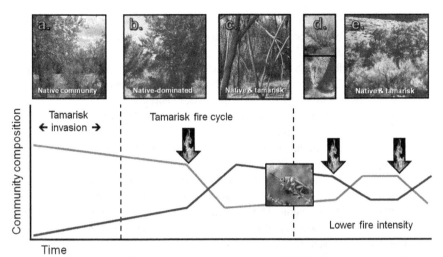

FIGURE 14.7 (a) Native riparian community composed of willows and cottonwoods. (b) Tamarisk colonizes openings in the native-dominated community increasing the proportion of flammable fine fuels. (c) Fires in native-dominated communities allow both natives and tamarisk to resprout. (d) The tamarisk leaf beetle reduces the resprouting ability of tamarisk, and increases tamarisk mortality over time. (e) The proportion of flammable fine fuels and fire intensity are reduced, allowing coexistence of tamarisk and natives.

Source: Photos courtesy of Gail Drus and Andrew Norton (beetle inset).

lowered water tables, and increased fire frequency associated with tamarisk invasion are likely to persist over the long term and to continue promoting tamarisk. However, tamarisk's superior physiological tolerance to environmental stresses (temperature extremes, salinity, drought and anoxia) may be diminished by leaf beetle herbivory. Synergisms between fire and herbivory may shift the system from one dominated by tamarisk to one where the coexistence of natives is possible, as increased tamarisk mortality over time will decrease flammable biomass and fire intensity in tamarisk monocultures, allowing native survival and recovery (see figure 14.7). Although eradication of tamarisk may not be possible, coexistence of native vegetation may restore biodiversity and habitat value to tamarisk-invaded areas.

Literature Cited

Alexieva, V., S. Ivanov, I. Sergiev, and E. Karanov. 2003. Interaction between stresses. Bulgarian Journal of Plant Physiology 29:1–17.

Bean, D. W., T. L. Dudley, and J. C. Keller. 2007. Seasonal timing of diapause induction limits the effective range *of Diorhabda elongata deserticola* (Coleoptera: Chrysomelidae) as a biological control agent for tamarisk (*Tamarix* spp.). Environmental Entomology 36:15–25.

Brooks, M. L., C. M. D'Antonio, D. M. Richardson, J. B. Grace, J. Keeley, J. M. DiTomaso, R. J. Hobbs, M. Pellant, and D. Pyke. 2004. Effects of invasive alien plants on fire regimes. BioScience 54:677–688.

Brooks, M., T. Dudley, G. Drus, and J. Matchett. 2008. Reducing wildfire risk by integration of prescribed burning and biological control of invasive tamarisk (*Tamarix* spp.). USGS Administrative Report, Project 05-2-1-18. El Portal, CA.

Busch, D. E. 1995. Effects of fire on southwestern riparian plant community structure. Southwestern Naturalist 40:259–267.

Busch, D. E., and S. D. Smith. 1993. Effects of fire on water and salinity relations of riparian woody taxa. Oecologia 94:186–194.

Busch, D. E., and S.D. Smith. 1992. Fire in a riparian shrub community: Postburn water relations in the *Tamarix-Salix* association along the lower Colorado River. Pages 52–55 *in* W.P. Clary, M. E. Durant, D. Bedunah, and C. L. Wambolt, comp. Proceedings of Ecology and Management of Riparian Shrub Communities. USDA-FS GTR-INT-289. USDA Forest Service. Fort Collins, CO

Busch, D. E., and S. D. Smith. 1995. Mechanisms associated with decline of woody species in riparian ecosystems of the southwestern US. Ecological Monographs 65:347–370.

Carreira J.A., and F. X. Niell. 1992. Plant nutrient changes in a semi-arid Mediterranean shrubland after fire. Journal of Vegetation Science 3:457–466.

Charudattan, R. 1986. Integrated control of waterhyacinth (*Eichhornia crassipes*) with a pathogen, insects, and herbicides. Weed Science 34:26–30.

Coffman, G. C., R. F. Ambrose, and P. W. Rundel. 2010. Wildfire promotes dominance of invasive giant reed (*Arundo donax*) in riparian ecosystems. Biological Invasions 12:2723–2734.

D'Antonio, C. M., and P. M. Vitousek. 1992. Biological invasions by exotic grasses, the grass/fire cycle, and global change. Annual Review of Ecology and Systematics 23:63–87.

Dalin, P., M. J. O'Neal, T. Dudley, and D. W. Bean. 2009. Host plant quality of *Tamarix ramosissima* and *T. parviflora* for three sibling species of the biocontrol insect *Diorhabda elongata* (Coleoptera: Chrysomelidae). Environmental Entomology 38:1373–1378.

DeLoach, C. J., R. I. Carruthers, J. E. Lovich, T. L. Dudley, and S. D. Smith, 2000. Ecological interactions in the biological control of saltcedar (*Tamarix* spp.) in the United States: Toward a new understanding. Pages 819–873 *in* N. R. Spencer, editor. Proceedings of the X International Symposium on Biological Control of Weeds, 4–14 July 1999. Montana State University, Bozeman.

DeSouza, J., P. A. Silka, and S. D. Davis. 1986. Comparative physiology of burned and unburned *Rhus laurina* after chaparral wildfire. Oecologia 71:63–68.

DiTomaso, J. M., M. L. Brooks, E. B. Allen, R. Minnich, P. M. Rice, and G. Kyser. 2006. Control of invasive weeds with prescribed burning. Weed Technology 20:535–548.

Drus, G. M. Unpublished data.

Dudley, T., M. Brooks, S. Ostoja, P. Shafroth, S. Roberts, G. Drus, M. Kuehn, I. Kuczynska, K. Hultine, K. Acharya, B. Conrad, D. Bean, and C. Deuser. 2011. Effectiveness Monitoring of Springfed Wetlands and Riparian Restoration Treatments. Pages 47–68, Final report for the Clark County Multiple Species Habitat Conservation Plan.

Dudley, T. L., and C. J. DeLoach. 2004. Saltcedar (*Tamarix* spp.), endangered species, and biological weed control: Can they mix? Weed Technology 18:1542–1551.

Dudley, T. L., C. J. DeLoach, J. E. Lovich, and R. I. Carruthers. 2000. Saltcedar invasion of western riparian areas: Impacts and new prospects for control. Transactions of the North American Wildlife and Natural Resources Conference 65:345–381.

Dwire, K. A., C. C. Rhoades, and M. K. Young. 2010. Potential Effects of Fuel Management Activities on Riparian Areas. Pages 181–205, Cumulative Watershed Effects of Fuel Management in the Western United States. USDA and US Forest Service, Rocky Mountain Research Station. Fort Collins CO. RMRS-GTR-231.

Dwire, K. A., and J. B. Kauffman. 2003. Fire and riparian ecosystems in landscapes of the western USA. Forest Ecology and Management 178:61–74.

Ellis, L. M. 2001. Short-term response of woody plants to fire in a Rio Grande riparian forest, Central New Mexico, USA. Biological Conservation 97:159–170.

Ellis, L. M., C. S. Crawford, and M. C. Molles Jr. 1998. Comparison of litter dynamics in native and exotic riparian vegetation along the middle Rio Grande of central New Mexico, USA. Journal of Arid Environments 38:283–296.

Everitt, B. L. 1980. Ecology of saltcedar: A plea for research. Environmental Geology 3:77–84.

Friedman, J. M., K. R. Vincent, and P. B. Shafroth. 2005. Dating floodplain sediments using tree-ring response to burial. Earth Surface Processes and Landforms 30:1077–1092.

Gregory, S. V., K. L. Boyer, and A. M. Gurnell. 2003. The ecology and management of wood in world rivers. American Fisheries Society. Bethesda, MD.

Hatcher, P. E. 1995. Three-way interactions between plant pathogenic fungi, herbivorous insects and their host plants. Biological Reviews 70:639–694.

Hay, M. E. 1996. Defensive synergisms reply to Pennings. Ecology 77:1950–1952.

Heywood, V. H. 1989. Patterns, extents and modes of invasions by terrestrial plants. Pages 31–60 *in* J. A. Drake, F. di Castri, R. H. Groves, F. J. Kruger, H. A. Mooney, M. Rejmanek, and M. Williams, editors. *Biological Invasions: A Global Perspective*. Wiley, New York, NY.

Horton, J. S. 1977. The development and perpetuation of the permanent tamarisk type in the phreatophyte zone of the Southwest. Pages 123–127 *in* R. R. Johnson and D. A. Jones, technical coordinators. Symposium on the Importance, Preservation, and Management of Riparian Habitat. 9 July 1977, Tucson, AZ. USDA Forest Service, General Technical Report RM 43. Fort Collins, CO.

Howard, S. W., A. E. Dirar, J. O. Evens, and R. D. Provenza. 1983. The use of herbicides and/or fire to control saltcedar (*Tamarix*). Proceedings of the Western Society of Weed Science 36:67–72.

Hudgeons, J. L., A. E. Knutson, K. M. Heinz, C. J. DeLoach, T. L. Dudley, R. R. Pattison, and J. R. Kiniry. 2007. Defoliation by introduced *Diorhabda elongata* leaf beetles (Coleoptera: Chrysomelidae) reduces carbohydrate reserves and regrowth of *Tamarix* (Tamaricaceae). Biological Control 43:213–221.

Knight, D. H. 1987. Parasites, lightning, and the vegetation mosaic in wilderness landscapes. Pages 59–83 *in* M. G. Turner, editor. *Landscape Heterogeneity and Disturbance*. Springer-Verlag, New York.

Knight, R. L. 1997. Wildlife habitat and public use benefits of treatment wetlands. Water Science and Technology 35:35–44.

Kulakowski D., T. T. Veblen, and P. Bebi (2003) Effects of fire and spruce beetle outbreak legacies on the disturbance regime of a subalpine forest in Colorado. Journal of Biogeography 30:1445–1456.

Lichtenthaler, H. K. 1998. The stress concept in plants: An introduction. Annals of the New York Academy of Sciences 851:187.

Lovich, J. E., T. B. Egan, and R. C. de Gouvenain. 1994. Tamarisk control on public lands in the desert of Southern California: Two case studies. Proceedings of the California Weed Conference 46:166–177.

McCullough, D. G., R. A. Werner, and D. Neumann. 1998. Fire and insects in northern and boreal forest ecosystems of North America. Annual Review of Entomology 43:107–127.

Menges, E. S., and M. A. Deyrup. 2001. Postfire survival in south Florida slash pine: Interacting effects of fire intensity, fire season, vegetation, burn size, and bark beetles. International Journal of Wildland Fire 10:53–63.

Ohmart, R. D., and B. W. Anderson. 1982. *North American Desert Riparian Ecosystems. Reference Handbook on the Deserts of North America.* Greenwood Press, Westport, CT.

Ohmart, R. D., B. W. Anderson, and W. C. Hunter. 1988. The ecology of the lower Colorado River from Davis Dam to the Mexico–United States international boundary: A community profile. Biological Report 85 (7.19), U.S. Fish and Wildlife Service, Washington, DC.

Owens, M. K., C. D. Lin, C. A. Taylor, and S. G. Whisenant. 1998. Seasonal patterns of plant flammability and monoterpenoid content in Juniperus ashei. Journal of Chemical Ecology 24:2115–2129.

Pattison, R. R., C. M. D'Antonio, and T. L. Dudley. 2010. Biological control reduces growth, and alters water relations of the saltcedar tree (*Tamarix* spp.) in western Nevada, USA. Journal of Arid Environments 75:346–352.

Pattison, R. R., C. M. D'Antonio, T. L. Dudley, K. K. Allander, and B. Rice. 2011. Early impacts of biological control on canopy cover and water use of the invasive saltcedar tree (*Tamarix* spp.) in western Nevada, USA. Oecologia 165:605–616.

Pettit, N. E., and R. J. Naiman. 2007. Fire in the riparian zone: Characteristics and ecological consequences. Ecosystems 10:673–687.

Pyne, S. J., P. L. Andrews, and R. D. Laven. 1996. *Introduction to Wildland Fire.* Wiley and Sons, New York.

Racher, B. 2003. Prescription development for burning two volatile fuel types. Doctoral thesis. College of Agricultural Sciences and Natural Resources, Texas Tech University, Lubbock.

Racher, B. J., R. B. Mitchell, C. Schmidt, and J. Bryan. 2001. Prescribed burning prescriptions for saltcedar in New Mexico. Research Highlights 1999: Noxious Brush and Weed Control 30:14–15.

Riggan, P. J., R. N. Lockwood, P. M. Jacks, C. G. Colver, F. Weirich, L. F. DeBano, and J. A. Brass. 1994. Effects of fire severity on nitrate mobilization in watersheds subject to chronic atmospheric deposition. Environmental Science and Technology 28:369–375.

Rood, S. B., L. A. Goater, J. M. Mahoney, C. M. Pearce, and D. G. Smith. 2007. Floods, fire, and ice: Disturbance ecology of riparian cottonwoods. Botany 85:1019–1032.

Saïdana, D., M. A. Mahjoub, O. Boussaada, J. Chriaa, I. Chéraif, M. Daami, Z. Mighri, and A. N. Helal. 2008. Chemical composition and antimicrobial activity of volatile compounds of *Tamarix boveana* (Tamaricaceae). Microbiological Research 163:445–455.

Sala, A., and S. D. Smith. 1996. Water use by *Tamarix ramosissima* and associated phreatophytes in a Mojave Desert floodplain. Ecology Applications 6:888–898.

Schindler, D. W. 1987. Detecting ecosystem responses to anthropogenic stress. Canadian Journal of Fisheries and Aquatic Sciences 44:6–25.

Shafroth, P. B., J. R. Cleverly, T. L. Dudley, J. P. Taylor, C. Van Riper III, E. P. Weeks, and J. N. Stuart. 2005. Control of *Tamarix* in the western United States: Implications for water salvage, wildlife use, and riparian restoration. Environmental Management 35:231–246.

Sih, A., G. Englund, and D. Wooster. 1998. Emergent impacts of multiple predators on prey. Trends in Ecology and Evolution 13:350–355.

Sosebee, R. E. 2004. Timing: The key to successful brush management with herbicides. Pages 85–98 *in* W. T. Hamilton, A. McGinty, D. N. Ueckert, C. W. Hanselka, and M. K. Lee, editors. *Brush Management: Past, Present, Future.* Texas A&M University Press, College Station.

Stromberg, J. C., and T. J. Rychener. 2010. Effects of fire on riparian forests along a free-flowing dryland river. Wetlands 30:1–12.

Subhadra, B. G. 2010. Overuse could leave Southwest high and dry. Science 329:1282.

Sugihara, N. G., J. van Wagtendonk, K. E. Shaffer, J. Fites-Kaufman, and A. E. Thode, editors. 2006. *Fire in California's Ecosystems.* University of California Press, Berkeley.

Taylor, J. P., and K. C. McDaniel. 1998. Restoration of saltcedar (*Tamarix* sp.)-infested floodplains on the Bosque del Apache National Wildlife Refuge. Weed Technology 12:345–352.

USDA. 2010. USDA APHIS PPQ Moratorium for Biological Control of Saltcedar (*Tamarix* species) using the biological control agent *Diorhabda* species (Coleoptera: Chrysomelidae). USDA, Animal and Plant Health Inspection Service.

van Wagtendonk, J. 2006. Fire as a physical process. Pages 38–73 *in* N. G. Sugihara, J. W. van Wagtendonk, J. Fites-Kaufman, K. E. Shaffer, and A. E. Thode, editors. *Fire in California Ecosystems.* University of California Press, Berkeley.

Weisenborn, W. D. 1996. Saltcedar impacts on salinity, water, fire frequency, and flooding. Pages 9–12 *in* Proceedings of the saltcedar management workshop. University of California Cooperative Extension, Rancho Mirage.

White, R. H., and W. C. Zipperer. 2010. Testing and classification of individual plants for fire behaviour: Plant selection for the wildland–urban interface. International Journal of Wildland Fire 19:213–227.

Zhang, D., L. Yin, and B. Pan. 2002. Biological and ecological characteristics of *Tamarix* L. and its effect on the ecological environment. Science in China Series D: Earth Sciences 45:18–22.

Zybach, B., M. Dubrasich, B. Brenner, and J. Marker. 2009. US Wildfire Cost-Plus-Loss Economics Project: The One-Page Checklist. Pages 1–20. Wildland Fire Lessons Learned Center. Advances in Fire Practice.

15

Tamarix

PASSENGER OR DRIVER OF ECOSYSTEM CHANGE?

Tyler D. Johnson

Nonnative species invasions have long been associated with changes in biological, ecological, and evolutionary trajectories. However, in recent years an increasing number of critiques of invasion ecology have suggested that the impacts of these species have been overstated or are simply incorrect (e.g., Gurevitch and Padilla 2004; Sax and Gaines 2008). One such genus receiving this scrutiny is *Tamarix* L. (e.g., Stromberg and Chew 2002a, b; Stromberg et al. 2009a). *Tamarix* invasion has occurred both in watersheds with dammed and/or channelized rivers and in rivers that are largely unaltered (Stromberg et al. 2007; Merritt and Poff 2010).

Dams and other structures that change the historic volume, velocity, or pattern of water discharge in the system are said to alter the hydrograph. The hydrograph for a given system can have varying effects on water quality and quantity, soil chemistry, wildlife habitat, floral and faunal communities, and fire regimes. However, there has been much debate in the literature about whether *Tamarix* invasion has similar influence and is the cause (Vitousek 1990; Busch and Smith 1993; Cleverly et al. 1997; Di Tomaso 1998; see also chapters 8, 9, and 14, this volume) or the consequence (Fenner et al. 1985; Anderson 1998; Stromberg 1998; Shafroth et al. 2002; Merritt and Poff 2010; see also chapters 5 and 16, this volume) of hydrograph changes. It is clear that the range of *Tamarix* has grown substantially in the century since its introduction (Brock 1994; Friedman et al. 2005) and has been predicted to continue to spread due to extent of suitable habitat (Morisette et al. 2006; Kerns et al. 2009) and its ability to adapt to colder climates (Sexton et al. 2002; Young et al. 2004; Friedman et al. 2008; Lehnhoff et al. 2011).

In areas that have experienced substantial ecosystem alteration associated with the establishment and spread of invasive species, the question arises whether the invading species caused the system to change (driver) or whether the invasion was facilitated by ecosystem change (passenger) (MacDougall and Turkington 2005; see also chapter 1, this volume). There is evidence that *Tamarix* has both direct and indirect interactions with factors that structure riparian plant communities. Riparian plant communities are generally structured by bottom-up controls that are functions of regional climate and local weather climate, hydrologic variables, fire dynamics and the salinity of the soil and groundwater (among rather a lot else). Here, I focus on the interaction that *Tamarix* has with these four most-studied ecological factors and how the models of passenger or driver fit with each.

Tamarix as a Passenger of Ecosystem Change

A passenger of ecosystem change achieves its dominance by tolerating or capitalizing on changes in the ecosystem that are outside the sphere of influence of the plant itself. The passenger is weakly interactive with the environment where it replaces native species. The invasive species' dominance is facilitated by ecosystem change that is not of the invader's making. The management of a passenger should focus on halting or reversing the change that allowed the invader to flourish (MacDougall and Turkington 2005).

The central tenet of the passenger model of ecosystem change is that the ecosystem is indeed changing, and it is clear that riparian areas of the American Southwest have undergone many alterations in the last century (Hastings and Turner 1965; Turner and Karpiscak 1980; Turner et al. 2003; Webb and Leake 2006; Webb et al. 2007). This period of change, beginning after 1880 (Hastings and Turner 1965), is concurrent with some of the early reports of *Tamarix* in the western United States (see chapter 16, this volume). Alteration in hydrology is clearly a driving mechanism of environmental degradation of regulated riparian areas. Evidence that *Tamarix* is a passenger of this degradation comes from its ability to tolerate the conditions, such as dry, salty soils and a vegetative community that is susceptible to fire, associated with intense hydrologic change. This suggests that where intensive severe hydrologic alteration has occurred, *Tamarix* is a passenger to that change.

TAMARIX AS A PASSENGER OF HYDROLOGY AND CLIMATE CHANGE

Hydrologic changes in southwestern riparian areas include complete or partial cessation of flow from dams built for storage, recreation or hydroelectricity, partial or complete diversion of stream flow for irrigation, and groundwater pumping (Webb and Leake 2006; Stromberg et al. 2007). These practices serve to reduce, or in some cases increase, base flow and alter natural flood regimes (Shafroth et al. 2002; Stromberg and Chew 2002a). Following river damming, there is a sudden and broad alteration in fluvial and alluvial dynamics that is generally followed by an adjustment in riparian vegetation, (Shafroth et al. 2002; Cooper et al. 2003; Webb and Leake 2006; Stromberg et al. 2007; Dewine and Cooper 2008). Below dams and impoundments in the southwestern United States a novel plant community, often dominated by *Tamarix*, has frequently developed (Merritt and Cooper 2000; Stromberg et al. 2007). In a region-wide study examining the influence of flow regulation and flow permanence on riparian vegetation, Stromberg et al. (2007) found that stream flow permanence is a strong determinant of riparian vegetation. They found that in naturally intermittent streams *Tamarix* was able to dominate at the expense of native trees, indicating that the traits described earlier also give *Tamarix* an advantage in unaltered riparian ecosystems. However, the effect of flow intermittency was amplified if the stream was intermittent from impoundment by a dam.

Region-wide patterns in atmospheric drivers have also been proposed to have caused a general expansion of woody plant habitat in the American Southwest, *Tamarix* among them. Climatic patterns that may have helped drive *Tamarix* range expansion include an increase in atmospheric CO_2 and a lengthening of the growing season, as well as an increase in both winter precipitation and total precipitation that influenced both the availability of water and flood disturbance (Turner et al. 2003). The increase in growing season length and disturbance frequency may have created or expanded unoccupied niches that *Tamarix* was able to exploit. Range-expansion models suggest that climatic variables are a key determinant of *Tamarix* distribution and potential for spread (Morisette et al. 2006; Bradley et al. 2009; Kerns et al. 2009).

Critical to understanding the traits that have allowed *Tamarix* to thrive as a passenger is its ability to tolerate conditions that are generally associated with fluvial modification or naturally dry environments. Evidence suggests that *Tamarix* is quite resistant to xylem cavitation induced by severe drought (Pockman and Sperry 2000; see chapter 9, this volume) and is a facultative phreatophyte, meaning that it can use both groundwater and water closer to the surface via capillary rise (Smith et al. 1998; Horton et al. 2001; Horton et al. 2003). *Tamarix* has also been demonstrated to access deeper ground water than some native riparian tree species (Horton and Campbell 1974). This is important because after impoundment, the downstream ground water table can drop below the rooting depth of some native trees. *Tamarix* can survive this change because it sustains contact with soil water, maintains tight stomatal control of transpiration, and uses vadose zone water when it is available (Horton et al. 2003; see also chapter 6, this volume). When water stress does occur in *Tamarix*, it is still able to operate at reduced water potentials that are inhibitory or lethal to native trees (Devitt et al. 1997; Horton et al. 2001; Horton et al. 2003). Such water stress can be due to precipitation patterns (i.e., atmospheric) and/or ground and surface water availability (i.e., hydrologic).

In addition to drought tolerance, seedling establishment patterns influence *Tamarix* dominance after flow alteration. In arid and semiarid riparian systems seeds of many native and exotic plant species require bare sediment created by natural flooding (Stromberg and Chew 2002a; Birken and Cooper 2006; Richardson et al. 2007). Native southwestern riparian trees have fairly brief periods of seed set, dispersion, and viability that coincide with receding winter and spring flood waters (Stromberg and Chew 2002a). In contrast, *Tamarix* can flower and set seed all summer long and can produce seedlings following late summer monsoons (Horton et al. 1960) or after historically late flooding events caused by flow regulation (see chapter 7, this volume). Both the native trees, such as willows and cottonwoods, and tamarisk, have relatively brief seed viability (see chapter 19, this volume). Flood regulation may therefore allow *Tamarix*, which is a reproductive generalist and is less dependent on the timing of flooding, to germinate and grow over a period of months in any growing season. However, where hydrologic conditions are favorable to native species establishment, they can out-compete *Tamarix* (Sher et al. 2000; Sher et al. 2002; Sher and Marshall 2003). This is further evidence that the condition of the ecosystem, rather than the mere presence of *Tamarix,* is most important in the initial invasion.

TAMARIX AS A PASSENGER OF FIRE

Fire in riparian areas is one of many ecosystem processes that can be altered by a change in flood regime and water availability. Reduced over-bank flooding allows fuels to accumulate on the ground as well as suspended in the canopy of plants. These fuels would normally have been removed or covered with sediment under an active flood regime (Stromberg and Chew 2002a). A change in fire frequency and severity can drive ecosystem change by altering competitive interactions that normally keep *Tamarix* in a less than dominant role (see chapter 14, this volume). *Tamarix*, as well as most of the common native riparian tree species, can resprout following fire, although this ability decreases in native trees with increasing age (see Glenn and Nagler 2005 for review of the subject; but see Stromberg and Rychener 2010). Some native riparian trees, such as Gooddings willow (*Salix gooddingii* C.R.Ball) resprout strongly after fire (Stromberg et al. 2009b).

   Tamarix may have several advantages over native trees when regenerating after and proliferating wildfire. Most *Tamarix* plants have shrublike characteristics (i.e., multiple stems, lower canopy branches, and a continuous vertical canopy) and a tendency to resprout vigorously and form stands with greater stem density than native trees (Bagstad et al. 2006). Resprouted stems can flower in their first year of growth, allowing *Tamarix* to colonize areas left bare by fire.

TAMARIX AS A PASSENGER OF SALINITY

In addition to altering fire regimes, a lack of flood pulsing can allow salts to accumulate in riparian soils and groundwater (Jolly et al. 1993; Merritt and Cooper 2000; Nagler et al. 2008; see also chapter 8, this volume). Soil salinity is another abiotic factor that moderates competitive interactions between *Tamarix* and native plants. *Tamarix* capitalizes on elevated salinity in soils that excludes native riparian plants not adapted to high salt conditions.

Tamarix as a Driver of Ecosystem Change

An invasive species that can drive ecosystem change must either out-compete native species or create vacant niche space that it can occupy. If an invasive species can dominate an area by its own alteration of the environment, it may perpetuate the alteration to the benefit of itself, and possibly other species. Regardless of the mechanism of establishment, its sheer numerical dominance, in many cases, will be enough to alter the system in some manner (MacDougall and Turkington 2005).

There are several instances where *Tamarix* is likely to drive changes in ecosystem function and biological diversity. These arise from the same mechanisms that affect it as a passenger: the alteration of hydrologic processes, fire regime alteration, and soil and groundwater salinization. Demonstrating that *Tamarix* is responsible for some of the change it is associated with is difficult, since correlation is not causation—but the mechanisms are very clear, and more compelling evidence is emerging.

TAMARIX AS A DRIVER OF HYDROLOGY

One way to identify where *Tamarix* was itself the agent of change in hydrology is to look at *Tamarix* on unregulated rivers, or before dams were built on that river. In an innovative study on the Green River in Utah, Birken and Cooper (2006) identified populations of *Tamarix* that established on bare in-stream sand deposits and actively invaded the floodplain *before* the river was regulated by the Flaming Gorge Dam. The study demonstrated that channel narrowing followed *Tamarix* establishment and spread. The channel narrowing was due to increased sediment catchment by dense *Tamarix* stands, followed by more seedling establishment on the newly created substrate (Birken and Cooper 2006). *Tamarix* stands have significantly higher stem density than native tree (but not shrub) species (Bagstad et al. 2006). The greater stem density of *Tamarix* changes the proportions of in- and near-stream alluvial structures, changing the sediment storage and subsequent vegetation pattern of the areas dominated by *Tamarix* (Cooper et al. 2003; see also chapter 7, this volume). A recent study concluded that *Tamarix* would have spread (though not necessarily dominated) throughout the west even without hydrologic alteration (Merritt and Poff 2010).

Much has been discussed and debated about *Tamarix* water use, and whether high evapotranspiration rates of this exotic plant are responsible for changes in the ecosystem (see chapters 4, 5, and 6, this volume). Despite many anecdotal accounts of *Tamarix* drying up streams and lowering reservoirs, it is exceedingly difficult to prove that *Tamarix* alone is the cause of the change. However, it is safe to say that when vegetation of any type occupies a niche that was previously bare (such as the sandbars and the highly arid environments that *Tamarix* can tolerate better than many native species), water in that system is affected.

TAMARIX AS A DRIVER OF FIRE

Tamarix has been described as a fire promoter, causing a frequent and high-intensity fire regime in some of the areas that it now dominates (Busch and Smith 1995; Zouhar 2003; see also chapter 14, this volume). There are many traits that may make *Tamarix* more prone to fire, including high stem density and many dead twigs. In thickets, dense and overlapping *Tamarix* canopies maintain continuous fuel loading (Zouhar 2003 and references therein). The vertical distribution of *Tamarix* foliage can function as a fuel ladder for fire to enter the crown of the thicket (see chapter 14, this volume). Further, mature *Tamarix* can produce significant fuel loads (dead twigs and leaves) over the course of a year. It has been shown that fire in riparian areas is more prevalent in *Tamarix*-dominated areas where fuel loadings were higher than in native vegetation (Busch 1995).

Most arid areas of potential *Tamarix* habitat have natural fire return intervals of decades to centuries, and some, such as the Sonoran Desert, seem not to have a historic fire regime at all (Zouhar 2003). The ecosystem response of any riparian system to fire is dependent on context. Fire may function as a natural disturbance in ecosystems with an evolutionary exposure to frequent fire, as is the case in the Chihuahuan Desert (Stromberg et al. 2009b). However, if fire is an exogenous disturbance (disturbance outside the evolutionary history of the system) to a riparian system, as it seems to be in the Sonoran Desert, the effects to the riparian ecosystem can be far-reaching. Following burning of *Tamarix*, the organic layer can be completely reduced to mineral ash (Bess et al. 2002), and the soil can become saline and very dry (Busch and Smith 1993).

While *Tamarix* may have traits to act as a fire promoter, a lack of seasonal flooding can lead to an accumulation of fuels that would have been removed under a natural flow regime (Stromberg and Chew 2002a). Thus, the prevalence of fire in riparian areas may be due more to river regulation than the fuel-loading capacity of *Tamarix*. However, *Tamarix* may speed the change in riparian ecosystems by shortening the fire return interval and capitalizing on the physical changes incurred by fire. If *Tamarix* acts as fire promoter, it stands to reason that in nonriparian areas, away from rivers with a changed hydrograph, it is altering fire regimes and driving a change in the ecology there.

TAMARIX AS A DRIVER OF SALINITY

Vegetation change driven by increased soil salinity is often attributed to *Tamarix* invasion. With its capacity to take up salts in the groundwater and to exude them via salt glands, there is certainly a visible mechanism (see chapter 8, this volume). High but variable soil salinities under dense *Tamarix* stands have been documented (e.g., Merritt and Cooper 2000; Shafroth et al. 2002), along with compelling evidence of dramatic increases in groundwater salinity due to some excretion of salts by the roots during water uptake (Nagler et al. 2008). Most of these studies are correlative, and do not demonstrate that *Tamarix* is responsible for the elevated salinity. One of the few ways cause is truly indicated is when salinity increases with time in a *Tamarix* stand to a greater degree than in native stands. Research designed to investigate this found that surface-soil salinity did indeed increase with the presence of *Tamarix* over time, but only outside levees, where flooding cannot flush accumulated surface salts (see chapter 8, this volume). The oldest stands also sometimes had very low salinity levels.

For native riparian trees, experimentally high levels of soil salinity have been shown to stunt cottonwood germination (Shafroth et al. 1995) and seedling growth both above and below ground (Rowland et al. 2004). This suggests strongly that if *Tamarix* does indeed salinize the soil or groundwater it could drive ecosystem change, although there is little field data available that directly tests this hypothesis. The accumulation of salts on floodplain soils and groundwater may be, like fire, context dependent. Under an active flood regime, floodwaters may remove accumulated salts thereby partially mitigating the driven change in salt accumulation.

Conclusion and Future Directions

To date, it appears that *Tamarix* in the American Southwest is both a passenger of and a driver of ecosystem change (Sher 2006; see also chapter 1 this volume). Research has documented that widespread vegetation changes below dams and other impoundments have allowed *Tamarix* to establish, likely as a result of the reduced competition from native species. However, *Tamarix* that was present on rivers before impoundment shows signs of driving ecosystem change; the impoundment may have accelerated change by altering the flood regime, after which *Tamarix* altered fire regimes and soil chemistry. In areas with little obvious ecosystem alteration, *Tamarix* can drive ecosystem change by exploiting vacant niches and tolerating, and in some cases proliferating, altered ecosystem functions (sediment, salt, and fire dynamics). It is clear, however, that most of the changes in areas dominated by *Tamarix* result from alteration of the natural hydrograph, both because of the direct effect of this change on the abiotic environment, and because these changes promoted the dominance of a new species, *Tamarix*. Thus, even where *Tamarix* removal is warranted, attention to the underlying hydrological issues is critical.

Whether this view of a combined passenger-driver model for *Tamarix* invasion holds for the leading edge of *Tamarix* expansion is unknown at this point. The influence that a changing climate will have on the expansion of *Tamarix* will change how the passenger/driver dichotomy describes *Tamarix* invasion. Within a changing climate regime and an expanding introduced range, the *Tamarix* syngameon will most likely continue to evolve to tolerate colder climates (Sexton et al. 2002; Friedman et al. 2008) and could exploit vacant niches at higher latitudes and elevations. Indeed, *Tamarix* has been observed as naturalized in British Colombia, Canada, and at 8500 feet in elevation in New Mexico.

There are limitations to describing exotic species invasions using a passenger/driver dichotomy. To have clear situations of passenger or driver, research must be able to show clear cause and effect relationships. Some experimental designs (i.e., control/treatment–before/after study designs) are not often feasible in the natural environment, especially in dynamic ecosystems such as riparian areas of the American Southwest. One recent observational study (Merritt and Poff 2010) concluded that *Tamarix* is mostly a passenger of the hydrologic change that reduced the competitive exclusion of *Tamarix* by native species. In general, it is simpler to demonstrate passenger effects than driver; the latter requires a very long experiment, since this is a long-lived species, and control of confounding variables.

There may be no definitive answer to the question of passenger or driver for *Tamarix*, but this dichotomy is a useful tool for describing what is known about an invasive species and its interaction with the ecosystem. Knowing the components and balance of passenger versus driver may suggest different management strategies and is an additional conceptual tool for managing invasive species in altered environments. It is likely that most invaders will be described as both a passenger and a driver (MacDougall and Turkington 2005), as I have done here with *Tamarix*. Whether *Tamarix* will continue its dual role in its new range is a matter for future research.

Literature Cited

Anderson, B. W. 1998. The case for saltcedar. Restoration and Management Notes 16:130–134.

Bagstad, K. J., S. J. Lite, and J. C. Stromberg. 2006. Vegetation, soils, and hydrogeomorphology of riparian patch types of a dryland river. Western North American Naturalist 66:23–44.

Bess, E. C., R. R. Parmenter, S. McCoy, and M. C. Molles. 2002. Responses of a riparian forest-floor arthropod community to wildfire in the middle Rio Grande Valley, New Mexico. Environmental Entomology 31:774–784.

Birken, A. S., and D. J. Cooper. 2006. Processes of *Tamarix* invasion and floodplain development along the lower Green River, Utah. Ecological Applications 16:1103–1120.

Bradley, B. A., M. Oppenheimer, and D. S. Wilcove. 2009. Climate change and plant invasions: Restoration opportunities ahead? Global Change Biology 15:1511–1521.

Brock, J. H. 1994. *Tamarix* spp. (saltcedar), an invasive exotic woody plant in arid and semi-arid riparian habitats of western USA. Pages 27–44 *in* L. C. de Waal, L. E. Child, P. M. Wade and J. H. Brock, editors. *Ecology and Management of Invasive Riverside Plants*. John Wiley and Sons Ltd. Chichester, UK, and New York.

Busch, D. E. 1995. Effects of fire on southwestern riparian plant community structure. The Southwestern Naturalist 40:259–267.

Busch, D. E., and S. D. Smith. 1993. Effects of fire on water and salinity relations of riparian woody taxa. Oecologia 94:186–194.

Busch, D. E., and S. D. Smith. 1995. Mechanisms associated with decline of woody species in riparian ecosystems of the southwestern US. Ecological Monographs 65:347–370.

Cleverly, J. R., S. D. Smith, A. Sala, and D. A. Devitt. 1997. Invasive capacity of *Tamarix ramosissima* in a Mojave Desert floodplain: The role of drought. Oecologia 111:12–18.

Cooper, D. J., D. C. Andersen, and R. A. Chimner. 2003. Multiple pathways for woody plant establishment on floodplains at local to regional scales. Journal of Ecology 91:182–196.

Devitt, D. A., J. M. Piorkowski, S. D. Smith, J. R. Cleverly, and A. Sala. 1997. Plant water relations of *Tamarix ramosissima* in response to the imposition and alleviation of soil moisture stress. Journal of Arid Environmnets 36:527–540.

Dewine, J. M., and D. J. Cooper. 2008. Canopy shade and the successional replacement of tamarisk by native box elder. Journal of Applied Ecology 45:505–514.

Di Tomaso, J. M. 1998. Impact, biology, and ecology of saltcedar (*Tamarix* spp.) in the southwestern United States. Weed Technology 12:326–336.

Fenner, P., W. W. Brady, and D. R. Patton. 1985. Effects of regulated water flows on regeneration of Fremont cottonwood. Journal of Range Management 38:135–138.

Friedman, J. M., G. T. Auble, P. B. Shafroth, M. L. Scott, M. F. Merigliano, M. D. Freehling, and E. R. Griffin. 2005. Dominance of nonnative riparian trees in western USA. Biological Invasions 7:747–751.

Friedman, J. M., J. E. Roelle, J. F. Gaskin, A. E. Pepper, and J. R. Manhart. 2008. Latitudinal variation in cold hardiness in introduced *Tamarix* and native *Populus*. Evoloutionary Applications 1:598–607.

Glenn, E. P., and P. L. Nagler. 2005. Comparative ecophysiology of *Tamarix ramosissima* and native trees in western US riparian zones. Journal of Arid Environmnets 61:419–446.

Gurevitch, J. and D. K. Padilla. 2004. Are invasive species a major cause of extinctions? Trends in Ecology and Evolution 19:470–474.

Hastings, J. R., and R. M. Turner. 1965. *The Changing Mile: An Ecological Study of Vegetation Change with Time in the Lower Mile of an Arid and Semiarid Region*. University of Arizona Press, Tucson.

Horton, J. L., S. C. Hart, and T. E. Kolb. 2003. Physiological condition and water source use of Sonoran Desert riparian trees at the Bill Williams River, AZ. Isotopes in Environmental and Health Studies 39:69–82.

Horton, J. L., T. E. Kolb, and S. C. Hart. 2001. Responses of riparian trees to interannual variation in ground water depth in a semi arid river basin. Plant, Cell and Environment 24:293–304.

Horton, J. S., and C. J. Campbell. 1974. Management of phreatophyte and riparian vegetation for maximum multiple use values. Paper RM-117. US Forest Service Research, Fort Collins, CO.

Horton, J. S., F. C. Mounts, and J. M. Kraft. 1960. Seed germination and seedling establishment of phreatophyte species. Rocky Mountain Forest and Range Experiment Station Paper 48. Ft. Collins, CO.

Jolly, I. D., G. R. Walker, and P. J. Thorburn. 1993. Salt accumulation in semi-arid floodplain soils with implications for forest health. Journal of Hydrology 150:589–614.

Kerns, B. K., B. J. Naylor, M. Buonopane, C. G. Parks, and B. Rogers. 2009. Modeling tamarisk (*Tamarix* spp.) habitat and climate change effects in the northwestern United States. Invasive Plant Science and Management 2:200–215.

Lehnhoff, E. A., F. D. Menalled, and L. J. Rew. 2011. Tamarisk (*Tamarix* spp.) establishment in its most northern range. Invasive Plant Science and Management 4:58–65.

MacDougall, A. S., and R. Turkington. 2005. Are invasive species the drivers or passengers of change in degraded ecosystems? Ecology 86:42–55.

Merritt, D. M., and D. J. Cooper. 2000. Riparian vegetation and channel change in response to river regulation: A comparative study of regulated and unregulated streams in the Green River Basin, USA. Regulated Rivers: Research & Management 16:543–564.

Merritt, D. M., and N. L. R. Poff. 2010. Shifting dominance of riparian *Populus* and *Tamarix* along gradients of flow alteration in western North American rivers. Ecological Applications 20:135–152.

Morisette, J. T., C. S. Jarnevich, A. Ullah, W. Cai, J. A. Pedelty, J. E. Gentle, T. J. Stohlgren, and J. L. Schnase. 2006. A tamarisk habitat suitability map for the continental United States. Frontiers in Ecology and the Environment 4:11–17.

Nagler, P. L., E. P. Glenn, K. Didan, J. Osterberg, F. Jordan, and J. Cunningham. 2008. Wide area estimates of stand structure and water use of *Tamarix* spp. on the lower Colorado River: Implications for restoration and water management projects. Restoration Ecology 16:136–145.

Pockman, W. T., and J. S. Sperry. 2000. Vulnerability to xylem cavitation and the distribution of Sonoran desert vegetation. American Journal of Botany 87:1287–1299.

Richardson, D. M., P. M. Holmes, K. J. Esler, S. M. Galatowitsch, J. C. Stromberg, S. P. Kirkman, P. Pyšek, and R. J. Hobbs. 2007. Riparian vegetation: degradation, alien plant invasions, and restoration prospects. Diversity and Distributions 13:126–139.

Rowland, D. L., A. A. Sher, and D. L. Marshall. 2004. Inter-and intra-population variation in seedling performance of Rio Grande cottonwood under low and high salinity. Canadian Journal of Forest Research 34:1458–1466.

Sax, D. F., and S. D. Gaines. 2008. Species invasions and extinction: The future of native biodiversity on islands. Proceedings of the National Academy of Sciences 105:11490.

Sexton, J. P., J. K. McKay, and A. Sala. 2002. Plasticity and genetic diversity may allow saltcedar to invade cold climates in North America. Ecological Applications 12:1652–1660.

Shafroth, P. B., J. C. Stromberg, and D. T. Patten. 2002. Riparian vegetation response to altered disturbance and stress regimes. Ecological Applications 12:107–123.

Shafroth, P. B., J. M. Friedman, and L. S. Ischinger. 1995. Effects of salinity on establishment of *Populus fremontii* (cottonwood) and *Tamarix ramosissima* (saltcedar) in southwestern United States. Western North American Naturalist 55:58–65.

Sher, A. A. 2006. Tamarisk: Passenger vs. driver of ecosystem change. ESA Special Session: Passengers versus Drivers of Ecosystem Change: Current Debate on *Tamarix* and Riparian Invasion. A. A. Sher and J. Stromberg, organizers. Ecological Society of America 91st Annual Meeting, Memphis, TN.

Sher, A. A., and D. L. Marshall. 2003. Seedling competition between native *Populus deltoides* (Salicaceae) and exotic *Tamarix ramosissima* (Tamaricaceae) across water regimes and substrate types. American Journal of Botany 90:413–422.

Sher, A. A., D. L. Marshall, and J. P. Taylor. 2002. Establishment patterns of native *Populus* and *Salix* in the presence of invasive nonnative *Tamarix*. Ecological Applications 12:760–772.

Sher, A. A., D. L. Marshall, and S. A. Gilbert. 2000. Competition between native *Populus deltoides* and invasive *Tamarix ramosissima* and the implications for reestablishing flooding disturbance. Conservation Biology 14:1744–1754.

Smith, S. D., D. A. Devitt, A. Sala, J. R. Cleverly, and D. E. Busch. 1998. Water relations of riparian plants from warm desert regions. Wetlands 18:687–696.

Stromberg, J. 1998. Dynamics of Fremont cottonwood (*Populus fremontii*) and saltcedar (*Tamarix chinensis*) populations along the San Pedro River, Arizona. Journal of Arid Environmnets 40:133–155.

Stromberg, J. C., and M. K. Chew. 2002a. Flood pulses and restoration of riparian vegetation in the American Southwest. Pages 11–49 *in* B. A. Middleton, editor. *Flood Pulsing in Wetlands: Restoring the Natural Hydrological Balance*. John Wiley and Sons, New York.

Stromberg, J. C. and M. K. Chew. 2002b. Foreign visitors in riparian corridors of the American Southwest: is xenophytophobia justified? Pages 195–219 *in* B. Tellman, editor. *Invasive Exotic Species in the Sonoran Region*. University of Arizona Press, Tucson.

Stromberg, J. C., M. K. Chew, P. L. Nagler, and E. P. Glenn. 2009a. Changing perceptions of change: The role of scientists in *Tamarix* and river management. Restoration Ecology 17:177–186.

Stromberg, J. C., S. J. Lite, R. Marler, C. Paradzick, P. B. Shafroth, D. Shorrock, J. M. White, and M. S. White. 2007. Altered stream flow regimes and invasive plant species: the *Tamarix* case. Global Ecology and Biogeography 16:381–393.

Stromberg, J. C., and T. J. Rychener. 2010. Effects of fire on riparian forests along a free-flowing dryland river. Wetlands 30:75–86.

Stromberg, J. C., T. J. Rychener, and M. D. Dixon. 2009b. Return of fire to a free-flowing desert river: Effects on vegetation. Restoration Ecology 17:327–338.

Turner, R. M., and M. M. Karpiscak. 1980. Recent vegetation changes along the Colorado River between Glen Canyon Dam and Lake Mead, Arizona. Professional Paper 1132. US Geological Survey, US Government Printing Office, Washington, DC.

Turner, R. M., R. H. Webb, J. E. Bowers, and J. R. Hastings. 2003. *The Changing Mile Revisited*. University of Arizona Press, Tucson.

Vitousek, P. M. 1990. Biological Invasions and ecosystem processes: Towards an integration of population biology and ecosystem studies. Oikos 57:7–13.

Webb, R. H., and S. A. Leake. 2006. Ground-water surface-water interactions and long-term change in riverine riparian vegetation in the southwestern United States. Journal of Hydrology 320:302–323.

Webb, R. H., S. A. Leake, and R. M. Turner. 2007. *The Ribbon of Green: Change in Riparian Vegetation in the Southwestern United States*. University of Arizona Press, Tucson.

Young, J. A., C. D. Clements, and D. Harmon. 2004. Germination of seeds of *Tamarix ramosissima*. Rangeland Ecology and Management 57:475–481.

Zouhar, K. 2003. *Tamarix* spp. Fire Effects Information System. US Forest Service, Washington, DC.

PART III

The Human Element

16

Tamarisk Introduction, Naturalization, and Control in the United States, 1818–1952

Matthew K. Chew

The history of tamarisk in the United States is one of scientists creating a monster. The first two phases of that process are summarized here, and a third one introduced. During phase one (c.1800–1938), tamarisks were celebrated and imported for qualities deemed beneficial. During phase two (1938–1952) the plants were disparaged for the same qualities and reframed as monsters. Phase three was the first era of federally organized tamarisk suppression, which continued into the 1970s. While looking for ways to conserve water, hydrologists and engineers became obsessed with eradicating tamarisk, a goal that offered some prospect of measurable success, while developing a strident antipathy toward the plants that overshadowed their original value. The participants were "men of their time" with jobs to accomplish and little patience for conflicting interests. Reporting the ongoing depredations of a monster offered a way to justify continued government action, and their own programs and employment.

Monsters "explode all of [our] standards for harmony, order and ethical conduct" (Gilmore 2003). They are "mistakes of nature," combining "characteristic components or properties of different kinds of living things or natural objects" and even (as here) artifacts and technologies (Gilmore 2003). They "signal borderline experience of uncontainable excess" (Kearney, 2003). Two centuries after the debut of Frankenstein, monsters still proliferate wherever science suffuses culture. "Each discipline...pushes back a whole teratology of knowledge beyond its margins...there are monsters on the prowl whose form changes with the history of knowledge" (Foucault 1970). But we need not rely on theories. A modern major newspaper declared tamarisk a "water-gulping, fire-feeding, habitat-ruining, salt-spreading monster" (Hartmann 2001) as if those were established botanical facts—even as continuing research rendered the claims increasingly untenable.

Phase One: Planting, Promotion, and Naturalization

A 1964 US Forest Service Research Note and a 1965 US Geological Survey (USGS) Professional Paper remain the best records of the American introduction and early spread of the genus *Tamarix*. These accounts provided anecdotes suitable to the needs of subsequent authors; but a historian cautioned, "Scientific bulletins from our experiment stations are a great source of folklore. When scientists write the results of their own experiments, they are clinical. But when they compose the "historical" sections of their bulletins, they are gullible as all get-out" (Isern 1997). The two papers appear well-researched as far as they go, but details of the arrival and earliest dispersal of tamarisks in North America remain obscure. So far, the earliest mention I have found of tamarisk growing in America is from Cambridge, Massachusetts (Peck 1818). Spotty documentation of commercial distribution survives in nursery catalogs from as early as 1823 on the Atlantic coast and about 1854 on the Pacific (Robinson 1965).

EARLY PROPAGATION

By 1868, six tamarisk species grew at the US Department of Agriculture (USDA) Arboretum in Washington, DC (Robinson 1965). Their shrubby habit and salt tolerance suggested utility for controlling beach erosion, and inspired the colloquial name "saltcedar." Army engineers hoping to stabilize a Texas barrier island planted saltcedars in the 1880s (Mansfield 1886). An Arizona rancher recalled that around 1898 his father stuck "a switch of a new plant he had found growing in a sandbar along the Gila River...in the moist soil of a ditch [where it became] a considerable clump of saltcedar" (Robinson 1965). Concurrently, a Rhode Island newspaperman echoed an earlier paean: "nothing can be more exquisitely graceful than the entire habit of this plant, and it is especially attractive in the early morning when its branches droop under the weight of silvery dew" (Davis 1899).

In 1897 the USDA's Section of Foreign Seed and Plant Introduction (SPI) made prospecting for useful plants a federal-government function (Pauly 2007). SPI staffer Mark Carleton's 1914 "special article" for *Science* was titled "Adaptation of the Tamarisk for Dry Lands." Carleton called tamarisk "the most drought resistant and otherwise hardy of all the trees and shrubs planted on [his own Texas farmstead]...there appears to be no limit in dryness of the soil on any usual Great Plains farm beyond which this plant will not survive." Furthermore, tamarisk was easily propagated by cuttings, so many plants could be realized from a small opening investment (Carleton 1914). A USDA ornithologist contributed a letter to *Science* noting that tamarisk flourished in wet Louisiana soils as well (McAtee 1914).

Shortly thereafter, botanist J. J. Thornber penned the University of Arizona's *Timely Hints for Farmers,* no. 121, summarizing tests of several tamarisk varieties in the Sonoran Desert. Recommending them for ornament, hedges, windbreaks and shade trees he wrote, "doubtless these plants will be put to other uses as they come to be better known" (Thornber 1916). Thornber is credited with introducing the evergreen athel tamarisk to America after procuring cuttings from an Algerian correspondent (Swingle 1924). Luna Leopold recalled that his father, Aldo, "planted a tamarisk in front of their house in Albuquerque, New Mexico about 1920" calling it "rather uncommon" (Robinson 1965). They lived "on a large lot with backyard access to the Rio Grande" (Meine 1988).

USGS geologist Kirk Bryan reported, "three plantations of the self-seeding saltcedar were set out [to retard gullying] in October, 1926 on the Rio Puerco by the middle Rio Grande Conservancy District... in July, 1927 [they] were growing vigorously" (Everitt 1980). *The California Cultivator,* a trade newspaper, reported, "Throughout Southern California there has been planted... quite a good deal of [athel] for windbreaks... it is a most rapid grower and is quite effective in checking the sweep of winds within three years" (Miller 1928). Tamarisk was gaining a reputation for screening sunlight, and mitigating wind and water flows under harsh southwestern conditions.

Mrs. Joseph Measures of Colorado extolled ornamental tamarisks in a 1933 article for *Flower Grower* magazine. But she also summarized the status of tamarisk in the southwest and hinted at the adaptability, dispersibility, and riparian habit that eventually characterized the genus. "In America, [tamarisk] has taken foothold in many places. In Colorado it is found in the alkali washes with the Willows and the native Cottonwoods, in waste areas; and as specimen shrubs in yards. In New Mexico, Arizona, and some parts of California it is used for windbreaks and division fences. Although it is a semi-tropical shrub [*sic*], it is adapting itself to places where the temperatures often reach ten to twenty degrees below zero" (Measures 1933).

NATURALIZED TAMARISK

The first naturalized American stand of tamarisks was reported on Galveston Island as "*Tamarix gallica*—I think" (Joor 1877). By 1900, tamarisks grew unsown and untended in Texas, Arizona, Utah, and California. T. H. Kearney collected tamarisk in Tempe, Arizona, calling it "common in river bottoms," while J. J. Thornber took a specimen in eastern Arizona's Safford Valley; populations were documented in New Mexico by the mid-1910s (Robinson 1965). Herbarium records accumulated, but these "escaped" plants otherwise attracted little interest. A Texas ornithologist included tamarisk as a component of the "Coast Prairie Thicket Association" without commenting on its origins or arrival (Oberholser 1925).

McMillan Dam was privately constructed on the Pecos River in 1893-1894. Lake McMillan was a sediment trap. A 1904 USGS survey found its water storage capacity diminished by over 22% after 10 years. Much of this sediment formed a delta at the reservoir inlet. By 1914, tamarisks were growing there. University of Texas dean of engineering T.U. Taylor evaluated Lake McMillan in 1930, finding that tamarisks occupied over 12,000 acres, one-fourth of the lake's original surface area. He quoted a 1928 letter from L. E. Foster of the Bureau of Reclamation describing a ten-square-mile tamarisk woodland occupying parts of the delta and upstream flood plain: "In many places the growth is so dense as to be almost impenetrable…[ranging] in height from twenty feet down to one foot or less. The diameter of the growth ranges from the size of a pencil to six or eight inches. Just how much effect this comparatively fine, dense growth has on the [water] velocities is not certain; but it is certain, however, that the heavier silt deposits are at the upper end of the area where such growth is more dense" (Taylor 1930). Foster saw the tamarisks as a useful trap within a trap, concentrating sediments at the upstream end of the reservoir. Taylor carried this idea forward, titling a subsection "Tamarisk on Guard" and writing that "the silt problems at McMillan Reservoir…have been materially lessened by the accidental propagation of this foreign evergreen shrub. Possibly the shrub can be planted at the upper limits of other reservoirs, located in regions of similar climatic conditions…to secure similar results." Taylor did not pretend to be a botanist. He was an engineer, searching for solutions to specific problems. He believed he had found a solution, rather than a problem, in tamarisk; but citing the Texas Board of Engineers, Taylor mentioned that at some points along the Pecos "the growth…has retarded the flow of the river to such an extent that…a flow of 2,000 cfs [now] produces the same…flood stages that 10,000 did before tamarisk appeared…it has become so dense…that it is often cheaper to construct a new canal than to clean the old one" (Taylor 1930).

Kirk Bryan introduced American tamarisk to the Association of American Geographers in 1932, concluding, "On the whole…the effects [of spontaneous tamarisk growth] on the works of man may be beneficial, and the general effect on minor streams while not entirely predictable will probably compensate for any disadvantage to other waterways"(Bryan and Hosea 1934). Meanwhile, a California zoologist wrote that Thornber's athel had "escaped from its original plantings on ranches and now occurs about many of the alkaline springs of Death Valley and the Salton Basin" (Jaeger, 1933). Also contemporaneously, a graduate student read a paper to the Illinois Academy of Sciences titled, "Some Influences of Man on Biological Communities." In his view, the advent of civilization encouraged the spread of "man tolerant" plants (Van Deventer 1934); unbeknownst to him, this idea had been proposed elsewhere in 1855 (Chew 2006). Van Deventer's work focused on invertebrates, and his vegetation analysis was strictly Clementsian. But an entomologist who studied insects in tamarisks and white sweet clover along Oklahoma's Canadian River observed that tamarisk seeds "find their way to favorable places, especially along rivers," and seedlings "appeared on the bare portions" of sediment banks near the water (Hefley 1937a). He mentioned that tamarisks were "grown locally for ornamental purposes in the upland section of the country surrounding the portion of the floodplain studied" but discounted the possibility that these could be the progenitors of plants in his study area. Elsewhere, Hefley (1937b) predicted that tamarisks would form a new type of riparian woodland in Oklahoma.

During the late 1930s, naturalized tamarisk remained a botanical curiosity. Michigan botanist Elzada Clover and her student Lois Jotter found isolated tamarisks growing along the Colorado River during a 1938 expedition that is remembered for making them the first women to traverse the Grand Canyon by boat (Clover and Jotter 1944; Cook 1987). In the 1940s and early 1950s, academic ecologists published botanically clinical views of tamarisk. A Kansas–Nebraska team observed that tamarisks defoliated by grasshoppers resprouted from their roots, surviving where some other trees failed (Albertson and Weaver 1945). Shinners (1948) of Southern Methodist University summed up the situation in Texas: "Salt cedar has evidently been cultivated in Texas for considerably more than half a century, and is probably to be found at the present time in every part of the state." Ware and Penfound (1949) revisited Hefley's Canadian River sites, and were the first Americans to record the important observation that salt cedar disperses seed throughout its growing season, phenology markedly different from that of spring-blooming cottonwoods and willows. They attributed tamarisk's tendency to form pure stands to its continuous breeding effort. Marks (1950) integrated tamarisk into his conceptual landscape, describing a *Pluchea–Tamarix* community "rich in species as well as in individuals." Later, Penfound (1953) called a "*Salix–Tamarix* association the most widespread of any of the plant communities found in Oklahoma lakes;" tamarisk went unmentioned in his summary of vegetation control methods.

EARLY ATTITUDES ABOUT WATER USE: PHREATOPHYTES

Around the time Aldo Leopold was planting his tamarisk in Albuquerque, USGS Ground Water Division Chief Oscar Meinzer (1920) proposed a functional category of plants that primarily use groundwater, listing a few species and genera as examples. He developed the concept with three further publications. In 1923 Meinzer established the term phreatophytes ("well plants") in an "Outline of Ground-Water Hydrology." There and in a sequel (1926), Meinzer defined a phreatophyte as "a species that habitually obtains its water supply" from groundwater rather than one that primarily derives its water from rainwater in the soil above the water table. In his view, phreatophytes were competing with farmers and others for shallow groundwater, an increasingly scarce resource.

Meinzer's 1927 USGS pamphlet *Plants as Indicators of Ground Water* seems to have been written prior to his 1926 paper, but it is more emphatic. He intensified his anthropomorphic imagery, subheading it "plants that habitually feed on ground water." His phreatophytes "habitually grow where they can send their roots down to the water table," and "such a plant is literally a natural well with pumping equipment, lifting water from the zone of saturation." Phreatophytes were machines; furthermore, they acted as if they had intentions.

Luna Leopold's assertion that tamarisk was uncommon helps explain its virtual absence from Meinzer's phreatophyte papers. He mentioned tamarisk only once, quoting South African geologist P. A. Wagner's 1916 list of "other plants that betray the presence of ground water at shallow depths" (Meinzer 1927). By 1927, most US publications associated tamarisk with drought tolerance and dry lands rather than riparian areas. Meinzer, based in Washington, DC, evidently relied on accounts of sites where tamarisk had not appeared. The omission was not his alone. Meinzer (1927) acknowledged he was "deeply indebted...for their critical examination of this paper and for many...valuable suggestions" to colleagues including J. J. Thornber, who apparently did not suggest tamarisks as candidate phreatophytes. Another near miss occurred when Meinzer cited three papers on rooting characteristics written between 1911 and 1923 by W. A. Cannon, who mentioned tamarisk only in a fourth work from the same era (Cannon 1912).

Meinzer has been acclaimed as the father of groundwater hydrology (Deming 2001), but he had no compunction about dabbling in plant ecology, and Thornber abetted him (Meinzer 1927). In his final phreatophyte paper, Meinzer (1926) used the subheading "Relation of Phreatophytes to Other Ecologic Groups," making a biological claim regarding his category's theoretical identity and significance. Meinzer (1927) suggested that hard disciplinary boundaries had previously prevented either botanists or hydrologists from "preempting" studies of groundwater use by plants. Where water availability is a limiting factor to plant growth, the ecohydrology of both surface water and groundwater persists as an interdisciplinary space occupied, sometimes uneasily, by both ecologists and hydrologists.

Sparse, variable precipitation makes the American Southwest ill suited to "dry" farming. By 1900, pre-Columbian stream water diversions and canals were being rehabilitated in Arizona deserts, and Mexican-era acequias continued to operate along the Rio Grande, but dependable, large-scale water storage and flood control were lacking (Pisani 1992, 2002). The Reclamation Act of 1902 began a massive public works project to fundamentally alter the region's hydrology (Pisani 2002). Over the ensuing six decades, the U.S. Bureau of Reclamation and its partners built 75 of the country's 100 largest reservoirs (Mermel 1958).

To understand tamarisk's impact on irrigation projects would take much longer. As federal dams rose, lowland rivers that had formerly flooded during spring thaws or briefly in midsummer became either regulated canals or dry washes that ran only when the new storage infrastructure was overwhelmed. Riparian trees phenologically adapted to spring flooding were left without the renewal of bare sediments their seeds needed for germination; seedlings might then be subjected to prolonged, drowning flows during summer agricultural irrigation. Tamarisks were equipped to occupy the new niche created when the dam builders shifted the "bare sediment season" from late spring to early autumn. The "bathtub ring" of damp silt exposed annually at the margin of every depleted reservoir further enhanced this opportunity for late-season establishment (Stromberg and Chew 2002).

During the Great Depression, make-work proposals grasped even at tamarisk. One appeared in *Science* magazine, replete with ideas for federal programs that could solve two or more problems simultaneously. Echoing Taylor's 1930 silting monograph, the author suggested "utilizing labor in planting Tamarisk or similar plants at the inflow of large reservoirs, such as Elephant Butte or Roosevelt Lakes" (Freudenthal 1933). He didn't realize that sufficient means for pan-regional distribution and propagation of tamarisk was already operating.

During the 1930s, managers employed tamarisk mostly as a panacea for erosion problems. A geographer wrote, "Tamarisk, introduced from Asiatic deserts, is transforming erosion and sedimentation processes along streams and lower deserts and is being planted in a few places in the lower parts of the [Navajo] reservation" (Hoover 1937). The Bureau of Reclamation continued erecting dams, while regional and local irrigation cooperatives continued diverting, pumping, and redistributing water. Biotic responses to these abiotic revisions went mostly unrecognized, or at least unreported. No one yet realized that reclamation amounted to a massive ecological subsidy for tamarisks.

Phase Two: New Science and a New Monster

Where water is scarce, farmers must compete for it with an array of other interests. Complicated water law is a signature theme of the American West. Three principles must be kept in mind. First, regardless of actual hydrological coupling, surface waters and closely associated shallow aquifers are not legally connected to deep groundwater. Second, there are statutorily defined "beneficial" uses for which surface water rights can be held; other "nonbeneficial" uses have no legal standing. Finally, rights to use surface water are governed via the doctrine of prior appropriation, that is, "first in time, first in right," where senior claims trump junior ones. Under some circumstances, a senior rights holder is allowed to transfer (usually sell) them to another user for new purposes, terminating the old use. Persuading someone to relinquish senior water rights is another matter. Where dry land is otherwise worthless, water transfers are controversial, expensive, and uncommon.

Nonbeneficial uses are like a businessman's fixed costs. If some nonbeneficial use can be curtailed, water will be liberated, and become claimable for beneficial use. Gauging channel flow is a tricky business. Estimating how flows might change under hypothetical conditions is far trickier. Oscar Meinzer proposed using phreatophytes to indicate shallow groundwater, and their water consumption as a factor to consider in calculating potential water yields. Nowhere in his phreatophyte papers did he speculate about managing vegetation to increase water yields, but that did not prevent his followers from trying.

RIO GRANDE

The National Resources Committee (NRC) and its successor Planning Board (NRPB) were creations of the Franklin Roosevelt presidency (Warken 1979). Following a national overview (NRC 1936) the NRC conducted a "joint investigation" of upper Rio Grande water issues in 1936-1937, and the NRPB investigated Pecos River issues in 1939-1940. Both responded to realized or feared water shortages and looming interstate water rights litigation. Both phreatophytes and tamarisks particularly were featured in both joint investigations.

As chairman of an ad hoc "Consulting Board," geographer Harlan Barrows supervised both projects, reputedly "knock[ing] bureaucratic heads together" to accomplish them (Colby and White 1961). Each investigation involved at least 10 bureaus from four cabinet-level agencies, plus the State Engineers of Texas, New Mexico, and (for the Rio Grande) Colorado (NRC 1938; NRPB 1942). Both called for unprecedented coordination among disciplines, agencies, scientists, and bureaucrats. The results and recommendations of the joint investigations were not seamless or internally consistent. Different chapters were prepared at different times, and persons other than the credited authors prepared many tables and figures. Nevertheless, it is significant that comprehensive, integrated scientific analyses of southwestern water issues were being attempted.

Many scientists took part in both projects. Notably, the USGS sent junior geologist C. V. Theis, who had recently proposed a new mathematical model of aquifer dynamics, to open a New Mexico office. The still-presiding Meinzer discouraged Theis from publishing, thinking a practical geologist unlikely to succeed where theoreticians were floundering (Bredehoft 2008). This created a troubled working relationship, but Theis had few employment options and busied himself with Rio Grande research. Soon the "Theis equation," describing groundwater movement by analogy with heat diffusion, "revolutionized the science of groundwater hydrology" (White and Clebsch 1994).

USDA personnel analyzed vegetation for the Rio Grande Joint Investigation (Rio JI). They never employed the term "phreatophyte," but neither did the USGS hydrologists, who referred to "plants that habitually feed on groundwater." Tamarisks were uniformly called "salt cedars," but noted only on vegetation maps. Some attempt was made to divide native (i.e., uncultivated) vegetation into categories such as "trees" and "brush" (NRC 1938) but salt cedar appeared in both. According to estimates published decades later, there were 5,500 acres of tamarisk in the river's Middle Valley in 1936 (Thomas and Gatewood 1963), but no such total appeared in the Rio JI data summaries.

Hoping to quantify water consumption, USDA personnel experimented with plantings in metal stock tanks where they could control water input, but salt cedar was not among the species tested for the Rio JI. The final summary reported that "data are not available to permit even an approximate determination of the total amount of water which might be feasibly and economically recovered from the million acre-feet or more by which the stream flow is annually depleted in supporting the growth of native vegetation in the upper Rio Grande basin. It seems unquestionable, however, that some fraction of this loss should be susceptible of economic recovery by proper drainage construction" (NRC 1938). They proposed salvaging water by draining the swamp—lowering the water table beyond root's reach of riparian phreatophytes.

PECOS RIVER

The Pecos River Joint Investigation (Pecos JI) emulated the Rio JI's organizational model, but recognized tamarisk at Lake McMillan. Nevertheless, its final report expressed differing opinions of the plants. A section detailing the history of irrigation by two USDA "Assistant Agricultural Economists" evidently relied on T. U. Taylor, claiming "the growth of tamarisk at the upper end of [McMillan] reservoir, starting about 1915, is credited with having appreciably arrested silting" (NRPB 1942). Meanwhile, USDA Irrigation Engineer Fred Scobey echoed Meinzer by classifying tamarisk and other "brush" as "non-beneficial but unavoidable users" of water. He estimated "about 50,000 acres of salt cedar (tamarisk) in the Basin, and each acre...consume[s] 5 acre-feet of water per year" (NRPB 1942). As for the Rio JI, Scobey lumped all uncultivated plants into the category "native vegetation" without distinguishing those of exotic origin. Where and when their ancestors originated was irrelevant to his water-use estimates. Although Scobey cited Meinzer for the Pecos JI, he substituted "ground-water plants" for phreatophytes (NRPB 1942). Scobey analyzed relative cover by cutting up aerial photographs along supposed boundaries between vegetation types and weighing the pieces. He identified 13,000 acres of tamarisk at Lake McMillan that "intercept silt and caused the delta formation in form and levels different from such a development without the growth. Also these cedars [*sic*] consume some 70,000 acre feet of water per annum. They constitute one of the major problems along the Pecos River" (NRPB 1942).

Silting was difficult to measure, but far easier to quantify than transpiration losses. Lake McMillan lost about half its original water-storage capacity to sedimentation during a half century of operation. "Consumptive" water use by "native" plants could only be estimated experimentally, in microcosm. For unexplained reasons, Pecos JI cooperators used different methods for evaluating different plants. They tried to measure water use by uncultivated phreatophytes in microcosm, but used field data for irrigated crops without caveats (NRPB 1942). It seems unlikely that any farmer would report using more water than he was entitled to or that verification would be a simple matter, and any conclusions drawn from comparing incommensurable data sets were already suspect.

The Pecos JI reported that native vegetation of all kinds routinely transpired more water per unit area than irrigated crops. "The normal annual consumptive use of water in the Carlsbad area [including Lake McMillan] by irrigated crops, salt grass and sacaton, and salt cedar is 2.8, 3.4, and 6.0 acre-feet per acre, respectively" (NRPB 1942). After factoring in local precipitation and the presumed replacement of floodplain grass by tamarisk, the investigation concluded that floodplain tamarisks were using an additional 2.6 acre-feet of water per acre from each of 2,000 floodplain acres, and Lake McMillan tamarisks were consuming 5.0 acre-feet of water from each of 11,000 delta acres. This put 60,200 acre-feet "as an estimate of the increased amount by which stream flow is depleted annually due to the salt cedars." The authors then waffled: "Actually the conditions of stream flow and the distribution and varying density of the salt cedars are such that the full depletion given by this estimate probably does not occur" (NRPB 1942). But the indictment was plain. The Pecos JI determined "the reduction or complete elimination of the salt-cedar [related water] losses offers a major opportunity for water saving in the Pecos River basin" (NRPB 1942).

Pecos JI findings regarding phreatophytes almost precisely parroted the Rio JI in its initial details. But this time, their recommendation was more specific. "From the data which have been presented and the analysis in the single instance of the salt cedar area above McMillan Reservoir, it seems unquestionable that a substantial fraction of this loss should be susceptible of economic recovery" (NRPB 1942). The death warrant had been drafted, if not yet signed. Thousands of acres of tamarisk had to be using lots of water, so eliminating them had to yield benefits, however unpredictable. Pecos JI personnel knew some ways to kill tamarisk, and they believed water salvage could be made manifest in acres of dead vegetation. A water-pumping, water-wasting monster was attacking the Pecos River. In some minds, confidence was high that it could be slain. However, there was less confidence that the stolen water could actually be recovered.

GILA RIVER

Another water shortage was developing in eastern Arizona. In 1939, prompted and financed by the War Department and Bureau of Indian Affairs, USGS hydrologists began inventorying the upper Gila River, where tamarisk had grown for a quarter century. One objective was quantifying phreatophyte water use in the Safford and Duncan-Virden Valleys. Despite Meinzer's omission, nothing suggests the investigators ever doubted that tamarisk was a phreatophyte. But a separate, preliminary paper by two of its principals admitted "there are no data on water use by cottonwood, tamarisk and baccharis, and such are being obtained" (Turner and Halpenny 1941). Using creative but primitive and incompletely described methods they estimated "completely killing... river bottom vegetation" in the Safford Valley would liberate 70,000 acre-feet of water annually (Turner et al. 1941); exactly the total predicted elsewhere by the Pecos JI.

Upstream of the Safford Valley, along the Gila's Eagle Creek tributary, are the Morenci copper mine and eponymous company town. The Phelps Dodge Corporation (PDC) acquired rights to the ore body in 1921 when copper demand was weak. Not until 1930 did PDC decide to pursue open-pit mining there, and it took seven more years to put together financing. The fortuitously timed expansion positioned PDC to supply the copper demands of a looming war. But PDC's Eagle Creek water was insufficient to supply the proposed operation, and water rights in the Safford Valley were fully allocated. PDC had to purchase existing rights or find "new" water. They began to investigate salvaging water by removing phreatophytes. After Pearl Harbor, the federal War Production Board directed PDC to "increase its overall copper production by 80 percent" (Jackson 1991). Supported by the Department of Commerce and the federal Defense Plant Corporation, USGS took over and expanded PDC's Safford Valley phreatophyte removal demonstration project. In 1950, the wartime research was published as *Use of Water by Bottom-Land Vegetation in Lower Safford Valley Arizona*. Its foreword stated, "This investigation, so far as is known, was the first attempt actually to measure the water consumed by a particular class of vegetation under natural conditions on a scale large enough to be of economic significance" (Gatewood et al. 1950). Its authors placed tamarisk conspicuously at the top of their phreatophyte inventory, misattributing its classification as such to Meinzer (who had retired 1946 and died in 1948). Then they shifted the subtext of tamarisk management with a series of unprecedented statements about the plants.

Saltcedar "thrived and spread at the expense of nearly all the native plant life" (Gatewood et al. 1950). In contrast with earlier categorical descriptions, this redeployed "native" to distinguish tamarisk from other riparian trees, and assertively cast it as a hyper-fecund cheater/parasite/usurper. Ironically, PDC's original plan (and its subsequent USGS iterations) had always included removing all nineteen phreatophyte species from the study reach; here they shifted eighteen of them to the list of victims.

Salt cedar formed "a dense jungle-like thicket that is difficult to penetrate" (Robinson 1952b). One study participant later revealed that they had attacked Safford Valley tamarisks with flamethrowers, an iconic weapon of the military's "island hopping" Pacific campaign, and a treatment of last resort for America's most intractable foreign enemies. Despite such measures, "shoots from an area cut over [regrew] to a height of 6 to 8 feet in about 12 months" (Gatewood et al. 1950). Like human aliens before and since, tamarisks became suspect because of their high fecundity, perceived aggressiveness, and unfair competitiveness. In summary, tamarisks were foreign, insular, recalcitrant, and unaccountably resilient. By standing in the way of mine expansion, they impeded commerce and security, flouting American values and natural propriety. USGS scientists had unreservedly described something very unplantlike. Tamarisk was a monster.

The lower Safford Valley study used six different methods to compute water use by phreatophytes and produced the highest-ever transpiration rate claimed for tamarisk: the annual equivalent of 7.2 acre-feet of water per acre of plants. The report was conspicuously silent about the prospects for sending that water to Morenci because PDC had already secured a much greater water supply by building the Horseshoe Dam on the distant Verde River in exchange for rights to relatively nearer Black River water (Johnson 1972). But the project had reunited tamarisk and one of its greatest detractors. Groundwater hydrologist Thomas Robinson, "fondly known to many as 'Mr. Phreatophyte,'" contributed to the Rio JI and then built a career on tamarisk-bashing (Johnson 1972). Robinson was the last person to supervise the Safford Valley study, and he coauthored the 1950 report (Gatewood et al. 1950). He followed it with a paper extrapolating from the Safford Valley findings (using "all available data"), claiming that at least 10.7 million acres of phreatophytes were wasting at least 15.7 million acre-feet of water annually in 13 states, and suggesting that data for all 17 western states would raise those numbers by 50%. He also alluded to a mythical monster, the Hydra: after bulldozing, "soon new growth appeared from each severed [tamarisk] root, so that two plants grew where one had grown before" (Robinson 1952b). Nevertheless, water salvage from phreatophyte removal remained hypothetical. It had neither been attempted on a significant scale for an extended period, nor demonstrated anywhere at any scale.

The Dawn of Phase Three: Slaying the Monster

Two new public projects devoted to phreatophyte (mainly tamarisk) control emerged around 1950. One was the ephemeral "New Mexico Salt Cedar Interagency Council," of which little evidence remains aside from a single report prepared by its subsidiary "Salt Cedar Interagency Task Force" composed of one state and three federal officials (Brown et al. 1951). The body of the report reviewed some of the published work on tamarisk by the USGS hydrologists and the USDA engineer discussed above. The document was classified "For Official Use Only– Not For Publication." A possible reason for suppression emerged in appendix 4, coauthored by the by-then celebrated hydrologist C. V. Theis, who would soon be promoted to coordinate USGS activities for the U.S. Atomic Energy Commission (White and Clebsch 1994).

Appendix 4 comprehensively criticized the assumptions and outcomes of previous USGS and USDA work on phreatophyte water consumption and disputed the premise of phreatophyte eradication efforts. Its language seems exasperated and didactic: "The primary reason that phreatophytes use water is because the water is available. Thus investigations of the means of salvage of water should include studies of the source of the water. Conversely, it seems that the most successful manner of saving water would be to deprive the plants of water, either by drains, pumps, or channels, rather than by taking the plants away from the water by eradication." It went on to point out that such draining would automatically collect the water and make it available for use, obviating uncertainties about net benefits (Brown et al. 1951). Fifteen years after the Rio JI, Theis's solution was to drain the swamp. Instead of assuming that killing tamarisk would yield water, he told managers to go after the water directly. Appendix 4's tamarisks were plants, not monsters.

The second public project was more substantial and longer lived, leaving a trail of detailed meeting minutes and other documents. The water-oriented Pacific Southwest Federal Inter-Agency Technical Committee appointed a Subcommittee on Phreatophytes, first chaired by a veteran of the Safford Valley studies. Before long, T. W. Robinson became a regular meeting attendee and occasional substitute secretary. By September 1952, he was its official secretary (Phreatophyte Subcommittee 1951-1959).

The Subcommittee persisted until 1970; only some of its earliest activities are covered here. The minutes of its first (1951) meeting reveal that the unpublished Salt Cedar Task Force report was already a topic of discussion. Appendix 4 was not mentioned, but the subcommittee had been organized to promote phreatophyte suppression, not to salvage water by engineered means. At its second meeting its members estimated that 40 species of tamarisk existed in the United States, but reduced its practical working inventory to two: the "marvelous shade tree" athel, and everything else, lumped together as *Tamarix gallica*. Later, they accepted their charge from the full Committee, adopted Meinzer's "phreatophyte" definition, and did other housekeeping (Phreatophyte Subcommittee 1951-1959).

At their seventh (June 1952) meeting, the Subcommittee considered a salient information request from the Arizona Underground Water Commission (AUWC), which wrote:

> [We] would appreciate very much any help your committee could give in obtaining for [us] information on the following points relating to phreatophytes in Arizona:
>
> 1. Maps showing areas of phreatophytes and amounts of water estimated to be used by them in each area.
> 2. Quality of water in each major area of phreatophyte use.
> 3. Possibilities and methods of destroying phreatophytes in each area.
> 4. Effect of destroying phreatophytes in each area on water supply. Where would water go which now is being used by phreatophytes?
> 5. Estimate costs of various methods of destroying phreatophytes per acre foot of water saved for beneficial use. (Phreatophyte Subcommittee 1951-1959)

The state of the Subcommittee's art and science were evident in its reply. It ignored all but item 3, assuring the AUWC that phreatophytes "use water in an uneconomical manner." It added some terse tabular statistics about survival rates of salt cedar "in the Gila River adjacent to Phoenix" after spraying with a mix of the new herbicides 2,4-D and 2,4,5-T (Phreatophyte Subcommittee 1951-1959), which later became notorious as "Agent Orange." Their bailiwick was slaying monsters, not salvaging water.

Later that year T. W. Robinson expanded the campaign against phreatophytes with the article "Water Thieves" in the trade magazine *Chemurgic Digest* (Robinson 1952a). Again he summarized Meinzer's work and added salt cedar to the phreatophyte list. He reiterated his 1951 maximum estimate that 25 million acre-feet of water was lost to "consumptive waste" by phreatophytes in the 17 western states: "twice the average annual flow of the Colorado River at Lees Ferry" (Robinson 1952b). The same year, a respected American geographer cited Robinson's symposium paper and the Safford Valley study, adding the accusation that "at the delta heads of reservoirs … [tamarisk] has become a jungle, and not only consumes large quantities of water, but clogs the inlet to the reservoir" (Meigs 1952). Thus, the very quality extolled by Taylor in 1930 and proposed for exploitation by Freudenthal in 1933 was transformed into a new indictment.

Summary and Conclusions

From 1877 to 1938 American academic botanists and ecologists displayed mild interest in *Tamarix* species without making them a research focus. During that era, horticulturists recommended tamarisks as drought and salt-tolerant ornamentals, windbreaks, and shade trees. Then, in less than a decade, from the inception of the Pecos JI in 1938 to the completion of the Safford Valley project in 1944, the opinions of scientists and scientifically trained natural resource managers about tamarisks changed drastically. Government employees (mostly affiliated with USDA) had imported, recommended, distributed, and planted tamarisks well into the Dust Bowl era. After seeing naturalized stands at Lake McMillan, other writers, led by USGS hydrologists, declared the plants worse than useless. Shrubs once extolled for erosion and sedimentation control became machine-like monsters pumping away scarce western water. On the heels of that assessment came a further redefining moment, the wartime recasting of Safford Valley tamarisks as alien aggressors. By 1950, a permanent interagency bureaucracy had sprung up, focused on slaying the beast rather than salvaging water. It would keep its members busily cutting, bulldozing, spraying, and reporting progress in terms of vegetation killed for another twenty years.

Naturalized tamarisk had no discernable public constituency and little prospect of gaining one. Few who advocated planting tamarisks would likewise have advocated the plants' unrestrained self-propagation. By the same token, there was little likelihood that a popular sentiment would develop against naturalized tamarisk. The rural West was sparsely populated, and relatively few citizens encountered tamarisk on private lands. The monstering of tamarisk required organization and impetus only the federal government could provide.

Lacking candid insider accounts, it is rarely obvious whether scientist-bureaucrats support the aims of the programs they are assigned to, or whether their rhetoric represents principled endorsement, sincere conviction, careerist pragmatism, or cynical manipulation. Tamarisk occupied major, mostly federal (or federalized) water projects. It was examined almost entirely through the lenses of hydrogeology and water politics, not botany or ecology. Tamarisk was a convenient scapegoat for the complex problems encountered by government water managers, whether they were true believers in the monster or not. Even so, it does not seem to have mattered strongly to the principals whether or not suppressing tamarisk ever made more "wet water" available. They could demonstrate productivity through acres of vegetation laid waste while suppressing or simply ignoring the lingering doubts over their theories, methods, and mandate. Monstering tamarisk was not a side effect. It established an attitude and perpetuated a program.

Acknowledgments

This chapter was condensed, with kind permission from Springer Science + Business Media B.V., from M. K. Chew, 2009. The Monstering of Tamarisk: How Scientists Made a Plant into a Problem. *Journal of the History of Biology* 42:231–266 © Springer 2009. Help and encouragement for this study came from Juliet C. Stromberg and the Arizona State University Center for Biology and Society and School of Life Sciences.

Literature Cited

Albertson, F. W., and J. E. Weaver. 1945. Injury and death or recovery of trees in prairie climate. Ecological Monographs 15:393–433.

Bredehoeft, J. D. 2008. An Interview with C. V. Theis. Hydrogeology Journal 16:5–9.

Brown A. F., J. G. Koogler, R. F. Palmer, and H. B. Elmendorf. 1951. Report to the Salt Cedar Interagency Council by the Salt Cedar Interagency Task Force. Albuquerque, NM.

Bryan, Kirk, and R. G. Hosea. 1934. Tamarisk: An Introduced Plant and Its Geographic Effect (1932 Conference Abstract). Annals of the Association of American Geographers 24:44–45.

Cannon, W. A. 1912. Some Features of the Root-Systems of Desert Plants. The Popular Science Monthly 81:90–99.

Carleton, M. A. 1914. Adaptation of the Tamarisk for Dry Lands. Science 39:692–694.

Chew, M. K. 2006. Ending with Elton: Preludes to Invasion Biology. Doctoral diss., Arizona State University School of Life Sciences, Tempe, AZ.

Clover, E. U., and L. Jotter. 1944. Floristic Studies in the Canyon of the Colorado and Tributaries. American Midland Naturalist 32:591–642.

Colby, C. C., and G. F. White. 1961. Harlan H. Barrows, 1877–1960. Annals of the Association of American Geographers 51: 395–400.

Cook, W. E. 1987. The *WEN*, the *Botany*, and the *Mexican Hat*: The Adventures of the First Women through Grand Canyon on the Nevills Expedition. Callisto. Orangevale, CA.

Davis, L. D. 1899. *Ornamental Shrubs for Garden, Lawn and Park Planting*. Putnam's, New York, NY.

Deming, D. 2001. *Introduction to Hydrogeology*. McGraw-Hill, Boston, MA.

Everitt, B. L. 1980. Ecology of saltcedar: A plea for research. Environmental Geology 3:77–84.

Foucault, M. 1970. The order of discourse. Translated by Ian McLeod. Pages 48–78 *in* Young, Robert, editor. 1981. *Untying the Text: A Post-Structuralist Reader*. Routledge. Boston, MA.

Freudenthal, L. E. 1933. Flood and erosion control as possible unemployment relief measures. Science 78:445–449.

Gatewood, J. S., T. W. Robinson, B. R. Colby, J. D. Hem, and L.C. Halpenny. 1950. Use of Water by Bottom-Land Vegetation in Lower Safford Valley Arizona. US Geological Survey. US Government Printing Office, Washington DC.

Gilmore, D. D. 2003. *Monsters: Evil Beings, Mythical Beasts, and All Manner of Imaginary Terrors*. University of Pennsylvania, Philadelphia.

Hartman, Todd. 2001.With luck, beetles may spill salt cedars. *Rocky Mountain News* (Denver, CO). Primary Source Media. HighBeam Research. 6 Feb. 2012. [Online]. http://www.highbeam.com.

Hefley, H. M. 1937a. Ecological studies on the Canadian River floodplain in Cleveland County, Oklahoma. Ecological Monographs 7:345–402.

Hefley, H. M. 1937b. The relations of some native insects to introduced food plants. Journal of Animal Ecology 6:138–144.

Hoover, J. W. 1937. Navajo Land Problems. Economic Geography 13:281–300.

Horton, J. S. 1964. Notes on the Introduction of Deciduous Tamarisk. US Forest Service Rocky Mountain Forest and Range Experiment Station, Fort Collins, CO.

Isern, T. D. 1997. Science or Myth? Plains Folk. North Dakota State University. [Online]. http://www.ext.nodak.edu/extnews/newsrelease/back-issues/000296.txt.

Jackson, D. C. 1991. Horseshoe Dam: Technical Report, HAER-AZ-24. US Bureau of Reclamation.

Jaeger, E. C. 1933. *The California Deserts: A Visitor's Handbook*. Stanford University Press. Palo Alto, CA.

Johnson, A. I. 1972. Groundwater hydrology: An introduction. Water Resources Bulletin 8:99–100.

Joor, J. F. 1877. The Tamarisk Naturalized. Bulletin of the Torrey Botanical Club 6:166.

Kearney, R. 2003. *Strangers, Gods and Monsters: Interpreting Otherness*. Routledge. London, UK.

Mansfield, S. M. 1886. Appendix S: Improvement of Rivers and Harbors in the State of Texas. Pages 1291–1336 *in* Annual Report of the Chief of Engineers, United States Army, to the Secretary of War, for the Year 1886. Part 2. US Government Printing Office, Washington, DC.

Marks, J. B. 1950. Vegetation and soil relations in the lower Colorado Desert. Ecology 31:176–193.

McAtee, W. L. 1914. Further Notes on Tamarisk [letter]. Science 39:906.

Measures, Mrs. J. A. 1933. Habits and bloom of the tamarisk. Flower Grower 19:331.

Meigs, P. 1952. Water problems in the United States. Geographical Review 42:346–366.

Meine, C. 1988. *Aldo Leopold: His Life and Work*. University of Wisconsin Press, Madison, WI.

Meinzer, O. E. 1923. Outline of Ground-Water Hydrology with Definitions. USGS Water-Supply Paper 494. US Government Printing Office, Washington, DC.

Meinzer, O. E. 1926. Plants as indicators of ground water. Journal of the Washington Academy of Sciences 16:553–564.

Meinzer, O. E. 1927. Plants as Indicators of Ground Water. USGS Water Supply Paper 577. Washington. US Government Printing Office, Washington, DC.

Meinzer, O. E. 1920. Quantitative methods of estimating ground-water supplies. Bulletin of the Geological Society of America 31:329–338.

Mermel, T. W. 1958. *Register of Dams in the United States*. McGraw-Hill, New York, NY.

Miller, C. C. 1928. Tamarix as a windbreak. *California Cultivator*, January 7, 1928, p.16.

National Resources Committee. 1936. *Drainage Basin Problems and Programs*. US Government Printing Office, Washington, DC.

National Resources Committee. 1938. The Rio Grande Joint Investigation in the Upper Rio Grande Basin in Colorado, New Mexico and Texas, 1936-1937. US Government Printing Office, Washington, DC.

National Resources Planning Board. 1942. *The Pecos River Joint Investigation: Reports of the Participating Agencies*. US Government Printing Office, Washington, DC.

Oberholser, H. C. 1925. The relations of vegetation to bird life in Texas. American Midland Naturalist 9:564–594.

Pauly, P. J. 2007. *Fruits and Plains: The Horticultural Transformation of America*. Harvard University Press. Cambridge, MA.

Peck, W. D. 1818. *A Catalogue of American and Foreign Plants, Cultivated in the Botanic Garden*. Harvard University Press, Cambridge, MA.

Penfound, W. T. 1953. Plant communities of Oklahoma lakes. Ecology 34:561–583.

Phreatophyte Subcommittee, 1951–1959. Minutes of Phreatophyte Subcommittee Meetings. Vol. 1. Pacific Southwest Inter-Agency Committee. (Private Collection.)

Pisani, D. J. 1992. *To Reclaim a Divided West: Water, Law and Public Policy 1848-1902*. University of New Mexico Press, Albuquerque.

Pisani, D. J. 2002. *Water and American Government: The Reclamation Bureau, National Water Policy, and the West*. University of California Press, Berkeley.

Robinson, T. W. 1965. *Introduction, Spread, and Areal Extent of Saltcedar (Tamarix) in the Western States*. USGS. US Government Printing Office, Washington, DC.

Robinson, T. W. 1952a. Water Thieves. Chemurgic Digest 11:12–15.

Robinson, T. W. 1952b. Phreatophytes and their relationship to water in western United States. Transactions American Geophysical Union 33:57–61.

Shinners, L. H. 1948. Geographic limits of some alien weeds in Texas. Texas Geographic Magazine 12:16–25.

Stromberg, J. C., and M. K. Chew. 2002. Foreign visitors in riparian corridors of the Southwest: Is xenophytophobia justified? Pages 195–219 *in* B. Tellman. *Invasive Exotic Species in the Sonoran Region*. University of Arizona Press, Tucson.

Swingle, W. T. 1924. The Athel, *Tamarix aphylla* (also called *T. articulata*), a Promising Windbreak for the Hot Irrigated Valleys of the Southwest. Indio, CA. Mimeograph. Walter Tennyson Swingle Collection, University of Miami, FL.

Taylor, T. U. 1930. Silting of Reservoirs. University of Texas Bulletin 3025, Austin.

Thomas, H. E., and J. S. Gatewood. 1963. *Drought in the Southwest, 1942-56*, USGS Professional Paper 372-D. US Government Printing Office, Washington, DC.

Thornber, J. J. 1916. *Tamarisks for Southwestern Planting: Timely Hints for Farmers*. University of Arizona Press, Tucson.

Turner, S. F., and L. C. Halpenny, 1941. Ground-water inventory in the Upper Gila River Valley in New Mexico and Arizona, scope of investigation and methods used. Transactions American Geophysical Union 22 (Part 3):738–744.

Turner, S. F., and others [*sic*]. 1941, Water Resources of Safford and Duncan-Virden Valleys, Arizona and New Mexico: USGS (unpublished mimeograph).

Van Deventer, W. C. 1934. Some Influences of Man on Biotic Communities. (Presentation abstract). Transactions of the Illinois State Academy of Sciences 26:137.

Ware, G. H., and W. T. Penfound. 1949. The vegetation of the lower levels of the floodplain of the South Canadian River in central Oklahoma. Ecology 30:478–484.

Warken, P. W. 1979. *A History of the National Resources Planning Board*. Garland Publishing, New York.

White, R. R., and A. Clebsch. 1994. C.V. Theis: The man and his contributions to Hydrogeology. Pages 45–56, *Selected Contributions to Ground-Water Hydrology by C. V. Theis, and a Review of His Life and Work*. USGS Water-Supply Paper 2415, US Government Printing Office, Washington, DC.

17

The Politics of a Tree

HOW A SPECIES BECAME NATIONAL POLICY

Tim Carlson

Tamarix occupies a unique niche among invasive species in the United States in that it has drawn bipartisan political interest in resolving an apparent problem at a time when many other invasive-plant issues get little or no attention. This chapter reviews the factors that have influenced federal and state legislative bodies, agencies, and a major private foundation. The dynamics of the *Tamarix* issue include the compounding effects of drought, wildfire, habitat, endangered species, agriculture, recreation, and anthropogenic actions and values. The chapter addresses how water concerns and the aesthetic perceptions of river systems affect public attitudes, while asking, why is the public engaged? Has the *Tamarix* issue, from a historical perspective, really met the requirements for public review and political decision making? Are the lessons learned from *Tamarix* transferable to other invasive species?

While most of the other chapters in this book are based on primary scientific data published in peer-reviewed journals, this chapter uses a "case study" approach that includes secondary literature review, interviews, social science, political science, philosophy, legal, and personal observations. Why this approach? The political reality is that science matters, but it is not the only thing that matters in decision making and policy making. We rely as much on our values (opinions) as on science (facts) in how we make decisions. A good example is the purchase of a car. From an environmental position, the Toyota Prius is currently the most fuel-efficient, environmentally friendly car now readily available—so why isn't everyone in the "conservationist community" driving one? We know the answer—we all take a slightly different set of factors into account in buying a car, such as price, the driving conditions we anticipate, load or passenger capacity, performance, safety, style, and even color and the number of cup holders. My graduate advisor in environmental engineering at Arizona State University drove a Checker taxi as his personal vehicle because from a strictly engineering standpoint it was the most durable, reliable, functional, and almost indestructible car available. I have never known anyone else who purchased a car based on such pure analysis.

This chapter shows the reader how the *Tamarix* problem has affected policy; several sections therefore intersect with other chapters in this text. Policy is not driven solely by science; it includes many other personal values. Thus, it is imperative to illustrate how science and values become intertwined to affect policy.

What Is Necessary for Policy to Be Adopted?

From both political science and social science perspectives, history shows that in the United States taking major actions to solve an environmental problem requires that the public believe three things:

1) The problem is critically important and deserves to be solved. Hans Bleiker of the Institute for Participatory Management and Planning, Systematic Development of Informed Consent wrote:

    In American society, to gain public support for an action the problem must be considered as SERIOUS . . . one that just HAS to be addressed. (Bleiker 2010)

2) The problem is directly linked to an important event or condition that causes a visceral reaction with the public. John Kingdon, in Agendas, Alternatives, and Public Policies, wrote:

    At any time, important people in and around government could attend to a long list of problems. . . . Those that gain prominence are tied to an issue. (Kingdon 1995)

3) Scientists and/or others have warned us about the problem and proposed solutions.

When all three conditions exist, the public will generally support a bipartisan approach to solving a problem. Past actions have been taken on such issues as water quality, bird population decline, forest health, and the building of an interstate highway system (see table 17.1). All but the last example have some relationship to a serious and "observable" environmental problem. But when doing policy analysis using case studies, it is often important to look beyond the general category that is of interest (i.e., environmental issues) to other important policy actions that can illustrate policy fundamentals. To emphasize this policy-making process, the interstate highway system example is more enlightening.

TABLE 17.1

Examples of public support for action

Problem	Linkage to an issue	Action supported by the public
Water quality of river, lakes, streams, wetlands, and estuaries.	Cuyahoga River in Cleveland catches fire.	Clean Water Act
Pesticide impacts to animals	Bald eagle decline	DDT banned
Hazardous wastes	Love Canal	Superfund legislation
Water quality of Chesapeake Bay	Decline in oyster and blue crabs	Chesapeake Bay restoration initiative
Decline in biodiversity	Whooping crane near extinction	Endangered Species Act
Decline in wildlife and wildlife habitat	Near extinction of megafauna such as Rocky Mountain elk, pronghorn, bighorn sheep, and numerous game birds and fish	Lacey Act of 1900, Migratory Bird Conservation Act of 1929, Wildlife Restoration Act of 1937, Sport Fish Restoration Act of 1950, and North American Wetlands Conservation Act of 1989
Declining health of American forests	Catastrophic forest fires of 2000–2003	Healthy Forest Restoration Act of 2003
Efficient transportation	World War II and the Cold War	National System of Interstate and Defense Highways Act of 1956

During World War II, Gen. Eisenhower saw the transportation advantages Germany enjoyed because of its Autobahn network, which also benefited the Allies by enhancing the mobility of their troops as they fought their way into Germany. These experiences, and his memory of the US Army's first transcontinental motor convoy from Washington, DC, to San Francisco in 1919, shaped Eisenhower's views on highways. They "started me thinking about good, two-lane highways, but Germany had made me see the wisdom of broader ribbons across the land." (FHA 1996; Ambrose 1990). For decades, it had been recognized in the United States that the country's highway system was in desperate need of modernization, but there simply wasn't sufficient political and public support to embark upon a massive public works program. That all changed in the early 1950s as the Soviet Union began expanding its atomic bomb program, and then in 1955 detonated its first hydrogen bomb. The Cold War had begun, and Americans' fear of nuclear engagement with the Soviet Union (evidenced by the interest in home bomb shelters and "duck and cover" training in schools) galvanized support for the National System of Interstate and Defense Highways Act of 1956. President Eisenhower's firsthand knowledge of Germany's Autobahn had added to an overwhelming acknowledgment by Congress that the United States should support a similar high-speed road network. On June 26, 1956, the Senate approved the bill by a vote of 89 to one. That same day, the House approved the bill by a voice vote, and three days later President Eisenhower signed the bill into law (Eisenhower 1948;

Ambrose 1990; FHA 1996), which has led to the creation of the interstate highway system we have today. This example illustrates well the three basic fundamentals of enacting major policy actions; that is, the public understood that the problem was serious (an inadequate highway system) and needed resolution, was linked to an important issue or event (recent memories of World War II and, most importantly, the Cold War fears), and there were solutions (German Autobahn design).

TAMARISK: A SERIOUS PROBLEM? CONNECTED TO AN ISSUE? HAS SOLUTIONS?

For over 50 years, many scientists and land managers have been telling the public and political leaders that tamarisk is a serious problem for a variety of reasons (see chapter 16, this volume). However, it was the drought in the West from 2002 through 2010 that focused public attention on the problem. In 2005 alone, *Tamarix* was mentioned in about 500 newspaper articles; over half also mentioned water (see chapter 1, this volume). When people in Denver are affected by watering restrictions and golf-course closures; when Lake Powell and Lake Mead are at all-time-low water-storage levels; and when states, especially in the Colorado River system, are experiencing severe water shortages, the issue goes public. Land and water managers in the West were thus compelled to investigate solutions to some of the water shortage problems by evaluating various methods, including controlling a nonnative invasive plant with a reputation for using a great deal of water. So, from a public and political perspective, the *Tamarix* issue certainly meets the first two criteria: it is considered a serious problem that deserves to be solved, and it is connected to an important issue, water availability. The question is, do we have the solutions to this perceived *Tamarix* problem? The answer is yes, for the most part. We have good approaches to planning appropriate riparian restoration (Shafroth et al. 2008), and Best Management Practices (BMPs) to control the species (Nissen et al. 2010), and for revegetation with other, presumably native plants (Sher et al. 2010). We have not always incorporated these tools in attempts at riparian restoration, but we are getting better at doing so.

BUT DOES TAMARIX REALLY USE A LOT OF WATER?

Two recent reports provide the most comprehensive assessment of the question of whether *Tamarix* really uses a lot of water. The U.S. Geological Survey (USGS) looked at the entire range for *Tamarix* in the western states, while the Tamarisk Coalition's assessment centered on the Colorado River Basin. The USGS report was based purely on literature review (USGS 2010). The Tamarisk Coalition used an independent peer panel, (ET Peer Panel) bringing together experts on riparian and floodplain ecosystems ecology, and using evapotranspiration (ET) rate measurement of the vegetation associated with these ecosystems; hydrologic interactions between vegetation, groundwater, and surface water; and tamarisk control and revegetation approaches (Tamarisk Coalition [TC] 2009; see also chapter 4, this volume). These two independent efforts came out with the same results: ET for tamarisk generally ranges between 700 to 1400 millimeters of water per year. By comparison, annual water requirements for high-yield corn range between 500 and 650 millimeters (Colorado State University [CSU] Cooperative Extension 2008); and for alfalfa, between 800 and 1500 millimeters per year (Hanson et al. 2008).

A key finding in both reports is that native streamside plants along western rivers, such as cottonwoods and willows, consume large but roughly similar amounts of water as *Tamarix*, though this was not new information. They are all phreatophytes, plants that obtain a significant portion of their water from the phreatic zone (zone of saturation) or the capillary fringe. However, both reports also found that tamarisk can persist on sites having greater depth to groundwater, where neither native cottonwoods nor willows can establish. In these higher terrace areas of a floodplain, *Tamarix* often dominates the vegetation community. Both reports draw a careful distinction between expected water savings and actual water availability. However, both conclude that water could be "saved" through appropriate and well-planned *Tamarix* control and restoration measures if

1) revegetation is a critical component of restoration;
2) replacement vegetation for tamarisk on upper floodplain terraces is composed of more xeric native grass and shrub species suitable for site-specific precipitation, soils, salinity, and groundwater depths; and
3) long-term maintenance occurs on the restoration areas.

If water savings are desired, would it not be prudent to understand what actual value we would gain from expensive *Tamarix* management?

The Economic Importance of Water Used by Tamarix

Of the values that affect policy making, one of the most important is financial. Economic benefit analysis almost always drives policy. When President Clinton established the National Invasive Species Council by Executive Order 13112, research on the environmental and economic impacts of invasive species (more than $137 billion per year in the United States to agriculture, forestry, public health, and endangered and threatened species) was used to develop policy (Pimentel et al. 2000). Executive Order 13112 defined an invasive species as "an alien species whose introduction does or is likely to cause economic or environmental harm or harm to human health" (NISC 2008).

If the recent findings of the USGS and the ET Peer Panel reports are correct in their assessment that sound *Tamarix* management conditions could lead to water savings, what would be the value of this "saved" water? As always with a difficult scientific, ecological, and economic issue, the answer is, it depends. The first step in describing the value of water is recognizing both how water is valued in the West and who would reap the benefit of the saved water.

HOW WATER IS VALUED

In the West, water is a commodity. Its value is dependent on state water law, water demand, and water availability. In Colorado, where more precipitation occurs on the western slope of the Continental Divide, yet most of the demand is on the eastern slope, water values can vary drastically as can how that water is used (for agricultural, municipal, or industrial purposes). In the American West, water use is governed by the law of prior appropriations, often referred to as "first in time, first in right." Water users with older, more "senior," water rights have priority for getting water when there are shortages, making these rights more valuable. Newer "junior" water rights are only filled after a senior water right is satisfied and therefore have less value. As Hal Simpson, Colorado's former state engineer, noted, "You could consider *Tamarix* to have the senior water right on the river; it always gets it" (Simpson pers. comm. 2003). It is thus reasonable to assume that *Tamarix* will always use water first because it can tap into deeper groundwater, or at least the associated capillary fringe, than can the native xeric plant community of the upper terraces of a river floodplain, even during the most severe drought. Therefore, the value of this lost water is relatively high.

Figure 17.1 illustrates the market value of water and its availability for purchase relative to seniority; that is, senior water rights, because they provide the highest probability for delivery and are finite in number, are the most expensive to purchase (Carlson 2005). It is also reasonable to argue that *Tamarix*'s water use value is at the apex of the pyramid shown by the arrow. Is it unreasonable to think that with this high value, *Tamarix* management efforts would be significantly expanding? They are not. Why? Understanding this apparent paradox requires a deeper look into western water law.

IF TAMARIX WATER USE HAS A HIGH VALUE, WHY AREN'T LAND MANAGERS REAPING THE REWARD?

If water is saved through good *Tamarix* management, meaning control and revegetation with natives, the unused water does not become a new water right for the land manager or landowner who performed the *Tamarix* management. It is not considered "new" water, but rather water in the system that remains available to existing water rights owners. In the 1974 water rights case *Southeastern Colorado Water Conservancy District v. Shelton Farms*, a farmer attempted to claim a new water right by removing tamarisk from his property. The Colorado Supreme Court disagreed: "This water is not properly considered as *developed* and instead, it should be legally classified as *salvaged*...for a fully appropriated river

Water Value
$/ac-ft

Always senior, even in extreme drought

Almost always senior, except in a drought

Normal water year

Usually called, except in wet years

Always called, except in extremely wet years

Water rights
Seniority

Water Rights Available for Purchase

FIGURE 17.1 Conceptual model of the market value of water and its availability for purchase relative to to seniority.
Source: Carlson 2005.

(Arkansas).... *[T]hirsty men cannot step into the shoes of a 'water thief'* (the non-native phreatophyte *Tamarix*)" (Colorado Supreme Court 1974). The Court held that if a wasteful practice is occurring and you remedy the situation, you are not entitled to the water that is recovered. The same legal concept applies to lining ditches to reduce seepage losses: the "saved" water is not yours to use for other purposes.

Although individual land managers cannot gain water rights from *Tamarix* management, management at the state level is important because all western states have water agreements with other states (e.g., Colorado River Compact of 1922) and want to ensure that every available water resource within their state is used efficiently. In the case of *Tamarix*, with its reputation for high water use, state water agencies see inefficiency.

Besides the *Tamarix* water-use issue, there are other considerations that affect policy decisions concerning this invasive species. Many landowners and land management agencies recognize other reasons to manage *Tamarix*, such as wildlife habitat, wildfire mitigation, stream carrying capacity, aesthetics, cultural values, and recreational enhancement.

Other Considerations in Tamarix Management

Both the USGS report and the Tamarisk Coalition's study recognize that tamarisk's impacts on wildlife are diverse, and depend on the species being considered (USGS 2010; TC 2009). It is commonly assumed that invasive plant species are detrimental to wildlife because they can displace the native plant species on which animals depend. This is true in many cases, especially when an animal has specialized on native plants for food or shelter (see chapter 10, this volume).

It is also possible that an invasive will serve as an acceptable substitute for the native plant it displaces and thus have little or no direct or immediate effects on the animal species. An example is the endangered southwestern willow flycatcher ([SWFL] *Empidonax traillii extimus*) in the lower Colorado River basin (see chapter 11, this volume). *Tamarix* provides suitable habitat for this bird, and some are concerned that tamarisk-control efforts might slow its recovery (USFWS 2005). In contrast, others have pursued legal action to have federal agencies engage in restoration efforts to reestablish native habitat where *Tamarix* has been suppressed by the *Tamarix* leaf beetle (*Diorhabda* spp.; USDCA 2009; see also chapter 22, this volume). There are several other species of birds and other animals that normally occur in riparian corridors but are absent from *Tamarix* stands (see chapter 10, this volume). It has been suggested that *Tamarix* removal in the upper Colorado River basin may benefit several endangered fish species, including the Colorado pikeminnow (*Ptychocheilus lucius*) and the razorback sucker (*Xyrauchen texanus*), by preventing the plant from increasing the sedimentation of backwaters these species use for breeding and rearing. (Krueger pers. comm. 2006). So the impacts of *Tamarix* on other species are not clearly black or white.

Both the USGS and the Tamarisk Coalition acknowledge that *Tamarix* removal can benefit many wildlife species in the long term when the project includes the revegetation of native plants, whether passive or active. Denuded areas rarely support much wildlife, and unless *Tamarix* is replaced by vegetation of equal or greater habitat value, wildlife will not benefit from control efforts, and may in fact be negatively affected. In cases where site characteristics or funding limitations make revegetation impracticable, it may be in the best interest of some wildlife species to leave *Tamarisk* in place.

Additional tamarisk impacts that are well documented include changes to stream morphology; increased wildfire frequency (see chapter 14, this volume); sedimentation (see chapter 7, this volume); soil salinity (see chapter 8, this volume); and cultural, recreational, and aesthetic impacts (Tamarisk Coalition 2009). For most of these, when *Tamarix* is dominant in a river system its impacts are negative or, at best, neutral. Also to be considered are the unknown impacts to the ecosystem from the *Tamarix* leaf beetle (see chapter 22, this volume).

Two other considerations that affect policy are the No-Action alternative as developed under the National Environmental Policy Act (NEPA), and climate change. Environmental legislation (NEPA; Clean Water Act [CWA]; Comprehensive Environmental Response, Compensation, and Liability Act [CERCLA], also referred as Superfund Act; and Endangered Species Act [ESA]) have long recognized the importance of discussing "not taking action." The No-Action alternative does not mean conditions will remain status quo if restoration decisions are not made. Impacts associated with *Tamarix* management versus impacts resulting from no action can be quantitatively or qualitatively evaluated based on research and historic evidence, and such information is cited in comprehensive *Tamarix* assessment reports (USGS 2010; TC 2009).

Determining whether an action is appropriate requires assessing the full range of impacts, not just the obvious ones (water and habitat), as well as the full range of alternatives including the No-Action option. The No-Action alternative is the de facto decision if no decision is made to undertake *Tamarix* management. Sometimes this is the appropriate course of action; sometimes it is not (see table 17.2).

A question asked of the ET Panel that directly relates to the No-Action alternative was: "If climate change occurs, what are the implications for ET rates from *Tamarix* as well as potential replacement vegetation?" The Panel responded that it is highly confident a region-wide rise in temperatures throughout the year due to climate change and that temperature increases could drive higher ET by increasing photosynthetic rates. Increased temperatures could also increase ET rates by extending the growing season and physical range of tamarisk. However, it should be noted that these effects are far from certain, in part due to the uncertainty about how climate change may affect precipitation rates and forms (i.e., snow versus rain) and about how increased CO_2 concentrations may alter plant water-use patterns (Tamarisk Coalition 2009).

TABLE 17.2

Comparison of active *Tamarix* management and the No-Action alternative

Issues	*Tamarix* management impacts	No-Action alternative impacts
Tamarix water consumption	The potential exists for water savings if *Tamarix* is replaced on floodplain terraces with xeric vegetation.	Increased water use by *Tamarix* will likely occur as *Tamarix* distribution increases and infilling occurs. Climate change may or may not lead to higher ET losses.
Tamarix spread	*Tamarix* could be reduced to a point that the species no longer has a significant negative impact on native plant species. With appropriate monitoring and maintenance these conditions can be sustained.	*Tamarix* will continue to spread up tributaries and along major river systems, except in those floodplain areas already fully occupied by *Tamarix* (ch. 3, this vol.). This spread and infilling will be aided by more frequent fires and climate change. Russian olive (*Elaeagnus angustifolia*), another nonnative tree, will gradually take over some areas *Tamarix* currently occupies because of its shade tolerance, and it could become the dominant species.
Wildlife and sensitive species	Some species may suffer habitat loss in the short term until revegetation efforts are successful. Habitat for many other terrestrial and aquatic species will benefit.	Some species may fare well in a *Tamarix* dominated floodplain. Other species, both terrestrial and aquatic, may suffer.

(Continued)

TABLE 17.2 (CONTINUED)

Issues	*Tamarix* management impacts	No-Action alternative impacts
Salinity and soil chemistry	Surface soil salinity would gradually decrease due to precipitation flushing. Where no overbank flooding is possible and capillary action raises leave salt residues due to evaporation, soil salinity will remain a problem.	Surface soil salinity will likely increase as distribution and infilling of *Tamarix* takes place. This increase can limit vegetation to species that are more salt tolerant.
Sedimentation	Erosion and transport could occur if revegetation does ameliorate this factor.	Stream morphology will likely continue to change to narrower and deeper channels.
Wildfire threat	Fire frequency and intensity would be reduced.	Fire frequency and intensity would increase.
Biological control	If biological control is accompanied by revegetation efforts, river system ecology should generally improve.	Biological control is occurring in the Upper Basin and will likely continue to spread into the Lower Basin as the *Tamarix* leaf beetle evolves or different ecotypes are introduced. Without native vegetation reestablishment, many species of wildlife including the endangered SWFL could lose habitat.
Cultural	Cultural impacts would improve with reestablishment of native plant communities. Short-term loss of shade for livestock. Possible archaeological damage due to bank erosion.	Overall, negative cultural impacts would increase.
Recreational	Generally, recreational impacts would improve with reestablishment of native plant communities. There would be short-term loss of shade for river users. Some types of river use may be affected if rivers become wider and shallower because of bank destabilization.	Overall, negative recreational impacts would increase.

Abbreviation: SWFL: southwestern willow flycatcher.

What Policies Have Actually Changed?

To answer the question of what policies have changed, it is critical for any policy maker to determine whether the three conditions stated at the beginning of this chapter have been or can be fulfilled.

Much has changed in the past ten years. To illustrate the degree to which policy has changed, shifts in national, state, regional, ecosystem, and endangered species recovery policies are summarized in the following sections. In one form or another they follow the prescription that a serious problem was recognized, was supported by public concern, and that solutions were available.

A NATIONAL APPROACH

Throughout the West, many roads and highways parallel rivers lined with solid stands of tamarisk—a constant reminder to travelers that the river's vegetation composition doesn't look the way they might expect it to (assuming they know what they're looking at, which is not always the case). This aesthetic perception is reinforced by state wildlife biologists, who generally favor native plant communities. This opinion that *Tamarix* simply did not belong in western river systems persuaded Congress to pass the *Salt Cedar and Russian Olive Control Demonstration Act* (Allard pers. comm.). The Act was signed into law as PL 109–320 by President George W. Bush on October 11, 2006. It was the only significant invasive-species legislation to pass in that session of Congress, and with nearly unanimous bipartisan support. How did this happen?

Congressional leaders believed that the *Tamarix* issue was a problem: the West was in the middle of a severe drought, and *Tamarix* had a reputation for using a large amount of water, as well as for being less desirable for wildlife than native vegetation. Rural constituents looked to Congress for help. I was asked by Congressional leaders Rep. Scott McInnis, Rep. (now Senator) Mark Udall, and Senator Pete Domenici to work with their staffs to develop legislation that could address their concerns. The legislation was vetted through the science advisor to the secretary of interior, land management agencies (Bureau of Land Management, National Park Service, Forest Service, USFWS, and Bureau of Reclamation), USGS, western farming and ranching interests, National Resources Conservation Service (NRCS), Native American tribes, environmental organizations, state agriculture and wildlife agencies, national laboratories, weed scientists, and political scientists. Through this process it became clear that more could be gained than through the traditional practice of funding individual pet projects scattered throughout Congressional districts in the West. We hoped to advance the understanding of the "state of the science," perform restoration with the objective of advancing the science on *Tamarix* and Russian olive, and finally, to help resolve the most vexing component of restoration, that is, sustainable funding. Accordingly, the legislation was structured with three fundamental components:

1. Assessment of the *Tamarix* and Russian olive problem, extent, and impacts
2. Funding of $80 million for large-scale demonstrations and associated research over a five-year period. These demonstration projects for control and revegetation were intended to serve as research platforms

for assessing restoration effectiveness, water savings, wildfire potential, wildlife habitat, biomass removal, and economics of restoration.

3.   Development of long-term management and funding strategies

Shortly after passage of the Act, the nation's economy went into a tailspin and only a small portion of authorized funding was appropriated for the demonstration projects or the identification of sustainable funding mechanisms. There was, however, appropriation for a scaled-down assessment of the *Tamarix* and Russian olive problem (USGS 2010). Although the demonstration projects and development of long-term strategies received little funding by Congress, these components of the new law are being picked up by others at the regional, state, and watershed levels.

REGIONAL EFFORTS

As of early 2012, the seven states of the Colorado River basin are considering jointly funding a small research project at the Cibola National Wildlife Refuge on the lower Colorado River. This research will use extensive groundwater-monitoring data and a hydrologic model to predict if reductions in ET on upper terraces of the floodplain will be converted to groundwater storage and streamflow. Remarkably, with all the research that has been done in the past, this study will be the first attempt to quantify water savings potential using actual research data modeled to site-specific conditions.

A major private foundation is investing in the Colorado River basin through its Freshwater Initiative program to improve streamside wildlife habitat affected by *Tamarix*. The Initiative grants funds to local partnerships working on riparian restoration in six different watersheds to complete *Tamarix* management demonstration projects, to pursue policy reforms at the federal and state levels, to support targeted scientific and economic analysis, and to advance outreach strategies. Examples include developing standard protocols for vegetation monitoring to assess watershed-wide shifts in plant communities; ecohydrological modeling to identify restoration sites that can be sustained by natural river dynamics; research on the influence of stressors on cottonwood recruitment and the means for reducing them; and reevaluating the role of the reestablishment of native plant communities to enable the recovery of endangered species. Additionally, the foundation recognized that that any long-term restoration effort must include a corresponding long-term management and sustainable funding strategy; thus, two significant studies were sponsored. One study developed a compendium of existing and newly emerging funding mechanisms (The Nature Conservancy [TNC] 2011), and the other focused on case-study analyses to identify sustainable mechanisms that have weathered the economic fluctuations of the past 80 years (Tamarisk Coalition 2011).

One aspect of regional policy shifts is the establishment of new organizations whose goals are to work specifically to solve the *Tamarix* problem. The Tamarisk Coalition based in Grand Junction, Colorado, is a good example. This nonprofit organization was formed in 2002 by volunteers representing local, state, and federal agencies, nonprofits such as The Nature Conservancy, and concerned private citizens. Their mission is to "restore riparian lands in the West through education and technical assistance." Interestingly, Tamarisk is not mentioned: control of this invasive plant is not the goal—restoration is the goal.

These are only a couple of examples of regional projects that focus on *Tamarix* management issues. Other significant efforts are beginning on the Missouri, Arkansas, and Rio Grande watersheds as well.

STATE INITIATIVES

Both policy making and funding for *Tamarix* management are occurring at state levels. The governor of Colorado issued an executive order to develop a state strategic plan to address the *Tamarix* problem (CO EO D002–03). Similar plans were developed for New Mexico (New Mexico Department of Agriculture [NMDA] 2005) and Kansas (Kansas Water Office [KWO] 2005). In California, the state assembly passed a bill authorizing the Department of Water Resources to work "in collaboration with the federal government, other Colorado River Basin states, and other entities to prepare a plan to control or eradicate tamarisk plants in the Colorado River watershed" (CAL 2006). To understand the scope of the problem, Colorado mapped and inventoried *Tamarix* throughout the entire state. This effort supported legislation for a $1 million pilot funding program was begun in 2008 that included monitoring and revegetation as a requirement for grant award (CO HB 07–1038). Likewise, the state NRCS modified its criteria for its Environmental Quality Improvement Program (EQIP) in Colorado to fund *Tamarix* control. The USFWS has also incorporated *Tamarix* control and revegetation into its *Partners for Fish & Wildlife* program. Another example is the state of Utah using wildfire mitigation funds for streamside restoration where *Tamarix* is considered a threat to property and human safety because of wildfire risk.

An ecosystem-specific example that is gaining funding is the revegetation efforts by NRCS Plant Materials Centers in New Mexico, Colorado, and Arizona. These specialized facilities are cooperating with local watersheds to provide training and ecotype-specific plant materials (trees, grasses, and shrubs) to enhance the success of *Tamarix* management.

WATERSHED PROBLEMS

Throughout the West there are active *Tamarix* management programs that are the primary focus of watershed work. This includes the Colorado, South Platte, North Platte, Big Horn, Republican, Arkansas, Purgatoire, Yampa, Green, White, Gunnison, Dolores, Escalante, San Rafael, San Juan, Upper Gila, Verde, and Virgin Rivers, the Rio Grande, and others. These projects are all local and driven by land managers' desire to improve wildlife habitat, reduce wildfire risk, and improve recreational enjoyment on the rivers. These are the primary values land managers see in *Tamarix* management and, if water savings occur, they are considered a secondary benefit.

Some of these smaller drainages have a significant advantage over rivers as they still have significant flows throughout much of the year, as well as pulse flows (higher runoff) from late-spring snow melt. Most are inventoried and mapped; thus, the watershed managers understand the extent of the problem and the costs of control and revegetation.

Training, education, and research are other components of these watershed programs, which include outreach to local communities, revegetation training, development of BMPs for *Tamarix* control (Nissen et al. 2010) and revegetation (Sher et al. 2010), specialized training on the role of grazing and riparian restoration, research on secondary invasives management, and research conferences that facilitate land managers' interaction with scientists (see chapter 21, this volume).

ENDANGERED SPECIES RECOVERY

The most interesting example of a policy shift is one now occurring with the recovery program for the endangered SWFL, and the potential impact of the *Tamarix* leaf beetle (*Diorhabda* spp.) on its nesting habitat. The presence of the beetle in the Colorado Plateau can be considered an active 5,500-kilometer-long (3,500 miles) biological control experiment along the Colorado, Green, Dolores, Gunnison, San Juan, Escalante, Virgin, and Yampa Rivers. There is good evidence of effectiveness of this biological control, and it could become the primary mechanism for *Tamarix* control and maintenance in the future. The advantages over other *Tamarix* control approaches might be significant, reducing use of herbicides and precluding more costly long-term solutions. However, the beetle has now spread into northern Arizona and New Mexico from release sites in Utah (see chapter 22, this volume). The defoliated tamarisk is directly affecting the endangered SWFL's critical habitat in the Virgin River watershed, because *Tamarix* had been providing suitable habitat for this bird (see chapter 11, this volume). There is growing concern that *Tamarix* removal or suppression might slow or even damage the recovery of the SWFL. Compounding this concern is the knowledge that three additional species of *Diorhabda* are becoming established in Texas and could spread, naturally or human-aided, into all of the SWFL's critical habitat.

Thus, we are left with the policy question, what are the next steps to protecting and enhancing the recovery of the SWFL in the presence of *Diorhabda*? Scientists and land managers working in the Colorado River system have grappled with this question, and firmly believe it requires a comprehensive program combining research, implementation, and monitoring to address urgent questions (TC 2010). Such changes to existing policy include the following:

- Revisiting the SWFL Recovery Program: Recognizing that *Diorhabda* is present or could be present in critical habitat areas in the near future, and using adaptive management to change the SWFL recovery plan's primary approach of retaining *Tamarix* to one that incorporates selective *Tamarix* management combined with the reestablishment of native plants (USFWS 2002).

- Implementing restoration: Begin native species revegetation efforts throughout SWFL critical habitat, especially in the "hot spot" area of the Virgin River system.
- Expanding monitoring: Monitoring has been vital to land managers in identifying the distribution and spread of *Diorhabda* and its effects on the ecosystem.
- Initiating new research: Research is necessary to understand how this watershed-scale experiment is responding to the unanticipated rate of the beetles' spread and effectiveness. Research areas are native vegetation response, secondary weed invasion, erosion, geomorphologic and hydrological changes, wildfire risk, soil salinity, *Diorhabda* dynamics, and wildlife abundance and diversity.

Conclusion

Has tamarisk affected policy? Of course it has. Research on the topic has exploded; watershed efforts have shifted from a water-savings focus to supporting the reestablishment of native plants; endangered species programs are being modified; legislation has been enacted; and funding is being expanded from the public sector to the private side. Why? Because the *Tamarix* issue met the three fundamentals of how policy is enacted/changed: (1) it is considered a serious problem that must be addressed—although research confirms that tamarisk can provide some valuable wildlife benefits, most scientists and especially the public recognize that for most wildlife species a native riparian habitat is superior; (2) it is connected to an issue—the drought of 2002–2010 and SWFL habitat defoliation by the *Tamarix* leaf beetle; and (3) there were solutions—state strategic plans, inventories, training, watershed organizations, BMPs, and funding from public and private sources.

Tamarix invasion is a unique problem that is considered important from diverse political and cultural perspectives, and it is being addressed in a nonpolarizing, cooperative spirit. If resolution of the common problem can be discussed among diverse groups with their unique biases, then other issues such as grazing, endangered species, other invasives, and water management can be brought into the solution.

The Escalante River watershed in Utah is a good example of how dominant nonnative species, *Tamarix* and Russian olive, focus on river functions. In 2010, various interests in the watershed elevated the discussion to include not only riparian restoration, but also watershed health and ecosystem processes, land uses, and the effects of climate change on species and system targets. These include the role of beaver as a tool for improving riparian function, river flows and water quality, native and nonnative fish, grazing management, aspen harvest, montane riparian and wet meadows, and Wild and Scenic designation. None of these actions could have gained the momentum that they now have without *Tamarix* species becoming emblematic of an impaired river system. These are real policy shifts.

LITERATURE CITED

Ambrose, S. E. 1990. *Eisenhower Soldier and President*. Simon & Schuster, New York .

Bleiker H. A. 2010. Citizen participation handbook and systematic development of informed consent. Institute for Participatory Management and Planning. [Online]. http://www.ipmp.com. Retrieved March 20, 2011.

CAL (California Legislation). 2006. California Legislation Assembly Bill 984 for Tamarisk Plant Control, signed into law by Governor Arnold Schwarzenegger, September 6, 2006.

Carlson T. 2005. Tamarisk and other invasive plants: Impacts to water resources in the West. Lecture at the Law of the Colorado River Conference, May 19, 2005. Las Vegas, NV.

Chesapeake Bay Program. 2009. History of the Chesapeake Bay Program. [Online]. http://www.chesapeakebay.net/about/how/history Retrieved August 9, 2012.

CO EO (Colorado Executive Order) D002-03. Colorado Governor William Owens Executive Order D 002 03 Directing State Agencies to Coordinate Efforts for the Eradication of Tamarisk on State Lands, January 2003.

CO HB (Colorado House Bill) 07-1038. Concerning Support for the Control of Non-native Plants from Colorado's Watersheds, February 2007.

Colorado State University (CSU) Cooperative Extension. 2008. Corn, Water Requirements. [Online]. http://www.extension.org/pages/14080/corn-water-requirements. Retrieved November 18, 2011.

Colorado Supreme Court, 1974. Southeastern Colorado Water Conservancy Dist. v. Shelton Farms. 529 P.2d 1321, at 1326–27 (Colo. 1974).

Comprehensive Environmental Response, Compensation, and Liability Act of 1980 (CERCLA) (42 U.S.C. 9601 et seq.).

Eisenhower D. D. (1948) 1997. *Crusade in Europe*. Johns Hopkins University Press, Baltimore, MD.

Endangered Species Act of 1973 (16 U.S.C. 1531–1544).

Executive Order 13122. *Invasive Species*. Federal Register, Volume 64, Number 25, February 8, 1999, Presidential Documents, pp. 6183–6186. [Online]. http://frwebgate.access.gpo.gov/cgi-bin/getdoc.cgi?dbname=1999_register&docid=99-3184-filed.pdf.   Retrieved March 20, 2011.

Federal Aid in Sport Fish Restoration Act of 1950 (16 U.S.C. 777–777k 64 Stat. 430).

FHA (Federal Highway Administration) Publications. Public Roads. Vol. 60, No. 1. Federal-Aid Highway Act of 1956: Creating the Interstate System, 2010. [Online]. http://www.fhwa.dot.gov/publications/publicroads/96summer/p96su1o.cfm.   Retrieved August 13, 2012.

Hanson B., K. Bali, S. Orloff, B. Sandon, D. Putman. 2008. How much water does alfalfa really need? Proceedings. California Alfalfa and Forage Symposium and Western Seed Conference, San Diego, CA, 2–4 December 2008. UC Cooperative Extension, Plant Sciences Department, University of California, Davis.

Healthy Forest Restoration Act of 2003 (84 U.S.C. 6501–6591).

KWO (Kansas Water Office). 2005. Ten-Year Strategic Plan for the Comprehensive Control of Tamarisk and other Non-native Phreatophytes, December 2005. Topeka, Kansas.

Kingdon J. W. 1995. *Agendas, Alternatives, and Public Policies.* 2nd ed. Adison-Wesley, New York, NY.

Lacey Act of 1910 (16 U.S.C. 3371–3378).

Migratory Bird Conservation Act of 1929 (16 U.S.C. 715 et seq.).

NISC (National Invasive Species Council) 2008. 2008–2012 National Invasive Species Management Plan, August 1. Department of the Interior. Washington DC. [Online]. http://www.invasivespeciesinfo.gov/council/mp2008.pdf.

National System of Interstate and Defense Highways Act of 1956 (23 U.S.C.).

Nissen S., A. Sher, and A. Norton, eds. 2010. Tamarisk best management practices in Colorado watersheds. Colorado State University, Ft. Collins.

NMDA (New Mexico Department of Agriculture). 2005. New Mexico Non-Native Phreatophyte/Watershed Management Plan. August 5, 2005. Tamarisk Coalition, Grand Junction, CO. Available online at http://www.sjwwii.org/NM%20Non-native%20 Phreatophyte%20and%20Watersh1%20%20.pdf.

North American Wetlands Conservation Act of 1989 (16 U.S.C. 4401–4412).

Pimentel D., L. Lach, R. Zuniga, D. Morrison. 2000. Environmental and economic costs of nonindigenous species in the United States. BioScience 50:53–65.

PL 109-320, U.S. Public Law 109th Congress. Salt Cedar and Russian Olive Control and Demonstration Act, 120 Stat. 1748. [Online]. http://frwebgate.access.gpo.gov/cgi-bin/ getdoc.cgi?dbname=109_cong_public_laws&docid=f:publ320.109.pdf.   Retrieved March 20, 2011.

Shafroth, P. B., V. B. Beauchamp, M. K. Briggs, K. Lair, M. L. Scott, and A. A. Sher. 2008. Planning riparian restoration in the context of *Tamarix* control in western North America. Restoration Ecology 16:97–112.

Sher, A. A., K. Lair, M. DePrenger-Levin, and K. Dohrerwend. 2010. Best management practices for revegetation after Tamarisk removal in the Upper Colorado River Basin. Denver Botanic Gardens, Denver, CO.

TC (Tamarisk Coalition). 2009. Colorado River basin tamarisk and Russian olive assessment, Report prepared for the Seven Basin States, December 2009. [Online]. http:// www.tamariskcoalition.org/ColoradoRiver.html. Retrieved August 9, 2012.

TC (Tamarisk Coalition). 2010. Internal communications of the outcomes of Phoenix SWFL meeting, FWS office, September 8–10, 2010.

TC (Tamarisk Coalition). 2011. Sustainable Funding Options for a Comprehensive Riparian Restoration Initiative in the Colorado River Basin, February 2011. [Online]. http:// www.tamariskcoalition.org/PDF/Sustainable%20Funding%20Options%20for%20 a%20Comprehensive%20Riparian%20Restoration%20Initiative%20in%20the%20 Colorado%20River%20Basin.pdf. Retreived August 9, 2012.

TNC (The Nature Conservancy). 2011. A Compendium of Financing Sources and Tools to Fund Freshwater Conservation, March 2011. [Online]. http://www.conservationgateway. org/file/compendium-financing-sources-and-tools-fund-freshwater-conservation.

USDCA (U.S. District Court for the District of Arizona). 2009. Center for Biological Diversity and Maricopa Audubon Society (plaintiffs) v. Animal and Plant Health Inspection Service and U.S. Fish and Wildlife Service (defendants)—Complaint for Declaratory and Injunctive Relief. Case 4:09-cv-00172-FRZ. Filed March 27, 2009. Note: Case dismissed due to compliance by Animal and Plant Health Inspection Service. [Online]. http://www.biologicaldiversity.org/news/press_releases/2009/south-western-willow-flycatcher-06-17-2009.html. Retreived August 10, 2012.

USFWS (U.S. Fish and Wildlife Service). 2002. Final Recovery Plan Southwestern Willow Flycatcher (*Exidonax trallii extimus*), August 2002. [Online]. http://www.fws.gov/southwest/es/Arizona/Documents/SpeciesDocs/SWWF/Final%20Recovery%20Plan/ExecSummary_Contents.pdf. Retrieved March, 20, 2011.

USFWS. 2005. Federal Register Volume 70, Number 201. Endangered and threatened wildlife and plant, Designation of critical habitat for the Southwestern Willow Flycatcher (*Exidonax trallii extimus*), Final Rule 50 CFR 17. October 19, 2005.

USGS (U.S. Geological Survey). 2010. Saltcedar and Russian Olive Control Demonstration Act Science Assessment, edited by Patrick Shafroth, Curtis Brown, and David Merritt. Prepared in cooperation with the Bureau of Reclamation, and the USDA Forest Service, Scientific. Investigations Report 2009–5247. [Online]. http://www.fort.usgs.gov/Products/Publications/pub_abstract.asp?PubID=22895.

18

A Philosophical Framework for Assessing the Value of Tamarisk

Naomi Reshotko

Is tamarisk good or bad? Or more specifically, does the continued existence of a species (such as tamarisk) have intrinsic value? Philosophers have been asking this kind of question for thousands of years, and many of them have endorsed theories that assume the notion of intrinsic value to be coherent and useful. All philosophers—whether they endorse the notion of intrinsic value or not—would agree that this notion of value requires a certain leap of faith; it requires the assumption of a foundational theory that does not have a basis in the natural world and that cannot be studied scientifically. It speaks of a "supernatural" notion of good (and bad) that must be assumed and *superimposed on*, rather than *studied through*, human experience.

This notion of supernatural goodness must be supported by arguments that do not rely on our experience of the world; despite this, proponents embrace such a notion because they find it necessary for dealing with questions regarding which human behaviors are right and which are wrong. They think that it is important for us to have definitive—and reasonably general—answers to such questions. Still, it would be nice if we could answer questions about the value of the tamarisk without having to make this leap of faith.

Fortunately, there is a way to assess the value of a species, population, or individual plant without making a leap of faith. It involves eliminating the notion of intrinsic value in favor of two other notions: conditional value and self-generated value. I will argue that the tamarisk has no value outside of a context. When we talk about the tree from some absolute, noncontextual, position, we will find that it is neither good nor bad. The tamarisk (like any other species) is a conditional good because it is good under some conditions, and bad under others.[1]

Before further examining these two notions of good, however, let's look at some reasons for why people believe that the tamarisk is good (or bad) in some more absolute sense. Interestingly, we will see that our reasons for saying that the tamarisk is intrinsically good won't actually be served by the notion of intrinsic goodness—even if we are willing to take that leap of faith.

The Importance of Maintaining a Human Scale for Goodness

Why do we care about preserving a species, or a community of species? Why do we want to preserve the particular ecosystem, or the kinds of ecosystems, of which these species are a part? Some obvious reasons are that they might be necessary to human survival or that they might enhance the human environment. Perhaps they enhance it aesthetically, or perhaps they give us food and other resources that we use for clothing or shelter. If they don't do these things directly, they might do them by providing habitat for other things that we eat and/or use as resources. Even if we don't use it currently, it is always possible that we will discover a use for a plant, or an ecosystem, whose utility escapes us at this time. Furthermore the ecosystems of which these species are a part might contribute to our well-being in more subtle ways. For example, there are many indications that the eradication of swamplands and tropical rainforests contributes to an unstable climate, and that we (human beings) will not be able to adapt to the future rate of climate change. In the case of tamarisk, it can be argued that it provides habitat (see chapter 10, this volume for its provision of bird habitat), that it is used to make honey (see chapter 19, this volume), and that it can provide shade for cattle and wildlife. It was planted extensively as a stream bank stabilizer (see chapter 16, this volume) and is sold as a drought-tolerant ornamental tree (chapter 19, this volume).

However, even if tamarisk enhances or is actually necessary for humans' (or other species') survival, this will not allow us to argue that it is *intrinsically* good. At the same time, if tamarisk does nothing but invade the habitat of other plants that enable or enhance human survival, it is still not *intrinsically* bad. Facts like those mentioned above can only convince us that the tamarisk is good or bad *for human beings*, and the fact that a thing is good or bad for human beings doesn't make it *intrinsically* good or bad. Vitamin C is good for human beings—we need it in order to survive. However, if we were to evolve in such a way that we no longer needed Vitamin C, it is unlikely that people would try to prevent it from going out of existence. We might still want oranges and broccoli—we have other *human-centered* interests in these foods. But all of the Vitamin C tablets in

---

[1] Of course, anything that is a "conditional good" could just as easily be called a "conditional bad." But I will follow the convention of using the term "conditional good."

the stores could be destroyed, and there would be no outcry. No one thinks that Vitamin C tablets are intrinsically good.

Some people do, however, argue that tamarisk is intrinsically bad or good—that it has to be completely eliminated or that it has to be maintained—not because it is valuable to people, but because it somehow has a value *in and of itself*. Let's look at what the basis for thinking this would have to be. Certainly, it's not that it is always catastrophic for a species to go out of existence. We have evidence that the loss of species and addition of new ones is part of evolution. We are probably happy that ancestors of our species made way for *Homo sapiens* in their ecosystems, allowing us to thrive. We have the idea, however (and we are probably right), that human behavior is currently eliminating species, resulting in fewer species in the world. In fact, many who support controlling tamarisk do so in the name of protecting a diversity of species (chapters 9, 21, and 25, this volume).

Decreased species diversity is *bad for human beings*. But that still doesn't make diversity *intrinsically* good—it makes it good *for* human beings. It also makes it good *for* most other species. It does not tell us why (or whether) diversity and particular species—like the tamarisk—that do or do not enhance it are intrinsically good or bad.

People who want to preserve diversity might claim that it needs to be preserved not because it is intrinsically good, but because it is good *for* the planet Earth. When they put it this way, they admit that this doesn't show that diversity is intrinsically good. If they can show that the survival of the Earth—even independent of the survival of people—is intrinsically good, however, that provides a nonhuman-centered reason for why diversity is good.

The supernatural character of intrinsic good makes it hard to determine whether Earth has this quality. Fortunately, we don't have to make that determination. Let's assume that the survival of Earth is an intrinsic good. It's not obvious that assuming Earth's survival or even the diversity of species on Earth is an intrinsic good means that protecting or eliminating the tamarisk (or behaving in any other particular way toward any particular plant, population, or species) is intrinsically good or bad.

Earth has a life span. There is evidence that there was a time when Earth did not exist, and there is evidence that there will come a time when Earth ceases to exist. We, as human beings, did not bring Earth into existence, and, even if we do nothing to hasten its demise, it will, someday, be destroyed by forces beyond our control. But Earth's lifespan is measured in geologic time, and its existence as a planet is not dependent on the survival of human beings. Even if we pollute the oceans with petroleum, cause the climate to change more rapidly and drastically than it otherwise would, or fill the atmosphere with radiation to the point that it is habitable only by ants and cockroaches, the planet will still exist. *Homo sapiens* will cease to exist, but Earth will not. However we want to characterize *Earth's* well-being, it should not depend on human existence. So when we care about an Earth (and solar system) that is habitable by human beings, we are not caring about maintaining the earth *for its own sake*. We are not caring about the goodness of the earth *on its own terms* or on a geologic scale. Rather, we are caring about the goodness of Earth on a human scale: we are caring about Earth insofar as it is good *for human beings*.[2]

There are those who argue that certain ecosystems have the "right to continued existence, and, at least in spots ... continued existence in their natural state" (Leopold 1949). They argue that we should care about these ecosystems for their own sake and not insofar as they are good for human beings. However, apart from celebrating the beauty and complexity of such ecosystems, they are unable to show from whence these rights arise. When push comes to shove, and they try to convince those who would not respect such rights, these same conservationists make human-centered arguments:

> A system of conservation based solely on economic self-interest is hopelessly lop-sided. It tends to ignore, and thus eventually to eliminate, many elements in the land community that lack commercial value, but that are (as far as we know) essential to its healthy functioning. It assumes, falsely, I think, *that the economic parts of the biotic clock will function without the uneconomic parts*. (Leopold 1949; emphasis added)

and

> Viewed from a deep ecocentric/ecological self level, the flourishing of wild species in wilderness and the best interests of humans coincide. Wilderness and wild species have to be preserved not only for planetary health and the inherent worth of biodiversity, but equally as well for who we are as human beings. (Sessions 1992)

They do this because there is no way to argue from first principles for the intrinsic value of ecosystems. An ecosystem's intrinsic value must be assumed and will only be granted by someone who already sympathizes with this notion. Certainly it is preferable to be in a position to educate opponents to the conservation of an ecosystem based on principles that can be observed and studied. The human-centered notions of good for which I argue, as well as the analysis of the functions of members of ecosystems that I offer, yield very similar conclusions to those sought by many preservationists who rail against an anthropocentric evaluation of our natural environment. Perhaps such thinkers fear that human-centered reasons for the preservation or elimination of a plant, species, or ecosystem will always be narrow and uneducated.[3] I contend that a human-centered approach that is educated and

---

[2] This argument is adapted from one made by Stephen Jay Gould. See David, Lenoir, and de Tonnac (2000).

[3] Leopold (1949) dismisses the notion of "enlightened self-interest" even as he makes the claim, cited above, that it is in our self interest to preserve the parts of the biotic clock that do not have economic value (208). Sessions seems to assume that any human-centered approach to the valuation of the natural world *must* be advocating an ignorant narrowly utilitarian or "Disneyland"

humbly aware of the limitations of our knowledge of the intricacies of the world's ecosystems will consider important long-term human goals that would be sacrificed were we to approach our human interests too narrowly. However, to be very clear, I do not put the long-term well-being of the planet or of any ecosystem on an equal footing with the long-term well-being of the human species as thinkers like Leopold and Sessions purport to do. Also, because of the arguments I've made in this section, I find their claims to this effect unsupportable.

Conditional Goods versus Unconditional Goods

It seems clear that, when we talk about whether or not the tamarisk (or anything else, other than human beings) is good, what we need to determine is whether it is good for human beings. How do we do that?

The tamarisk gets its value from the role that it plays in the systems in which it is a participant. We need to evaluate which roles it plays in which systems, and then we have to decide whether or not those systems benefit human beings. The analysis doesn't end there: we may find that the same system benefits human beings under some circumstances, but has a neutral—or even negative—effect in others. Each of these determinations can be made by using scientific methods to study the tamarisk, the systems in which it plays a role, and the way in which those systems interact with people in both the short and the long (but not geologic) term.

The philosophical framework I advance regards tamarisk and the systems in which it plays a role as conditional goods: they are a means to an end, and they bring that end about under some conditions and not under others. A conditional good that brings about a beneficial end is good; one that brings about a detrimental end is bad. Many of us will voluntarily have a root canal performed on one of our teeth at some point in our lives—but we don't do it just for fun. We choose to have a root canal because it is a means to having teeth that function well. When the nerve in my tooth is dying, a root canal is beneficial and therefore good—a root canal is good under those conditions. Having a root canal on a healthy tooth is not beneficial: it unnecessarily weakens a healthy tooth, and therefore is bad. So a conditional good is a means to an end, and it is good or bad depending on whether the end it leads to is good or bad. It's important to note that the end to which a conditional good leads may be a conditional good itself. We do many things as a means to improve our health, but even health can be a conditional good: consider

mentality where humans unthinkingly "manage" and exploit the natural world as "theme parks" and "roadside attractions" (96–105). Interestingly, Curtin, 2005, and Slater, 1995, argue that it is American conservationists like Leopold and Sessions who have placed a Disney-like "Edenic Narrative" on the notion of wilderness and on particular ecosystems, treating areas like the Amazonian rain forest and the Yosemite Valley as if they have no native populations and no history of functioning as a resource for those populations. For a more comprehensive notion of enlightened self-interest, see Reshotko (2006, 57–73).

the people who were not drafted during the Vietnam War because they had poor eyesight, or were missing the index ("trigger") finger of their right hand. These people might feel, all things considered, that they were lucky to have had health issues that kept them out of Vietnam. So both the penicillin shot that I get in order to improve my health, and the improved health that it gives me, are conditional goods. For riparian ecosystems, removal of tamarisk may be beneficial because it is a means to reducing wildfire risk, thus making tamarisk a conditional bad in those ecosystems (see chapters 14 and 21, this volume). Conversely, killing this tree may result in loss of habitat for an endangered bird we are trying to protect, and therefore the tamarisk would be a conditional good for that endangered bird in that environment (see chapter 11, this volume). Since the tamarisk has conditional value, the tree itself has no value outside of its impact and uses.

Are there unconditional goods? Note than an unconditional good is not an intrinsic good. An unconditional good is also a means—it is good only because it brings about a beneficial end; it is not good *in itself* but rather is only (even if always) good *for something else*. The preceding discussion makes it look like the continued existence of Earth is an unconditional good *for human* beings. No matter what the circumstances, the human species will be better off if Earth continues to exist than if it does not.[4] We also made an argument that appears to imply that a diversity of species is unconditionally good for each and every species.[5] It is easy to see that, even if there are some unconditional goods, they will be vastly outnumbered by conditional goods. Every species or population of plant or animal—other than human beings—will turn out to be conditionally good or bad for human beings. Each will be a means to some human-centered end where that end is only sometimes beneficial.

The above is certainly true of tamarisk; its presence cannot be considered unconditionally positive for humans' interests, even if it can be argued that its negative impacts have been overstated or misinterpreted in the past (see chapters 8 and 5, this volume). Rafters who don't have river access or places to beach because of miles-long tamarisk thickets and birders who expect to see woodpeckers and eagles in the shade of cottonwood trees—only to find a monoculture of tamarisk—are justified in calling the stands of trees that they encounter "bad" as they prevent them from accomplishing their goals. The problem comes when they generalize and claim that all tamarisk trees are unconditionally bad, and don't recognize that,

---

[4] This is just an example, and it could be wrong. We might be able to imagine a science fiction scenario in which the last human beings escape a deteriorating and dangerous (*for them*) planet earth and blow it up in order to ensure their survival as they migrate to another planet.

[5] Note that the fact that something is unconditionally good for a *species* does not make it unconditionally good for populations or organisms that are *members* of that species. While it could benefit a species to eliminate all of the carriers of a certain genetic illness, it would not benefit those members who were the carriers who would not have experienced any symptoms of the illness they carried and might not have passed it down to their offspring. Species also exist in a time frame that has to be considered apart from what is good *for* their particular members. Still, concerns about the extinction of our species can have an impact on the well-being of members of our species, so that it might be reasonable to forgo some current benefits in the interest of our having more generations of descendents.

while tamarisk prevent rafting and bird watching in some environments, they might further some other legitimate goals in these same environments (or in others).

A clarification is important before we continue: neither the conditional goods nor any other goods I am talking about in this essay are "relative" or "subjective." Such theories of goodness claim that goodness is *relative to* a subject's or observer's beliefs. They would say that root canals and tamarisks are good *for* those who *think* them good and bad for those who *think* them bad. Conditional goods are *objectively* good in certain determinate circumstances and objectively bad in others. The fact that I want a root canal or think it is good is not what makes it good for me. It is the objectively determinable condition that my teeth (and the rest of me) are in that makes it good or bad, and that is why I go to the endodontist to be advised on which choice to make. A person might like the way the tamarisk looks and think that planting some in his yard will enhance his life, but whether or not it actually enhances his life will depend on how it interacts with his property values, septic field, the health of his family and pets, and so on. And these interactions and their consequences are objective and independent of what any individual thinks they are or will be. The claim that tamarisk is good or bad for some particular species of animals must be an objective claim about how it inhibits or enhances their survival or thriving—not a claim about whether or not they like them (e.g., see chapter 10, this volume; Bateman et al. show objective evidence that the tamarisk is bad for some species and good for others).

Self-Generated versus Other-Generated Goods

Conditional goods are only good or bad as *means* to some other thing that brings benefit or harm. Self-generated and other-generated goods are different from conditional and unconditional goods: they are ends. In particular, self-generated goods are the ends that help us to evaluate the goodness or badness of conditional goods. It may help to consider this in the context of what philosophers have discussed with respect to the self-generated good of individual human beings.[6] Many philosophers throughout the ages[7] (and some contemporary psychologists: Czikszentmihalyi 1990; Seligman 2002; Gilbert 2006; Lyubomirsky 2008) argue that a positive and optimal condition is the terminus of all human endeavors.[8] Choose anything you want to do or want to obtain

---

[6] "Self-generated" and "other-generated" are my own terms for describing human ends. I think they are the terms that need to replace "intrinsic" and "extrinsic" in a naturalized, empirical framework. See Reshotko (2006), for arguments to this effect.

[7] Many interpreters find that Plato's early dialogues argue for this thesis. See Reshotko (2006). Aristotle begins his *Nicomachean Ethics* stating that happiness is the end of all human pursuits. The utilitarians Bentham and Mill, and later, Sidgewick argue that this is the case.

[8] This condition is often identified with the English term "happiness," but see Nettle (2005, 17–30) for reasons to avoid using "happiness" to refer to the kind of ultimate human experience discussed by many psychologists. Still I will use the term "happiness" here at times with the stipulation that it refers to whatever optimal human experience turns out to be.

and ask yourself why you want it; when you have your answer, ask in turn why you want to do or obtain *that*. Continue to do this, and you will find that you have a long iteration of questions and answers until you finally give the answer "because I want to be happy." But, if you ask yourself why you want to be happy, you will not be able to give a further reason for wanting *that*. It is happiness that makes each thing you use as a means to becoming happy, good. Each of those means has *other-generated*, conditional, value. But, happiness generates its own (unconditional) value—it is always good, and its goodness derives from itself rather than from something else.

In order for this notion of self-generated value to work within the world of science and empiricism, we need the condition that constitutes the self-generated value to be objective and determinate. So, if we are going to use the happiness of individual human beings in our equation, we will have to assume that it is objective, determinate, and—at least theoretically—measurable. This is a controversial thesis, but it is one that many philosophers and psychologists have supported with arguments (Czikszentmihalyi 1990; Gilbert 2006; Lyubomirsky 2008).[9] Note that the English word "happiness" might have connotations that make this claim hard to swallow. It might be best to think of what has self-generated value for an individual human being as that person's "thriving," or "flourishing," rather than "happiness." The idea would then be that human beings are built in such a way that they inevitably try to thrive, and everything they do is done in an effort to make themselves thrive. Furthermore, we are used to the idea that we defer to an expert who studies us objectively, and who determines whether or not we are thriving. I might think myself happy, but I might still defer to a doctor who determines that I have such high blood pressure that I risk having a stroke while doing most of my favorite activities. Because of this expert's determination, I might grant that—despite my beliefs to the contrary—I am not thriving. I might think that I am happy, but might still defer to a psychologist who claims that my addiction to cocaine and my lack of interest in my family is evidence that I am not thriving. Similarly, whether or not a stand of trees is part of a thriving ecosystem is also something that can be determined objectively and a point on which the casual observer should be willing to defer to an ecologist.

To evaluate the goodness of the tamarisk for human beings, we need to think of a self-generated good for the *human species,* or for particular populations of human beings, rather than for particular, individual, human beings. What helps individual human beings thrive could be different from what allows us to thrive as a species or as a population.

You probably recognize that even worrying about your own individual happiness isn't straightforward. You have probably noticed that worrying about your own happiness isn't all that conducive to being happy. To use a simple example,

---

[9] Also see White (2006), for a discussion of this tactic and also the alternative of viewing happiness as a "regulative concept."

suppose that tonight is your last opportunity to read the last few chapters of *War and Peace,* and you really want to know how it ends. But suppose that it is also your best friend's birthday; she lives out of town and you really miss her, and she has come to visit you, hoping to spend her birthday with you. If you thought that all that mattered to your happiness was your own happiness, you might read *War and Peace,* while sitting in the same room as your friend. But this is an unlikely choice. Each activity, reading *War and Peace,* and spending time with your friend, is less fulfilling when you try to do them at the same time. One reason for this is that how much you enjoy spending time with your friend is partly dependent on how happy she is when you are with her. So, making ourselves happy is a highly contextual endeavor—as is our thriving. Conditional goods like reading *War and Peace,* and hanging out with our friends, have to be knit together in very particular ways if they are to yield beneficial results. The thriving of a population or species also involves integrating lots of conditional goods in unique ways. The values of those conditional goods that result in a species thriving will almost certainly be different from the values of those conditional goods that result in the thriving of only particular individuals.[10]

For the purpose of evaluating the tamarisk, let's begin with the context created by the self-generated good of the thriving of the human species. In other words, let's assume that there is something objective that constitutes the thriving of the species, and let's assume that our concerns for maintaining a planet that is habitable by human beings are driven by this goal. Let's also assume that it is a goal that we want either because it is built into us to care about our species' continued existence,[11] or because it is a means to the thriving of each individual in our species.[12] It is important that we understand that what constitutes this thriving is objective and, at least theoretically, determinable by scientific investigation. So, we have established the thriving of the human species as an unconditional and self-generated good toward which all human beings naturally strive. We are now able to evaluate our endeavors with respect to the tamarisk in terms of that goal.

---

[10] For example, it promotes the survival of a species for its individual members to experience inordinate fear and anxiety over conditions that threaten their well-being and to experience very little satisfaction from conditions that promote it, but this is a source of distress to individual members of the human species and often inhibits their happiness. See Nettle (2005) on the "life/dinner problem."

[11] If this were the case than we would just consider it a contingent part of human nature to worry about—either consciously or behaviorally—the survival of the human species; such concerns would be "hard-wired" into us, perhaps because they have been adaptive in the past.

[12] In this case, my own happiness or thriving is enhanced by my confidence that the human species is not likely to become extinct—this might be borne of my natural feelings of attachment to my children and, by extension, all of my future descendents. Even if I am content that the world will outlast me, I want to be assured that my children and grandchildren don't have to worry about apocalypse in their lifetimes and that might mean being assured that my grandchildren don't have to worry about there being apocalypse in their grandchildren's lifetimes, and so on.

Natural Functions

In order to assess the value of the tamarisk, we will look at how it functions in the systems in which it participates, and how those functions affect human thriving. We will find that, since it participates in different systems differently, and because the same system interacts with human activities differently in different contexts, there are some populations that should be controlled or eradicated, and others that should be left alone or supported (see chapter 17, this volume).

The tamarisk does a lot of things: among others, it uses water, it performs photosynthesis, it provides shade, food, and other things relevant to the habitat of species with whom it coexists. We need to objectively determine which of those things constitutes its objective, natural, function.[13] We will find that whether the tamarisk has a function, and what *objective* natural functions it has, varies with the systems in which it participates. Consider the human heart: it pumps blood and it makes noise—which of these two activities is *the* natural function of the human heart? We all want to say that its function is to pump blood. We could justify that by noting that we are evaluating it with respect to the thriving of an individual human being, or of the human species. A human being's thriving is supported by the organized interaction of a number of systems, each of which plays a role in keeping her alive. It is the function of the circulatory system to circulate the blood so that each part of the body receives the nutrients necessary to perform its function. The heart is part of the circulatory system, and its pumping blood allows the circulatory system to do its job. Thus, we are justified in singling out the pumping of the blood as the natural function of the human heart, considered in the context of maintenance of the normal functioning of the human organism.[14] On the other hand, a doctor might feel my pulse, or listen to my heart with a stethoscope; in this case, the sound that my heart makes plays a role in a different system—a diagnostic system concerning what is going on in the human organism. In that system, the sound my heart makes has an objective, natural function as an indicator of my health. While the circulatory system is embedded in systems whose ultimate goal is the thriving of humans in one way, the diagnostic system that makes use of the heart's sound is embedded in those systems in a different way. They are two different systems that ultimately serve the same goal, but in very different fashions. There may be times when the sound my heart makes obstructs certain systems that are also trying to contribute to my thriving; a doctor might listen to my lungs to see if I have bronchitis, and have trouble hearing the relevant noises due to the beating of my heart. While I, of course, want my heart to continue to beat, it could

---

[13] There are two competing philosophical theories of what it is for something to be a natural function: that developed by Wright 1973 and that developed by Cummins 1975. I will make use of the systems theory developed by Cummins as defended by Davies (2001). Davies arguments have convinced me that it is the more coherent of the two theories.

[14] This must be underwritten by the assumption that normal functioning fulfills objective criteria, where one of those objective criteria is staying alive without external technological support.

be that the sound it makes plays no useful, objective, role in diagnosing problems with my respiration. So, when we ask what the natural function of the heart or the tamarisk is, there is no single answer; it will differ with the context—with the systems within which human beings are interacting with each one. Even so, each natural function that an object has is objectively *the* natural function of the thing in question *in* the particular system in which it has that function.

Because we must assess tamarisk in the context of good for human beings (even if that good is the general one of helping to maintain ecosystems that support the continued existence of the human species on the planet), the natural function of tamarisk in various contexts may range from preventing sand from blowing onto train tracks to providing wildlife habitat. However, with regard to tamarisk management and valuation of the tamarisk, it may be more relevant to consider instead the natural function of riverside (i.e., riparian) plant communities where tamarisk grows, and whether the presence of tamarisk there contributes to or detracts from that function. Riparian ecosystems serve many functions for humans, and are especially important in arid environments as they are where the highest concentration of plants and other biodiversity will be found (Gregory et al. 1991). The primary complaint about tamarisk is that it alters the historic functioning of these ecosystems. Thus, the debate about whether tamarisk is good or bad might be best framed in terms of what the various functions of riparian habitat are with respect to its allowing the human species to thrive, or at least, what the value of tamarisk is in these riparian systems.

Assessing the Value of the Tamarisk, Population by Population

Using the model described above, we can conclude that the tamarisk is a conditional, other-generated good. Thus, its value *for human beings* can only be assessed within a system in which it has an objective function, where that system is embedded in systems that interact in ways that support the thriving of the human species on Earth. That is, wherever we find the tamarisk, we have to see how its functions are connected to the continued existence of our species. This might not be straightforward, but it gives us something concrete that is subject to empirical investigation, upon which to base our arguments. Currently, some tamarisk populations are creating a fire risk for nearby residential and business developments (see chapter 14, this volume). These particular populations of tamarisk have woven themselves into the interactions between certain riparian systems and particular residential and economic systems. These systems—and their particular instances—are also conditional, other-generated goods. If those tamarisk populations inhibit the proper functioning of these systems *and* these same residential and economic systems play objective, natural functions in further systems that promote the thriving of the human species, then perhaps the tamarisk are conditionally bad in these contexts. Thus, those particular tamarisk populations might need to be controlled or eliminated. But this tells us nothing about what to do with populations of tamarisk that are not inhibiting the proper functioning of residential and business developments.[15]

Some tamarisk provide habitat for the endangered southwestern willow catcher (see chapter 11, this volume). The continued survival of this bird species is also a conditional, other-generated good that needs to be evaluated with respect to the systems in which it is embedded in order to determine if it makes a positive contribution, even if only remotely, to the thriving of the human species. As discussed earlier, the promotion of continued diversity might indeed be a functional component of the goal of preservation of the human species. Although it also has its limits: it matters that there are many and diverse species, but it's not clear how important it is that any particular species be among them. The argument that we don't know enough to determine exactly what the effects on diversity of the eradication of any particular species could mean that we need to promote their continued existence wherever possible.[16] It could easily be the case, however, that tamarisk populations that are near developments, but provide habitat for endangered species, will have a different (conditional, other-generated) value than those that are in remote areas, or on a private ranch, or are intermixed with native trees. That is: it is impossible to state uniformly that this species is good or bad, or even that it has value or not, because its conditional value will differ with each stand of trees.

Using This Model in the Absence of Full Knowledge

Even if we are convinced that this is an appropriate model to use in order to evaluate the tamarisk, and that there are objective criteria that will help us to delineate what it is for the human species to thrive, it is unclear how, if we don't right now know what those objective criteria are, we can use this model in practice. If this is the correct model to use, then the responsible thing to do is to form educated hypotheses concerning criteria for the thriving of the human species. It is relatively easy to propose *some* criteria, even if they do not fill in the entire picture.

One likely criterion for the thriving of the human species is its continued existence. Thriving is, of course, something beyond survival. For example, thriving might require survival in certain numbers—enough that we maintain sufficient diversity within our species that we can continue to undergo biological evolutionary adaptation. It might include the ability to experience our lives as something

---

[15] Note that we need to evaluate the way *particular* residences and businesses fit into *particular* larger systems. It could turn out that a particular house was poorly located so begin with. It would make sense to evaluate the tamarisk on such lots with respect to other systems in which they play a role, even if we deem then conditionally bad for these residences.

[16] Stephen Jay Gould alludes to this sort of reasoning (David et al. 2000).

other than burdens and mere opportunities for procreation. Our ability to adapt to our environments through culture and technology, and to enhance our lives—or ruin them—using these tools is arguably unique to the human species, and might be considered essential to the identity of members of our species. Proposals concerning what constitutes our thriving will be controversial, and may be hard to justify or debunk. It could be that the answers to these questions require philosophical rather than empirical or scientific investigation.

However, assessing the value of populations of tamarisk (and other species) relative to the survival of the human species might be enough to form a baseline for the engagement in research concerning in which contexts the tamarisk is a conditional good, and in which it is a conditional bad. For example, it seems safe to propose that having the Earth's human habitats and their climates change no faster than humans can adapt to them is either a good under many conditions, or an unconditional good.[17] There might be instances in which the tamarisk (or another kind of plant) supports the maintenance of habitat in ways that ultimately serve us, and other instances where it doesn't. Our theory will require us to treat the tamarisk differently in these different cases, as illustrated above. It is likely that we can arrive at a certain amount of consensus on the questions of the conditional value of various populations of a wide variety of plants and animals keeping only these criteria in mind, even while efforts are made to figure out what it is for the human species to thrive, and which particular instances of systems—in which particular places—support that flourishing and should be maintained, or inhibit it and should be altered.[18]

Literature Cited

Csikszentmihalyi, M. 1990. *Flow: The Psychology of Optimal Experience*. HarperCollins, New York, NY.

Cummins, R. 1975. Functional analysis. Journal of Philosophy 72:741–760.

Curtin, D. 2005. *Environmental Ethics for a Post-Colonial World*. Rowman and Littlefield, Lanham, MD.

David, C., F. Lenoir, and J. de Tonnac. 2000. *Conversations about the End of Time*. Fromm International, New York.

Davies, P. 2001. *Norms of Nature: Naturalism and the Nature of Functions*. MIT Press, Boston, MA.

Gilbert, D. 2006. *Stumbling on Happiness*. Random House, New York.

---

[17] It might not be an unconditional good; we could imagine a scenario in which human beings escape the earth and find a habitat that is even more conducive to their thriving than the earth is. The possibility that this would actually happen, however, seems vanishingly small.

[18] I am grateful to Anna Sher for several helpful suggestions regarding the examples in this article. I am grateful to Thomas Nail for much discussion and several helpful suggestions regarding this article.

Gregory, S.V., F. J. Swanson, W. A. McKee, and K. W. Cummins. 1991. An ecosystem perspective of riparian zones. BioScience 41:540–551.

Leopold, A. 1949. *A Sand County Almanac*. Oxford University Press, New York.

Lyubomirsky, S. 2008. *The How of Happiness: A Scientific Approach to Getting the Life You Want*. Penguin Press, New York.

Nettle, D. 2005. *Happiness: The Science behind Your Smile*. Oxford University Press, Oxford.

Reshotko, N. 2006. *Socratic Virtue: Making the Best of the Neither-Good-nor-Bad*. Cambridge University Press, Cambridge, UK.

Seligman, M. 2002. *Authentic Happiness*. Free Press, New York.

Sessions, G. 1992. Ecocentrism, wilderness, and global ecosystem protection. Pages 90–130 *in* M. Oelschlaeger, editor, *The Wilderness Condition*. Sierra Club Books, San Francisco, CA.

Slater, C. 1995. Amazonia as edenic narrative. Pages 114–131 *in* W. Cronon, editor. *Uncommon Ground: Toward Reinventing Nature*. Norton, New York.

White, Nicholas, 2006. *A Brief History of Happiness*. Blackwell, Oxford, UK.

Wright, L.1973. Functions. *Philosophical Review* 82:139–168.

19

Botany and Horticulture of Tamarisk

Martin F. Quigley

Plants of the genus *Tamarix* L. (Tamaricaceae, one of four families in the order Caryophyllales) are large shrubs or small trees comprising more than 40 species and hybrids. Species in the genus are native to mostly arid environments from North Africa and the Mediterranean basin eastward to northern India and west-central China. Its center of distribution appears to be in the dry valleys of the Middle East and northwest to Pakistan, but it has been widely planted for centuries, so native ranges are indeterminate (Bailey 1976). Given *Tamarix*'s wide geographic distribution and very long history of human use and propagation, the naming and identification of species is convoluted and difficult, with many redundant synonyms in common use (see table 19.1; see also chapter 2, this volume). In English, its common names are tamarisk, saltcedar (generically), or athel (usually only for *T. aphylla*, the tallest-growing, most treelike species of the genus).

Botanical Description

The tamarisk plant has fine-textured, often very dense twiggy branching; the plant is multistemmed and woody, with thick, peeling to deeply fissured bark (Bailey 1976). It is generally fast-growing when young, slowing with age, and can have a life span of many decades (Dirr 1990). Its deciduous leaves are alternate, sessile, often sheathing, small, and scalelike, with salt-secreting glands (see chapter 8, this volume). The tiny leaves are mostly needlelike as in *T. aphylla*, or can be completely reduced to scales. The dense, shrubby, and irregular form of most species can be very attractive, and the foliage is among the most finely textured of the large, nonconiferous shrubs. It is tolerant of a wide range of soils, moisture,

319

FIGURE 19.1 The tamarisk has a spectacular floral display lasting for many weeks.
Source: Courtesy of Tim Carlson.

temperature, and altitude, but grows particularly well in dry, hot and well-drained sites (see chapter 5, this volume; Dirr 1990).

Even its harshest detractors cannot deny that when flowering, tamarisk is spectacular (see figure 19.1). In full bloom, every branch and twig is covered with flowers, almost masking its bright or gray-green foliage. Flowers are small, short-pedicelled or sessile, and borne in dense racemes, usually in terminal panicles; in some species the tiny flowers are clustered densely in the leaf axils. In most species the flowers are perfect (having both sexes), but *T. dioica* is dioecious, with male and female flowers on separate pants (Bailey 1976). Flower petals are usually a pale to medium pink, though the color ranges from almost white to a magenta red depending on species and site conditions. Sepals and petals number 4 to 5; stamens usually 4 to 5, rarely 8 to 12, sometimes slightly connate at the base; the ovary is one celled, surrounded at the base by a more or less deeply 5- or 10-lobed disk; styles are 2 to 5, clavate or short and thick; the fruit is a capsule, dehiscent into 3 to 5 valves; seeds are numerous, minute, and bear tufts of hair at the apex (Bailey 1976; see also chapter 2, this volume).

Most tamarisk species have fairly long periods of bloom, lasting many weeks to months, and the various species have distinctly different flowering phenologies: some are early spring bloomers, some peak in summer, and others bloom into autumn (see table 19.1). Though tamarisk seeds are very numerous and equipped for dispersal by wind and water, their very small size precludes the visible drifts of seeds often seen with phreatophyte species of *Populus* and *Salix,* both in the family Salicaceae, though they all share the characteristic brief seed viability common to riparian woody plants. Reproduction in tamarisk is prolific (*T. gallica* has up to 1,000,000 seeds/kg), and in addition to seed production, the plant can spread vegetatively from multiple root-crown sprouts, and by adventitious rooting from submerged or buried stems, or from lateral roots. In Italy, the Istituto Superiore

TABLE 19.1

Summary of *Tamarix* species names and characteristics (cf. Bailey 1976 and others)

Species*	Form	Flower season	Native origin
T. africana Poir.	Shrub/tree	Spring	W.Europe, NW Africa, Canary Islands
****T. aphylla* (L.) Karst., "athel"	Tree	Summer/autumn	N. Africa, E. Mediterranean. Synonym: *auriculata*
T. aralensis Bunge.	Shrub/tree	Spring/summer	South-central Russia, Iran
T. canariensis Willd.	Tree	Spring/summer	Mediterranean
****T. chinensis* Lour.	Tree	Spring	East Asia. Synonyms: *caspica, elegans, japonica, juniperina, libanotica, plumosa*
T. dioica Roxb. Ex Roth.	Tree	(Bloom time not noted)	India, Pakistan, Iran, Afghanistan
****T. gallica* L.	Small tree/shrub	Spring/summer	S. Europe. synonyms: *algeriensis, anglica*
T. hispida Willd.	Small tree/shrub	Summer/autumn	Central Asia
T. indica Willd.	Small tree/shrub	Summer/autumn	Northern India (not cultivated)
****T. parviflora* DC	Tree/shrub	Spring	Turkey, Greece, Yugoslavia, Albania. Synonym: *tetrandra*
****T. ramosissima* Ledeb.	Small tree/shrub	Spring/summer	East. Europe to Central/ East Asia Synonyms: *aestivalis, amurensis, indica, odessana, pentandra*

*Of these verifiable species, those most commonly found naturalized in the American Southwest are *T. chinensis*, *T. parviflora* (often misidentified as *T. pentandra*), and *T. ramosissima*. There is also evidence of hybridization among the invasive species (see Gaskin and Schaal 2002; see also ch.2, this vol.). Other recorded species names include *T. androssowii, arceuthoides, autromongolica, boveana, canariensis, dalmatica, elongata, gansuensis, gracilis, hampeana, hohenackeri, jintaenia, karelinii, laxa, leptostachys, meyeri, mongolica, sachuensis, smyrnensis, taklamakanensis, tarimensis*, and *tenuissima*. Baum (1978) also records a suite of *Tamarix* species found in Israel: *T. arvensis, aplexicaulis, gennessarensis, hampeana, jordanis, negevensis, nilotica, passerinoides, parviflora, palestina*, and *tetragyna*.

** denotes species found in the United States.

per la **Protezione** e la **Ricerca Ambientale**, or ISPRA, formerly known as Agency for the Protection of the Environment and Technical Services (APAT 2003), in a pamphlet about the environmental use of tamarisk as a soil stabilizer, states that under natural conditions, the viability of the tiny seeds is quickly lost (Young et al. 2004). Sowing without any pretreatment should immediately follow collection. The seed may be stored at a temperature of +3°C to +4°C for one to two years, but with a considerable loss of viability. Germination occurs rapidly, within 24 hours of seed imbibation (Dirr 1990), though subsequent early development of the seed-lings can be slow. Vegetative propagation by means of cuttings taken in autumn/winter is also widespread in horticulture (Dirr 1990).

In general, tamarisk seedlings require a moist substrate for germination and during early establishment, which is why most tamarisk populations begin in riparian or in seasonal drainage bottoms, from which they can spread vegetatively to drier upland sites (see chapter 1, this volume). Tamarisk root structure is particularly dense in the riparian zones of arid ecosystems, often with deep roots to the water table, and extensive fibrous, fine roots near the soil surface. The density of tamarisk stands may preclude germination or establishment of other species' seeds through competition for water and light, and also because the foliar salt glands, sequestering and excreting salts that are drawn up from the substrate, promote a saline environment of the soil surface below the plants (see chapter 8, this volume).

Gaskin and Schall (2002) note that in North America there are tamarisk hybrids that are not found in the native Eurasian ranges of the species. These are crosses between *Tamarix aphylla* and *T. parviflora*, between the morphologically similar *T. canariensis* and *T. gallica*, and the most widespread colonizer, a hybrid of *T. chinensis* and *T. ramosissima* (chapter 2, this volume).

Ethnobotany

In its native range from west-central Asia, the Middle East, and the Mediterranean, various *Tamarix* species have long been exploited by humans for various uses, including fuelwood, tools, basketry (UNEP 2003), hedges and fencing for livestock enclosure, and also for shade for humans and animals. Of course, in extreme environments where there are limited numbers of woody species and sparse cover, most trees are exploited in many different ways. Tamarisk's very dense and fibrous wood makes it difficult to work, and cutting it quickly dulls traditional tools; however, modern alloys and machinery can produce bowls and other objects from boles large enough to work (see figure 19.2). For centuries, in very arid environments, tamarisk has also been widely planted for windbreaks for its dense

FIGURE 19.2 The dense, fibrous wood of tamarisk can be shaped into bowls and tool handles.

branching, and soil stabilization for its extensive root systems (see chapter 16, this volume). Conversely, the shrub has also been planted for de-watering of marginal fields in the way that *Salix* and *Populus* have been used elsewhere in the north temperate zone (UNEP 2003).

Bailey (1976) cites ancient use of tamarisk for medicine, dyestuff, and food. Rustomjee and Nanabhai (1903) recorded that tamarisk bark is astringent, bitter, and tonic; they also document that ashes of tamarisk may contain large quantities of sulphate of soda, an additive to silica in preindustrial glassmaking. Rustomjee and Nanabhai (1903) also document medicinal uses of exudate (manna) from *T. gallica* (with synonyms *T. indica*, *T. orientalis*, and *T. articulata*), prepared from the *C. manniparus* galls, in decoctions, tinctures, or ointments for treatment of diarrhea and dysentery, as a mild laxative, and as an expectorant. They also state that the soft twigs are used in rural areas as pessaries (cervical contraceptives). Investigations of medicinal uses of *T. gallica* also include the treatment of kidney stones (calcium oxalate; Bensatal and Ouahrani 2008) and as a liver tonic and tumor inhibitor (Sehrawat and Sultana 2006). *T. nilotica* has been used since pharaonic times as an antiseptic, for fever and headache relief, as an anti-inflammatory, and aphrodisiac (Abouzid and Sleem 2011). Newly discovered tannins nilotinin D8 (9) and heirtellin A (1) isolated from *T. nilotica* killed human tumor cells (Ouabi et al 2010).

In the Judeo-Christian bible (Gen. 21:22–34) Abraham planted tamarisk at Beersheba, "the well of the oath"; tamarisk was widespread and common in bibilical lands (Baum 1966). Several plants of the Middle East are known to produce manna (Exod. 16:4; *maan* in Arabic, *gaz* in Farsi); these include *Tamarix manifera*, found in northern Iraq and Iran (Grami 1998). The manna is created when a boring insect, most often *Coccus manniparus* (related to the Cochineal beetle), penetrates the tamarisk stem, stimulating gall formation, from which exudes a sweet, white, somewhat nutritious sap that can be gathered without difficulty after it congeals (Anderson 1984).

A dark, strong-flavored honey is made by wild and domesticated bees that gather nectar from stands of tamarisk; there is currently some commercial production in the American Southwest. In parts of its native range, Libya for example, the honey is considered low grade (Hisham el-Waer, pers. comm.) With tamarisk's extended and prolific blooming over many months, particularly at times when few other nectar sources are available in the desert, it is easy for honeybees to produce a relatively pure honey, with a very distinct flavor. There are also unverified claims by some apiarists that tamarisk honey can prevent or cure minor pathogens in bee hives, such as chalk brood and sac brood.

Horticulture and Distribution

It has been many centuries since agriculture and horticulture diverged; that is, since plant aesthetics and plant productivity took distinctly different paths in human manipulations. Farmers are conservative, crop focused, and tend to reinforce existing habits rather than to experiment with new plants or methods. Gardeners, on the other hand, will plant anything that promises even an ephemeral visual reward, or uniqueness in the garden, even when the plant demands special conditions or extra care. Avid gardeners have always sought the newest, the most different, the showiest plant—and in some cases have gone to great lengths to obtain something that the neighbors will envy (Taylor and Gordon 2000). To get a plant that is unobtainable legitimately, a gardener will nip off a cutting in a botanic garden or park, grab some seed in someone else's garden, or dig plants from public land—even in the dark of night if necessary. This is to illustrate that some gardeners may disregard environmental consequences of their personal plantings: it is only recently that the landscape and horticulture industry has begun to limit propagation and sale of invasive species.

Horticulture also has a gentler side: most plant enthusiasts freely share their treasures with others, just for the asking. But there is also a cutthroat aspect of one-upsmanship in gardening that began in early Victorian England, and persists today, in which any new introduction is jealously guarded for as long as possible (Taylor and Cooper 2000). Once a plant has gone public, designers and landscape architects often wait until the more risk-taking gardeners have proven its reliability, and until the nursery industry has proven its durability and propagated it for general sale. Once tamarisk had proved itself in the hot, arid conditions of Texas to Southern California, it was freely traded, and became a staple in the landscape and nursery trade (see chapter 16, this volume).

Whatever the motive for plant introduction—agricultural, horticultural, medicinal, or simply accidental arrival—the North American landscape is full of plants, both woody and herbaceous, productive, ornamental, or weedy, from around the globe—and especially from temperate Eurasia. It remains unexplained why the converse is not true: very few temperate North American plants have naturalized to the point of invasiveness in Eurasia.

Despite the scattered biblical and historical references to tamarisk planted to beautify oases or settlements, the propagation of tamarisk as an ornamental landscape plant really began when Western Europeans began to plant it in gardens, which probably dates from the time of the Moorish conquest of the Iberian peninsula. Beginning in the early nineteenth century, various species of tamarisk were introduced many times to North America and South America (see chapters 2 and 16, this volume). The route of introduction probably originated in the arid biomes of the Middle East and North Africa, but no records remain of which species went where. Propagated by dormant-season stem cuttings, tamarisk was easily carried, along with other food and ornamental plants, such as the apples and roses of Persia, through Western Europe and Great Britain, and thence to the new world (Joor 1877).

The earliest Atlantic coast North American nursery catalogs to mention tamarisk, during the 1820s, extolled its virtues not for drought tolerance, but for salt tolerance: it was promoted as a tough, reliable, and decorative windbreak for beachfront properties (Dirr 1990). It is probable that these plant merchants were unaware of the arid origins of the plant, and knew it only as a plant that could thrive in sandy, salty, nutrient-limited soil.

While shrubby species of tamarisk can reach small tree stature by structural pruning and training to a single leader, as an ornamental the plant is usually top-pruned severely to encourage foliage density and lateral spread, and to promote new growth for maximum flowering display. Hard pruning immediately after flowering is the recommended timing for optimal bloom in the next growing season. In the early twentieth century, tamarisk had found its supporting, but never starring, role in eastern North American gardens as a large, colorful, but unruly novelty plant (see Davis 1899 and Measures 1933). By the end of the nineteenth century, tamarisk was planted both as a windbreak and as an ornamental, from Texas to Southern California (Horton 1977).

Tamarix resembles other Eurasian genera that have overtaken large swaths of the North American landscape (e.g., *Lonicera, Elaeagnus, Rhamnus, Ailanthus*) in that it is extremely prolific, with large seed output, efficient dispersal, and, under the right conditions, a very high rate of germination. Like these other successful woody invaders, it also spreads vegetatively, sprouting vigorously from basal shoots and from an extensive root system. But unlike many widespread woody invaders in North America, in the moist, temperate climate of the eastern United States tamarisk has never become a problem either in gardens or in natural woodlands. (Gaskin [chapter 2, this volume] reports that *T. tetragyna* has naturalized in Georgia, but that it is rare.) Despite its hardiness in arid conditions, tamarisk often shows decline in growth under conditions of high fertility and constant moisture. When grown in nutrient-rich soils with abundant moisture, tamarisk tends to have weaker wood, more open growth, and a shortened life span (Dirr 1990), limiting its ability to naturalize or be competitive in the benign circumstances of North America east of the Mississippi. Today, at least five of its species, and some hybrids of those, are occasionally found in gardens and parks across the North American continent, though never in great numbers or very common use. Unlike some globally successful animal dispersals, invasiveness in plants is situational, with very few exceptions. That is, ranges are limited at some point by edaphic factors such as moisture, soil, or winter temperatures, and biotic factors such as competition. A plant may be a very aggressive colonizer after introduction but only in a specific set of environmental conditions—and therefore may be perfectly suited for ornamental horticulture in those areas where it will not naturalize. It is ironic that this vigorous plant is not always particularly competitive in benign conditions, even in the Southwest where it otherwise has established very successfully (Sher et al. 2002).

In both Texas and in Southern California, tamarisk was first introduced as a garden novelty, a large flowering shrub that could withstand extreme heat and prolonged drought (Thornber 1916; Swingle 1924; Tellman 2002). Its ability to flourish without irrigation was quickly noted, and the shrub was subsequently promoted by land and water managers for stream and reservoir bank stabilization (chapter 16, this volume).

Tamarisk naturalizes very quickly in some riparian areas, and it is now extremely common and widely recognizable throughout the desert southwest, where, in contrast to its behavior in more mesic climates and weathers, it has become a dominant, invasive plant in both riparian and upland ecosystems (see chapter 1, this volume). It now forms almost pure stands along significant reaches of the Colorado River and Rio Grande basins, and their tributaries, as far north as Wyoming (see chapter 9, this volume).

Tamarisk's greatest biological success in North America, therefore, is not as a garden plant, but as a naturalized colonizer in the canyons and deserts of the American Southwest, where the soils, climate, and altitudes resemble those of its native ranges in Eurasia. These stands can be extremely competitive, excluding most other trees, and their longevity has not yet been determined, as tamarisk has been present only about a century, and its range appears to be still expanding (see chapter 3, this volume). Thus, while tamarisk is not particularly invasive in its native range, or in moist, benign temperate zones, it has acquired an important ecological role in the New World, far from the oases of the Middle East and the Central Asian deserts.

Though tamarisk is currently listed as a noxious plant by many state and federal agencies in the United States and Canada, it is still offered for sale in the landscape trade not only in many large and small nursery catalogs east of the Mississippi (e.g., New York, Ohio, Indiana, Minnesota) but also west of the river (Nebraska and central Canada, for example) In particular it is listed online for sale by many midwestern companies, for example by Dayton Nurseries in Ohio. One catalog (*Rob's Plants*) says of *T. ramosissima* "Pink Cascade" that it has "sprays of pink flowers from July through the end of summer, with deciduous, needlelike foliage. It is salt tolerant, but luckily it will grow in our nonsalty soil as well." Dayton is even more eloquent describing the cultivars "Pink Cascade" and "Summer Glow": "Masses of deep pink flowers create an ostrich plume effect in summer on this hardy shrub. Foliage is soft and feathery. Cut back almost to the ground each winter to improve its flowering display and growth habit. Ideal for creating a colorful splash in any sunny border!" And even better, tamarisk is "deer and rabbit resistant." Though these horticultural selections are listed as cultivars of straight species, it is likely that some are also undocumented hybrids, as found by Gaskin and others in the western United States (chapter 2, this volume).

Conclusion

After almost two centuries in North America, tamarisk remains a horticultural curiosity and has become an environmental issue of national proportion. Its traditional ethnobotanical uses have not been promulgated here. Nor has tamarisk gained wide popularity as a garden shrub in eastern North America, probably because of its large size, rather dead-looking winter aspect, and its somewhat wild, unsymmetrical habit, requiring more drastic annual pruning than most modern

FIGURE 19.3 Athel trees (*Tamarix aphylla*) with trunk diameters of up to one meter at the Furnace Creek Visitors Center, Death Valley, California.

gardeners dare to do—all making it unsuitable for small suburban yards. Tamarisk's reputation as an invader has lessened its appeal for gardeners, even in areas where it has never naturalized or become a political issue. It remains a seldom-seen ornamental east of the Mississippi, one of the now "old-fashioned" introduced shrubs that are considered too large for small urban and suburban gardens, and are relatively unfamiliar to parks and garden personnel. In the Southwest, tamarisk is still used occasionally as an ornamental shrub or tree, despite its well-publicized invasiveness there. Large, long-lived specimens can be found in both private and public landscapes from Southern California to central Texas. For me, a classic horticultural and environmental dichotomy can be seen at Furnace Creek in Death Valley: a very old, long-tended row of athel trees (*T. aphylla*) frames the entry to the visitors center (see figure 19.3), while just to the north a vast stand of dense and tangled naturalized salt cedar (probably *T. chinensis*), clearly invasive and supremely competitive, provides shade and shelter to human and avian visitors.

Literature Cited

Abouzid, S., and A. Sleem. 2011. Hepatoprotective and antioxidant activities of *Tamarix nilotica* flowers. Pharmaceutical Biology 49:392–395.

Anderson, F., editor. 1984. "Herbals through 1500." *The Illustrated Bartsch*, vol. 90. Abaris, New York.

APAT. 2003. Seed propagation of Mediterranean trees and shrubs. Agency for the Protection of the Environment and for Technical Services. Rome, Italy. [Online]. http://www.apat.gov.it/site/en-GB/APAT/The_Institute_-_ex_APAT.

Bailey, L. H. (L.H.Bailey Hortorium), 1976. *Hortus Third*. New York: Macmillan.

Baum, B. R. 1966. Monographic revision of the genus *Tamarix*. Final Results Report for USDA Proj. # A10-FS-9. Department of Botany, Hebrew University, Jerusalem, Israel.

Baum, B. R. 1978. *The Genus Tamarix*. The Israel Academy of Sciences and Humanities, Jerusalem.

Bensatal, A., and M. R. Ouahrani. 2008. Inhibition of crystallization of calcium oxalate by the extraction of *Tamarix gallica* L. Urological Research 36:283–287.

Davis, L. 1899. *Ornamental Shrubs for Garden, Lawn and Park Planting*. G. Putnam's Sons, New York:.

Dirr, M. 1990. *Manual of Woody Landscape Plants*. 4th ed. Stipes Publishing Company. Champaign, IL.

Gaskin, J. F., and B. A. Schaal, 2002. Hybrid *Tamarix* widespread in U.S. invasion and undetected in native Asian range. Proceedings of the National Academy of Sciences 99 (17):11256–11259.

Grami, B. 1998. Gaz of Khunsar: The manna of Persia. Economic Botany 52(2): 183–191.

Joor, J. F. 1877. The tamarisk naturalized. Bulletin of the Torrey Botanical Club 6(32):166.

Horton, J. S. 1977. The development and perpetuation of the permanent Tamarisk type in the phreatophyte zone of the Southwest. Presented at the Symposium of the Importance, Preservation, and Management of the Riparian Habitat, 7/9/77, Tucson, AZ.

Measures, J. A. 1933. Habit and Bloom of the Tamarisk. Flower Grower 19:331.

Orabi, M. A. A., S. Taniguchi, M. Yoshimura, T. Yoshida, K. Kishino, H. Sakagami, T. Hatano. 2010. Hydrolyzable tannins of tamaricaceous plants. III.(1)Hellinoyl- and macrocyclic-type ellagitannins from *Tamarix nilotica*. Journal of Natural Products 73:870–879.

Rustomjee, N. K., and N. K. Nanabhai. 1903. *Materia Medica of India and Their Therapeutics*. Caxton Works, Bombay.

Sehrawat, A., and S. Sultana. 2006. *Tamarix* gallica ameliorates thioacetamide-induced hepatic oxidative stress and hyperproliferative response in Wistar rats. Journal of Enzyme Inhibition and Medicinal Chemistry 21:215–223.

Sher, A. A., D. L. Marshall, and J. Taylor. 2002. Spatial partitioning within southwestern floodplains: Patterns of establishment of native *Populus* and *Salix* in the presence of invasive, non-native *Tamarix*. Ecological Applications 12:760–772.

Swingle, W. T. 1924. The athel, *Tamarix aphylla* (also called *T. articulata*), a promising windbreak for the Hot Irrigated Valleys of the Southwest, Indio, CA. Mimeograph. Walter Tennyson Swingle Collection, University of Miami, FL.

Taylor, G., and G. Cooper. 2000. *Gardens of Obsession*. Seven Dials, Cassell & Company, London.

Tellman, B. 2002. Human introduction of exotic species. Pages 25–46 *in* B. Tellman, editor. *Invasive Exotic Species in the Sonoran Region*. University of Arizona Press and Arizona-Sonora Desert Museum, Tucson.

Thornber, J. J. 1916. Tamarisks for southwestern planting. *Timely Hints for Farmers*. University of Arizona Agricultural Experiment Station, Tucson.

UNEP, Parker, C. 4/28/2003. *Tamarix: A natural resource*. New York: United Nations Environmental Programme (1997).

Warren, D. K. and R. M. Turner. 1975. Saltcedar (*Tamarix chinesis*) seed production, seedling establishment, and response to inundation. Journal of the Arizona Academy of Sciences 10:135–144.

Young, J. A., C. D. Clements, and D. Harmon. 2004. Germination of seeds of *Tamarix ramosissima*. Rangeland Ecology and Management 57:475–481.

PART IV

Management

20

Tamarisk Management

LESSONS AND TECHNIQUES

Cameron H. Douglass, Scott J. Nissen, and Charles R. Hart

We have been managing tamarisk for almost as long as it has been naturalized in the southwestern United States. Historical reports show that as early as the 1930s—when many were still actively planting tamarisk for its purported benefits—water engineers and farmers began to remove tamarisk trees when they were found to constrict flows in irrigation ditches (Taylor 1930). While our current management options for tamarisk are more sophisticated than those of 80 years ago, most are no more than variations on the techniques and strategies that have been tried over the years. Our advantage today, as managers of tamarisk, is the experience we have collectively gained, the many lessons that have been learned from the long history of tamarisk management.

History of Tamarisk Control

The first widespread introductions of tamarisk began in the 1800s (see chapter 16, this volume) though a few authors say that introduction occurred as far back as the mid-seventeenth century by Spanish settlers and travelers (Hefley 1937; Tellman 2002a). By the 1830s, many prominent US nurseries frequently sold several *Tamarix* spp., including *T. pentandra, T. gallica, T. chinensis,* and others (Tellman 2002a). In 1870, the US Department of Agriculture released *T. pentandra* for sale to the public, following earlier plantings at the National Arboretum in Washington, DC (Robinson 1965). Into the 1930s, tamarisk was still considered by many to have beneficial purposes, particularly large-scale plantings to enhance **333** erosion control, and the sedimentation projects that were considered necessary at the time to support burgeoning agricultural operations in the region (Chew 2009). For example, after observing that the presence of dense tamarisk in the floodplain of Lake McMillan had prevented silting of the reservoir by slowing inlet flow, Dr. T. Taylor of the University of Texas, Austin, suggested that tamarisk should be used similarly at other reservoirs (Taylor 1930). At the same time, however, tamarisk was being labeled a problematic plant (Taylor 1930).

The oldest cases of ornamental plantings escaping cultivation were actually much earlier in 1880 and 1897, in Utah and Texas, respectively (Tellman 2002a; Shafroth et al. 2005). Not until several decades later, though, were naturalized populations widespread, such as dense infestations along the beaches and banks of the Pecos River in Texas that were cited by Texas Board of Water engineers as having reduced river flows as much as fivefold (Taylor 1930). Farmers northwest of Barstow, Texas, began to abandon densely infested irrigation canals, saying that it was cheaper to construct an entirely new canal than attempt to clean the infested ones (Taylor 1930). Dr. Taylor expressed the conundrum posed by tamarisk in the 1930s: "In thirty years the tamarisk in the Pecos Valley has spread from one lone tree near Roswell to a rather heavy growth along the river from Roswell to Barstow. The tamarisk can be a blessing to the reservoir, but it can be a troublesome factor to the canal" (Taylor 1930).

In addition to responding to public pressure to provide water for expanding agriculture in the Southwest, New Deal work projects during the Great Depression brought a renewed focus on water issues throughout the Southwest. Dam and reservoir construction that had begun at the turn of the twentieth century progressed unabated, altering natural flow regimes throughout regional river systems (Stromberg and Chew 2002; Chew 2009). New, artificial flow regimes in southwestern waterways contributed to an overall lack of management of, and disinterest in, rapidly expanding tamarisk infestations (Dudley et al. 2000; Stromberg and Chew 2002; Tellman 2002b). For example, along a 170-mile stretch of the Brazos River in north-central Texas, an analysis of historical aerial photos shows that tamarisk acreage increased 52% between 1940 and 1969 (Busby and Schuster 1973). Up until the 1960s, tamarisk was still being planted in the region (West and Nabhan 2002).

Concerted efforts to control tamarisk appear to have begun after 1942, when federal irrigation engineers in the Pecos River Joint Investigation first formally accused the species, correctly or not, of being a disproportionate consumer of water, and therefore a target for management (Robinson 1952; see also chapter 16, this volume). By 1948, the US Bureau of Reclamation (BOR) staff were treating hundreds of acres of tamarisk in the delta of the McMillan Reservoir using 2,4-D herbicide, at a cost of $4/acre (Subcommittee 1951–1970). About the same time, in a project sponsored by the US Department of Commerce and the Defense Plant Corporation, US Geological Survey (USGS) staff attempted to control tamarisk along a tributary of the Gila River in Arizona using bulldozers and military-issued flamethrowers (Robinson 1952). Unfortunately, within 12 months the treated trees regrew to be six to eight feet tall (Robinson 1952). Tamarisk management within the Gila River and Salt River drainages of Arizona gathered steam in 1951 following construction of the Gila and Salt River Floodway by the US Army Corps of Engineers. In the spring and early summer of that year 40 acres of tamarisk were bulldozed and the resulting debris burned, and then in September 1951, the regrowth was used for the first large-scale trial comparing the effectiveness of 10 herbicide treatments (Subcommittee 1951–1970). Foliar applications of 2,4-D and 2,4,5-T at two concentrations were tested, as were cut-stump and basal-bark applications of 2,4-D in diesel fuel at three concentrations.

Despite these early control projects, tamarisk has continued to spread, and the scale and scope of programs to control tamarisk have grown rapidly as well. There are now improved chemical options for killing tamarisk trees, and a biological control agent (tamarisk leaf beetles, *Diorhabda* spp.; see chapter 22, this volume) has been successfully used to control tamarisk populations. This chapter presents an overview of current, commonly used management strategies, and also presents several examples that illustrate how the described methods have been successfully implemented. While we detail a variety of removal methods, we know that none of these alone will successfully or sustainably control tamarisk. As with other invasive plant species, control of tamarisk will only be effective over the long term if individual, site-specific management strategies are integrated into a comprehensive plan (Shafroth et al. 2008; see also chapter 23, this volume). Toward this end, we also aim to demonstrate that sustainable tamarisk management can address the preservation of ecosystem functioning and biotic integrity, and also ensure that the economic value and productivity of land is maintained and even enhanced.

Management Options

FLOODING

One of the primary reasons tamarisk has become so abundant throughout the southwestern United States is the alteration of flooding patterns and reduced peak flows in natural waterways (Stromberg et al. 2007; see also chapter 7, this volume). It has been shown that restoration of natural flooding regime can suppress tamarisk by promoting native species' growth (Sher et al. 2002; see also chapter 21, this volume). But many sites where tamarisk is problematic are no longer hydrologically connected to a flooding source that would allow this treatment option to be feasible. That said, inundation of mature tamarisk trees seems only to kill them reliably with at least two to three years of flooding (Wiedemann and Cross 1979). On the other hand, flooding of infested sites during seed germination, or during the first year of seedling growth, has been found to cause consistently high (> 90%) tamarisk mortality (Smith and Kadlec 1983, Gladwin and Roelle 1998, Sprenger et al. 2001).

PRESCRIBED FIRE

A small but growing literature is emerging regarding the effects of fire on tamarisk (see chapter 14, this volume). With regard to fire as a management tool, it appears that prescribed burns alone do not effectively control tamarisk, as it is well adapted to all but the hottest and longest fires (Busch 1995; Dudley et al. 2000; Racher and Britton 2003; Racher 2009). Following fires tamarisk plants will resprout vigorously from unaffected roots, and can regrow up to several meters in a year (Brock 1994). There are examples of successful prescribed fires for the control of tamarisk, but only in combination with herbicide treatments. For example, Harms and Hiebert (2006) documented a 95% reduction in tamarisk foliar cover at about 30 sites in southwestern states where prescribed burns were followed by herbicide application. Racher (2009) found that although burning mature stands does reduce canopy cover substantially, mortality is very inconsistent and generally low, especially for sites with a history of burns. It has been proposed that prescribed burns are probably best used for biomass or debris removal rather than as a primary control for tamarisk (Racher and Britton 2003).

BIOLOGICAL CONTROL

Biological control in the form of defoliation by *Diorhabda* spp. leaf beetles is the newest tool for tamarisk management and will likely be increasingly significant in years to come (see chapter 22, this volume). Here, we discuss observations that are particularly relevant to the integration of biological control with other management options. It takes several years of defoliation by *Diorhabda* species to kill mature tamarisk trees, and some estimates suggest that it would take at least a decade to ensure 75% to 80% mortality in a given stand (Dudley et al. 2000). Damage to tamarisk trees by leaf beetle defoliation tends to be fatal mostly on younger trees and seedlings that do not have large enough root systems to recover from the repeated stress of herbivory (Dudley et al. 2000).

While this rate of impact is slower compared to chemical or mechanical control, it also presents an opportunity to combine biological control with other methods. For example, Brooks et al. (2008) found that tamarisk trees that had been first weakened physiologically by repeated *Diorhabda* beetle defoliation were more likely to be killed by prescribed burns than trees that experienced no herbivory. Similarly, we have observed that establishment of *Diorhabda carinulata* occurs more rapidly on trees regrowing following mechanical removal of aboveground biomass, and that the effects of herbivory are more immediate. Presumably, the overall physiological effect of beetle herbivory on these trees—which are already stressed from biomass removal—would also be more severe. This suggests that the integration of aboveground biomass removal and biological-control releases timed to curtail regrowth could be promising for long-term tamarisk management.

MECHANICAL TREATMENTS

Many mechanical strategies to reduce or remove aboveground tamarisk biomass have been tried over the years, including bulldozing, shredding/mulching (brush mowing), chaining, disking, grubbing, knifing, roller chopping, and root plowing (Subcommittee 1951–1970; Brock 1994; Smith et al. 2002). However, most mechanical strategies that only remove the aboveground stems of mature trees do not kill the plants, as resprouting can be vigorous. There is some evidence that young (first-year) tamarisk seedlings are particularly sensitive to mechanical removal, even disking at a shallow depth of five inches (Smith et al. 2002). For the same reasons that trees respond positively to fire, mature tamarisk plants tend to regrow vigorously after any aboveground biomass removal, unless there is also damage or extraction of the root crown and lateral root system (McDaniel and Taylor 2003b).

Killing tamarisk trees using only mechanical approaches requires the physical removal of trunks, plant crowns, and lateral roots (Hart 2009). Track hoes or excavators equipped with specialized grubbing attachments are effective at removing most above- and belowground biomass, but this must be followed by the raking and sifting of the soil to remove lateral roots. The latter can be done with root rakes mounted on bulldozers (see chapter 21, this volume). Other equipment used to clear large areas of aboveground tamarisk biomass are site-preparation tractors or skid steers with forestry mulching attachments mounted on the front (see figure 20.1; Nissen et al. 2010). Smaller versions of these mulching attachments can also be mounted onto the swing arms of excavators, allowing for more selective

FIGURE 20.1 An example of a typical site preparation tractor with a forestry mulching head (hydro axe) that can be used to shred the aboveground portion of tamarisk trees.

Source: Photo courtesy of Cameron Douglass.

removal of tamarisk trees, which is useful if native woody species are growing among the tamarisk, or vice versa. Complete removal of root crowns and lateral roots can cause fairly severe soil disturbances that may require mediation before revegetation (Scifres 1980; Sher et al. 2010).

Whole plant removal (e.g., by using an excavator with a thumbed bucket; see figure 20.2) can successfully kill tamarisk if conditions allow for removal of the actual root mass (Nissen et al. 2010). If excavation is planned for winter months, it is important to ensure that the soil is not frozen, as this can prevent complete removal of the root crown and enable resprouting (Scifres 1980). With any mechanical removal, it is particularly important to plan for follow-up monitoring and any necessary re-treatments, since the success of many long-term tamarisk management programs relying primarily on mechanical treatments is incumbent on annually re-treating sites (Brock 1994).

Tamarisk regeneration following removal of top growth alone follows a certain pattern: (sometimes dramatically) increased stem densities; shorter, more erect stems with smaller diameters; and a full, bushy canopy (see figure 20.3). Tamarisk regrowing after aboveground mechanical treatments has often increased vigor, with plants producing more stems per unit area than is typical of undisturbed plants. The shorter-stature tamarisk regrowth allows for more efficient application

FIGURE 20.2 A track hoe equipped with a thumbed bucket can be used to excavate entire tamarisk trees. Note the root crown attached to the tree on the ground; it is essential to remove this whole structure in order to minimize the chance of resprouting.

Source: Photo courtesy of Cameron Douglass.

FIGURE 20.3 This image shows a site one year after two types of mechanical tamarisk were completed. To the left of the dotted line the aboveground portion of trees was shredded using a site preparation tractor equipped with a Hydro Ax™. To the right of the same line whole trees were excavated using a track hoe and then shredded on site. In the background mature, untreated trees can be seen.

Source: Photo courtesy of Cameron Douglass.

of herbicides to individual plants, with reduced risk to other nearby vegetation. Alternatively, prescribed burning can be more effectively used as a follow-up treatment at sites where mature tamarisk is first cleared mechanically and the trees windrowed (Taylor and McDaniel 1998; Racher 2009).

CHEMICAL TREATMENTS

Over the past half century many herbicides have been used in attempts to control tamarisk, though chemical control was not reliably effective until the introduction of imazapyr in the mid-1980s. These included various formulations of 2,4-D; 2,4,5-T (silvex); picloram; dicamba; triclopyr; glyphosate; and finally, imazapyr (Scifres 1980; Brock 1994). The only products that are environmentally compatible with application in riparian environments that tamarisk frequently infests are 2,4-D, glyphosate, triclopyr, and imazapyr. Until the introduction of triclopyr and imazapyr, the most widely used product was 2,4-D, which did not consistently control tamarisk trees (Kerpez 1987; Brock 1994). Glyphosate is still commonly used but is not very effective when used alone (Fick and Geyer 2010). For example, (Duncan 2010b) found that foliar applications of glyphosate resulted in only 32% mortality, whereas combinations of glyphosate and imazapyr resulted in 99% mortality. Mixtures of glyphosate and imazapyr are very common, and reduce the required amount of the more expensive and environmentally persistent imazapyr (Carpenter 1998; Duncan and McDaniel 1998).

There are various methods of herbicide application to tamarisk, but generally the methods are targeted at foliage ("foliar" treatments); the bark of an intact tree ("basal" bark refers to applications that target the bottom 18–24 inches of a tree's bark); and a freshly cut stump surface. The part of the tree targeted by a given herbicide is important to consider because this determines what time of year treatments can be made (e.g., foliar treatments can only be done when trees have leaves), and can influence the difficulty and cost of the treatments. Generally, herbicide applications that target the bark or cut surfaces are more expensive and take longer than foliar treatments because they require the applicator to directly apply the product to individual trees. However, these types of applications greatly reduce negative off-target impacts to desirable vegetation. All herbicides currently used on tamarisk are systemic, which means that once applied to the foliage, a stump, or after penetrating the tree's bark, the chemicals move (translocate) within the plants' vascular system to the roots.

Imazapyr was registered by the Environmental Protection Agency (EPA) for noncrop use in July 1984 and first sold shortly thereafter (EPA 2006). This herbicide belongs to a chemical family called imidazolinones that kill plants by inhibiting the synthesis of an enzyme critical to the production of branched-chain amino acids, and ultimately, proteins (Shaner and O'Connor 1991). Imazapyr has relatively long-lived residual soil activity (25–142 days), and once in soil it is broken down mostly by microbial metabolism (Senseman 2007). This means that it can persist in the soil at levels toxic to plants for several months to a year after it is applied, and consequently inhibits establishment of other plants, both unwanted and desirable (Sher et al. 2010). The duration of effective weed control from imazapyr soil residues depends on factors that affect soil microbial activity (e.g., soil temperatures and moisture), as well as soil organic matter content and pH levels (Vizantinopoulos and Lolos 1994; Nissen et al. 2010). On the other hand, imazapyr is very water soluble, and will degrade quickly (2–3 days) in water when exposed to sunlight (Mallipudi et al. 1991). Most products containing imazapyr are labeled primarily for use as foliar applications (e.g., Habitat®, Arsenal®, Arsenal® Powerline™); exceptions include Chopper® Generation II, which is formulated for dormant season basal bark applications.[1]

Triclopyr, introduced early in the 1970s, is a synthetic auxin herbicide that mimics the physiological effects of a natural hormone ubiquitous in plants (Howard et al. 1983). At naturally occurring concentrations these hormones are essential to proper plant functioning, but at high concentrations the compounds disrupt plant growth and cause death in susceptible broadleaved plants (Nissen et al. 2010). Commercial herbicides that contain triclopyr generally have relatively limited soil residual activity (10–46 days), and like imazapyr are rapidly degraded (less than 48 hours) when in water and exposed to sunlight (Woodburn et al. 1993; Senseman 2007). Triclopyr can be formulated for basal bark and cut-stump

---

[1] Any reference to product trade names is made with the understanding that no discrimination is intended and that the authors, their employers, or the publishers of this book, endorse no products. Always consult product labels for the most current information. Any person using a pesticide is responsible for reading, understanding, and following the product label.

applications (e.g. Remedy Ultra™, Garlon® 4 Ultra™, Pathfinder II™) or can be formulated for cut surface applications (Garlon® 3A).

Imazapyr is the most consistently effective choice for application to tamarisk by aircraft in aerial applications (see figure 20.4; Duncan and McDaniel 1998; Nissen et al. 2010). Both fixed-wing aircraft and helicopters are used to aerially treat tamarisk, though fixed-wing planes are only advantageous for treating large, monotypic stands (McDaniel and Taylor 2003a). Helicopters are better for aerial applications because of their slower air speeds, closeness to the ground, and ability to use higher application volumes (Hart et al. 2005). Furthermore, their maneuverability allows for treatment of variably shaped and sized tree stands, and avoidance of desirable natives, notably *Populus* and other riparian tree species (McDaniel and Taylor 2003a; Nissen et al. 2010).

Ground-based, individual plant herbicide applications can take many forms, but are either foliar and applied during the growing season or targeted to the bark or cut surfaces during the winter when trees are dormant (Hart 2009). Individual plant treatments with imazapyr can be very effective (more than 95% control), and such applications will reduce the herbicide amount needed to treat a site and minimize the potential for overspray into desirable vegetation (Duncan and McDaniel 1998; Nissen et al. 2010). Application timing is important for all foliar tamarisk treatments, which are in August and September, before plants begin senescing and shunting foliar resources to their root systems (Duncan and McDaniel 1998; McDaniel and Taylor 2003a).

Low-volume basal bark or cut-surface applications using higher concentrations of either herbicide can also be effective at controlling tamarisk, and such applications can be carried out year round (Howard et al. 1983, Parker and

FIGURE 20.4 A helicopter being used to aerially apply herbicides to a remote stand of tamarisk.
Source: Photo courtesy of Charles Hart.

Williamson 2000). This application is most effective on smaller trees, partly because young trees have smoother, thinner bark that is more easily penetrated by herbicides (Parker and Williamson 2000). Applicators can use backpack sprayers, all terrain vehicles (ATVs), or horse-mounted sprayers. The added value of following prescribed burns of mature tamarisk trees with individual plant chemical control has been demonstrated in a few instances (Fox 2001; Racher and Britton 2003). Alternatively, McDaniel and Taylor (2003a) found more than 93% long-term control with an aerial application of imazapyr, followed three years later by a summer burn.

It is important when selecting any herbicidal option for controlling tamarisk to consider habitat in which applications will be taking place, and specifically how close to any perennial stream the application will occur (Carpenter 1998). Also, while it is often desirable to remove dead materials after trees have been killed using chemical controls, it is critical that any follow up treatments be delayed for two years (Duncan and McDaniel 1998, Hart 2009). This period allows for complete translocation of a systemic herbicide throughout the entire root system of a tamarisk tree and will ensure that the entire plant is killed. Dead or dying trees that are disturbed before herbicides are completely translocated will often resprout from surviving root fragments.

Finally, it is important to understand the trade-offs when using herbicides to control tamarisk. Imazapyr, glyphosate, and other nonselective chemistries will kill nearly any vegetation they contact, the exception being some weed species such as *Bassia scoparia* (kochia) that have developed resistance to these herbicides (Shafroth et al. see chapter 23). Kochia is a widespread exotic weed in crop fields where herbicides with the same mode of action as imazapyr are frequently used. The selection pressure applied in these fields led to the development of resistance by kochia, a plant species which has since spread widely into natural areas (Primiani et al. 1990). Also, its persistence in the soil can pose a challenge for revegetation (Sher et al. 2010). Herbicides such as triclopyr and other synthetic auxinic chemicals—which only kill broad-leaved plants—will be more selective and only affect sensitive plants (Howard et al. 1983). However, we should be concerned that shifts in functional groups of native plant communities have been documented in rangelands due to the repeated use of selective herbicides (Pearson and Ortega 2009, Ortega and Pearson 2011).

Case Studies

PECOS RIVER ECOSYSTEM PROJECT, TEXAS: A STORY ABOUT WATER

The Pecos River Ecosystem Project was proposed by the Red Bluff Water and Power Control District in 1997 to reduce tamarisk populations along the Pecos River (Hart et al. 2005; Hatler and Hart 2009; McDonald 2010). The objectives of the project were to increase the efficiency of water delivery from the river to irrigation districts in the Red Bluff District and to improve the quality of the water by decreasing salinity. In the planning meetings, several major concerns emerged: that the treatment method selected should provide a high rate of tamarisk mortality while minimizing detrimental effects on existing native vegetation, that it should be as economical as possible, and that soil loss from stream banks should be minimized. Revegetation of riverbanks with native plants to complete the ecosystem restoration was a major concern.

The project was set up in two phases: (1) treatment of tamarisk, and (2) debris removal and maintenance. Before treatment could begin, a section 24(C) special-use label was requested for Arsenal™ herbicide use on tamarisk in rangeland and aquatic areas in Texas, and in 1999 was approved by the Texas Department of Agriculture. Without the 24(C) label, Arsenal™ could only be used on noncrop and right-of-way applications, so applications along the Pecos River were deemed "off label." Since then, imazapyr has received a federal rangeland and aquatic use designation, and the 24(C) is no longer necessary. Another daunting task was to obtain permission from private landowners along the river. A "spray easement" or landowner contract (a legal document signed by both parties and filed in the appropriate county offices) was developed between the Project and private landowners to grant the Project access onto the property for tamarisk treatment and follow-up management over a 10-year period. More than 800 easements were signed by private landowners, with a rejection rate of less than 1%. Bids were solicited from aerial applicators in late summer 1999, and the project was awarded to North Star Helicopters. With funding, landowner permissions, and the applicator contract in hand by August 1999, treatments began the following month.

Applications were made with a helicopter outfitted with a sectioned boom and special nozzles that allowed for large droplets and high total spray volume. The combined effect of the sectioned boom (which meant that certain portions could be turned off if desired) and modified nozzles meant that herbicide overspray on nontarget vegetation was minimized. From 1999 through 2005, 289 river miles (13,497 acres) along the Pecos River and various tributaries and springs within the basin were treated, and about $2.7 million in local, state, and federal funds was spent. The average mortality of tamarisk after aerial applications was estimated at 80% to 90% in ground surveys. Prescribed burning and the establishment of *Diorhabda* beetle populations for maintenance of still-living trees are currently following up the chemical control.

As part of the Project, a research site was established in 2001 along the river to document and characterize water salvage as a result of tamarisk management. The study monitored a series of shallow wells equipped with highly sensitive pressure transducer water level loggers to detect diurnal groundwater fluctuations, and to estimate evapotranspiration and water salvage from tamarisk stands in treated and adjacent untreated areas. Hays (2003) estimated significant potential for water salvage as a result of tamarisk control. Subsequent analysis by Hatler and Hart (2009) found that water salvage from this site initially peaked at 82% two years after treatment then steadily declined to a low of 7% five years after treatment, following some tamarisk regrowth. Imperative to understanding these results is that (1) the study defined water salvage as "water available for other ecological functions" and should not be interpreted as additional stream flow; (2) much of the water salvage would presumably be used over time by desirable native replacement vegetation; (3) during the study period, both drought and flood conditions influenced results; and (4) vegetation replacing tamarisk during the study period was mostly shallow-rooted grasses and forbs that have comparatively fewer direct effects on groundwater fluctuations. Such results would not be expected in systems where phreatophytes again dominated riverside vegetation after tamarisk control (see chapter 4, this volume). Furthermore, the study concluded that the salvage would be short-lived if a strategy for vegetation maintenance was not implemented.

In 2007, to verify water salvage along the Pecos River, Texas, McDonald (2010) designed a study to characterize surface and groundwater interactions following tamarisk removal and to "locate" previously estimated water salvage. Plot-scale evaluations did not detect any increase in base flow of the river nor any increased groundwater recharge directly attributable to tamarisk control. However, a cursory analysis of base flow at a larger scale did indicate significant increases several years into the project. McDonald concluded, however, that additional studies were needed to positively link these increases with the tamarisk management program. These studies have been unable to verify that tamarisk removal increases stream flow or groundwater recharge or "salvage" water for use by other desirable native plants. Other researchers have reached similar conclusions (chapters 5 and 6, this volume). Projects that remove saltcedar and Russian olive with the intention of reducing evapotranspiration (ET) and increasing flow in streams have produced mixed results, with most studies failing to demonstrate significant long-term changes (Archer et al. 2011).

Overall, it is clear from the many attempts at quantifying water salvage from tamarisk removal that changes to stream flow or groundwater recharge are very site and drainage specific (Shafroth et al. 2005) and are linked to variable weather, tamarisk removal effectiveness and the composition of replacement vegetation (Hatler and Hart 2009).

DINOSAUR NATIONAL MONUMENT: THE POWER OF VOLUNTEERISM

Not every project attempting large-scale tamarisk management revolves around water salvage or water delivery. Sometimes tamarisk removal is undertaken to improve wildlife habit, increase biodiversity, or enhance recreational value. The US national parks and monuments systems were established to preserve and protect wild and scenic areas so that they could be enjoyed by every generation. Dinosaur National Monument, in northwestern Colorado and northeastern Utah, is a wild and scenic area particularly prized by white water rafters and boaters. The site encompasses the confluence of the Yampa and Green Rivers and provides recreational opportunities to an estimated 14,000 river rafters each year (Utah PCD n.d.).

The riparian corridors along the Yampa and Green Rivers were once dominated by tamarisk to the point that trees interfered with the recreational experience of visitors by overrunning landing sites and camping areas; however, there was very limited funding for management. Land managers for the monument discovered a secret weapon: public volunteers. For the past 15 years, Dinosaur National Monument has had a very active weed management program with massive inputs of volunteer assistance. Almost 5,500 volunteers have contributed over 20,000 hours assisting with weed management projects (Utah, PCD n.d.). This program includes many of the common control strategies that have been discussed, but there have been no aerial herbicide spraying operations or large-scale mechanical control in the Monument. Tamarisk removal has targeted important river campsites and beaches using cut stump treatments of imazapyr, as well as a unique individual plant removal strategy of winching large tamarisk stumps out of the ground using only hand tools. The success of the Weed Warriors program in Dinosaur is just one example of the contribution volunteers can make to tamarisk management and invasive plant management in general (Duncan 2011). By engaging with the public, the park and the volunteers both benefit. The Monument has been able to improve the aesthetics of the river and enhance the recreational experience for visitors far beyond what could have been achieved with the limited funding provided to the park for invasive plant management, and volunteers become advocates for invasive species management.

CENTRAL UTAH PUBLIC LANDS: INTERAGENCY COOPERATION

One of the most difficult problems in large-scale tamarisk removal is coordination between all the entities—private, state and federal—responsible for the land where the targeted infestation occurs. Landowners or state and federal agencies do not always have compatible land management objectives, making planning and implementing invasive species management difficult. One of the best examples of cooperation between public government agencies is the 22-member consortium of city, county, state, and federal agencies that collaborated on the removal of tamarisk from 13,850 acres of US Forest Service (USFS) and Bureau of Land Management (BLM) property in Buckhorn Wash in central Utah. Buckhorn Wash is infamous as the hideout of Butch Cassidy and the Wild Bunch, but it is also home to spectacular scenery and Native American petroglyphs (rock art). Many of the canyons in central Utah were heavily infested with tamarisk, and Manti-La Sal National Forest (USFS) administrators were concerned about the impact that tamarisk was having on natural and recreational resources. The USFS partnered with the Order of the Arrow, the leadership corps of the Boy Scouts of America (BSA), to implement a large-scale tamarisk control program (Duncan 2010a).

In 2007 and 2008, BLM and Utah State University employees mapped tamarisk infestations in the three canyons. Five hundred sixty volunteers joined forces on this project and treated 46 linear miles of tamarisk—with an average of 97.5% control one year after treatment. BSA members were responsible for the removal of smaller diameter branches and tamarisk stems while USFS and BLM sawyers used chainsaws to remove the largest stems, and licensed herbicide applicators then treated trees with a modified cut stump method. The combined efforts of the BSA and other local, state and federal land management agencies controlled over 13,000 acres of tamarisk and improved the riparian environment of the Buckhorn Wash using 17,793 volunteer service hours at a value of nearly $350,000.

Costs, Impacts, and Trade-Offs

While there are many factors that influence the selection of tamarisk control tools—including characteristics of the tamarisk infestation, site constraints, understory vegetation, and project goals—cost is usually the most important determinant of which strategy will be feasible (Shafroth et al. 2008). Management of any invasive species is inherently costly, but management of woody invaders such as tamarisk is even more costly because it requires two equally important phases: control (killing trees) and biomass management (removing and disposing or reusing branches, trunks and stumps). Biomass management options can include burning, raking into piles, chipping to create mulch, conversion into wood pellets for stoves, and even use as a fuel for downdraft gasification (Sher et al. 2010; Nielsen et al. 2011). Regardless of the strategy, biomass management can be more costly than the control phase itself (Taylor and McDaniel 2004; Coalition 2008). Although occasionally it may not be possible or desirable to remove standing dead trees, in most cases such management will promote desirable replacement vegetation (Sher et al. 2010).

There are two scales of tamarisk management projects: a larger, watershed or drainage-wide scale; and a more localized, site-specific approach. Many factors determine the scale of a management effort, but cost will likely be the most important, and will particularly influence the choice of management tools. For example, large projects usually use lower cost/area, highly efficient management tools such as biocontrol or aerial herbicide applications. There are more options for tamarisk removal available to smaller projects, such as removing a patch of trees from an urban riverside park. At these higher value sites, more selective, intensive, and expensive strategies can be used to completely remove trees and actively reestablish desirable native plant communities. Such methods can include targeted removal by excavators or cut stump methods that reduce exposure of desirable vegetation to herbicide.

Furthermore, available funds will often define the area that can be treated and how management will be carried out. There are implicit trade-offs between the financial costs of a management project, its ecological impacts or benefits, and the control tools that are chosen. The goal of tamarisk management is frequently to restore some degree of pre-invasion functioning to a site, but the impacts resulting from removal or control may delay or prevent benefits from ecosystem restoration. For example, while aerial applications of an herbicide such as imazapyr are very cost effective, off-target soil residues mean that relatively expensive active revegetation efforts are likely necessary. Conversely, targeted individual tree removal and control strategies—which are generally much more costly in the short term—may enable the preservation of existing understory plant communities and ecosystem functioning, and mean that intensive restoration activities will not be needed.

There have been many reports summarizing the relative costs of many of the management options presented here (e.g. Coalition 2008; Barz et al. 2009), but expenses such as fuel and herbicides will vary year to year, almost always increasing. Herbicidal control of tamarisk can no longer be accomplished for $4/acre as it was in 1948, but the cost for brand name formulations of imazapyr has fallen considerably over even the past five years (Barz et al. 2009).

Integrated Tamarisk and Ecosystem Management

It is clear that single-method strategies for managing tamarisk are simply not effective over time. It is vital when designing tamarisk management programs to use the suite of tools that will have the highest chance of both maintaining tamarisk control and conserving the inherent ecological resilience of a treated site (Jorgensen 1996; Pearson and Ortega 2009). Selecting appropriate strategies and implementing them in the proper sequence are keys to long-term success. Tamarisk control can be thought of as the "designed disturbance" component of Sheley et al.'s (1996) successional weed management framework, and in this context tamarisk should be removed using strategies that positively redirect the successional trajectory of a site. The use of complementary or even synergistic control methods can actually accelerate the rate of natural ecosystem recovery (Masters and Nissen 1998). To ensure the long-term success of management projects it is essential that monitoring and maintenance are planned for and carried out—meaning in particular that secondary invaders or surviving tamarisk plants are actively controlled (Shafroth et al. 2008).

For a truly sustainable approach to managing tamarisk, we must recognize that (1) management method has a direct impact on the capacity of sites to recover, and (2) that controlling tamarisk is only a small part of the comprehensive program needed to attempt management of an ecosystem (see chapter 23, this volume). First, management of invasive species begins when a land manager decides that a certain species has a disproportionately negative impact on a site. The techniques that are chosen to remove the targeted species, and the order in which they are implemented, directly affect ecosystem recovery, both from the management action itself, and the sudden absence of the dominant plant species (Taylor and McDaniel 2004; Vincent et al. 2009). Strategies that remove a dominant invasive plant such as tamarisk may indirectly facilitate a secondary invasion of an equally noxious species (Pearson and Ortega 2009), or may cause other negative outcomes such as severe erosion during flooding (Vincent et al. 2009).

Second, invasive species removal will not automatically result in a positive outcome if other underlying problems with ecosystem processes (e.g., hydrology, nutrient cycling or excessive biotic impacts use such as grazing) are not also addressed. The success of tamarisk removal and site restoration has been closely tied to larger scale changes in natural resources (particularly hydrology) that enhance the capacity for sites to self-repair, such as the removal of grazing pressure or the restoration of natural hydrologic flows and flooding patterns (Stromberg and Chew 2002; Taylor and McDaniel 2004). For example, an important part of the successful management of tamarisk infestations along the Mojave River at Barstow Resource Area in California was the installation of a "riparian management fence" to exclude both grazing and off-road recreational vehicles (Lovich et al. 1994; Chavez 1996). This allowed the understory plant community to regenerate without being trampled and prevented erosion and soil disturbance associated with heavy use. In this context, it is important for managers to consider whether larger-scale modifications in watershed hydrology and other ecosystem processes are possible to ensure that tamarisk removal and site restoration projects will be successful (Stromberg and Chew 2002). Sites that can be restored to natural ecosystem processes, or that have a high revegetation capacity or especially desirable understory plant community, should be prioritized as especially strong candidates for tamarisk management (Taylor and McDaniel 2004; Parker et al. 2005).

Conclusions

Some methods used today to control tamarisk are very similar to those that were first used eighty years ago, such as mechanical extraction. Others are more recent, most notably the availability of selective and less environmentally harmful herbicides. The use of *Diorhabda* beetles as biological control agents for tamarisk, however, is ironically perhaps the most advanced and modern means we have to manage tamarisk. Arguably, what has changed the most over the years we have been tackling tamarisk—and what will hopefully continue to improve—is our understanding of ecosystems and how our management of individual components affects all the others. Similarly, our history of managing tamarisk has taught us many lessons, principally the importance of using integrated suites of tools rather than relying on single methods. We have also learned through experience the value of using community volunteers and interagency collaborations, since invasive species such as tamarisk are an issue that we must deal with collectively.

The past has taught us a lot about how the ways in which we manage a plant such as tamarisk can affect the condition, economic value, and long-term ecological vitality of invaded sites. Recent research in particular allows us to know with a greater degree of certainty how ecosystems and certain organisms, such as secondary invaders, will respond to specific management tools. By developing comprehensive and adaptive management plans, using integrated and intentional suites of management tools, and learning our lessons from past efforts we might be able to begin pushing back the tide of tamarisk that has quietly crept across the southwest for almost two hundred years.

References

Archer, S. R., K. W. Davies, T. E. Fulbright, K. C. McDaniel, B. P. Wilcox, and K. I. Predick. 2011. Brush Management as a Rangeland Conservation Strategy: A Critical Evaluation. Pages 105–170 *in* D. D. Briske, editor. Conservation Benefits of Rangeland Practices: Assessment, Recommendations, and Knowledge Gaps. U.S. Department of Agriculture, Natural Resources Conservation Service, Washington DC.

Barz, D., R. P. Watson, J. F. Kanney, J. D. Roberts, and D. P. Groeneveld. 2009. Cost/benefit considerations for recent saltcedar control, Middle Pecos River, New Mexico. Environmental Management 43:282–298.

Brock, J. H. 1994. *Tamarix* spp. (saltcedar) an invasive exotic woody plant in arid and semiarid habitats of western USA. Pages 27–44 *in* L. C. de Wall, editor. Ecology and Management of Invasive Riverside Plants. J. Wiley & Sons, New York, NY.

Brooks, M., T. R. Dudley, G. Drus, and J. Matchett. 2008. Reducing wildfire risk by integration of prescribed burning and biocontrol of invasive tamarisk (*Tamarix* spp.), p. 40. U.S. Geological Survey, El Portal, CA.

Busby, F. E., and J. L. Schuster. 1973. Woody Phreatophytes along the Brazos River and Selected Tributaries Above Possum Kingdom Lake, 168. Texas Water Development Board, Austin.

Busch, D. E. 1995. Effects of fire on southwestern riparian plant community structure. Southwestern Naturalist 40:259–267.

Carpenter, A. T. 1998. Element Stewardship Abstract: *Tamarix* spp. The Nature Conservancy, Arlington, VA.

Chavez, R. A. 1996. Integrated weed management: concept and practice. Pages 32–36 *in* J. D. DiTommaso and C. E. Bell, editors. *Proceedings of the Saltcedar Management Workshop. June 12, 1996, Rancho Mirage, CA.* University of California Cooperative Extension Service, Holtville, CA.

Chew, M. K. 2009. The monstering of tamarisk: How scientists made a plant into a problem. Journal of the History of Biology 42:231–266.

Coalition, T. 2008. Assessment of Alternative Technologies for Tamarisk Control, Biomass Reduction and Revegetation. Tamarisk Coalition, Grand Jucntion, CO.

Dudley, T. R., C. J. DeLoach, J. E. Lovich, and R. I. Carruthers. 2000. Saltcedar invasion of western riparian areas: Impacts and new prospects for control. Pages 345–381 *in* 65th North American Wildlife and Natural Resources Conference. Wildlife Management Institute, Indiana, IL.

Duncan, C. 2010a. Scouts team-up against tamarisk. Techline Newsletter, Western Range and Wildlands Edition. Weed Management Services, Helena, MT. [Online]. http://www.techlinenews.com/index.php?option=com_content&view=article&id=96:scouts-team-up-against-tamarisk&catid=57:western-rangeland-news&Itemid=32.

Duncan, C. 2011. Weed inventory key to measuring program success at Dinosaur National Monument in Utah. Pages 6–9, Techline Newsletter, Western Range and Wildlands Edition. Weed Management Services, Helena, MT.

Duncan, K. W. 2010b. Inidividual plant treatment of saltcedar. Pages 106–110 *in* Saltcedar and Water Resources in the West Symposium. Texas Agricultural Experiment Station, Texas Cooperative Extension, San Angelo.

Duncan, K. W., and K. C. McDaniel. 1998. Saltcedar (*Tamarix* spp.) management with imazapyr. Weed Technology 12:337–344.

EPA (U.S. Environmental Protection Agency). 2006. Reregistration Eligibility Decision for Imazapyr. Page 108. U.S. Environmental Protection Agency, Washington, DC.

Fick, W. H., and W. A. Geyer. 2010. Cut-stump treatment of saltcedar (Tamarix ramosissima) on the Cimarron National Grasslands. Transactions of the Kansas Academy of Science 111:223–226.

Fox, R. B. 2001. Saltcedar Management Following a Summer Wildfire at Lake Meredith National Recreation Area. Master of Science thesis. Texas Tech University, Lubbock.

Gladwin, D. N., and J. E. Roelle. 1998. Survival of plains cottonwood (*Populus deltoides* subsp. *monilifera*) and saltcedar (*Tamarix ramosissima*) seedlings in response to flooding. Wetlands 18:669–674.

Harms, R. S., and R. D. Hiebert. 2006. Vegetation response following invasive tamarisk (*Tamarix* spp.) removal and implications for riparian removal. Restoration Ecology 14:461–472.

Hart, C. R. 2009. *Saltcedar Biology and Management.* Texas A & M University, College Station.

Hart, C. R., L. D. White, A. McDonald, and Z. Sheng. 2005. Saltcedar control and water salvage on the Pecos River, Texas, 1999–2003. Journal of Environmental Management 75:399–409.

Hatler, W. L., and C. R. Hart. 2009. Water loss and salvage in saltcedar (*Tamarix* spp.) stands on the Pecos River, Texas. Invasive Plant Science and Management 2:309–317.

Hays, K. 2003. *Water Use by Saltcedar (*Tamarix *spp.) and Associated Vegetation on the Canadian, Colorado and Pecos Rivers in Texas.* Texas A & M University, College Station.

Hefley, H. M. 1937. Ecological studies on the Canadian River floodplain in Cleveland County, Oklahoma. Ecological Monographs 7:345–402.

Howard, S. W., A. E. Dirar, J. O. Evans, and F. D. Provenza. 1983. The use of herbicides and/or fire to control saltcedar (*Tamarix*). Pages 65–72 *in* Western Society of Weed Science Annual Meeting. Western Society of Weed Science, Las Vegas, NV.

Jorgensen, M. C. 1996. The use of prescribed fire and mechanical removal as means of control of tamarisk trees. Pages 28–29 *in* J. D. DiTommaso and C. E. Bell, editors. *Proceedings of the Saltcedar Management Workshop. June 12, 1996, Rancho Mirage, CA.* University of California Cooperative Extension Service, Holtville, CA.

Kerpez, T. A. 1987. *Saltcedar Control for Wildlife Habitat Improvement in the Southwestern United States.* U.S. Department of Interior, Fish and Wildlife Service, Washington, DC.

Lovich, J. E., T. B. Egan, and R. C. de Gouvenain. 1994. Tamarisk control on public lands in the desert of southern California: two case studies. Pages 166–177 *in* 46th Annual California Weed Conference. California Weed Science Society, San Jose, CA.

Mallipudi, N. M., S. J. Stout, A. R. daCuntha, and A.-h. Lee. 1991. Photolysis of imazapyr (AC 243997) herbicide in aqueous media. Journal of Agricultural Food Chemistry 39:412–417.

Masters, R. A.,and S. J. Nissen. 1998. Revegetating leafy surge (*Euphorbia esula*)–infested rangeland with native tallgrasses. Weed Technology 12:381–390.

McDaniel, K. C. and J. P. Taylor. 2003a. *Aerial Spraying and Mechanical Saltcedar Control.* Pages 100–105 *in* Saltcedar and Water Resources in the West Symposium. Texas Agricultural Experiment Station and Texas Cooperative Extension, San Angelo, TX.

McDaniel, K. C., and J. P. Taylor. 2003b. Saltcedar recovery after herbicide-burn and mechanical clearing practices. Journal of Range Management 56:439–445.

McDonald, A. K. 2010. *Hydrologic Impacts on Saltcedar Control Along a Regulated Dryland River.* Texas A&M University, College Station.

Nielsen, J. M., T. Walton, J. Diebold, and R. Walt. 2011. Converting riparian rstoration waste to energy: Testing tamarisk (*Tamarix* spp.) woody biomass as fuel for donwdraft gasification. 2011 Tamarisk Research Conference. Tamarisk Coalition, Tucson, AZ.

Nissen, S. J., A. A. Sher, and A. Norton. 2010. *Tamarisk: Best Management Practices in Colorado Watersheds.* Colorado State University, Fort Collins.

Ortega, Y. K., and D. E. Pearson. 2011. Long-term effects of weed control with picloram along a gradient of spotted knapweed invasion. Rangeland Ecology and Management 64:67–77.

Parker, D., M. Renz, A. Fletcher, F. Miller, and J. Gosz. 2005. *Strategy for Long-term Management of Exotic Trees in Riparian Areas for New Mexico's Five River Systems, 2005–2014.* USDA Forest Service, Southwestern Region, Albuquerque, NM.

Parker, D., and M. Williamson. 2000. *Low-Impact, Selective Herbicide Application for Control of Exotic Trees in Riparian Areas: Saltcedar, Russian-Olive and Siberian Elm.* USDA Forest Service, Southwestern Region, Albuquerque, NM.

Pearson, D. E., and Y. K. Ortega. 2009. Managing Invasive Plants in Natural Areas: Moving Beyond Weed Control. Pages 1–21 *in* R. V. Kingely, editor. *Weeds: Management, Economic Impacts and Biology.* Nova Science Publishers, Hauppage, NY.

Primiani, M. M., J. C. Cotterman, and L. L. Saari. 1990. Resistance of kochia (*Kochia scoparia*) to sulfonylurea and imidazolinone herbicides. Weed Technology 4:169–172.

Racher, B. 2009. Saltcedar: Is burning an option? Fire Science Brief 50:1–6. [Online]. http://www.firescience.gov/projects/briefs/00-2-29_FSBrief50.pdf.

Racher, B. J., and C. M. Britton. 2003. Fire in saltcedar ecosystems. Pages 94–99 *in* Saltcedar and Water Resources in the West Symposium. Texas Agricultural Experiment Station and Texas Cooperative Extension, San Angelo.

Robinson, T. W. 1965. *Introduction, Spread and Areal Extent of Saltcedar (*Tamarix*) in the Western States.* U.S. Geological Survey, Washington, DC.

Robinson, T. W. 1952. Phreatophytes and their relationship to water in Western United States. Transactions American Geophysical Union 33:57–61.

Scifres, C. J. 1980. *Brush Management Principles and Practices for Texas and the Southwest.* Texas A & M Universeity Press, College Station.

Senseman, S. A., editor. 2007. *Herbicide Handbook.* 9th ed. Weed Science Society of America, Lawrence, KS.

Shafroth, P. B., J. R. Cleverly, T. R. Dudley, J. P. Taylor, C. V. Riper, E. P. Weeks, and J. N. Stuart. 2005. Control of Tamarix in the western United States: Implications for water salvage, wildlife use, and riparian restoration. Environmental Management 35:231–246.

Shafroth, P. B., V. B. Beauchamp, M. K. Briggs, K. Lair, M. L. Scott, and A. A. Sher. 2008. Planning riparian restoration in the context of *Tamarix* control in western North America. Restoration Ecology 16:97–112.

Shaner, D. L., and S. L. O'Connor. 1991. *The Imidazolinone Herbicides*. CRC Press, Boca Raton, FL.

Sheley, R. L., T. J. Svejcar, and B. D. Maxwell. 1996. A theoretical framework for developing successional weed management strategies on rangeland. Weed Technology 10:766–773.

Sher, A. A., D. L. Marshall, and J. Taylor. 2002. Spatial partitioning within southwestern floodplains: Patterns of establishment of native *Populus* and *Salix* in the presence of invasive, non-native *Tamarix*. Ecological Applications 12:760–772.

Sher, A. A., K. Lair, M. Deprenger-Levin, and K. Dohrenwend. 2010. *Best Management Practices for Revegetation in the Upper Colorado River Basin*. Denver Botanic Gardens, Denver, CO.

Smith, L. M., and J. A. Kadlec. 1983. Seed banks and their role during drawdown of a North American marsh. Journal of Applied Ecology 20:673–684.

Smith, L. M., M. D. Sprenger, and J. P. Taylor. 2002. Effects of discing saltcedar seedlings during riparian restoration efforts. Southwestern Naturalist 47:598–601.

Sprenger, M. D., L. M. Smith, and J. P. Taylor. 2001. Testing control of saltcedar seedlings using fall flooding. Wetlands 21:437–441.

Stromberg, J. C., and M. K. Chew. 2002. Foreign visitors in riparian corridors of the American Southwest: Is xenophytophobia justified? Pages 195–219 *in* B. Tellman, editor. *Invasive Exotic Species in the Sonoran Region*. University of Arizona Press and Arizona-Sonora Desert Museum, Tucson, AZ.

Stromberg, J. C., S. J. Lite, R. Marley, C. Paradzick, P. B. Shafroth, D. Shorrock, J. M. White, and M. S. White. 2007. Altered stream-flow regimes and invasive plant species: the *Tamarix* case. Global Ecology and Biogeography 16:381–393.

Subcommittee, P. 1951–1970. Minutes of the Phreatophyte Subcommittee Meetings. Page 440 *in* Phreatophyte Subcommittee, Pacific Southwest InterAgency Committee (PSIAC) of the Water Resources Council, editor, Various. Washington, DC.

Taylor, J. P., and K. C. McDaniel. 1998. Restoration of saltcedar (*Tamarix* sp.)-infested floodplains on the Bosque del Apache National Wildlife Refuge. Weed Technology 12:345–352.

Taylor, J. P., and K. C. McDaniel. 2004. Revegetation strategies after saltcedar (*Tamarix* spp.) control in headwater, transitional, and depositional watershed areas. Weed Technology 18:1278–1282.

Taylor, T. U. 1930. *Silting of Reservoirs*. University of Texas, Austin.

Tellman, B. 2002a. Human Introduction of Exotic Species. Pages 25–46 *in* B. Tellman, editor. *Invasive Exotic Species in the Sonoran Region*. University of Arizona Press and Arizona-Sonora Desert Museum, Tucson.

Tellman, B. 2002b. The range of control methods. Pages 319–330 *in* B. Tellman, editor. *Invasive Exotic Species in the Sonoran Region*. University of Arizona Press and Arizona-Sonora Desert Museum, Tucson.

Utah PCD (Partners for Conservation and Development), Saving Utah's Landscape: Dinosaur National Monument. Utah State University Cooperative Extension, Logan, UT. [Online]. http://extension.usu.edu/files/publications/publication/pub__6803066.pdf.

Vincent, K. R., J. M. Friedman, and E. R. Griffin. 2009. Erosional consequence of saltcedar control. Environmental Management 44:218–227.

Vizantinopoulos, S., and P. Lolos. 1994. Persistence and leaching of the herbicide imazapyr in soil. Bulletin of Environmental Contamination and Toxicology 52:404–410.

West, P., and G. P. Nabhan. 2002. Invasive Plants: Their occurrence and possible impact on the central Gulf Coast of Sonora and the Midriff Islands in the Sea of Cortes. Pages 91–111 *in* B. Tellman, editor. *Invasive Exotic Species in the Sonoran Region.* University of Arizona Press and Arizona-Sonora Desert Museum, Tucson.

Wiedemann, H. T., and B. T. Cross. 1979. *Saltcedar Control along Shorelines of Lakes.* Texas Agricultural Experiment Station, Texas A & M University, College Station, TX.

Woodburn, K. B., F. R. Batzer, F. H. White, and M. R. Schultz. 1993. The aqueous photolysis of triclopyr. Environmental Toxicology and Chemistry 12:43–55.

21

Tamarisk Management at Bosque Del Apache National Wildlife Refuge

A RESOURCE MANAGER'S PERSPECTIVE

Gina Dello Russo

Over the past 65 years Bosque del Apache National Wildlife Refuge (hereafter "the Refuge") has been working to control tamarisk and other weeds as a part of its overall management. In the 1980s, the Refuge began aggressive tamarisk removal and supported research directed at fine-tuning control techniques while improving understanding of competition between tamarisk and native woody plant species. Because of this work, we have become a leader in control of this aggressive invasive species and in reestablishing the processes and native plant communities associated with southwest riparian ecosystems. All aspects of the work, including planning, site preparation, implementation, and monitoring have changed over time, but we have learned some basic lessons that we carry out into the field with us each day. This chapter tells the story of *Tamarix* management on the Refuge from the perspective of one of its resource managers.[1]

We learned from experience that the most important consideration for effective tamarisk control and successful plant replacement is always considering the project as a whole; tamarisk control is only one step in implementation (see chapter 23, this volume). Understanding how historic river processes maintained native plant communities in the past plays a role in successful establishment and effective control of tamarisk now, just as knowing how alterations to these processes will inform site selection, site analysis, control techniques, and expected riparian health resulting from any project. The steps of planning, evaluation, implementation,

---

[1] The findings and conclusions in this article are those of the author(s) and do not necessarily represent the views of the U.S. Fish and Wildlife Service.

FIGURE 21.1 Wetlands at Bosque del Apache National Wildlife Refuge, near Socorro, New Mexico.
Source: Photo courtesy of Bosque del Apache.

and maintenance cannot be rushed. Each step has its value in ensuring successful native plant reestablishment and future sustainability of restored riparian wildlife habitat.

The Refuge is located on the Rio Grande in central New Mexico, the middle Rio Grande of the state (Cochiti Reservoir downstream to Elephant Butte Reservoir). It was established in 1937 as a wintering area for migratory birds, primarily the sandhill crane (*Grus canadensis*). Refuge management focuses on improving the riparian ecosystem, including mimicking the creation and succession processes of native riparian forests, cottonwood (*Populus deltoides* spp. *wislizeni*) and willows (*Salix gooddingii* and *S. exigua*), wetlands dominated by emergent native and naturalized grasses, forbs, and sedges, and wet meadows (*Distichlis spicata* and *Sporobolus airoides* dominate; see figure 21.1). To maintain habitat for the thousands of birds and other wildlife that depend on the Refuge, tamarisk had to be controlled and if possible eradicated from floodplain areas.

Bosque Del Apache National Wildlife Refuge: Early Experiences (1942–1992)

The Refuge sits at the lower end of the middle Rio Grande of New Mexico below two major tributaries, the Rio Salado and Rio Puerco. These tributary watersheds contribute a substantial seasonal water and sediment supply to the Rio Grande main stem above the Refuge. Tamarisk became established on the middle Rio Grande beginning in the 1920s, and by 1950 there were thousands of acres established in the area of the Refuge (US Bureau of Reclamation unpublished data).

Elephant Butte Reservoir, located downstream of the Refuge, was filled in 1942 for the first time (Crawford et al. 1993). The resulting pool caused a change in slope on the river upstream. This altered hydrology during high river flows limited the river's ability to move sediment, to establish a defined channel in places, and to establish native plants (Harrison 2011). In the 1950s, the drainage system to move water off the floodplain was rehabilitated. A large conveyance channel, the Low Flow Conveyance Channel (LFCC), and spoil levee were constructed on the eastern portion of the floodplain at that time as well. This divided the floodplain into an "active" portion, which is connected to and dependent on river flows, and a "historic" floodplain, now disconnected from the river. By 1960, the LFCC was completed, and from 1960 to 1984, it carried river flows of up to 2,000 cubic feet per second (cfs) from the San Acacia Diversion Dam downstream to Elephant Butte Reservoir. Tamarisk spread during this period, reducing the diversity and patchy nature of the riparian plant communities found on the Refuge and surrounding floodplain (Bureau of Reclamation unpublished data). Once tamarisk was established, river flows were unable to remove it, so historic patterns of plant succession in native riparian plant communities were disrupted. By 1992, the middle Rio Grande was a highly regulated river reach with numerous dams and diversions, groundwater depletions, and large areas of tamarisk (Everitt 1998).

TAKING ON TAMARISK CONTROL AND THE EARLY PROCESS OF DEVELOPING EFFECTIVE TECHNIQUES

Early Refuge documents note that tamarisk was "taking root but can't be removed because of soft silt sometimes waist deep" (1941 Refuge Narrative). By 1942, staff considered tamarisk control the "most serious problem" facing the Refuge (Harrison 2011). Tamarisk formed dense monotypic stands in areas with a high groundwater table (Campbell and Dick-Peddie 1964), and its growth pattern was assumed to provide limited wildlife benefit. At this time, water use by tamarisk was also considered higher than that of native tree species (Shafroth et al. 2005). Tamarisk fire behavior was extreme, limiting wildfire suppression and resulting in the loss of native riparian forests (Stuever 1997).

Control of tamarisk began in 1947, first with experimental small-scale application of 2,4-D herbicide; results were "not too encouraging" (1947 Refuge Narrative). From 1948 to 1987, we hand sprayed and mowed tamarisk at various sites, even chaining some sites using a ship's anchor chain pulled between two bulldozers (Vicente pers. comm.). In 1960, the U.S. Bureau of Reclamation adopted the use of a root plow to remove tamarisk, improving control (Horton 1960). By the 1980s, the Refuge was using this method, augmented by the use of a root rake to increase root material removal from the soil (Vicente pers. comm.; see figure 21.2).

In 1985, the Refuge hired wildlife biologist John P. Taylor, who quickly developed an adaptive management approach to tamarisk control. In 1987, we had the first trial of applying herbicide by fixed-wing aircraft, called an "aerial application."

FIGURE 21.2 Bulldozer equipped with root rake removing tamarisk below ground biomass at the Refuge.
Source: Photo courtesy of Bosque del Apache.

This application proved effective; the 50-hectare plot showed 95% control after three years. This began a period of applied research with Dr. Kirk C. McDaniel of New Mexico State University and others focused on comparing mechanical (i.e., removing trees with a root plow and rake attached to a large bulldozer) and chemical (aerial application of imazapyr) control for both efficacy and cost/benefit ratios (Horton 1960; Taylor and McDaniel 1998). The Refuge also made a substantial investment by dedicating staff and buying bulldozers and specialized equipment necessary to mechanically control tamarisk. Although both approaches seemed to work well under some conditions, efficacy varied with soil texture and moisture, depth to water table, and size of tamarisk (Refuge unpublished data). In particular, areas of high soil moisture and fine soil texture favored continued tamarisk resprouting after initial control.

As a result of this work, we became adept at killing and removing tamarisk. Between 1987 and the early 1990s, large monotypic tamarisk tracts were controlled by a combination of aerial herbicide and bulldozer mechanical treatments. After the trees were sprayed and appeared dead (approximately three years), they were bulldozed into piles. Then, root plows and rakes were used to cut and then pull out what remained below ground. All this material was burned in large piles (see figure 21.3). This mechanical treatment disturbed the soil and vegetation considerably, so it was used solely in areas where tamarisk was nearly the only species present ("monotypic stands"). Given how frequently we saw tamarisk resprout after the first treatment, the importance of follow-up treatment could not be overstated. Generally this meant one or two mechanical or chemical treatments after initial

FIGURE 21.3 Burning piles of tamarisk at the Refuge.
Source: Photo courtesy of Bosque del Apache.

mechanical removal. We learned it is best to ensure that tamarisk was completely controlled before attempting to establish native plants.

THE PROCESS OF LEARNING ABOUT NATIVE PLANT ESTABLISHMENT

The Refuge began planting cottonwood, Goodding's willow, and riparian shrub areas, but these early tamarisk control/native plant restoration projects were less than ten hectares on average and considered experimental. Early studies on native plants, soil and water requirements (Anderson and Ohmart 1984; Swenson and Mullins 1985; Fenchel et al. 1987; Bayley 1995) informed planting approaches at the Refuge. Native species were occasionally planted from seed, but rooted plants and poles (limbs or boles without roots) of cottonwood and willow proved to be most effective (Swenson and Mullins 1985). With sufficient groundwater, both cottonwood and willow poles are capable of sprouting roots. Early revegetation efforts were generally successful, but we found that many aspects of the soil and water availability strongly influenced survival of plantings. Site factors that influenced these early project designs (and that continue to determine native plant species selection if plantings are included in restoration projects) were plant salinity tolerance, site soil texture and salinity, and depth to groundwater (Anderson and Ohmart 1984; Shafroth et al. 1995). Testing soil salinity in particular correlated with cottonwood and willow pole planting success, since these species are sensitive to the saline conditions often associated with tamarisk stands (Rowland et al. 2004; see also chapter 8, this volume). Early on, we collected soil samples taken from different depths of the soil horizon to test salinity; later, we used an electromagnetic instrument (EM 38) to measure soil conductivity, which was correlated to soil salinity (Sheets et al. 1994). Wells for sampling groundwater (piezometers) were installed and monitored to determine the range in yearly groundwater fluctuation. Planting cottonwood and Goodding's willow poles deeply enough to ensure successful establishment was labor intensive. Restoration sites ranged from less than one-hectare (windrows along Refuge farm units) to approximately ten-hectare plots. By the early 1990s, we were considering native plant restoration on a larger scale using alternative techniques.

SUMMARY OF THE EARLY YEARS

Over the first 50 years of tamarisk establishment and control on the Refuge, researchers and Refuge staff developed an understanding of the effort required to control this species and establish desirable vegetation in its place. We determined some of the requirements for successful control including chemical mixes and mechanical control techniques. By 1992, we had a systematic process to control historic floodplain tamarisk and a skilled maintenance crew committed to the effort. Early Refuge efforts to restore desirable plant communities were limited to active plantings of woody plants and impounding water to promote grass species. This approach proved successful in various site conditions depending on water availability and careful selection of planting locations based on soil conditions. However, these restored cottonwood and willow stands contained only a few shrub species, and trees were usually planted in rows (see figure 21.4); there was limited plant diversity, food, or nest structure for birds. But this work served as a catalyst to future efforts, showing stakeholders such as private landowners, policy

FIGURE 21.4 Rows of cottonwood trees planted in 1986, after tamarisk removal at the Refuge.
Source: Photo courtesy of Bosque del Apache.

FIGURE 21.5 Tamarisk resprouting after fire.
Source: Photo courtesy of Anna Sher.

makers, and the general public how tamarisk control could be a successful step toward native plant establishment.

Ever since tamarisk was introduced to the area, the Refuge has focused on its control, with some lags in effort due to budget restrictions (Harrison 2011). We did this because we saw habitat diversity diminishing over time; this diversity was crucial to our mission of providing for winter water birds, addressing endangered species needs, and benefiting the overall ecosystem. Large-scale tamarisk control became a Refuge priority following a human-caused wildfire, the Volley Fire, which burned on the Refuge in 1986. During this fire, native trees were killed by the extreme heat, but tamarisk resprouted aggressively (Harrison 2011; see figure 21.5). After tamarisk control, cleared areas were managed to provide seasonally flooded grasslands or emergent wetlands for water birds in winter. For example, United States wintering greater sandhill crane numbers had become very low by 1960, there were only 1,700 birds nationwide, and 1,200 of those were wintering at the Refuge. We focused on providing habitat for sandhill cranes and other migratory winter water birds, and combined with lesser sandhill crane, by 1989 numbers had reached 17,200 at the Refuge alone. Tamarisk removal was an important component of creating this habitat, and following this and other desirable outcomes of tamarisk removal, the Refuge turned to larger projects.

Adaptive Management and Implementation of Larger-Scale Projects (1992–2000)

By the early 1990s, the Refuge and others were starting to explore and discuss the physical and biological complexity of functioning native riparian ecosystems and the potential for rehabilitation of those systems under altered conditions (Crawford et al. 1993). Research published during this time began to focus on how alterations to western rivers, such as damming, suppressed native riparian plants and promoted tamarisk (Fenner et al. 1985; Auble et al. 1994; Busch and Smith 1995). Many working on Southwest rivers began to link the importance of natural overbank flooding and gradual drawdown of the water table to physical processes, survival, succession, and diversity of riparian plant communities and their ability to compete with tamarisk (Segelquist et al. 1993; Shafroth et al. 1995; Ellis et al. 1999; Sher and Hyatt 1999). Such perspectives had a great influence on Refuge managers in the context of the larger-scale projects that were tackled during the 1990s.

LARGE-SCALE RESEARCH PROJECTS IMPLEMENTED

One Refuge experiment established on the active floodplain evaluated the ability of native plant seedlings to compete with tamarisk seedlings under historic high flow conditions (Taylor et al. 1999). John Taylor and Loren Smith developed a replicate study in which plots were mechanically cleared of mature, dense tamarisk, with control plots left standing. All plots were adjacent to the Rio Grande main channel. Following site preparation and spring, peak river flows timed to mimic the historic hydrograph, native and nonnative woody plant establishment was measured on both treatment and control plots. Anna Sher (2002) tagged each of the hundreds of seedlings in the treatment plots, recorded their survival and growth over the next three years, and correlated them to environmental variables. With only natural seed dispersal and overbank flooding at the site, abundant native cottonwood and willow were established alongside tamarisk seedlings (Taylor et al. 1999). Initial densities of tamarisk were much higher than that of the native trees, but there was high mortality of tamarisk seedlings where cottonwood seedlings were densest (Sher et al. 2002). This research was important because it was the first to show that native woody vegetation could easily compete with tamarisk if the timing of river flows was favorable and mature tamarisk were removed. Native plant response on cleared plots showed promise for "letting the river" assist in restoration. After ten years, these sites were revisited and the cottonwood-dominated stands were measured and evaluated (Taylor et al. 2006). Tamarisk is still present in these plots but is a minor understory component of the young cottonwood gallery forest.

Understanding river processes and a method for mimicking those processes at control/riparian restoration sites became the platform for future planning and implementation on the Refuge. Site evaluation included past and present surface and groundwater hydrology, river reach geomorphology and sediment movement, and site soil chemistry and texture through the soil horizon to groundwater table. In terms of hydrology, subsequent research focused on water required for successful native tree and shrub germination, growth, and competition. This included thresholds for surface water recession rates that encouraged native and/or nonnative woody plant responses and competitive interactions (Sher et al. 2000; Sprenger et al. 2001; Sher and Marshall 2003; Bhattacharjee et al. 2008).

Other research conducted on the Refuge at this time included extensive surveys of bird, rodent, and insect use of tamarisk relative to native trees, and a multiyear investigation of the role of flooding in nutrient cycling in both tamarisk and native stands by a group from the University of New Mexico (Ellis 1995; Molles et al. 1995; Ellis et al. 1997; Ellis 2000). Papers based on this work made the case that river flooding and irrigation water management at the Refuge were powerful tools for ecosystem restoration (e.g., Molles et al. 1998).

TAMARISK CONTROL METHODS REFINED AND EXPANDED

McDaniel and Taylor (2003) continued to evaluate tamarisk control techniques. Their experimental designs focused on comparing aerial chemical application and mechanical control in large monotypic stands of tamarisk, including both efficacy and costs of each practice. Prescribed fires were used to remove above-ground biomass of tamarisk that had been treated with herbicide in study plots (see figure 21.6). In mechanical control plots, piles of trees were burned, and roots were plowed, raked to the surface, and burned; this technique is still used today (see chapter 20, this volume). Abiotic conditions at the sites (soil texture, chemistry, and groundwater levels) were also measured to evaluate possible correlations between percent control and these factors. One of the most important findings of this research was that both chemical and mechanical control techniques rarely achieved 100% kill of tamarisk after one application; both required follow-up control to varying degrees (Duncan and McDaniel 1998; McDaniel and Taylor 2003). We also used other tamarisk control techniques, such as cut stump for isolated larger trees, foliar and basal bark herbicide application on

FIGURE 21.6 Standing dead tamarisk burning after being killed by herbicide, at the Refuge.
Source: Photo by John Taylor, Bosque del Apache.

resprouts, and mechanical removal with an excavator equipped with a hydraulic thumb in areas of large mixed stands. These techniques were effective as well.

During this period, maintenance staff at the Refuge observed that burning the trees after they had been sprayed with herbicide was problematic. Tamarisk tended to resprout following the fires, requiring added mechanical follow-up treatment. Also, because these prescribed fires left all tamarisk root crowns intact, the herbicide-burn technique required additional plowing and raking to remove the troublesome root crowns. Therefore, they recommended that mechanical removal by a bulldozer with a brush blade be done first; when pushing over live trees, root crowns stay attached to the tree and come out of the ground with it. This reduced the need for subsequent passes with the root plow and rake (Reeves and Tafoya pers. comm.). A combination of formal research and suggestions by maintenance staff resulted in continuing improvements in tamarisk control at the Refuge.

MIMICKING NATURAL PROCESSES OF THE RIO GRANDE

Advances in tamarisk control were matched in advances in passive revegetation, as the role of hydrology was more fully realized. Taking what was learned on the active floodplain experiment (Taylor and Smith 1999; Sher et al. 2002), we developed a new technique for native plant establishment on the historic floodplain that produced a much more diverse native plant species assemblage, mimicked some of the crucial river processes linked to healthy riparian ecosystems, and lessened the staff time and equipment required to establish these plants. The steps toward restoration at these new sites used the same techniques for tamarisk control, but infrastructure and water delivery became more important.

Once tamarisk was cleared, site topography was measured and berms were built to pond water to a given elevation, creating management units (see figure 21.7). Inflow and outflow water-control structures were installed in these units and, finally, timed with the spring high flows, surface water was applied. This flooding left saturated soils and mud flats as water receded. If timed correctly, native woody species germination was effective, as studies at the Refuge and elsewhere have shown (Poff et al. 1997; Auble and Scott 1998; Friedman et al. 1995; Mahoney and Rood 1998; Sprenger et al. 2001; Rood et al. 2005; Bhattacharjee et al. 2008). The first successful establishment of native trees and shrubs using this technique was accomplished in 1996, my first involvement with a restoration project on the Refuge. Subsequent Refuge projects have restored approximately 400 hectares of riparian forest using this technique. We learned that maintenance floods on these restored sites, again mimicking subsequent periodic high flows, stimulated nutrient cycling, lessened fuel loading, and flushed surface salts from the soil (Refuge unpublished data). Monitoring on early flood establishment sites found that a return interval of four to six years for these maintenance floods would keep soil salinity in the range measured at time of plant establishment.

FIGURE 21.7 Berms defining management units at the Refuge.
Source: Photo courtesy of Bosque del Apache.

NATIVE PLANT ESTABLISHMENT: SUCCESSES AND CHALLENGES ON LARGER PROJECTS

It was never the goal of the Refuge to simply control tamarisk. Early on, Refuge staff realized the importance of native plant competition with tamarisk to achieve lasting control. It was also never the goal of Refuge riparian restoration to predetermine which plant community was appropriate for each site once we moved from installed plantings. The Refuge set the stage by clearing the tamarisk and providing water, one component of the historic river processes. Once flooded, the seed bank, soil texture, and position on the floodplain facilitated a certain mosaic of plants. Monitoring at these sites showed plant diversity that reflected soil characteristics and soil moisture gradients (Sprenger et al. 2001).

One of the benefits of this approach was that both riparian forest and low-lying emergent wetlands were created. This patchy pattern provided greater structural diversity, shelter, and food resources for wildlife species (Refuge unpublished data). It also provided fuel breaks where wetlands occurred, reducing the potential extent of wildfire. Finally, this restoration technique provided us with greater water management flexibility. When drought conditions limited surface water supply, we reduced riparian forest restoration or maintenance water and waited for a more favorable year.

Some failures in forest restoration also taught us an important lesson: because passive establishment of native species depended on an ability to mimic historic flooding, and because our ability to predict or mimic those hydrographs depended on snowpack runoff or ditch water delivery, scheduling restoration was not always predictable. Forest restoration sites that had been prepared by removing tamarisk were kept waiting for the right conditions, either in river flows or ditch water supply. If in the first establishment year, the moisture regime was unfavorable to native woody species, the site was cleared of weeds and tamarisk seedlings by using a shallow disk attached to a farm tractor. In this way, the site was maintained until a more favorable spring flood or water supply was possible. This technique of first year tamarisk sprout control became a standard Refuge tool.

SUMMARY OF ADAPTIVE MANAGEMENT AND LARGE-SCALE PROJECTS

Research and monitoring on and off the Refuge produced crucial information on river ecosystem dynamics (Stromberg et al. 1991; Naiman et al. 1993; Ellis et al. 1999), on altered river-system characteristics, including changes in flood timing, magnitude, and duration and riparian forest response (Ritter 1993; Scott et al. 1996; Friedman et al. 1996; Auble and Scott 1998; Richter and Richter 2000), and on riparian wildfire effects, which were becoming more common on the middle Rio Grande (Stuever 2000; Ellis 2000; Refuge unpublished data). Reduced river water and sediment delivery changed sediment movement and channel dimensions, motivating a larger scale effort to determine longer-term outcomes (Crawford et al. 1993; Friedman et al. 1996; see also chapter 7, this volume). Policy documents provided voice to societal interest in tamarisk control, riparian restoration, and overall river health on the middle Rio Grande (Crawford et al. 1993).

As we moved to larger restoration projects on the Refuge, implementation became standardized. Planning, site analysis and monitoring, invasive plant control and maintenance, infrastructure placement, and native plant establishment were now the basic steps. Project design and site analysis still included the evaluation of soil chemistry and texture, laboratory analysis of representative soil samples, groundwater measurements (at larger sites), and general topographic information. Site monitoring included checking tamarisk and native-plant stem density, plant diversity, and growth and competition between native plants and tamarisk. Our maintenance staff became advisors for other efforts to remove large tamarisk forest stands. This work and knowledge paved the way for the natural next step of considering restoration on a watershed rather than Refuge scale, and reaching out to those seeking to do similar work in other locales.

Moving Outside the Refuge Boundary: Landscape Level Tamarisk Control (2000–2010)

By the end of 1990s, we were beginning to assist community efforts and to consider the Refuge in the context of a larger landscape. After John Taylor's unexpected death in 2004, this off-refuge work became a new focus, augmenting the previous focus on riparian research with outreach and technical assistance. The Save Our Bosque Task Force, a local nonprofit organization started in 1993 and made up of agency representatives (including Refuge staff), private landowners, and community members took the lead in this effort to analyze such issues as invasive species control, fire danger, ecosystem health, endangered species, and recreation opportunities along the river. The Task Force began to address community issues including degradation of the bosque (or riparian forest) adjacent to Socorro, San Antonio, Bosquecito, and other villages.

SENSE OF PLACE: THE FIRST COORDINATED WORK BEYOND REFUGE BOUNDARIES

Focusing first on river parks, illegal dumping, and human-caused fires (Kernberger 2005), the Task Force recognized that the river corridor would have limited value to wildlife if upstream and downstream habitat continued to be affected by wildfires, tamarisk, or development on the floodplain. An unimpeded river corridor serves many wildlife species, and the open land allows flood waters to pass downstream safely, a process crucial to the survival of cottonwood and willows and endangered vertebrate species including Rio Grande silvery minnow and southwestern willow flycatcher.

In 1999, area government agency personnel and nonprofit organizations met with private landowners on the active floodplain to discuss protection and riparian restoration on their lands and the issue of floodplain encroachment. These discussions resulted in a conceptual restoration plan (CRP) for the 73 kilometers of river from San Acacia Diversion Dam to the San Marcial railroad bridge, including the 16 kilometers on the Refuge (TetraTech 2004).

Following completion of the CRP, the Refuge developed a programmatic restoration plan and environmental assessment for the active floodplain under our jurisdiction (Bosque del Apache NWR 2005). This document laid out general techniques to be used on the active floodplain to maintain and improve lateral connectivity between the river channel and floodplain, control invasive plant species, and promote river processes within the limited floodplain. At about the same time, Refuge staff became involved in other landscape-level projects such as the Middle Rio Grande Endangered Species Collaborative Program.

COMMUNITY PARTNERSHIPS ON THE ACTIVE FLOODPLAIN THROUGHOUT THE SOCORRO VALLEY

With planning documents in place and a group of landowners interested in conservation easements and habitat restoration, the Task Force and Rio Grande Agricultural Land Trust, with assistance from the Refuge and other agencies, secured funding and protected 222 hectares of private flood-prone lands, cleared tamarisk from these lands, and began the first phase of native plant establishment.

Private landowners have been active in follow-up tamarisk control and in improving wildlife habitat diversity on their lands. We have provided native plant materials to not-for-profit groups restoring riparian areas.

In 2001 and again in 2004, we partnered with two Middle Rio Grande Pueblos, state and local agencies, and private landowners to restore and improve wetland habitat for bird species on and off the Refuge (NAWCA proposals 1999, 2003). Refuge projects were designed to treat approximately 728 hectares of mature tamarisk and restore native forest, wetlands, and grasslands. The Marcial Fire, another human-caused wildfire, burned across chemical treatment Refuge NAWCA sites in 2006, changing the direction and timing of work for this project. New funding for burned-area rehabilitation (Parametrix 2008) allowed the Refuge to improve infrastructure, purchase much-needed replacement equipment, and partner with adjacent landowners to control tamarisk. One research project done in conjunction with the Marcial Fire was designed to address the difficult task of xeric site restoration following tamarisk control (Beauchamp et al. 2009). The first two years of monitoring saw primarily nonnative weed species established on seeded sites. These xeric sites are some of the hardest to restore because, although they are within the historic floodplain, they are disconnected from surface and shallow groundwater.

Efficiency in implementing tamarisk control/native plant restoration projects varies depending on staff, equipment, and funding, and with the NAWCA/Marcial Fire Project, we moved beyond that efficient work schedule. The result was noxious weed encroachment, necessary modifications to original plans, and additional maintenance of the project area. Progress was still made on the 700-hectare site, with the restoration of approximately 300 hectares, but Refuge staff has determined that approximately 15 to 25 hectares of tamarisk control per year, followed by native plant establishment, is a more appropriate implementation rate for Refuge resources.

Other collaborative tamarisk work included the NM FWS Fire Program's installation of large fuel breaks on the Refuge to limit the wildfire threat on the active floodplain. We also began coordinating a US Fish and Wildlife Service Invasive Species Strike Team (ISST) program on six National Wildlife Refuges in the state, focusing on early detection/rapid response and the comprehensive mapping and monitoring of invasive species, including tamarisk (Refuge unpublished reports). The ISST program along with other programs like the New Mexico State Forestry Inmate Work Camp has provided assistance in follow-up tamarisk control.

In all these ways, we had a role in large-scale land management efforts and gained influence with stakeholders. Concurrently, the research that was done on the Refuge laid the groundwork for ever-larger projects that would influence the scientific community.

RESEARCH BROADENS ITS FOCUS

The Refuge increasingly became a desirable location for restoration and tamarisk research, in part because of the careful records that were kept for each restored area (Cleverly and Dello Russo 2007; see also chapter 24, this volume). Several universities used our "living laboratory" for both small and large projects, most with the objective of improving restoration methods or understanding tamarisk's impact on the ecosystem. In some cases, the Refuge was one replicate in a much larger study (e.g., Finch et al. 2006; see also chapter 8, this volume). The Refuge became the first Land Management Research and Demonstration National Wildlife Refuge (LMRD) in 2004. The purpose of LMRDs is to foster research and information transfer; this designation facilitated application of the research that had been done at the Refuge to projects outside its borders. In 2006, the Tamarisk Coalition, a southwestern nonprofit organization that provides education and technical assistance for riparian restoration, held a national research conference devoted specifically to tamarisk in the Southwest. Projects conducted partially or entirely at the Refuge (including by Refuge staff) were well-represented: 17 presenters in such categories as restoration, biological control, distribution and ecology, water use, chemical control strategies, research needs, and modeling. It was clear that the Refuge had become an important stage for larger-scale tamarisk research projects.

The Refuge Evapotranspiration Tower Transition Project on the active floodplain was another major research project to occur at the Refuge; the purpose to evaluate water use by nonnative and native plants (Cleverly et al. 2002; Shafike et al. 2007). Long-term monitoring by a consortium of universities and national laboratories to measure the environmental conditions and water vapor coming from a tamarisk forest site began in 1999. The University of New Mexico and New Mexico State University continued this monitoring following 25 hectares of tamarisk removal adjacent to one tower (2005–2007). After tamarisk removal, a dense patch of primarily native woody plants and open grassland were established through natural processes (Dello Russo 2010). Water use of these newly established habitats was compared to that of mature tamarisk (Samani et al. 2007; Shafike et al. 2007). This work revealed that water use in the cleared area declined at first, but within the second year of establishment, native plant water use was increasing, as expected (see chapter 6, this volume). However, to date this use has not reached the rates of the neighboring mature tamarisk (Bawazir pers. comm.).

SUMMARY OF LANDSCAPE LEVEL WORK

Because of staff changes and competing priorities during this period, we had to reevaluate the focus and direction of the tamarisk removal program, shifting our role from leading tamarisk/riparian restoration research to playing a supporting role. The Refuge's transition is reflected in the shift from publishing scientific articles to creating reach-wide management plans (Tetra Tech 2004; Dello Russo and Najmi 2004; Bosque del Apache NWR 2005), land-management wetland reviews, private-lands project designs, and advocacy papers to decision makers (Dello Russo 2007). Working with diverse interests, we have been a part of an important dialogue on landscape-level analysis (Dello Russo 2009) and how we adapt to changes predicted for the Rio Grande (Millar et al. 2007; Enquist and Gori 2008; Davies 2010).

2010 and Beyond: Working within the Context of a Changing Environment

Looking back, our history might seem to reflect a well-paced transition from small-scale to large-scale projects, from artificial to passive restoration techniques, from trials to established protocol for tamarisk removal. The scale of projects has increased; the Refuge focus has grown to include a river reach of 70 kilometers or more. And in the process Refuge staff, along with other practitioners, have learned the importance of place in determining restoration goals (Hummel and Caplan 2007). But that is a simplification of the commitment made by the Refuge managers, biologists, and maintenance crews to follow through and learn as a team. It was an opportunity taken and the changes on the landscape from our restoration efforts are visible and impressive (see figure 21.8).

Even more impressive are the concerted efforts of many along the middle Rio Grande and other river systems who are also learning a sense of place; each making progress toward greater biodiversity through tamarisk control. In some cases, tamarisk offers wildlife habitat that no longer exists with native plants because of alterations to southwest rivers (Anderson and Barrows 1998; Stromberg 1998; Sogge et al. 2008). However, the Refuge and other preserves in the middle Rio Grande have shown that is not the case in central New Mexico, and they are diligently working to successfully control tamarisk.

ASSESSING FUTURE NEEDS: TOOLS AND COMMUNICATION

Three decades of extensive tamarisk control and revegetation have brought to light several gaps that need filling as we continue to improve habitats at the Refuge and beyond. The most important of these are for ecosystem evaluation/assessment tools and a framework for communication between researchers and land managers that facilitates adaptive management.

In a dynamic river ecosystem where natural fluctuation is expected and human-caused change inevitable, tools to evaluate proposed changes to middle Rio Grande water management will be critical. Currently, climate change is predicted to decrease snowpack water in the Rio Grande watershed, and possibly result in more severe summer rainfall (Enquist and Gori 2008). Competing for this limited water supply will be a growing human population joining other current water users. Our responsibility to provide quality wildlife habitat is all the more important if the river corridor is further fragmented and natural areas begin to decline in wildlife value (Mueller et al. 2005; Stewart et al. 2005; Haney 2007; Seavy et al. 2009). Research and restoration priorities will focus on how to provide for riparian ecosystem

(a)

(b)

(c)

FIGURE 21.8 (a) Tamarisk before removal. (b) Restoration site after tamarisk removal. (c) Site after revegetation with native species.

Source: Photo courtesy of Bosque del Apache.

sustainability under these conditions. A cooperative effort between us, universities and other government agencies is developing a decision support system tool (Brand et al. 2011; Hansen et al. 2010); this tool will simulate future trends in river condition, evaluate potential threats, and help us predict river resiliency (Coonrod, Roach and Tashjian pers. comm.). The introduced tamarisk beetle, found on the middle Rio Grande in 2011 (New Mexico Department of Game and Fish unpublished data), will be one factor to consider in this context (see chapter 22, this volume).

Even with all of the work the Refuge has done in outreach and technology transfer, a criticism it has received over the years has been that our observations, techniques and lessons have not been fully communicated to the larger restoration community. It is a valid criticism, and true of most land managers. With new technology, new avenues may now open to land managers for communicating restoration work. Establishing a dialogue to share experiences, discuss problems, and learn as a group will be a priority for us. Both evaluation tools and improved communication will require state-of-the-art technology, highlighting the importance of multidisciplinary skills in resource managers.

SUMMARY OF FUTURE STEPS

Unlike research, where removing the subjectivity or bias of human impression is critical to gaining knowledge, land management has a human component that is reflected best in the experience and vision of practitioners. We hope that by implementing meaningful restoration, research, and monitoring projects and involving many disciplines and partners, the Refuge has improved our understanding of Southwest river systems. Thanks to managers, biologists, technicians, and maintenance workers who have taken ownership of this comprehensive land management program, restoration work at Bosque del Apache National Wildlife Refuge will continue. Advocating for and involving stakeholders in the evolution of landscape-level planning and implementation of riparian lands protection and restoration will also continue throughout the river reach, with opportunities for the Friends of Bosque del Apache National Wildlife Refuge, other nonprofit organizations, private landowners, volunteers, and local citizens to partner on projects. With improved technology in modeling and analysis, continued partnerships with experts in related fields, and expected limited resources and added stresses to the system, prioritizing tamarisk removal and native habitat restoration will need to be more comprehensive and strategic.

Literature Cited

Anderson, B. and C. Barrows. 1998. The debate over tamarisk. Restoration and Management Notes 16(2): 129–139.

Anderson, B. W., and R. D. Ohmart. 1984. Final Report. Vegetation Management Study for the Enhancement of Wildlife along the Lower Colorado River. U.S. Department of the Interior, Bureau of Reclamation, Albuquerque, NM.

Auble, G. T., J. M. Friedman, and M. L. Scott. 1994. Relating riparian vegetation to present and future streamflows. Ecological Applications 4(3):544–554.

Auble, G. T., and M. L. Scott. 1998. Fluvial disturbance patches and cottonwood recruitment along the upper Missouri River, Montana. Wetlands 18(4):546–556.

Bayley, P. B. 1995. Understanding large river-floodplain ecosystems. Bioscience 45:153–158.

Beauchamp, V. B., C. M. Pritekel, and P. B. Shafroth. 2009. Evaluation of revegetation methods for saltcedar-dominated xeric riparian areas along the Rio Grande, New Mexico: USGS Science Support Partnership progress report for FY2008. Fort Collins, CO.

Bhattacharjee, J., J. P. Taylor Jr., L. M. Smith, and L. E. Spence. 2008. The importance of soil characteristics in determining survival of first-year cottonwood seedlings in altered riparian habitats. Restoration Ecology 16(4):563–571.

Bosque del Apache NWR. 2005. Control of non-native plant species and reestablishment of native riparian forest, wetlands, grasslands, and in-channel habitats on the active floodplain of the Rio Grande at the Bosque del Apache NWR, Socorro County, New Mexico. US Fish and Wildlife Service. Bosque del Apache National Wildlife Refuge, San Antonio, NM.

Brand, L. A., J. C. Stromberg, D. C. Goodrich, M. D. Dixon, K. Lansey, D. Kang, D. S. Brookshire, and D.J. Cerasale. 2011. Projecting avian response to linked changes in groundwater and riparian floodplain vegetation along a dryland river: A scenario analysis. Ecohydrology 4:130–142.

Busch, D. E., and S. D. Smith. 1995. Mechanisms associated with decline of woody species in riparian ecosystems of the southwestern U.S. Ecological Monographs 65:347–370.

Campbell, C. J., and W. A. Dick-Peddie. 1964. Comparison of phreatophyte communities in New Mexico. Ecology 45:492–505.

Cleverly, J., and G. Dello Russo. 2007. Salt cedar control: Exotic species in the San Acacia Reach. Pages 76–79 in L.G. Price P. Johnson, and D. Bland editors. *Water Resources of the Middle Rio Grande: San Acacia to Elephant Butte.* New Mexico Bureau of Geology and Mineral Resources, Socorro.

Cleverly, J. R., C. N. Dahm, J. R. Thibault, D. J. Gilroy, and J. E. A. Coonrod. 2002. Seasonal estimates of actual evapotranspiration from *Tamarix ramosissima* stands using 3-dimensional eddy covariance. Journal of Arid Environments 52(2):181–197.

Crawford, C. S., A. C. Culley, R. Leutheuser, M. S. Sifuentes, L. H. White, and J. P. Wilber. 1993. Middle Rio Grande ecosystem: Bosque biological management plan. U.S. Fish and Wildlife Service, Albuquerque, NM.

Davies, P. M. 2010. Climate change implications for river restoration in global biodiversity hotspots. Restoration Ecology 18(3): 261–268.

Dello Russo, G. 2010. Evapotranspiration Tower Transition Project Final Report. Middle Rio Grande Endangered Species Collaborative Program, Albuquerque, NM.

Dello Russo, G. 2007. Opportunities for long-term bosque preservation in the San Acacia reach. Pages 84–88 in L.G. Price, P. Johnson, and D. Bland editors. *Water Resources of the Middle Rio Grande: San Acacia to Elephant Butte.* New Mexico Bureau of Geology and Mineral Resources, Socorro.

Dello Russo, G. 2009. Restoration in New Mexico: the floodplains. Southwest Hydrology 8(2): 31–33.

Dello Russo, G., and Y. Najmi. 2004. Planning for large-scale habitat restoration in the Socorro Valley, New Mexico. Pages 86–91 in C. Aguirre-Bravo, P. J. Pellicane, D. P. Burns, and S. Draggan editors. Monitoring Science and Technology Symposium

Proceedings RMRS-P-42CD. USDA Forest Service Rocky Mountain Research Station, Fort Collins, CO.

Duncan, K. W., and K. C. McDaniel. 1998. Saltcedar (*Tamarix* spp.) management with imazapyr. Weed Technology 12:337–344.

Ellis, L. M. 1995. Bird use of *Tamarix* and cottonwood vegetation in the Middle Rio Grande Valley of New Mexico, U.S.A. Journal of Arid Environments 30:339–349.

Ellis, L. M. 2000. Flooding and fire as disturbance mechanisms in riparian areas. Pages 44–46 *in* J. P. Taylor, editor. Proceedings from the conference on fire in riparian areas sponsored by the Middle Rio Grande Bosque Initiative. Albuquerque, NM.

Ellis, L. M., C.S. Crawford, and M.C. Molles. 1997. Rodent communities in native and exotic riparian vegetation in the Middle Rio Grande Valley of central New Mexico. Southwestern Naturalist 42:13–19.

Ellis, L. M., M. C. Molles Jr., and C. S. Crawford. 1999. Influence of experimental flooding on litter dynamics in a Rio Grande riparian forest, New Mexico. Restoration Ecology 7(2):193–204.

Enquist, C., and D. Gori. 2008. A Climate Change Vulnerability Assessment for Biodiversity in New Mexico, Part I: Implications of Recent Climate Change on Conservation Priorities in New Mexico. The Nature Conservancy of New Mexico, Santa Fe.

Everitt, B. L. 1998. Chronology of the spread of tamarisk in the central Rio Grande. Wetlands 18(4):658–668.

Fenchel, G., W. Oaks, and E.A. Swenson. 1987. Selecting desirable woody vegetation for environmental mitigation and controlling wind erosion and undesirable plants in the Rio Grande and Pecos River valleys of New Mexico. 5-year interim report (1983-1987). USDA-SCA-Plant Materials Center, Los Lunas, NM.

Fenner, P., W. W. Brady, and D.R. Patton. 1985. Effects of regulated water flows on regeneration of Fremont cottonwood. Journal of Range Management 38(2):135–138.

Finch, D. M., J. Galloway, and D. Hawksworth. 2006. Monitoring bird populations in relation to fuel load treatments in riparian woodlands with tamarisk and Russian olive understories. Pages 113–120 *in* C. Aguirre-Bravo, P. J. Pellicane, D. P. Burns, and S. Draggan, editors. Monitoring Science and Technology Symposium Proceedings RMRS-P-42CD. USDA Forest Service Rocky Mountain Research Station, Fort Collins, CO.

Friedman, J. M., W. R. Osterkamp, and W. M. Lewis, Jr. 1995. Restoration of riparian forest using irrigation, artificial disturbance, and natural seedfall. Environmental Management 19(4): 547–557.

Friedman, J. M., W. R. Osterkamp, W. M. Lewis Jr. 1996. The role of vegetation and bed-level fluctuations in the process of channel narrowing. Geomorphology 14: 341–351.

Haney, J. 2007. Rivers and water management in the Southwest. Southwest Hydrology 6(3):22–35.

Hansen, L., J. Hoffman, C. Drews, and E. Mielbrecht. 2010. Designing climate-smart conservation: Guidance and case studies. Conservation Biology 24:63–69.

Harrison, R. J. 2011. Bosque del Apache: A brief history. Friends of Bosque del Apache National Wildlife Refuge, San Antonio, NM.

Horton, J. S. 1960. Use of a root plow in clearing tamarisk stands. Research Note 50. United States Department of Agriculture, Rocky Mountain Forest and Range Experiment Station, Fort Collins, CO.

Hummel, O. C., and T. Caplan. 2007. Saltcedar control and restoration—be careful with generalizations. Southwest Hydrology 6(6):26–27.

Kernberger, C. 2005. What's "Saving our Bosque" all about? Save Our Bosque Task Force, Socorro, NM.

Mahoney, J. M., and S. B. Rood. 1998. Streamflow requirements for cottonwood seedling recruitment-an integrative model. Wetlands 18(4):634–645.

McDaniel, K. C., and J. P. Taylor. 2003. Saltcedar recovery after herbicide-burn and mechanical clearing practices. Journal of Range Management 56(5):439–445.

Millar, C. I., N. L. Stephenson, and S. L. Stephens. 2007. Climate change and forests of the future: Managing in the face of uncertainty. Ecological Applications 17: 2145–2151.

Molles, M. C. Jr., C. S. Crawford, and L. M. Ellis. 1995. Effects of an experimental flood on litter dynamics in the middle Rio Grande riparian ecosystem. Regulated Rivers: Research and Management 11(3–4):275–281.

Molles, M. C. Jr., C. S. Crawford, L. M. Ellis, H. M. Valett, and C. N. Dahm. 1998. Managed flooding for riparian ecosystem restoration. Bioscience 48(9):749–756.

Mueller, R. C., C. M. Scudder, M. E. Porter, R. T. Trotter III, C. A. Gehring, and T. G. Whitham. 2005. Differential tree mortality in response to severe drought: Evidence for long-term vegetation shifts. Journal of Ecology 93: 1085–1093.

Naiman, R. J., H. Decamps, and M. Pollock. 1993. The role of riparian corridors in maintaining regional biodiversity. Ecological Applications 3(2):209–212.

Parametrix. 2008. Restoration analysis and recommendations for the San Acacia reach of the Middle Rio Grande, NM. Prepared by Parametrix. Albuquerque, NM.

Poff, N. L., J. D. Allan, M. B. Bain, J. R. Karr, K. L. Prestegaard, B. D. Richter, R. E. Sparks, and J. C. Stromberg. 1997. The natural flow regime: A paradigm for river conservation and restoration. BioScience 47:769–784.

Richter, B. D., and H.E. Richter. 2000. Prescribing flood regimes to sustain riparian ecosystems along meandering rivers. Conservation Biology 14(5): 1467–1478.

Ritter, A. M. 1993. Channel dynamics in the context of cottonwood recruitment on the Middle Rio Grande, New Mexico. Master's thesis. Colorado State University, Fort Collins.

Rood, S. B., G. M. Samuelson, J. H. Braatne, C. R. Gourley, F.M. R. Hughes, and J. M. Mahoney. 2005. Managing river flows to restore floodplain forests. Frontiers in Ecology and the Environment 3:193–201.

Rowland, D, A. A. Sher, and D. L. Marshall. 2004. Cottonwood population response to salinity. Canadian Journal of Forest Research 34: 1458–1466.

Samani, Z. A. S. Bawazir, R. K. Skaggs, M. P. Bleiweiss, A. Pinon, and V. Tran. 2007. Water use by agricultural crops and riparian vegetation: An application of remote sensing technology. Journal of Contemporary Water Research and Education 137:8–13.

Scott, M. L., J. M. Friedman and G. T. Auble. 1996. Fluvial process and the establishment of bottomland trees. Geomorpholoogy 14:327–339.

Seavy, N. E., T. Gardali, G. H. Golet, F. T. Griggs, C. A. Howell, R. Kelsey, S. L. Small, J. H. Viers, and J. F. Weigand. 2009. Why climate change makes riparian restoration more important than ever: Recommendations for practice and research. Ecological Restoration 27(3):330–338.

Segelquist, C. A., M. L. Scott, and G.T. Auble. 1993. Establishment of *Populus deltoides* under simulated alluvial groundwater declines. American Midland Naturalist 130:274–285.

Shafike, N., S. Bawazir, and J. Cleverly. 2007. Native versus invasive: Plant water use in the middle Rio Grande basin. Southwest Hydrology 6(6):28–29.

Shafroth, P.B., J.M. Friedman, and L.S. Ishinger. 1995. Effects of salinity on establishment of *Populus fremontii* (cottonwood) and *Tamarix ramosissima* (saltcedar) in southwestern United States. Great Basin Naturalist 55:58–65.

Shafroth, P. B., J. R. Cleverly, T. L. Dudley, J. P. Taylor, C. van Riper III, E. P. Weeks, and J. N. Stuart. 2005. Control of *Tamarix* in the western United States: Implications for water salvage, wildlife use, and riparian restoration. Environmental Management 35:231–246.

Sheets, K. R., J. P. Taylor, and J. M. H. Hendricks. 1994. Rapid salinity mapping by electromagnetic induction for determining riparian restoration potential. Restoration Ecology 2:242–246.

Sher, A. A., and D. L. Marshall. 2003. Competition between native and exotic floodplain tree species across water regimes and soil textures. American Journal of Botany 90: 413–422.

Sher, A. A., D.L. Marshall, and J. Taylor. 2002. Spatial partitioning within southwestern floodplains: Patterns of establishment of native *Populus* and *Salix* in the presence of invasive, non-native *Tamarix*. Ecological Applications 12:760–772.

Sher, A. A., D.L. Marshall, and S.A. Gilbert. 2000. Competition between native *Populus deltoides* and invasive *Tamarix ramosissima* and the implications for reestablishing flooding disturbance. Conservation Biology 14:1744–1754.

Sher, A. A., and L. H. Hyatt. 1999. The Disturbance-Invasion Matrix: A new framework for predicting plant invasions. Biological Invasions 1(3–4):109–114.

Sogge, M. K., S. J. Sferra, and E. H. Paxton. 2008. Saltcedar as habitat for birds: Implications to riparian restoration in the Southwest. Restoration Ecology 16:146–154.

Sprenger, M. D., L. M. Smith, and J. P. Taylor. 2001. Testing control of saltcedar seedlings using fall flooding. Wetlands 21:437–441.

Stewart, I. T., D. R. Cayan, and M. D. Dettinger. 2005. Changes toward earlier streamflow timing across Western North America. Journal of Climate 18: 1136–1155.

Stromberg, J. C. 1998. Functional equivalency of saltcedar (*Tamarix chinensis*) and Fremont cottonwood (*Populus fremontii*) along a free-flowing river. Wetlands 18(4):675–686.

Stromberg, J. C., D.T. Patten and B.D. Richter. 1991. Flood flows and dynamics of Sonoran riparian forests. Rivers 2(3):221–235.

Stuever, M. C. 2000. Bosque fires and fire mortality of cottonwood. Pages 30–43 *in* J. P. Taylor, editor. *Proceedings from the Conference on Fire in Riparian Areas sponsored by the Middle Rio Grande Bosque Initiative*. Albuquerque, NM.

Stuever, M. C. 1997. Fire induced mortality of Rio Grande cottonwood. Master's thesis. University of New Mexico, Albuquerque.

Swenson, E. A., and C.L. Mullins. 1985. Revegetating riparian trees in southwestern floodplains. Pages 135–138 *in* R. R. Johnson, C. D. Ziebell, D. R. Patton, P. F. Ffolliott, and R.H. Hamre, editors. Riparian Ecosystems and Their Management: Reconciling Conflicting Uses. U.S. Forest Service. General Technical Report RM-120. Albuquerque, NM.

Taylor, J. P., D. B. Wester, and L. M. Smith. 1999. Soil disturbance, flood management, and riparian woody plant establishment in the Rio Grande floodplain. Wetlands 19(2):372–382.

Taylor, J. P., and K. C. McDaniel. 1998. Riparian management on the Bosque del Apache National Wildlife Refuge. New Mexico Journal of Science 38:219–232.

Taylor, J. P., L. M. Smith, and D. M. Haukos. 2006. Evaluation of woody plant restoration in the Middle Rio Grande: Ten years after. Wetlands 26(4):1151–1160.

Tetra Tech, Inc. 2004. Conceptual restoration plan, active floodplain of the Rio Grande, San Acacia to San Marcial, New Mexico. Save Our Bosque Task Force, Socorro, NM.

22

Bring on the Beetles!

THE HISTORY AND IMPACT OF TAMARISK
BIOLOGICAL CONTROL

Dan Bean, Tom Dudley, and Kevin Hultine

On a hot July afternoon in 2007, we watched small yellow- and black-striped beetles flying up into the wind and being carried upstream along the Dolores River in Utah, near the confluence with the Colorado River. The surroundings were spectacular: deep-cut canyons of red and beige sandstone and green ribbons of tamarisk, willows, and an occasional cottonwood lining the river. The insects weren't spectacular though; they were lazy fliers and appeared to have no direction except to be taken where the wind blew them. By mid-August beetles had moved far upstream, defoliating all tamarisks from the confluence of the Dolores to the Colorado state line 30 kilometers away. It was one of the signs of success for a tamarisk biological control program that had been over 20 years in the making (DeLoach et al. 2004; Carruthers et al. 2008). At the same time, it was an unsettling display of the capacity of a small, herbivorous insect to change western North America's riparian ecosystems, much as tamarisk itself already had (Hultine et al. 2010a). Most biological control programs fly under the radar of the popular press and the public and are important to only farmers, ranchers, and weed managers, but the tamarisk biocontrol program has developed a much larger audience. The beetles have become controversial even as they take their place among the most spectacular of biological control success stories (Van Driesch et al. 2010; see also chapter 1, this volume). This chapter is about the beetles and their biology and does not address the controversies surrounding their use (see chapters 11, 17, and 25, this volume). Our hope is that by understanding the biology of tamarisk leaf beetles we can have more productive discussions concerning their current and future impact on riparian ecosystems, and their use as a tamarisk management tool.

Background

A BRIEF HISTORY OF WEED BIOLOGICAL CONTROL

Biological control, or "biocontrol," is the use of natural organisms such as predators, herbivores, and pathogens to control pest organisms (Bellows and Fisher 1999). Biocontrol depends on the establishment of new ecological relationships. The balance between the pest and the biocontrol agent is determined by a complex array of factors including the biology of pest and agent, the interaction with other of organisms in the system such as predators or pathogens, and the impact of abiotic factors such as weather, soil chemistry, water availability, elevation, and latitude. The goal of biocontrol is to maximize suppression of the pest (the target) while minimizing negative effects on other organisms (nontargets) and the environment. By achieving this goal the modern practice of biocontrol has become a safe, sustainable, and inexpensive pest control strategy.

Weed biocontrol in North America started in the mid-1940s with the introduction of the *Chrysolina* beetles to control an introduced rangeland weed, *Hypericum perforatum* (Saint John's wort, or Klamath weed; Huffaker and Kennett 1959). The results were a spectacular reduction of *H. perforatum* populations, returning tens of thousands of acres of rangeland from a nearly useless condition back to grazing land, and at a minimal cost. The success of the project stimulated the development of other weed biocontrol projects in North America. Some have been very successful and some not (for a summary of weed biocontrol in the United States, see Coombs et al. 2004).

The safety of weed biocontrol came under question from within the biocontrol community and from the broader community of ecologists who were concerned that introduced organisms were not being properly evaluated prior to release into the environment (Louda et al. 2003). There were a few notable instances where an introduced biocontrol agent was not host-specific enough; they were able to feed and develop on nontarget plants in the field. The most notorious of these is the thistle seed head weevil *Rhinocyllus conicus*, released against musk thistle (*Carduus nutans*), despite having a broad host range enabling it to feed on nontarget native thistles (Louda et al. 2003). Since that time there have been changes in attitudes toward native species, changes in regulatory procedures and the development of a best practices code within the biocontrol community (Balciunas 2000) all of which drastically reduce the possibility of "another *R. conicus*" in modern weed biological control. In fact the practice of weed biological control has been remarkably safe and effective (Pemberton 2000) and the successes worldwide are impressive (Van Driesch et al. 2010).

SELECTION OF TAMARISK AS A BIOCONTROL TARGET

A weed must be nonnative, have serious and widespread economic and ecological impact, and be resistant to other control methods before it is targeted for biological control. It is best if the weed has no close native relatives in North America, especially congeners (Pemberton 2000). Tamarisk fits these criteria well. There are no North American native plants in the tamarisk family (Tamaricaceae), tamarisk is capable of rapid spread and difficult to control (Dudley et al. 2000; DeLoach et al. 2000; Shafroth et al. 2005) and is widespread throughout the western United States (Friedman et al. 2005). The tamarisk biological control program was begun by Lloyd Andres, Robert Pemberton, and their colleagues at the US Department of Agriculture Agricultural Research Service (USDA ARS) in the 1970s; the program was further developed in the 1980s by Jack DeLoach and his colleagues, also of the USDA ARS. It began with gathering information on tamarisk and the insects that feed on it in the native range of the plant, which stretches from West Africa to Central China (DeLoach et al. 2003; Tracy and Robbins 2009). DeLoach and colleagues, including cooperating scientists from many countries, compiled a list of about 300 insects and mites that feed on tamarisk, and selected a few that appeared to be host specific and that seemed promising as potential biological control agents (DeLoach et al. 2003).

SELECTION OF THE TAMARISK BEETLE

Robert Pemberton (USDA ARS) saw beetles feeding on tamarisk in Mongolia, and he was convinced that this insect, the tamarisk beetle *Diorhabda elongata*, family Chrysomelidae (leaf beetles), could do substantial damage to tamarisk and was worth pursuing as a biocontrol agent. The beetle, along with several other insects found on tamarisk in the native range, was selected for further evaluation, including taking a close look at the host range of the insect in the native range. If a biocontrol agent is specific for the target plant, then it will be found on that plant alone in the native range. Insects are rejected as potential biocontrol agents if they are found to be generalist feeders; that is, if they feed on nontarget plant taxa in the native range. *D. elongata* was found on plants in the genus *Tamarix* and the closely related genus *Myricaria* and no others in the native range. It was known that *D. elongata* were closely associated with tamarisk along rivers and streams in the native range (Lopatin 1977), which is a good indication that they are coevolved specialists. *D. elongata* was imported into the United States and held in quarantine at USDA facilities first in Temple, Texas, starting in the early 1990s, and then at both the Temple facility and the USDA ARS quarantine facility in Albany, California, in the late 1990s (DeLoach et al. 2003). Extensive research then followed, to elucidate the beetle's biology, taxonomy, and host specificity in its new range.

Biology of the Tamarisk Beetle

AN INTRODUCTION TO THE TAMARISK BEETLE

When an insect is used as a biocontrol agent, many aspects of its biology are more closely scrutinized than they would otherwise be. For weed biocontrol agents, this always includes numerous studies of host-plant use. Research may also include a detailed analysis of development, behavior, physiology, and taxonomy. This was the case with *D. elongata*, which is much better understood now than when it was first selected as a tamarisk biocontrol agent.

The tamarisk beetle goes through three caterpillar-like stages (instars), and during each instar it clings to the host plant and feeds almost constantly, even at night if temperatures are warm. The first two instars are dark bodied while the third and final larval instar is dark with yellow stripes running longitudinally down each side of the insect (see figure 22.1). When the third instar is finished feeding, it descends into the leaf litter, where it glues a loose collection of plant material and sand around itself, forming a casing within which it pupates. The pupa is delicate and pale yellow; it is easily damaged if removed from the protective casing. Within about 12 days at summer temperatures the adult is formed, remaining within the pupal case for about a day while it darkens and hardens (sclerotization) and acquires the distinctive black stripes and spots that characterize a mature adult. The adult climbs out of the leaf litter and up into the canopy, where it feeds on foliage for about four days before becoming reproductively mature (Bean et al. 2007b). Although the larval stages are the most destructive of tamarisk, the adults inflict some damage on the plants; adults are thought to inflict less than 20% of the overall herbivory. Reproductive maturity is marked by development of oocytes within the ovarioles of females and the development of the accessory glands in males (Bean et al. 2007b). Once reproductive maturity is reached, insects mate many times. Females lay eggs either singly or, more often, in clusters of about 5 to 15 eggs that are deposited on the host plant. Hatchling larvae

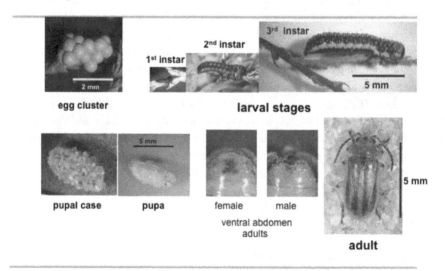

FIGURE 22.1 Life stages of the tamarisk beetle (*Diorhabda* spp).
Source: Photos courtesy of Dan Bean.

immediately feed and begin to grow. The entire life cycle takes about 30 to 40 days (Lewis et al. 2003a).

The tamarisk beetle life cycle shifts with the approach of fall. Adult beetles cease reproducing and instead enter dormancy known as "diapause" (Bean et al. 2007b). Insect diapause is a genetically determined dormant state marked by a buildup of metabolic reserves, changes in behavior, and the acquisition of freeze tolerance in temperate climates (Beck 1980). Diapause often begins in response to environmental cues such as photoperiod and temperature indicating that stressful conditions will soon prevail (Beck 1980; Saunders 2002). This is the case with the tamarisk beetle: the shortening day lengths of late summer and fall are used as a signal to enter diapause prior to the onset of winter (Lewis et al. 2003a; Bean et al. 2007a,b). Tamarisk beetles' behavior changes as they enter diapause. Instead of flying from plant to plant, adult beetles descend into the leaf litter beneath plants, where they overwinter (Bean et al. 2007). In spring they emerge from the litter, begin feeding, and then climb to the ends of branches where using their antennae they sense pheromones and host-plant chemicals and move to join aggregations of reproductive beetles (Cossé et al. 2005, 2006). Soon thereafter they mate and lay eggs. Beetles usually produce two or more generations per season, with the decreasing photoperiod in late summer acting as a cue for the cessation of reproductive development and entry into diapause.

TAXONOMY OF TAMARISK-FEEDING *DIORHABDA*

The extant taxonomy of a biocontrol agent has often been constructed using external features that do not distinguish closely related and morphologically similar species (cryptic species) differing in key characteristics that may be important to their use in biocontrol, such as host-plant specificity (Toševski et al. 2011). The taxonomic revision of the tamarisk-feeding *Diorhabda* is another excellent example of a study clarifying our understanding of a biocontrol agent, enabling more effective use of the agent.

The beetles used in the first host range studies, caged studies, and for the first experimental releases were classified as *Diorhabda elongata deserticola* and were collected in central Asia (DeLoach et al. 2003). It became apparent in 2001 soon after the first field releases of *D. elongata* in the United States that different ecotypes (genetically distinct populations adapted to different sets of environmental conditions) of *D. elongata* would be required to match climate and host-plant species, as well as other ecological features across the vast geography of *Tamarix* distribution in North America (DeLoach et al. 2004). Fortunately *D. elongata* was found in association with *Tamarix* across a wide swath comprising three continents stretching from West Africa to eastern Asia (DeLoach et al. 2004; Tracy and Robbins 2009; see figure 22.2). Throughout their range *D. elongata* forms were known to be specialists on the genus *Tamarix* except in the case of *D. e. deserticola* which was also known to feed on members of the related genus *Myricaria* (Tracy and Robbins 2009).

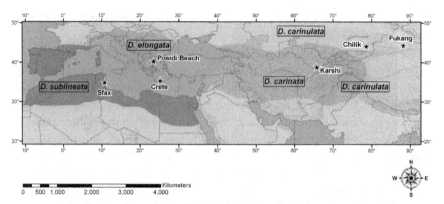

FIGURE 22.2 Distribution of tamarisk-feeding *Diorhabda* species in Eurasia and North Africa. Stars indicate collection sites.

Source: Adapted from Tracy and Robbins (2009).

In 2002, additional collections of *D. elongata* were made from many sites in Eurasia and North Africa (DeLoach et al. 2004; Milbrath and DeLoach 2006a) and processed through the USDA ARS quarantine facility in Albany, California. When the first new ecotype of *D. elongata,* from the island of Crete, was brought into the quarantine, we noticed that it appeared to be slightly larger and more greenish than the Central Asian (Fukang) ecotype. Our first studies of the Crete ecotype showed that the cues for diapause induction were different from those measured in the original Fukang ecotype (Dudley et al. 2006; Dalin et al. 2010).

Other ecotypes were subsequently collected and brought into quarantine at the Albany facility. These included populations from near Sfax, Tunisia; Posidi, Greece; Karshi and Buchara, Uzbekistan; and Turpan, China (Milbrath and DeLoach 2006a). Among these ecotypes there were variations in the photoperiodic induction of diapause (Dudley et al. 2006; Dalin et al. 2010), in overwinter survival, and possibly in the timing of diapause break in the spring (Milbrath et al. 2007), variations in host-plant selection and use (Milbrath and DeLoach, 2006a, b; Dalin et al. 2009; Herr et al. 2009; Dudley et al. 2012), and in the number of generations per season (Hudgeons et al. 2007; Milbrath et al. 2007; Dalin et al. 2010). These characteristics were important in matching beetle ecotypes with geographic regions in North America, and specific targets in the genus *Tamarix* (Carruthers et al. 2006; Dalin et al. 2009; Tracy and Robbins 2009), which made them valuable in the tamarisk biocontrol program. Some of these ecotypes have established in areas where the original *D. e. deserticola* failed (Hudgeons et al. 2007; Carruthers et al. 2008), and establishment has been correlated with the physiological characteristics that vary among ecotypes (Milbrath et al. 2007; Dalin et al. 2009; 2010). A taxonomic revision of the tamarisk-feeding members of the genus *Diorhabda* became necessary in the context of the biocontrol program.

The tamarisk-feeding beetles in the genus *Diorhabda* were previously recognized as a single species with three or four subspecies, although some authors have disagreed with this taxonomic hierarchy (reviewed in Tracy and Robbins 2009). As a consequence of the tamarisk biocontrol program, James Tracy and Tom Robbins of the USDA ARS in Temple, Texas, began a major study of the morphology, taxonomy, and biogeography of *Diorhabda elongata* that included detailed morphological analysis of reproductive structures (Tracy and Robbins 2009). They divided the tamarisk-feeding *Diorhabda* into five species, based primarily on concealed reproductive structures of both males and females (Tracy and Robbins 2009).

What was formerly *Diorhabda elongata* is now classified as a species complex consisting of five tamarisk feeding specialists. The biological control program has used four of these, and three of them have become established in North America (see figure 22.3). This includes two populations of the Central Asian species *Diorhabda carinulata* (northern tamarisk beetle) that have become widespread in Colorado, Utah, Nevada, and Wyoming and have established in Oregon and Idaho. Populations have also moved into northern New Mexico and northern Arizona. *Diorhabda carinata* (larger tamarisk beetle) was originally collected in Uzbekistan and has been released in Texas but is not well established, possibly due to early spring emergence, which makes it susceptible to late freezes (Milbrath et al. 2007). *Diorhabda elongata* (Mediterranean tamarisk beetle) has been released in Texas and California and is established at multiple sites in Texas (Hudgeons et al. 2007, Milbrath et al. 2007) and in the Cache Creek drainage of central California (Carruthers et al. 2006). *Diorhabda sublineata* (subtropical tamarisk beetle) was collected in Tunisia, released in Texas along the Rio Grande in 2009, and has established there, defoliating up to 100 kilometers of riverside tamarisk as of 2010, but may have physiological limitations preventing northward expansion (Milbrath et al. 2007).

DIORHABDA HOST RANGE AND SAFETY

Safety is one of the first considerations in the process of selecting and evaluating potential biocontrol agents (Balciunas 2000). Safety testing requires the evaluation of the host specificity of the potential agent; that is, will it feed and develop on plants other than the plant intended for control (the target)? All other plants are considered nontargets and may include agricultural crops, native plant species, and nonnative species that have ornamental value. If the candidate agent is shown to feed on a nontarget, the question becomes would those nontarget plants be at risk under field conditions? If so, further testing is done to evaluate risk and decisions are made based on the results (e.g., Lewis et al. 2003b; Dudley and Kazmer 2005). The procedures for evaluating safety of a potential agent have been described in detail by several authors (e.g., Coombs et al. 2004) and the host range testing for *Diorhabda* was carefully planned under contemporary standard procedures (DeLoach et al. 2003).

Risk of damage to nontarget plants is highest with plants that are most closely related to the target species (Pemberton 2000), and so testing focused on the closest relatives of *Tamarix*, in addition to an array of more distantly related plants, including

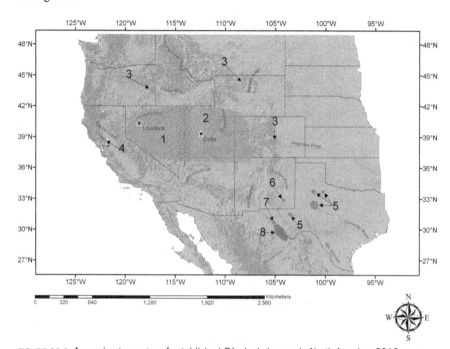

FIGURE 22.3 Approximate ranges of established *Diorhabda* spp. in North America, 2010: areas 1, 2, and 3 = *D. carinulata*, areas 4, 5, 6 and 7 = *D. elongata*, area 8 = *D. sublineata*. Area 1 represents the distribution of *D. carinulata*, Fukang ecotype, that came from release sites near Lovelock, Nevada. Area 2 represents the distribution of *D. carinulata* (Chilik ecotype) coming from an initial release site near Delta, Utah. Cross-hatching between areas 1 and 2 indicates a likely hybridization zone between ecotypes. More-localized established populations are indicated by 3 and are found in eastern Oregon, northern Wyoming, and central Colorado along the Arkansas River. There are other known smaller areas of establishment in Wyoming and South Dakota, which are not shown. Area 4 is an established population of *D. elongata* (Crete ecotype) along Cache Creek, central coast range of California. This population is notable since it has established on *T. parviflora*. Areas denoted by 5 are *D. elongata* established along several drainages in west Texas, and 6 shows a single site in eastern New Mexico. Area 7 denotes a small establishment of *D. elongata* along the Rio Grande; area 8 indicates the approximate boundaries of *D. sublineata* establishment along the Rio Grande.

Sources: Approximate boundaries of known established populations were drawn using Arc Map 10.0 (Esri, Redlands, CA). The approximate boundaries for *D. carinulata*, areas 1 and 2, were obtained in part from the Tamarisk Coalition (www.tamariskcoalition.org) and from T. Dudley. Establishment of *D. elongata* along Cache Creek in California has been described (Carruthers et al. 2008), and current boundaries were obtained from John Herr (USDA ARS, Albany). Establishment of *D. carinulata* in eastern Oregon was verified by Eric Coombs at the Oregon Department of Agriculture, Salem, Oregon. Establishment of *D. sublineata* and *D. elongata* on the Rio Grande was verified by Chris Ritzi, Sul Ross University, Alpine, Texas. Establishment of *D. elongata* in western Texas was described by Hudgeons et al. (2007), and further spread was verified by Allen Knutson at the Texas AgriLife Research and Extension Center, Dallas, Texas. Establishment of *D. elongata* in New Mexico was verified by Kevin Gardner and David Thompson, New Mexico State University, Las Cruces.

habitat associates and agriculturally and economically important nontarget species (Wapshere 1974; Balciunas 2000; DeLoach et al. 2003). The first round of evaluation was done using no-choice tests in which larvae were placed on the test plant with no alternatives. No-choice tests are considered the most stringent tests of host range since a negative no-choice result means that insects starve in the presence of green plant material—an emphatic sign of rejection. In no-choice tests the only plants that supported *Diorhabda* were in the genera *Tamarix*, *Myricaria*, and *Frankenia* (DeLoach et al. 2003; Lewis et al. 2003b). The genus *Frankenia* has members native to North America, and so was selected for further testing (Lewis et al. 2003b).

The genus *Frankenia* is in the family Frankeniaceae, which shares the order Tamaricales with the Tamaricaceae, making members of the genus *Frankenia* the closest North American relatives of the genus *Tamarix* (Spichiger and Savolainen 1997; DeLoach et al. 2003). It wasn't surprising that this was the only North American genus that could support feeding and also be a suitable substrate for egg-laying (oviposition) by a *Tamarix* specialist. The species *F. salina* and *F. palmeri* supported larval development to adulthood and were considered physiologically acceptable host plants, although not optimal (DeLoach et al. 2003; Lewis et al. 2003b). The risk to *Frankenia* was then tested under conditions in which insects had a choice between *Frankenia* and other plants and, finally, under field conditions where insects may encounter *Frankenia* in a natural setting.

Since host selection is made primarily by females during choice of oviposition sites, the next round of tests was designed to measure the acceptability of *Frankenia* as a substrate for oviposition. In these tests caged beetles oviposited on *Frankenia* but at a significantly lower rate than on *Tamarix* (Lewis et al. 2003b). In an open field study oviposition and feeding damage to *Frankenia* were measured at two widely separate field sites where *Frankenia* spp. plants were exposed to very high *Diorhabda* populations in the process of defoliating tamarisk. Despite the high beetle populations no eggs were laid on *Frankenia* and feeding damage to *Frankenia* was very low (less than 4% leaf damage compared to tamarisk less than two meters away, which was almost completely defoliated). These experiments led the researchers to conclude that the risk to *Frankenia* spp. in the field was very low (Dudley and Kazmer 2005). This is a classic example, common in weed biological control, where there is some risk to nontarget species but it is small enough; and the potential benefits of the program, large enough, that the project is allowed to continue.

When new collections of *Diorhabda elongata sensu lato* were imported for use in the biocontrol program, they were tested using the same procedures used for the original populations, even though at that time they were considered to be a single species (DeLoach et al. 2003; Milbrath and DeLoach 2006a; Herr et al. 2009). The ecotypes (some now separate species) all showed a similar pattern; host preference for *Tamarix* but with some oviposition and development possible on *Frankenia*, especially *F. salina*.

DIORHABDA PHENOLOGY

Phenology is the study of the seasonal timing of life history events such as flowering or bud break in plants, and reproduction, migration and diapause in insects. The correct seasonal timing of life history events is critical for success of a population and mistiming can be fatal. Growth, development and reproduction must occur during favorable times, when temperature, moisture, and other ecological factors allow it, and dormancy must be timed to enable avoidance of harsh conditions such as freezing temperatures or a lack of moisture. In herbivorous insects the timing of life history events is also tied to host-plant phenology. If the insect emerges from dormancy too soon or enters dormancy too late in the season it will starve. Likewise, if it remains dormant when the host plant is available, it will not be able to optimize resources. The tamarisk beetle presents an interesting example of phenology mismatches in new habitats in North America.

Tamarisk beetles collected in Central Asia were genetically encoded to respond to the photoperiods, temperatures, and tamarisk phenology encountered there. Both Central Asian collection sites are at northern latitudes (44°N for Fukang, China; 43°N for Chilik, Kazakhstan) with continental climates. The first releases of the northern tamarisk beetle *D. carinulata* failed to establish in Texas, at latitudes of 33°N and 29°N (Lewis et al. 2003a). Beetles ceased reproduction early in the summer and failed to overwinter, although they became reproductively active when brought into the laboratory under constant light (Lewis et al. 2003a). The early cessation of reproduction and the failure to establish were thought to be the result of a mismatch between insects adapted to central Asia and the environmental conditions they encountered in Texas. Early-summer day lengths are nearly 1.5 hours shorter at the Texas field sites than in Fukang, China, which could cause insects to misread the season and enter diapause prematurely (Lewis et al. 2003a; Bean et al. 2007a). To test this hypothesis and better predict where and when beetles would be reproductive, additional measurements of their developmental responses to photoperiod and temperature were made.

The developmental response of insect populations to day length can be quantified and expressed as critical day length (CDL) for diapause induction, which is the day length at which 50% of individuals in a population enter diapause (Beck 1980; Tauber et al. 1986; Saunders 2002; see figure 22.4). The CDL for the original Fukang population of *D. carinulata* was measured under controlled laboratory conditions by rearing beetles at different temperatures and photoperiods and evaluating reproductive status. In a separate experiment, beetle populations were sampled at regular intervals throughout the spring and summer at sites located along a latitudinal gradient. Field-collected beetles were evaluated for reproductive status (diapause vs. reproductive), which was plotted against day lengths 13 days prior to the field-site collection dates. The 13-day subtraction was necessary to account for the timing of the developmental processes leading to the expression of diapause traits in *D. carinulata* (Bean et al. 2007a). The CDL measured in the field populations averaged 14 hours 39 minutes and varied by only five minutes across five field sites ranging from 37°N to 44°N latitude (Bean et al. 2007a). Under controlled conditions at an average temperature of 25°C, the CDL was slightly longer at 14 hours 53 minutes (Bean et al. 2007a) and varied between 15 hours 8 minutes and 14 hours 50 minutes at temperatures between

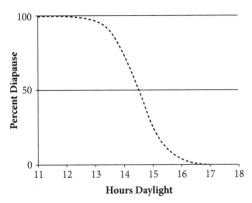

FIGURE 22.4 Developmental response of a hypothetical insect population to day length. The day length at which half of the population will enter diapause is 14.5 hours.

22°C and 31°C. The temperature effect on CDL was small compared with those measured in other insect species (cf. Tauber et al. 1986). When day lengths fell below about 14 hours 40 minutes at field sites across North America, the tamarisk beetle populations ceased reproducing and entered diapause, and at some sites this happened during the hottest days of summer, when the likelihood of survival through fall and winter to emerge in the spring was low. Because of latitudinal variation in day length patterns this put more southern beetle populations out of synchrony with the new environments into which they had been introduced.

Day lengths are a function of latitude; northern hemisphere day lengths increase until June 20, and then decrease until December 21. As a consequence the CDL for diapause in *D. carinulata* occurs earlier in the summer at more southern latitudes. For instance, the day length was 14 hours 39 minutes on July 3 at the Bishop, California, field site (37° 05'), and 50% diapause was measured 13 days later, on July 16. At a higher latitude site near Lovell, Wyoming (44° 50'), day length was 14 hours 38 minutes on August 2, and 13 days later, on August 15, the *D. carinulata* population reached 50% diapause (Bean et al. 2007a). The timing of diapause induction in the field occurred earlier in the season at more southern locations; whereas the onset of inhospitable temperatures and the senescence of tamarisk occur later in the season at more southern locations (Dalin et al. 2010). Therefore, beetle phenology became more out of synchrony with tamarisk phenology, the more southward the beetles were introduced. The asynchrony became more severe as a function of decreasing latitude until at some southern sites, such as the Texas sites, or the Bishop, California, site, beetles failed to establish self-sustaining populations. This was a problem for the biocontrol program since the largest stands of tamarisk were at southern locations, where beetle phenology was unsuitable. There were two possible solutions: first, import beetles adapted to more southern climates and second, wait for natural selection to act on the CDL of established *D. carinulata* populations, driving evolution of beetle phenology toward an optimal state for more southern areas.

Less than a year after the 2001 open field releases of the northern tamarisk beetle, *D. carinulata*, other more southern-adapted *Diorhabda* were collected from North Africa and Eurasia and tested as tamarisk biocontrol agents (DeLoach et al. 2004). *D. elongata*, collected from the island of Crete, Greece (35.8°N) and known as the Crete ecotype of *D. elongata*, has become established at several locations in Texas (Hudgeons et al. 2007a; Carruthers et al. 2008). The Crete ecotype of *D. elongata* has been shown to remain reproductive late into the season (September) at latitudes south of 36°N where the original releases of *D. carinulata* failed (Milbrath et al. 2007; Dalin et al. 2010). The CDL of the Crete ecotype of *D. elongata* was 13 hours 16 minutes, about 1.5 hours shorter than the CDL for the Fukang ecotype of *D. carinulata* (Dalin et al. 2010), which accounted for the late season reproductive activity measured in this ecotype. The subtropical tamarisk beetle, *D. sublineata*, has been released along the Rio Grande in southern Texas, where it is established and has defoliated tamarisk along 100 miles of the river (figure 22.3). *D. sublineata* has a CDL of 12 hours 50 minutes (Dudley et al. 2006; Bean unpub. data) which allows populations to remain reproductive into late summer, even in southern Texas.

D. carinulata (Fukang ecotype) was introduced into the field in 2001, and by 2005, the beetles were reproducing later in the season at two widely separate field sites; Lovelock, Nevada (Dudley pers. obs.), and Pueblo, Colorado (Debra Eberts, US Bureau of Reclamation, pers. obs.). At the Pueblo field site larvae were found late into August where formerly beetles had ceased reproduction and larvae were gone by the beginning of August. At the Lovelock site larvae were observed in early September where prior to 2005 they had not been seen past mid-August. The shift in *Diorhabda* phenology suggested adaptation of the critical day length for diapause induction. A reevaluation of CDL in these and three other populations showed that CDL had decreased by as much as 54 minutes in field populations of *D. carinulata* (Fukang ecotype; Bean et al. 2012). It has long been known that CDL can change over time (Beck 1980; Tauber et al. 1986), but the rate of evolution has been surprisingly rapid in *D. carinulata*. This has meant the successful establishment of beetles in more southern areas where they now coexist with an endangered bird subspecies, the southwestern willow flycatcher *Empidonax traillii extimus*, which is known to nest in tamarisk (Bateman et al. 2010). The change in CDL has made *D. carinulata* a more effective biological control agent (Dalin et al. 2010; Bean et al. 2012), but it has also brought into play conflicts between the tamarisk biocontrol program and the management of an endangered species (Bateman et al. 2010; Dudley and Bean unpub data; see also chapter 11, this volume).

In the spring northern tamarisk beetles emerge from the leaf litter and begin to feed immediately, soon after the appearance of the first tamarisk foliage, regardless of latitude or day length (Lewis et al. 2003a). Sensitivity of diapausing adult beetles to photoperiod ends during the early winter (Bean unpub. data), and after that, they are held in diapause by low temperatures. Emergence of adults in the spring coincides with the greening of tamarisk and the termination of diapause most likely requires several warm days to complete. This is a pattern often seen in temperate-zone insects (Beck 1980; Tauber et al. 1986), and it gives flexibility to the life cycle, allowing insects to become active as soon as rising temperatures permit. Although this is a common theme in insect phenology, work is still needed on the thermal requirements for spring emergence, so that it can be better predicted.

BEHAVIOR: HOW BEETLES CONGREGATE TO DEFOLIATE TAMARISK

During the first observations of *Diorhabda* population expansion in North America it was noted that when beetles fed on tamarisk, they did so in large numbers and often completely defoliated the plants (Pattison et al. 2010). In areas where beetles were active it was common to find thousands of adults on a single tree and subsequently to find the tree covered with larvae. Aggregation behavior is a key to the beetles' effectiveness in defoliating tamarisk, and chemically mediated communication plays a major role in this behavior. The underlying chemistry was explicated in the laboratories of Robert Bartelt and Allard Cossé, USDA ARS, Peoria Illinois, with the goal of better understanding beetle behavior as well as developing chemical baits for monitoring *Diorhabda* in the field.

D. *carinulata* were tested for sensitivity to chemicals released from other *D. carinulata* (pheromones), as well as for sensitivity to volatile compounds (kairomones) released by the host plant, tamarisk. The detection method used was gas chromatographic electroantennographic detection (GC-EAD) in which electrical impulses from the antennae were measured in response to chemical mixes, which were separated into individual components and blown gently over the antennae to mimic odors encountered in nature (see Cossé et al. 2005, 2006). Insect antennae are very sensitive to semiochemicals, chemicals used by the insect to alter behavior. The antennae of *Diorhabda* were no exception. Two compounds produced by male beetles got major electrical responses from male or female *D. carinulata* antennae (Cossé et al. 2005), and at least 15 compounds produced by tamarisk plants also got electrical responses from antennae (Cossé et al. 2006). However, field testing was needed to show if the compounds had any effect on beetle behavior in a natural setting.

The chemical compounds that had stimulated the antennae of *D. carinulata* were tested under field conditions in western Nevada. Small vials containing the compounds were punctured with a pin to slowly release the compounds and hung next to yellow sticky cards in tamarisk trees or shrubs (see figure 22.5). Insects attracted to the test compound(s) got stuck on the paper and were counted. Two male-produced antennally active compounds, (2E, 4Z)-2, 4-heptadienal and (2E, 4Z)-2, 4-heptadien-1-ol, attracted male and female beetles during the main flight period, from early afternoon until dusk (Cossé et al. 2005). Emission of a male-produced aggregation pheromone was considered unusual, but now there are a

FIGURE 22.5 David LaClergue counts *D. carinulata* adults trapped on a sticky card baited with a combination of GLVs and the *Diorhabda* pheromone blend. Beetles attracted to the odors alight on the sticky card and are trapped. Cards with no baits attract fewer than 10 beetles.

growing number of similar examples in chrysomelids, including a major agricultural pest, the Colorado potato beetle (Dickens et al. 2002), and in *Galerucella* spp., a biocontrol agent for the noxious weed purple loosestrife (Bartelt et al. 2008).

Field tests in Nevada also showed that four tamarisk-produced compounds were attractive to adult beetles, and that these were from a group of compounds known as green leaf volatiles ([GLVs]; Cossé et al. 2006). The GLVs smell like a newly mown lawn to the human nose; vials containing them attract thousands of adult beetles in the late spring before tamarisk leafs out, and again after beetles defoliate tamarisk in the summer. It seemed clear that GLVs signaled the presence of food to hungry beetles, but there is possibly more to the story, since feeding by beetles increased GLV emissions from foliage by ten- to twentyfold (Cossé et al. 2006). High levels of the GLVs may signal the presence of feeding adult beetles, which would in turn attract more beetles, turning GLVs into another aggregation signal for adult beetles. GLVs or other volatile plant compounds may also be involved in host-plant selection, which could explain the preference of *D. carinulata* adults for *T. ramosissima* over *T. parviflora* (Dalin et al. 2009; Dudley et al. 2012).

The combination of GLVs and pheromone blend was more attractive than the additive effect of each alone, suggesting a synergy between the semiochemicals (Cossé et al. 2006). We proposed that a combination of pheromone emission by males and GLV emission by plants on which beetles were feeding would provide the most powerful attractive signals for colonization (Cossé et al. 2006). Beetles captured in sticky traps baited with GLVs were mostly males, ranging from 77% to 96% of trap catch (Cossé et al. 2006). This caused us to postulate that males are more active in the search for suitable host plants. They alight on plants, sample the foliage, and either remain on that plant, where they feed and emit pheromones, or move on. If males remained on a plant the combination of male-produced phero-mones and GLVs would be a powerful attractant, calling in males and females alike, resulting in colonization, mating, egg laying, and eventual defoliation (Cossé et al. 2006).

Movement of adult beetles through the habitat is episodic and may take place on a small-scale (tens of meters) or a much larger-scale (tens of kilometers). These movements can be surprising considering the poor flight abilities of these insects (beetles appear to float into the breeze rather than show strong directed flight). There are times of day and times during the season when beetles are more likely to move short distances or to make long-range dispersal flights.

Short-range dispersal flights occur daily during the spring and summer, when beetles are reproductively active. Large swarms of adults fly to the tops of trees, particularly in the late afternoon, in what has been described as swarming. These movements are local, with beetles moving only a few meters or tens of meters at a time. Flight activity is greatest in the mid to late afternoon with mixed-sex swarms moving from tree-top to tree-top, breaking up and reforming dynamically until activity comes to a halt at sunset (Cossé et al. 2005). In the morning, beetles remain in the trees, where they form mating pairs, and they rarely fly until the afternoon. Trees that have such aggregations receive numerous egg masses; in many cases, they become defoliated within about two weeks as the larvae feed and mature. This behavioral pattern changes in the mid to late summer as beetles begin to enter diapause. Diapause-destined beetles avoid flying and are more likely to remain on the same green tamarisk trees until they complete feeding. Aggregations of adult beetles are often found together on one or a few trees, where they can be observed feeding but not mating. Once they are fully fed, they descend to the duff beneath the tamarisk where they overwinter.

Beetle behavior has also been studied on caged plants in a laboratory setting, under controlled temperature and photoperiod. When beetles were held under long days (16 hours), they showed dispersal activity in the afternoon and mating pair formation in the morning, the same as seen during the early summer in the field (Bean et al. 2007b). When beetles were held under 12-hour days, they rarely flew or left the plant and did not mate or lay eggs; after about 15 days they left the plants and crawled down into the leaf litter (Bean et al. 2007b). Males were shown to emit aggregation pheromone under 16-hour days, but they did not emit phero-mone when days were 12 hours (Bean et al. 2007b). These results indicate two dis-tinct photoperiodically controlled behavioral patterns: the pattern of reproductive adults and the pattern of diapause-destined adults.

Long-range dispersals are more difficult to track, and the long-range move-ment of individuals has not yet been measured, but we have observations suggesting that when beetle populations grow large, and resources become locally depleted, beetles will fly up into the wind currents and alight at distances several kilometers from the origin. Although we haven't observed the movement of the same individual beetle over long distances, we have measured the movement of *D. carinulata* populations over 40 kilometers in one season. We also have field observations from different areas and different seasons suggesting large-scale dispersal flights.

On a clear, warm afternoon at the Lovelock, Nevada, field site in early July of 2004 Bartelt and Bean were working in two different areas, separated by over two kilometers, when we witnessed a major dispersal event. Bartelt noted adult beetles taking flight from the tops of tamarisk trees and moving straight up into the air until they disappeared from view. I noted beetles "raining" out of the sky and alighting on some of the tamarisk trees near to where I was working, about two kilometers away. There were no large beetle populations in the area between us, so we presumed that the beetles flying from one site were the same ones landing at the other. In eastern Utah in 2007 we set up yellow sticky cards baited with pheromone on the top of a mesa that was over 1,500 feet higher than the nearest tamarisk plant. We captured over 20 adult *D. carinulata* that had flown up from the tamarisk thickets far below. These observations indicate long-distance dispersal behavior when resources become locally depleted, and they also indicate that dispersing beetles will end up distributed over the landscape. It seems likely that once dispersed, adult *Diorhabda* use both pheromone and kairomones signals to find and coalesce on isolated tamarisk, as has been seen with other leaf beetles (Grevstad and Herzig 1997; Bartelt et al. 2008). The isolated beetles will be attracted to a male or males that have found the tree and started emitting aggregation pheromone, as well as to the plants giving off GLVs (Cossé et al. 2006). This would lead to colonization of tamarisk in remote areas as dispersed beetles become reaggregated. It could also lead to "hot spots" where beetles aggregate on a few trees within large stands of tamarisk, beyond areas where main beetle populations are moving. Aggregations on remote tamarisk stands or individual trees are often observed, and "hot spots" within larger tamarisk stands are common, both here and in their native range in Asia.

The Tamarisk Beetle in North America

The tamarisk program is unique among weed biocontrol programs. Instead of simply releasing the insects and making them available to weed managers for use as quickly as possible (program implementation releases), in the tamarisk program the release of beetles took place in three stages. The tamarisk beetles were first released into secure cages (Dudley et al. 2001, then into the open at experimental sites where they were closely monitored and finally general implementation releases were made. The reasons and rationale for this approach are discussed elsewhere (DeLoach et al. 2003, 2004; Caruthers et al. 2008), but in short, the extra measures were taken out of concern for the habitat of the southwestern willow fly-catcher and because postrelease monitoring of the biocontrol impacts has become a high priority in biocontrol programs.

EXPERIMENTAL RELEASES

The northern tamarisk beetle was first released at 10 field sites, into large, secure field cages enclosing several tamarisk plants. These caged releases were made in 1998 and 1999 at seven sites, and in 2000 at three more. These releases enabled beetles to feed on tamarisk under natural conditions so that their ability to develop and survive the winter could be quantified. In spring 2001, beetles were taken from the cages and released into the open field at seven of the 10 sites (DeLoach et al. 2003, 2004). Three sites where the beetle populations had not done well were not used for open field releases. The seven open field release sites were named for nearby towns: Bishop, California; Delta, Utah; Pueblo, Colorado; Lovell, Wyoming; Seymour, Texas; Lovelock, Nevada; and Schurz, Nevada, (DeLoach et al. 2004). Beetles established in the open field (which means that they had growing populations that survived winter) at five sites but failed to establish at the Bishop, California, and Seymour, Texas, sites.

Beetles started to defoliate tamarisk trees at the Lovell, Wyoming, and Pueblo, Colorado, sites by midseason in 2002; the results were promising, but the first large-scale defoliation wasn't seen until the late summer 2002. In August 2002 we got word from Jeff Knight at the Nevada Department of Agriculture that "beetles were everywhere" and were turning plants from green to light brown in a tamarisk thicket in western Nevada. The site was near Lovelock, Nevada (figure 22.3), and was the first big tamarisk biocontrol success. Ray Carruthers (USDA ARS) from the Albany labs went out to photograph the beetles and see the damage to tamarisk first hand. He brought back stories of masses of black larvae dripping from the trees and more intense defoliation than we had expected. By the end of 2002, the beetles had defoliated the first hectare; by the end of 2003, 300 hectares; and by the end of 2004, about 10,000 hectares had been defoliated (DeLoach et al. 2004, Carruthers et al. 2008, Pattison et al. 2010). This expansion was dramatic and has become indicative of what beetles can do under optimal environmental conditions.

Similar beetle population expansions were seen at the Delta, Utah, and Schurz Nevada, sites, but they lagged behind the Lovelock site by about a year (Carruthers et al. 2008). At the Pueblo Colorado, and Lovell, Wyoming, sites, the defoliation of tamarisk has been far less expansive and dramatic. This has been attributed to predation of beetles by ants and other predators, and weather patterns (DeLoach et al. 2004, Carruthers et al. 2008). At the two southernmost experimental field-release sites, Bishop, California, and Seymour, Texas, beetles were never established, probably due to the shorter day lengths there (DeLoach et al. 2004; Bean et al. 2007a). At two of the California caged-release sites northern tamarisk beetles did not establish within the cages and were never released into the open. The tamarisk species at these sites is *T. parviflora,* which is not an optimal host for *D. carinulata* (Dalin et al. 2009), and host-plant mismatch was probably a factor in the failure of populations to establish (Dudley et al. 2012). The first experiments, caged- and open field releases of *D. carinulata,* were very instructive, helping to define the limitations of the species, and to inform the selection of other *Diorhabda* species to be used where *D. carinulata* had failed (DeLoach et al. 2004).

In Texas, where short day lengths were thought to inhibit establishment of *D. carinulata* (Lewis et al. 2003), the three other *Diorhabda* species were tested. All three have shorter CDLs than is measured in *D. carinulata* (Dudley et al. 2006, Dalin et al. 2010), and this translated to an increased period of reproductive activity and four to five generations per season instead of one generation, as is seen with *D. carinulata* (Milbrath et al. 2007). *D. sublineata* and *D. elongata* have become established at experimental field sites in Texas but *D. carinata* was thought to have failed to establish (Carruthers et al. 2008), probably due to high overwintering mortality (Milbrath et al. 2007). Although they went undetected for about five years, small populations of *D. carinata* survived and were discovered in west Texas in 2012 where they have now proliferated (J. Tracy, personal communication).

In California, *D. elongata* are now well established at an experimental release site where *D. carinulata* had failed (Carruthers et al. 2008). The site is on Cache Creek in the coast range where the tamarisk species is *T. parviflora.* The success of *D. elongata* at that site, where *D. carinulata* failed, may be a consequence of *T. parviflora* being a better host match for *D. elongata* than for *D. carinulata* (Dalin et al. 2009).

IMPLEMENTATION OF TAMARISK BIOCONTROL

Following the experimental releases, the tamarisk biocontrol program shifted to the implementation phase in which beetles were widely distributed to weed and natural-resource managers to control tamarisk infestations. The largest implementation project was directed by Richard Hansen of the United States Department of Agriculture Animal and Plant Health Inspection Service, Plant Protection and Quarantine (USDA APHIS PPQ). This project resulted in the distribution of beetles to sites in Wyoming, Colorado, Oregon, Idaho, Washington, Montana, South Dakota, Nebraska, Kansas, and Iowa. Overall about a half million adult *D. carinulata* were collected from western Nevada by personnel from the USDA APHIS and the Colorado Department of Agriculture. These were sorted and distributed out of the Colorado Department of Agriculture's Palisade Insectary in a program that started in 2005 and lasted until 2008. This was the only major program designed for the interstate distribution of tamarisk beetles. Individual states had programs for intrastate implementation, including Texas, where *D. elongata* was used for intrastate redistribution (Knutson and Muegge 2008), and Utah where intrastate distributions were made from large *D. carinulata* populations near Delta, Utah.

CURRENT RANGE OF *DIORHABDA* IN NORTH AMERICA

Four tamarisk beetle species, *D. carinata, D. carinulata, D. elongata,* and *D. sublineata* have been released as part of the tamarisk biocontrol program, but *D. carinata* are not known to be established, probably because of poor overwintering survival rates (Milbrath et al. 2007). The other three species are established and have spread from the initial release sites. The distribution of *D. carinulata* is the most widespread, and the Fukang, China, and Chilik, Kazakhstan, ecotypes are now found in the intermountain regions of Colorado, Utah, Wyoming, Nevada, Idaho, and Oregon. Recently *D. carinulata* have moved into New Mexico from Utah along the San Juan River. They have also moved into northern Arizona from southern Utah, following the Virgin River drainage (Bateman et al. 2010). *D. carinulata* have been introduced into Texas and California, but thus far they have not established in those two states (Carruthers et al. 2008). The Fukang ecotype is established in Nevada, Wyoming, Oregon, Idaho, and eastern Colorado on the Arkansas River while the Chilik ecotype is established in Utah, northern New Mexico, northern Arizona, and western Colorado. A third *D. carinulata* ecotype from Turpan, China, was tested in cages and found to thrive at the Bishop, California site (Dudley et al. 2006) but was never released into the open there. One small release of the Turpan ecotype along the Arkansas River in Colorado failed to establish. *D. elongata* were introduced into Texas, California, and New Mexico, but in all three instances they were very slow to establish defoliating populations. They now seem to be well established in all three states and are defoliating substantial areas in California along Cache Creek, in west Texas on several drainages including the Rio Grande, and in eastern New Mexico in the Pecos drainage. *D. sublineata* was the last species to be introduced into the field, and all introductions have been made along the lower Rio Grande starting in 2009. Population expansion and *Tamarix* defoliation occurred rapidly, and at the end of 2010 tamarisk had been defoliated along approximately 100 kilometers of the Rio Grande and Rio Conchos. Populations of *D. elongata* and *D. sublineata* are now sympatric or will soon become so at one site on the Rio Grande (see areas 6 and 7, figure 22.3) and may soon become sympatric in other parts of west Texas. *D. carinata* were released in northwestern Texas but self-sustaining populations do not seem to be carrying over from year to year (Milbrath et al. 2007; Carruthers et al. 2008).

IMPACTS OF *DIORHABDA* ON TAMARISK

Diorhabda feeds exclusively on tamarisk leaves, leading to periods of defoliation that last several weeks or longer. Specifically, the beetles scrape leaf waxes and the cuticle before attacking the mesophyll and the vascular system resulting in leaf desiccation and subsequent leaf drop (Dudley 2005). Repeated defoliation events (i.e., over several consecutive growing seasons) result in carbon starvation that, in turn, reduces leaf production and growth, leading in some cases to tamarisk mortality (Hudgeons et al. 2007; see also chapter 9, this volume). Following the implementation releases of the beetle, several monitoring programs were initiated to track tamarisk condition, beetle populations, and plant community changes following biocontrol. An evaluation of all monitoring results is beyond the scope of this chapter, so we focus on tamarisk mortality as a consequence of biological control. The need for detailed monitoring programs and a description of some that are already in place are discussed elsewhere (Carruthers et al. 2008, Hultine et al. 2010a; Bateman et al. 2010).

RATES OF TAMARISK MORTALITY

One of the most frequent questions from weed managers is, can the tamarisk beetles kill the host plant? The answer is that the beetles can kill tamarisk plants but the rate and percentage of mortality can be highly variable. In western Colorado we have monitored 10 sites, each with 25 permanently marked tamarisk plants, set up in either 2006 or 2007. Beetles began defoliating tamarisk at the sites starting in 2008, and at the end of 2011 site mortality ranged from 0% to 56% of the 25 permanently marked trees (see figure 22.6). At the Humboldt Sink in northern Nevada, where the first establishment of *D. carinulata* occurred in 2002 (DeLoach et al. 2004; Dudley et al. 2006; Pattison et al. 2010), a series of 32 to 40 trees radiating in the four cardinal directions from the initial release site at points two kilometers and five kilometers from the site (where beetles colonized in subsequent years), has been regularly monitored. Almost no mortality was observed until the third season of defoliation, after which a steady increase in mortality was recorded in subsequent years (see figure 22.7). However, the rate of mortality was variable, with about 80% mortality at the original release site by 2009, where trees received repeated defoliation two and even three times per season; whereas at greater

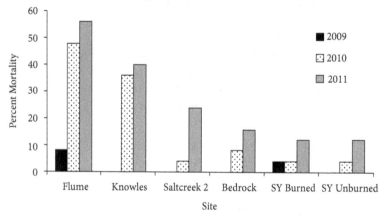

FIGURE 22.6 Tamarisk mortality at six field sites in western Colorado measured over three seasons. Sites were defoliated by beetles for the first time in either 2007 or 2008, and all sites had been defoliated at least four times by 2011. Of the 10 sites in this study, four have had no mortality and are not shown.

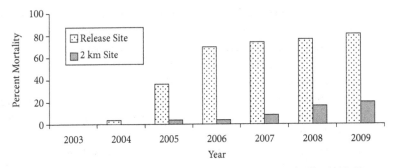

FIGURE 22.7 Mortality of regularly monitored *T. ramosissima* trees at the Humboldt River, Nevada, *Diorhabda* research release location. Death is indicated for the immediate release site, which had repeated defoliation starting in 2002 following release in 2001, and at our monitoring station 2 kilometers away from the release point where biocontrol insects established in 2004; here, defoliation was less intense.

distances, where beetles colonized in subsequent years (Pattison et al. 2010) cumulative mortality has reached roughly 20%. Some trees were lost to fire or the bulldozer, making precise rate measurements difficult. There has been no recovery of any tree previously noted as "dead" (no green foliage observed over the course of a full season), while there has been some moderate recovery of live canopy volume in the surviving trees as herbivory intensity has declined over time.

Why some tamarisk stands experience more or faster rates of mortality than others is still an open question. Mortality likely hinges on several factors, including the number of defoliation events, the timing of defoliation, plant access to resources, stand age and population genetics. The more often a given tamarisk plant is defoliated, the more likely that it will be carbon "starved" (Hudgeons et al. 2007b) to the point that it no longer has the resources to maintain basic metabolism. The number of events that are required to starve and kill a plant (if the plant is killed at all) can vary from about three to eight, likely depending on other factors outlined below. However, some plants have been killed after a single event. Another critical factor may be the seasonal timing in which defoliation takes place. Just prior to senescence, plants reallocate mobile nutrients and photosynthates from the canopy to internal storage compartments (usually in the roots) in order to construct new tissues the following spring. If plants are defoliated in fall or late summer just prior to reallocating internal resources to storage, they may be unable to produce enough new leaves and subsequent photosynthetic compounds the following spring to support a minimal amount of metabolic activity. The result is a downward-spiraling resource limitation that ends in mortality.

Mortality rates may also be connected to the availability of abiotic resources such as water, mineral nutrients, and sunlight. Relative to many other riparian tree species, tamarisk is fairly drought tolerant so that tamarisk stands often establish and grow over broad groundwater depth, soil moisture, and salinity gradients (Pockman and Sperry 2000; Glenn and Nagler 2005; Hultine and Bush 2011; see also chapter 9, this volume). It is plausible that tamarisk trees and stands along drier margins of riparian floodplains are more likely to succumb to repeated defoliation events than are trees occurring where water is more plant-available. To our knowledge, there are currently no published reports that relate tamarisk mortality rates from beetle defoliation to water- and/or nutrient-availability gradients. However, relationships between tree mortality, episodic disturbance, and resource availability have been reported in other tree species (Manion 1991; Ogle et al. 2000; McDowell et al. 2010).

Rates of beetle-induced mortality may also be related to stand age. Tamarisk trees/shrubs are relatively short-lived and are rarely found older than 80 years. As tamarisk stands age, they lose vigor, produce fewer leaves per unit ground area and use fewer resources (Hultine et al. 2010b). We therefore suspect that older stands are also more susceptible to beetle-induced herbivory and yield higher rates of mortality compared to younger stands if all else is equal.

Finally, given selection pressures related to resource allocation, beetle-induced mortality is almost certainly related to plant genetics. Riparian tree species, such as tamarisk, are typically selected for fast rates of growth following recruitment. This is due to the intense competition for sunlight and space that can emerge among individual riparian trees despite the high disturbance frequency that is common in these systems. Tamarisk has also evolved under intense pressure from herbivory. In its home range, tamarisk is attacked by 325 species of insects and mites from 88 genera (Kovalev 1995). This plant–insect coevolution has likely resulted in a diverse set of defense strategies across its range. Among these is one in which plants maintain relatively high storage-metabolite pools to draw on to replace damaged tissues following herbivory (Bloom et al. 1985). However, resource allocation to metabolite storage comes with the trade-off of reduced allocation to growth. Therefore, this inevitable trade-off has likely resulted in a diverse set of strategies among and within populations, for allocating resources to balance growth with surviving episodic disturbance. If so, then one could hypothesize that plants expressing relatively high growth rates are more likely to succumb to repeated defoliation events than plants with lower growth rates. Radial growth rates of tamarisk trees that were killed following several years of defoliation in southeastern Utah were in fact significantly higher than for co-occurring trees that survived an equal amount of defoliation (Hultine unpublished data). If these results are common over large areas, then it is plausible that the *Diorhabda* biocontrol program will reduce the overall productivity and genetic diversity of surviving tamarisk populations.

Although *Tamarix* eradication may be the goal of many management efforts, biocontrol by itself will never lead to that result. The goal of biological control, after all, is not eradication but suppression of the target pest plant to levels at which damage to natural or economic resources is acceptable. It is very likely that the desired level of control will not be achieved with *Diorhabda* spp., requiring the introduction of additional biocontrol agents. Two other insects (a weevil, *Coniatus tamarisci*, and a mealy bug, *Trabutina mannipara*) have already been approved through the USDA for eventual release, once the questions regarding *Diorhabda* introduction have been resolved. In combination with herbicide application (O'Meara et al. 2010) or prescription fire (see chapter 14, this volume), biocontrol may play an important role in reducing *Tamarix* densities to tolerable levels, even to the point of local extirpation, if desired. Nonetheless, benefits will accrue even from heavy damage or moderate mortality levels, for example, reductions in the transpiration losses of groundwater to the atmosphere (Pattison et al. 2010), enhancement of food resources for wildlife, and increased productivity of desired vegetation previously suppressed by dense tamarisk infestations.

CONCLUSIONS AND LONG-TERM OUTLOOK FOR *DIORHABDA* IN TAMARISK CONTROL.

The long-term projection for beetle activity and tamarisk decline depends largely on the equilibrium established between beetles and *Tamarix*. It is certain that beetles will not drive tamarisk to extinction in North America, but it is equally certain that in some areas the beetles have compromised tamarisk health on a scale that will have major ecological impacts. The death of tamarisk plants or the decline of green biomass will make resources available for other plants that were outcompeted by tamarisk. We also believe that tamarisk beetles will eventually be found in most tamarisk stands in North America, as species and ecotypes move through the system and evolve to match environmental conditions found across the vast range of *Tamarix*. Although there will be no "victory" over tamarisk in the West, there may be a decline in density that will greatly lower the impact of the plant on native ecosystems.

Literature Cited

Balciunas, J. K. 2000. Code of best practices for biological control of weeds. Pages 435–436 *in* N. R. Spencer, editor. *Proceedings of the X International Symposium on Biological Control of Weeds*, July 1999, Montana State University, Bozeman.

Bartelt, R. J., A. A. Cossé, B. W. Zilkowski, R. N. Wiedenmann, and S. Raghu. 2008. Early-summer pheromone biology of *Galerucella calmariensis* and relationship to dispersal and colonization. Biological Control 46:409–416.

Bateman, H. L., T. L. Dudley, D. W. Bean, S. M. Ostoja, K. R. Hultine, and M. J. Kuehn. 2010. A river system to watch: Documenting the effects of saltcedar (*Tamarix* spp.) biocontrol in the Virgin River valley. Ecological Restoration 28:405–410.

Bean, D. W., P. Dalin, and T. L. Dudley. 2012. Evolution of critical day length for diapause induction enables range expansion of *Diorhabda carinulata*, a biological control agent against tamarisk (*Tamarix* spp.). Evolutionary Applications 5:511–523.

Bean, D.W., T. Wang, R.J. Bartelt, and B.W. Zilkowski. 2007b. Diapause in the leaf beetle *Diorhabda elongata* (Coleoptera: Chrysomelidae), a biological control agent for tamarisk (*Tamarix* spp.). Environmental Entomology 36:531–540.

Bean, D. W., T. L. Dudley, and J. C. Keller. 2007a. Seasonal timing of diapause induction limits the effective range of *Diorhabda elongata deserticola* (Coleoptera: Chrysomelidae) as a biological control agent for tamarisk (*Tamarix* spp.) Environmental Entomology 36:15–25.

Beck, S. D. 1980. *Insect Photoperiodism*. Academic Press, NY.

Bellows, T.S. and T.W. Fisher. 1999. *Handbook of Biological Control*. Academic Press, San Diego, CA.

Bloom A. J., F. S. Chapin III, and H. A. Mooney. 1985. Resource limitation in plants-and economic analogy. Annual Review of Ecology and Systematics 16:363–392.

Carruthers, R. I., C. J. Deloach, J. C. Herr, G. L. Anderson, and A. E. Knutson. 2008. Saltcedar areawide pest management in the western United States. Pages 271–299 *in* O. Koul, G. Cuperus, and N. Elliott, editors. *Areawide Pest Management: Theory and Implementation*. CAB International. Cambridge, MA.

Carruthers, R. I., J. C. Herr, J. Knight, and C.J. DeLoach. 2006. A brief overview of the biological control of saltcedar. Pages 137–140 *in* M. S. Hoddle and M. W. Johnson, editors. *Proceedings of the fifth California conference on biological control*. Riverside, CA.

Coombs, E. M., J. K. Clark, G. L. Piper, and A. F. Cofrancesco, editors. 2004. *Biological Control of Invasive Plants in the United States*. Oregon State University Press, Corvallis.

Cossè, A. A., R. J. Bartelt, B. W. Zilkowski, D. W. Bean, and E. A. Andress. 2006. Behaviorally active green leaf volatiles for monitoring the leaf beetle, *Diorhabda elongata*, a biocontrol agent for saltcedar, *Tamarix* spp. Journal of Chemical Ecology 32:2695–2708.

Cossè, A. A., R. J. Bartelt, B. W. Zilkowski, D. W. Bean, and R. J. Petroski. 2005. The aggregation pheromone of *Diorhabda elongata*, a biological control agent of saltcedar (*Tamarix* spp.): Identification of two behaviorally active components. Journal of Chemical Ecology 31:657–670.

Dalin, P., D. W. Bean, T. Dudley, V. Carney, D. Eberts, K. T. Gardner, E. Hebertson, E. N. Jones, D. J. Kazmer, G. J. Michels, S. A. O'Meara, and D. C. Thompson. 2010. Seasonal adaptations to day length in ecotypes of *Diorhabda* spp. (Coleoptera: Chrysomelidae) inform selection of agents against saltcedars (*Tamarix* spp.). Environmental Entomology 39:1666–1675.

Dalin, P., M. J. O'Neal, T. Dudley, and D.W. Bean. 2009. Host plant quality of *Tamarix ramosissima* and *T. parviflora* for three sibling species of the biocontrol insect *Diorhabda elongata* (Coleoptera: Chrysomelidae). Environmental Entomology 38:1373–1378.

DeLoach, C. J., P. A. Lewis, J. C. Herr, R. I. Carruthers, J. L. Tracy, and J. Johnson. 2003. Host specificity of the leaf beetle, *Diorhabda elongata deserticola* (Coleoptera: Chrysomelidae) from Asia: A biological control agent for saltcedars (*Tamarix*: Tamaricaceae) in the Western United States. Biological Control 27:117–147.

DeLoach, C. J., R. Carruthers, T. Dudley, D. Eberts, D. Kazmer, A. Knutson, D. Bean, J. Knight, P. Lewis, J. Tracy, J. Herr, G. Abbot, S. Prestwich, G. Adams, I. Mityaev, R. Jashenko, B. Li, R. Sobhian, A. Kirk, T. Robbins, and E. Delfosse. 2004. First results for control of saltcedar (*Tamarix* spp.) in Cullen, editor. *The Open Field in the Western United States*. Eleventh International Symposium on Biological Control of Weeds, Canberra, Australia

DeLoach, C. J., R. I. Carruthers, J. E. Lovich, T. L. Dudley, and S. D. Smith. 2000. Ecological interactions in the biological control of saltcedar (*Tamarix* spp.) in the United States: Toward a new understanding. Pages 819–873 *in* N. R. Spencer, editor. *Proceedings for the X International Symposium on Biological Control of Weeds,* 4–14 July 1999. Montana State University, Bozeman.

Dickens, J. C., J. E. Oliver, B. Hollister, J. C. Davis, and J. A. Klun. 2002. Breaking a paradigm: Male-produced aggregation pheromone for the Colorado potato beetle. Journal of Experimental Biology 205:1925–1933

Dudley, T. D., and D. J. Kazmer. 2005. Field assessment of the risk posed by *Diorhabda elongata*, a biocontrol agent for control of saltcedar (*Tamarix* spp.), to a nontarget plant, *Frankenia salina*. Biological Control 35:265–275

Dudley, T.L. 2005. Progress and pitfalls in the biological control of saltcedar (*Tamarix* spp.) in North America. *Proceedings of the 16th U.S. Department of Agriculture interagency research forum on gypsy moth and other invasive species;* 18–21 Jan. 2005. Technical Report NE-337. USDA Forest Service General, Annapolis, MD, Morgantown, WV.

Dudley, T. L., C. J. DeLoach, J. E. Lovich, and R. I. Carruthers. 2000. Saltcedar invasion of western riparian areas: Impacts and new prospects for control. Pages 345–381 *in* R. E. McCabe, and S. E. Loos, editors. *Transactions of the 65th North American Wildlife & Natural Resources Conference,* 24–28 March 2000, Chicago, IL. Wildlife Management Institute, Washington, DC.

Dudley, T. L., C. J. DeLoach, P. A. Lewis, and R. I. Carruthers. 2001. Cage tests and field studies indicate leaf-eating beetle may control saltcedar. Ecological Restoration 19:260–261.

Dudley, T. L., D. W. Bean, R. R. Pattison, and A. Caires· 2012. Selectivity of a biological control agent, *Diorhabda carinulata* (Chrysomelidae) for host species within the genus *Tamarix*. Pan Pacific Entomologist. 88:319–341.

Dudley, T. L., P. Dalin, and D. W. Bean. 2006. Status of biological control of *Tamarix* spp. in California. Pages 137–140 *in* M. S. Hoddle and M. W. Johnson, editors. *Proceedings of the fifth California conference on biological control*. Riverside, CA.

Friedman, J. T., G. T. Auble, P. B. Shafroth, M. L. Scott, M. F. Merigliano, M. D. Freehling, and E. R. Griffin. 2005. Dominance of nonnative riparian trees in western USA. Biological Invasions 5:747–751.

Glenn, E. P., and P. L. Nagler. 2005. Comparative ecophysiology of *Tamarix ramosissima* and native trees in western U.S. riparian zones. Journal of Arid Environments 61:419–446

Grevstad, F.S., and A. L. Herzig. 1997. Quantifying the effects of distance and conspecifics on colonization: experiments and models using the loosestrife leaf beetle, *Galerucella calmariensis*. Oecologia 110: 60–68.

Herr, J.H., R.I. Carruthers, D.W. Bean, C. Jack DeLoach and J. Kashefi. 2009. Host preference between saltcedar (*Tamarix* spp.) and native nontarget *Frankenia* spp within the *Diorhabda elongata* species complex (Coleoptera: Chrysomelidae). Biological Control 51:337–345.

Hudgeons, J. L., A. E. Knutson, C. J. DeLoach, K. M. Heinz, A. McGinty, and J. L. Tracy. 2007a. Establishment and biological success of *Diorhabda elongata elongata* on invasive *Tamarix* in Texas. Southwestern Entomologist 32:157–168

Hudgeons, J. L., A. E. Knutson, K. M. Heinz, C. J. DeLoach, T. L. Dudley, R. R. Pattison, and J. R. Kiniry. 2007. Defoliation by introduced *Diorhabda elongata* leaf beetles (Coleoptera: Chrysomelidae) reduces carbohydrate reserves and regrowth of *Tamarix* (Tamaricaceae). Biological Control 43:213–221.

Huffaker, C.B., and C.E. Kennett. 1959. A ten year study of vegetational changes associated with biological control of Klamath weed. Journal of Range Management 12: 69–82.

Hultine, K. R., J. Belnap, C. van Riper III, J. R. Ehleringer, P. E. Dennison, M. E. Lee, P. L. Nagler, K. A. Snyder, S. E. Uselman, and J. B. West. 2010a. Tamarisk biocontrol in the western United States: ecological and societal implications. Frontiers of Ecology and the Environment 8:467–474

Hultine, K. R., P. L Nagler, K. Morino, S. E. Bush, K. G. Burtch, P. E. Dennison, E. P. Glenn, and J. R. Ehleringer. 2010b. Sap flux-scaled transpiration by tamarisk (*Tamarix* spp.) before, during and after episodic defoliation by the saltcedar leaf beetle (*Diorhabda carinulata*). Agricultural and Forest Meteorology 150:1467–1475.

Hultine, K. R., and S. E. Bush. 2011. Ecohydrological consequences of non-native riparian vegetation in the southwestern U.S: A review from an ecophysiological perspective. Water Resources Research 47W07542. doi:0.1029/2010WR010317.

Knutson, A., and M. Muegge. 2008. Biological control of saltcedar using a leaf-feeding beetle in Texas: 2006–2008. Final Report to NRCS Conservation Innovation Grant Program, Project #68-7442-5-467.

Kovalev, O.V. 1995. Co-evolution of tamarisk (Tamaricaceae) and pest arthropods (Insecta: Arachnida: Acarina), with special reference to biological control prospects. Pensoft, Sofia.

Lewis, P. A., C. J. DeLoach, A. E. Knutson, and J. L. Tracy. 2003a. Biology of *Diorhabda elongata deserticola* (Coleoptera: Chrysomelidae), an Asian leaf beetle for biological control of saltcedars (*Tamarix* spp.) in the United States. Biological Control 27:101–116.

Lewis, P. A., C. J. DeLoach, J. C. Herr, T. L. Dudley, and R. I. Carruthers. 2003b. Assessment of risk to native Frankenia shrubs by the Asian leaf beetle, *Diorhabda elongata deserticola* (Coleoptera Chrysomelidae) introduced for biological control of saltcedars (*Tamarix* spp.) in the western United States. Biological Control 27:148–166.

Lopatin, I. K. 1977. *The Leaf-Beetles of Central Asia and Kazakhstan*. Nauka Publishing, Leningrad. (1984 English translation from the Russian, Amerind Publishing Company, New Delhi).

Louda, S. M., R. W. Pemberton, M. T. Johnson, and P. A. Follett. 2003. Nontarget effects: The Achilles' heel of biological control? Annual Review of Entomology 48:365–396.

Manion P. D. 1991. *Tree Disease Concepts*. 2nd ed. Prentice-Hall, Upper Saddle River, NJ.

McDowell N. G., C. G. Allen, and L. Marshall. 2010. Growth, carbon isotope discrimination, and mortality across a ponderosa pine elevation transect. Global Change Biology 16:399–415.

Milbrath, L.R., and C. J. DeLoach. 2006b. Acceptability and suitability of athel, *Tamarix aphylla*, to the leaf beetle *Diorhabda elongata* (Coleoptera: Chrysomelidae), a biological control agent of saltcedar (*Tamarix* spp.) Environmental Entomology 35:1379–1389.

Milbrath, L. R., and C. J. DeLoach. 2006a. Host specificity of different populations of the leaf beetle *Diorhabda elongata* (Coleoptera: Chrysomelidae), a biological control agent of saltcedar (*Tamarix* spp). Biological Control 36:32–48.

Milbrath, L. R., C. J. DeLoach, and J. L. Tracy. 2007. Overwintering survival, phenology, voltinism and reproduction among different populations of the leaf beetle *Diorhabda elongata* (Coleoptera Chrysomelidae). Environmental Entomology 36:1356–1364.

O'Meara, S., D. Larsen, and C. Owens. 2010. Methods to control saltcedar and Russian olive. Pages 65–102 *in* P. B. Shafroth, C. A. Brown, and D. M. Merritt, editors. Saltcedar and

Russian olive control demonstration act science assessment: U.S. Geological Survey Scientific Investigations Report 2009–5247.

Ogle, K., T. G. Whitham, and N. S. Cobb. 2000. Tree-ring variation in pinyon predicts likelihood of death following severe drought. Ecology 81:3237–3243

Pattison, R. L., C. M. D'Antonio, T. L. Dudley, K. K. Allander, and B. Rice. 2010. Early impacts of biological control on canopy cover and water use of the invasive saltcedar tree (*Tamarix* spp.) in western Nevada, USA. Oecologia. doi:10.1007/s00442-010-1859-y.

Pemberton, R.W. 2000. Predictable risks to native plants in weed biocontrol. Oecologia 125:489–494

Pockman, W.T., and J. S. Sperry. 2000. Vulnerability to xylem cavitation and the distribution of Sonoran Desert vegetation. American Journal of Botany 87:1287–1299

Saunders, D. S. 2002. *Insect Clocks*. 3rd ed. Elsevier Science, Amsterdam, The Netherlands.

Shafroth, P. B., J. R. Cleverly, T. L. Dudley, J. P. Taylor, C. Van Riper III, E. P. Weeks, and J. N. Stuart. 2005. Control of *Tamarix* in the western United States: Implications for water salvage, wildlife use, and riparian restoration. Environmental Management 35:231–246.

Spichiger, R., and V. Savolainen.1997. Present state of Angiospermae phylogeny. Candollea 52:435–455.

Tauber, M. J., C. A. Tauber, and S. Masaki. 1986. *Seasonal Adaptations of Insects*. Oxford University Press, New York.

Toševski, I., R. Caldara, J. Jović, G. Hernández-Vera, C. Baviera, A. Gassmann, and B. C. Emerson. 2011. Morphological, molecular and biological evidence reveal two cryptic species in *Mecinus janthinus* Germar (Coleoptera, Curculionidae), a successful biological control agent of Dalmatian toadflax, *Linaria dalmatica* (Lamiales, Plantaginaceae). Systematic Entomology 36: 741–753

Tracy J. T., and T. O. Robbins. 2009. Taxonomic revision and biogeography of the *Tamarix*-feeding *Diorhabda elongata* (Brullé, 1832) species group (Coleoptera: Chrysomelidae: Galerucinae: Galerucini) and analysis of their potential in biological control of tamarisk. Zootaxa 2101:1–152

Van Driesche, R. G., R. I. Carruthers, T. Center, M. S. Hoddle, J. Hough-Goldstein, L. Morin, L. Smith, D. L. Wagner, et al. 2010. Classical biological control for the protection of natural ecosystems. Biological Control, supplement 1, S2–S33.

Wapshere, A. J. 1974. A strategy for evaluating the safety of organisms for biological weed control. Annals of Applied Biology 77:201–211.

23

Riparian Restoration in the Context of *Tamarix* Control

Patrick B. Shafroth, David M. Merritt, Mark K. Briggs, Vanessa B. Beauchamp, Kenneth D. Lair, Michael L. Scott and Anna Sher

River corridors are highly susceptible to the colonization and spread of nonnative species in part because of relatively high natural disturbance rates, the capacity for rapid, long-distance propagule dispersal, and various anthropogenic perturbations (Richardson et al. 2007). Proliferation of nonnative plants in riparian corridors is relatively common globally, and invasions are associated with substantial ecological, economic, or management costs (Webb and Erskine 2003; Holmes et al. 2005; Shafroth et al. 2010a). In cases where large-scale alien plant control has been undertaken, ecological restoration is often cited as a desired outcome. The case of *Tamarix* spp. in western North America serves as an excellent example of the challenges, potential conflicts, and information needs associated with restoration in the context of invasive species control (Shafroth and Briggs 2008a; Shafroth et al. 2008b; Shafroth et al. 2010a).

Rationales for controlling or eliminating *Tamarix* from sites, river reaches, or entire watersheds include assumptions that natural recovery or the active restoration of native plant communities will follow exotic plant removal (McDaniel and Taylor 2003; Quimby et al. 2003). Although it is often assumed or implied that *Tamarix* removal alone constitutes "restoration," to meet restoration goals and objectives additional measures are often required to facilitate the establishment of desired replacement vegetation. In this chapter, the term "restoration" refers to conversion of *Tamarix*-dominated sites to a replacement vegetation type that achieves specific management goals and helps return parts of the system to a desired and more natural state or dynamic. Removal of *Tamarix*, even coupled with the replanting of replacement vegetation, may restore a desired appearance of a riparian area and some functions, but reestablishing the processes that maintain such form is usually necessary to sustainably restore form and function. The degree to which a site is restored following removal of *Tamarix* depends on many factors, such as the site's potential for the desirable replacement vegetation, direct and indirect effects of removal (e.g., effects of herbicides on nontarget vegetation), the efficacy of passive and active restoration measures, and the maintenance of processes that support native vegetation and prevent recolonization by nonnative communities.

In this chapter we synthesize published research related to restoring native riparian vegetation following *Tamarix* control or removal. We begin with a brief discussion of objective setting and the planning of such efforts. Next, we discuss the importance of considering site-specific factors and of context in selecting and prioritizing sites for restoration. We conclude with a discussion of costs and benefits associated with active, passive, and combined restoration approaches, providing examples from the literature and presenting an overview of the key issues to consider in carrying out restoration projects at a range of scales. Although our focus is on restoration following *Tamarix* control, most of the material in this chapter also applies to situations in which the removal of different nonnative riparian taxa is emphasized (e.g., Russian olive; Shafroth et al. 2010a) or, more generally, to riparian restoration in the western United States.

Restoration Planning and Objective Setting

Shafroth et al. (2008b) proposed a process for developing restoration projects for *Tamarix*-dominated bottomland sites to encourage resource managers, restoration practitioners, and policy makers to plan for restoration up front when contemplating *Tamarix* control. The process consists of seven sequential steps: (1) goal identification; (2) development of objectives, including evaluation of important ecological and nonecological site factors; (3) site prioritization; (4) development of a site-specific plan; (5) project implementation; (6) post-implementation monitoring and maintenance; and (7) adaptive management (see figure 23.1).

The first steps in a restoration planning process include a clear articulation of both the general goals and more-specific objectives (Briggs 1996; Clewell et al. 2005; Shafroth et al. 2008b). Goals of *Tamarix* control and subsequent restoration may include decreasing site-level transpiration loss in an attempt to increase water yield, increasing lateral channel instability, replacing *Tamarix* with native vegetation, improving wildlife habitat, preventing further spread, reducing plant biomass, and reducing the intensity and frequency of riparian wildfires (McDaniel and Taylor 2003; Shafroth et al. 2005; Bateman et al. 2008; see also chapter 1, this volume). For a project with an overall goal of improving wildlife habitat, objectives

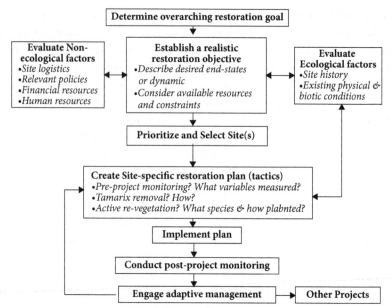

FIGURE 23.1 Seven-step process for planning restoration in the context of *Tamarix* removal.
Source: Adapted from figure 1 in Shafroth et al. (2008b).

for restoration might include specifying the type of plant community or habitat structure desired to replace *Tamarix*, determining the spatial scale of the effort, and identifying the wildlife taxa expected to respond positively (or negatively) to the restoration effort.

When setting objectives, it is important to identify specific metrics that will be monitored to determine the degree of project success. Recognizing and considering the trade-offs and conflicts among different goals is also important in the project's planning phase. For example, a floodplain fuels-reduction project might have entirely different goals and objectives than one aimed at enhancing wildlife habitat, potentially resulting in short-term losses of habitat structure to achieve beneficial reductions in fuel loading (McDaniel and Taylor 2003; Bateman et al. 2008).

Site Factors and Context for Restoration

Developing realistic objectives for the composition of replacement vegetation following *Tamarix* removal depends on an evaluation of the ecological and nonecological (e.g., legal and institutional) factors associated with a candidate restoration site or river reach. Ecological evaluations can also help prioritize restoration sites. The outcome of restoration efforts is influenced by site conditions and processes prior to and following removal of *Tamarix*. Some river reaches are good candidates for *Tamarix* removal followed by restoration, but other sites may be too degraded or have little potential for restoring physical processes and recovery of native riparian vegetation (Taylor and McDaniel 2004). Factors such as saline or desiccated soils, changes in the historic hydrograph of a river, or the presence of nonnative seed banks may hinder replanting and restoration.

In addition to the physical and biological aspects of restoration sites and river reaches, many nonecological factors can influence site selection and restoration success. Nonecological factors include permits and other legal requirements, funding for the project and post-project monitoring, site access and logistics, and stakeholder interactions (including community involvement, education, and conflict resolution). Fostering partnerships and collaborations, leveraging funds, sharing resources, and involving volunteers will facilitate restoration efforts.

In this section, we focus on the most important ecological factors, including valley and bottomland geomorphology; flow regimes; groundwater dynamics; soil chemistry and texture; and the structure, composition, and relative abundance of native and nonnative vegetation present at a site, and we consider spatial and tem-poral scale in planning restoration projects (Bay and Sher 2008; Shafroth et al. 2008b). Taking into account the effects of perturbations upstream, downstream, or in the uplands adjacent to a particular project, along with anticipated future changes in the watershed may enhance a project's likelihood of success, decrease unexpected outcomes, and provide context, facilitating the transfer of information from one site to the next.

The importance of project scale is evident in biological control of *Tamarix*. Leaf-eating beetles in the genus *Diorhabda* have been released and have spread around the western United States for biological control of *Tamarix* (Hudgeons et al. 2007; Hultine et al. 2010; see also chapter 22, this volume). The beetles defoli-ate *Tamarix*, sometimes more than once during each growing season, significantly reducing canopy and increasing plant mortality along long river reaches (Pattison et al. 2011). This pattern of *Tamarix* death presents a unique situation in which the scale and scope of control and restoration are not known in advance, in contrast to most restoration efforts. Such unprescribed patterns could require more proactive restoration approaches, such as the planting of replacement vegetation in advance of biological control to mitigate negative effects on wildlife (Paxton et al. 2011; see also chapter 11, this volume).

VALLEY AND BOTTOMLAND GEOMORPHOLOGY

Most *Tamarix* control and restoration occurs in river bottomlands, though this work also occurs along reservoir margins, vernal pools, and springs. In a river-ine setting, the physical foundation upon which restoration efforts occur begins with valley geomorpholgy, which can range from a narrow bedrock canyon to a wide alluvial valley. In a valley, the bottomland is the mosaic of alluvial or collu-vial features, including channel bed, banks, bars, floodplains, and terraces. These geomorphic surfaces generally differ in their elevation above and lateral distance from the active channel; the differences are associated with gradients of inundation frequency, shear stress and disturbance intensity, depth to groundwater, and soil characteristics. In turn, these physical variables significantly influence plant communities (Richards et al. 2002; Merritt et al. 2010). For example, in contrast to dry terraces, bars are characterized by frequent inundation, relatively high levels of physical disturbance, shallow groundwater, and low soil salinity, making them suitable for disturbance-adapted, mesic species. Narrow canyons along flow-regulated (dammed) rivers may have few geomorphic surfaces and may have had only sparse riparian vegetation growth prior to dam construction and *Tamarix* colonization (Webb et al. 2007). Thus, an important early step in the site-assessment process is to identify the range of geomorphic surfaces present and to evaluate replacement vegetation potential (Kondolf and Piegay 2003). It is also crucial to understand how different perturbations in the watershed create either sediment deficit or sediment surplus at restoration sites and how current and future sediment dynamics may influence site conditions. Finally, site conditions can be dynamic over space and time, in part because of feedbacks among vegetation, hydrology, and sediment erosion and deposition (see chapter 7, this volume).

FLOW REGIME, FLUVIAL DISTURBANCE, AND WATER AVAILABILITY

River flow regime and associated fluvial disturbance (floods, low flows, rates of stage change) interact with sediment regime and effects of *Tamarix* removal to exert tremendous influence on site restoration potential (Stromberg et al. 1991; Mahoney and Rood 1998; Hughes 1997; Stromberg et al. 2007a; Vincent et al. 2009). In the western United States, aspects of flow regime that may favor native pioneer trees (cottonwoods and willows, genera *Populus* and *Salix*) over *Tamarix*, or allow a mix of native species with *Tamarix*, include (1) floods that are large enough to create bare, moist germination sites; (2) flood recession timing that is synchronized with the seed dispersal period of native pioneer trees; (3) a flood recession rate that is slower than seedling root growth (1–4 cm per day); (4) low flows that provide sustained surface water and groundwater availability; and (5) a lack of subsequent floods until desirable plants are large enough to resist flood-induced physical damage (Mahoney and Rood 1998; Hughes et al. 2001). Where these conditions have been met, native seedlings and saplings have been able to establish successfully in the presence of *Tamarix* (Shafroth et al. 1998; Sher et al. 2002; Nagler et al. 2005), ultimately dominating relatively unregulated river reaches (Stromberg et al. 2007b; Merritt and Poff 2010; Mortenson and Weisberg 2010). Conversely, *Tamarix* is usually dominant along flow-regulated rivers (Stromberg et al. 2007b; Merritt and Poff 2010; Mortenson and Weisberg 2010; Nagler et al. 2011). The frequency of flows suitable for native seedling establishment strongly influences the heterogeneity and age-class diversity of riparian forests (Mahoney and Rood 1998) and thus their functions, values, and physiognomic characteristics.

It is important to consider low flows and alluvial groundwater dynamics that limit plant establishment and growth, as different plant species and communities are associated with specific ranges of depth to groundwater (Meinzer 1927; Stromberg et al. 2007a). For example, canopy dieback and mortality of native *Populus* are associated with sustained groundwater declines in well-drained soils (Scott et al. 1999; Shafroth et al. 2000; Cooper et al. 2003). Other studies have shown that native riparian woodland species (*Populus fremontii* and *Salix goodingii*) dominate *Tamarix* at sites along rivers with perennial flow, where water tables fluctuate less than 0.5 meters over the season, groundwater depths are less than 2.6 meters, and surface water persists for more than three-fourths of the year (Lite and Stromberg 2005). A study of sites that were revegetated following *Tamarix* control revealed that those with characteristics favorable for mesic vegetation (e.g., shallow water tables, low salinity, high precipitation) contained a lower density and cover of *Tamarix* than other revegetated sites (Bay and Sher 2008; see also chapter 24, this volume). Where depth to groundwater is extreme and overbank flooding is infrequent or absent, vegetation depends on precipitation, which is sparse and highly variable in the arid and semiarid western United States. In these situations, revegetation may require planting xeric native species or irrigating mesic taxa until well established.

SOIL SALINITY AND TEXTURE

Soil salinity can greatly influence site restoration following *Tamarix* removal, because plants vary in their tolerance of soil and water salinity (see table 23.1; Vandersande et al. 2001; Shafroth et al. 2008b; Beauchamp et al. 2009; Beauchamp and Shafroth 2011). Soil salinity on many bottomland surfaces in the western United States has increased to levels that permit growth of salt-tolerant plants only (see chapter 8, this volume). Salts may be concentrated in floodplain soils because of evaporation from shallow water tables, agricultural runoff, and the reduced frequency of flushing flows and elevated water tables that result from river regulation (Jolly et al. 1993; Anderson 1995, Glenn et al. 2012; Merritt and Shafroth in press; see also chapter 8, this volume). Floodplain soil salinity also varies due to natural variation in the salt content of different geologic formations, substrates, and water sources. Determining soil salinity as part of the site evaluation process can allow planners to develop lists of site-appropriate species for restoration or revegetation (table 23.1; Beauchamp et al. 2009).

Soil texture is an important consideration when selecting restoration sites because it affects soil moisture, salinity, cation exchange capacity, nutrient availability, aeration, and the height of the capillary fringe above the water table, which may affect competitive interactions between *Tamarix* and replacement species. *Tamarix* grows on a wide range of substrate textures ranging from gravel to clay; replacement species are often more specialized, having a narrower range of soil-texture requirements (Sher and Marshall 2003; Beauchamp and Shafroth 2011).

TABLE 23.1

Descriptions of generalized site and plant community types, associated soil electrical conductivities, and representative plant genera characteristic of *Tamarix* removal sites in the western United States.

Salinity-moisture regime	Vegetation community	Electrical conductivity		Representative genera
Mesic, lower salinity sites with seasonally shallow water tables or surface flows	High proportion of nonchenopod trees, shrubs, grasses, and annual and perennial forbs	< 4 dS/m	Trees	*Populus, Salix, Celtis, Prunus, Forestiera, Juglans, Robinia*
			Shrubs	*Salix, Amorpha, Baccharis, Pluchea, Ephedra, Lycium, Shepherdia, Rhus, Ericameria/Chrysothamnus*
			Grasses	*Distichlis, Sporobolus, Paspalum, Leymus, Spartina, Panicum*
			Forbs	*Anemopsis, Sphaeralcea, Corydalis, Eriogonum*
Ephemerally mesic, highly saline sites receiving periodic groundwater and/or surface flow (for example, alkali sinks)	High proportion of halophytic chenopod species; few grasses	>12 dS/m	Shrubs	*Suaeda, Atriplex, Allenrolfea, Sarcobatus*
			Grasses	*Distichlis, Puccinellia, Sporobolus, Muhlenbergia*
			Forbs	*Salicornia, Heliotropium, Atriplex (herbaceous)*
Xeric, moderately to highly saline sites	Mixture of shrubs, forbs, and grasses; dominated by halophytic species within the Chenopodiaceae	> 8 dS/m	Trees	*Acacia, Prosopis, Parkinsonia/Cercidium*
			Shrubs	*Atriplex, Allenrolfea, Suaeda, Isocoma, Sarcobatus*
			Grasses	*Sporobolus, Elymus, Pascopyrum, Leptochloa, Pleuraphis, Panicum*
			Forbs	*Sphaeralcea, Heliotropium, Frankenia*
Xeric, less saline sites	Mixture of shrubs, forbs (including legumes), and grasses; higher proportion of forbs and grasses	< 8 dS/m	Trees	*Chilopsis, Forestiera*
			Shrubs	*Lycium, Ephedra, Krascheninnikovia, Rhus, Prosopis, Fallugia, Lesquerella*
			Grasses	*Achnatherum, Bothriochloa, Bouteloua, Elymus, Eragrostis, Pleuraphis, Panicum*
			Forbs	*Oenothera, Sphaeralcea, Anemopsis, Ambrosia, Baileya, Frankenia, Chrysopsis/Haplopappus*

Sources: Adapted from synthesis of Bernstein (1958), Federal Water Pollution Control Administration (1968), Dick-Peddie (1993), Ogle (1994), Natural Resources Conservation Service (1996),. Lair and Wynn (2002a, b), Tangee and Kielen (2002), Beauchamp et al. (2009), and Swift (2010). For each salinity-moisture regime, it is generally best to use local, native species. Modified from table 3 in Shafroth et al. (2008b). Units are in deciSiemens per meter (dS/m).

BIOTIC FACTORS

In addition to the physical factors described earlier, several biotic factors influence site restoration potential. These include the availability of native and nonnative plant propagules, presence of mycorrhizal fungi, and competitive interactions among species. With or without the active planting of desired replacement vegetation, seeds or vegetative propagules of native or exotic species often occur on site or nearby (Goodson et al. 2001), or they may be dispersed to the site (Merritt and Wohl 2002). Sites that have been occupied by *Tamarix* can be recolonized by *Tamarix* unless conditions have changed considerably. Invasions of other nonnative species (secondary weeds) can also follow *Tamarix* removal. Within *Tamarix*'s range, these species include kochia (*Bassia scoparia*), bromes (*Bromus* spp.), perennial pepperweed (*Lepidium latifolium*), Russian knapweed (*Acroptilon repens*), Bermudagrass (*Cynodon dactylon*), and Russian thistles (*Salsola* spp.; Weeks et al. 1987; McDaniel and Taylor 2003). Sites occupied by such ruderal (disturbance-adapted) weeds may inhibit native species establishment.

Mycorrhizae in the soil can influence the establishment and growth of replacement vegetation (see chapter 13, this volume). Studies in Arizona indicate that sites dominated for many years by monotypic stands of *Tamarix* lack arbuscular mycorrhizal fungi ([AMF]; Beauchamp and Stutz 2005). Arbuscular mycorrhizal fungi associate symbiotically with the roots of many plant species and help plants acquire relatively immobile soil nutrients, particularly phosphorus, in exchange for the carbon produced in photosynthesis (Smith and Read 1997). Many native riparian plants have AMF associations; however, *Tamarix* and many weed species are nonmycorrhizal (Allen 1991; Titus et al. 2002; Beauchamp and Stutz 2005; see also chapter 13, this volume).

Approaches to Restoration Following *Tamarix* Control

Once *Tamarix* has been successfully controlled as part of a restoration plan, a site can be (1) left to be naturally colonized by replacement vegetation, (2) modified to facilitate establishment of native taxa, (3) reseeded or replanted with desirable replacement vegetation, or (4) some combination of the above. *Passive restoration* refers to facilitating the return of desirable system dynamics and species composition by removing one or more underlying stressor(s). *Active restoration* approaches include manipulating a site to prepare it for restoration; revegetating the site by introducing seeds, transplant stock, or cuttings; or irrigating or otherwise manipulating the site to enhance recovery. The decision to employ active or passive approaches depends on restoration objectives and expected outcomes, site characteristics, watershed attributes, practical constraints, and available resources. The success of both active and passive restoration approaches can also be influenced by effects of *Tamarix* control on the site.

PASSIVE RESTORATION

In riparian systems, approaches to passive restoration include removing invasive species, eliminating structures that control channel form, and restoring natural processes, such as flooding, sediment fluxes and associated fluvial dynamics (Stromberg 2001). Passive restoration can also involve removing stressors that might inhibit native species establishment, including herbivores (such as livestock), other plants that might compete for moisture or sunlight, or recreational activities (Mortenson et al. 2008). In some cases, simply removing *Tamarix* may allow for the natural recovery of native vegetation, but only if site conditions and physical processes that support native vegetation are intact or restored following removal (Sher et al. 2010).

Included in passive restoration is the maintenance or restoration of physical factors and processes that enhance establishment of desirable native species (e.g., see chapter 21, this volume). Restoring natural or seminatural flow regimes can provide hydrologic conditions that favor native species or inhibit *Tamarix* along extensive river reaches (Molles et al. 1998; Taylor et al. 1999; Nagler et al. 2005; Rood et al. 2005; Stromberg et al. 2007a; Shafroth et al. 2010b), flush salts accumulated in floodplain soils (see chapter 8, this volume), and generate spatial and temporal variability in riparian forest structure typical of natural systems (Hughes et al. 2005; see figure 23.2).

Natural recovery of riparian vegetation following *Tamarix* control is often the only option, and can be successful if the primary underlying stressor is the presence

FIGURE 23.2 Managing flow releases from dams has been used on several rivers in western North America to promote restoration of riparian vegetation. This photograph shows an experimental flood release on the Bill Williams River in March 2006, which was designed, in part, to reduce the density of *Tamarix* seedlings and promote regeneration of native *Populus* and *Salix* seedlings. See Shafroth et al. (2010b) for more details.

Source: Photo courtesy of Patrick B. Shafroth.

or dominance of *Tamarix*. Natural recovery requires that native species exist on a site as remnant vegetation, as propagules (i.e., seed or plant parts that can sprout) present in the soil, or as propagules dispersed from elsewhere. The presence of remnant native vegetation also can be important in providing microsites favorable for plant establishment. The absence of remnant native taxa suggests that natural recovery will likely be difficult unless other physical processes are restored, stressors removed, or propagules added. For these reasons, it is important that native species are not eliminated or significantly reduced as a part of *Tamarix* removal efforts. Finally, successful natural recovery following *Tamarix* control requires that soil and climatic and hydrologic conditions during the recovery period (1–5 years following treatment) are suitable for the colonization and establishment of native species and expansion of remnant native vegetation. Because weather in the western United States is highly variable between years, re-treatment or other site maintenance may be necessary for years before successful natural recovery will occur. Harms and Hiebert (2006) noted increases in native plant cover on only a few of 33 natural recovery sites across the western United States. Merritt and Johnson (2006) reported diminished species diversity on sites after *Tamarix* removal and slow recovery of native species during the first five years following treatments, compared to sites from which *Tamarix* was not removed. DeWine and Cooper (2008) found that the superior shade tolerance of *Acer negundo* (box elder) allows it to outcompete and overtop *Tamarix*, eventually replacing it in the absence of any control measures.

ACTIVE RESTORATION

When sites are severely degraded or disconnected from the active floodplain, opportunities to restore natural processes are not available, or other constraints prevent the implementation of passive methods, active revegetation may be considered. Active restoration measures require considerable inputs of labor, funding, and time. Key to revegetation success is selecting plant species that are adapted to current site conditions, which depends on understanding their ecophysiological characteristics as well as the characteristics of the revegetation site. In addition, the success of active revegetation can be influenced by effects of *Tamarix* removal, soil preparation, types of propagules used, and planting method, as well as irrigation (Bunting et al. 2011) and the control of other weeds that may invade subsequent to *Tamarix* removal.

Soil preparation. Revegetation following control of dense and/or mature stands of *Tamarix* is often difficult without soil preparation (Herbel et al. 1973; Pinkney 1992; Taylor et al. 1999; Anderson et al. 2004; Merritt and Johnson 2006). Soil surface treatments can be used to (1) create soil surface microrelief to enhance precipitation capture and infiltration; (2) reduce, redistribute, and/or dilute salts in the *Tamarix* leaf litter and upper soil profile; (3) improve soil texture characteristics for improved seed germination and establishment; and (4) ensure proper depth placement and incorporation of seed and/or mycorrhizal inoculum. Although mechanical seedbed preparation can be accomplished during *Tamarix* removal with implements such as root plows and root rakes, other machinery such as roller choppers, land imprinters, and pitter-seeders cause less environmental disturbance (see figure 23.3).

After *Tamarix* is cleared, the material may be taken off site, or shredded and left on site as a groundcover mulch (Dixon 1990; Lair and Wynn 2002a, b). A sufficient quantity of mulch can suppress secondary weed flushes, buffer adverse environmental extremes (wind, temperature, erosion), and enhance moisture retention. However, very deep mulch may prevent the establishment of desirable species by creating a barrier to seeds reaching the soil or the emergence of sprouts from soil seed banks (Merritt and Johnson 2006).

In areas where the abundance and vigor of mycorrhizal spores are low, amending the soil may improve the performance of natives over nonmycorrhizal exotics (Allen and Allen 1984; Allen and Allen 1986; Hanson 1991; Beauchamp et al. 2009). Mycorrhizal inoculum (spores) can be obtained either commercially or by harvesting and incorporating soil from adjacent native stands, a preferred method as it does not use nonnative inoculum. Most commercial mycorrhizal inocula contain spores of *Glomus intraradices, G. mosseae, G. aggregatum,* and/or *G. fasciculatum.* All these species have been found in riparian areas dominated by cottonwood or willow (*Populus* or *Salix*) and on riparian terraces dominated by

FIGURE 23.3 Active restoration following *Tamarix* removal often requires extensive site manipulation, including grading the soil surface, seeding, or amending the soil, all of which were done on this site along the Rio Grande in Bosque del Apache National Wildlife Refuge, New Mexico.

Source: Photo courtesy of Vanessa B. Beauchamp.

sacaton (*Sporobolus wrightii*) in Arizona (Kennedy et al. 2002; Beauchamp et al. 2006).

High levels of soil salinity may be reduced by the mechanical creation of microtopographic relief on the soil surface, or with commercial soil amendments. The most commonly used products promote (1) a chemical reaction whereby soluble salts are converted to neutral or acidic compounds, or (2) physical adsorption of sodium (Na^{++}) via colloidal attachment and sequestration (Richards 1954). Although these products may reduce salinity or sodicity, their effectiveness is limited, first, by the cost of the higher application rates required in soils with high electrical conductivity, and second, by the need to be incorporated via tillage or irrigation for maximum efficacy, which is often unfeasible. Flood irrigation with fresh water can also reduce soil salinity (see chapter 8, this volume).

Plant material and species selection. When selecting plant materials for restoration projects, it is advisable to use container stock or seed collected locally (Burton and Burton 2002; McKay et al. 2005). At a minimum, plant materials should be adapted to the soils, water availability, elevation, and climate characteristic of the project site. Other considerations when selecting plants include germination rates, seedling vigor, seedbed preparation, seeding methods for field establishment, and the ability of planted species to reproduce and spread without further management.

Transplanting rooted plants often establishes plants more quickly than seeding. Seed banks or planted seeds can be negatively affected by seed predation, and young transplant stock can be affected by herbivory. Clusters of transplants, spaced and planted throughout a restoration site, can develop into seed-source "islands," long-term sources of propagules. Planning for the needs of transplant stock well in advance of installation is necessary to grow stock to the necessary size and maturity for persistence, and to prepare for necessary maintenance, such as irrigation.

Ecology of seeded species and seeding approaches. Following *Tamarix* removal, ruderal or weedy species may dominate the site for the first one to five years (McDaniel and Taylor 2003; Merritt and Johnson 2006). Long weed-dominated or bare-ground phases following *Tamarix* removal can be shortened by planting mixes of early-, mid-, and late-successional perennial species, in concert with sound integrated weed management measures. Some sites may first require the establishment of species that are better adapted to less than ideal environmental conditions until the site stabilizes.

The species composition of plants that become established following *Tamarix* removal strongly influences subsequent plant establishment and successional dynamics ("initial floristics" sensu Egler 1954; Gilpin 1987; Kline and Howell 1987; Allen 1995). For example, including vigorously reproducing species like quailbush (*Atriplex lentiformis*) in seeding of xeric *Tamarix* control sites commonly results in quailbush dominance for extended periods, inhibiting establishment of other species that were concurrently seeded (Pinkney 1992; Bay and Sher 2008).

Establishing cottonwoods (*Populus* spp.) or other shade canopy species can effectively suppress co-establishing *Tamarix* (Sher et al. 2002).

Seeding may promote competitive displacement of pioneering species by later-establishing, stress-tolerant plants ("facilitation models"; Grime 1979; Kline and Howell 1987). On highly disturbed substrates, native species establishment may be delayed, or desired successional trajectories may be altered when late successional species are planted exclusively in attempts to accelerate succession (Gilpin 1987; Allen 1995). Instead, the seeding and establishment of less competitive species can be followed by subsequent interseeding or overseeding of more competitive species (Romney et al. 1987; Redente and Depuit 1988).

Suppressing competition from *Tamarix* or secondary weeds may help determine the composition and sequence of initial and subsequent plantings. For example, along the upper Pecos River in southeastern New Mexico, long-term (50- to 60-year) chemical and mechanical *Tamarix* control converted many riparian habitats to monotypic stands of kochia (*Bassia scoparia*). Native grasses were then seeded on some sites, and once they were sufficiently established to suppress kochia (in concert with herbicidal kochia control measures), the grass community was augmented by interseeding desirable forbs and shrubs.

COMBINING ACTIVE AND PASSIVE RESTORATION

In cases where passive approaches alone may not be able to completely restore key ecological functions, combinations of passive and active approaches that seek to mimic natural processes have proven to be effective. For example, several projects have been successful in establishing desirable riparian vegetation by providing favorable hydrology, sometimes in combination with *Tamarix* control or native seed augmentation (Friedman et al. 1995; Roelle and Gladwin 1999; Bhattacharjee et al. 2006; see also chapter 21, this volume). Native species establishment can be enhanced by first amending site conditions and then broadcasting seeds, planting seedlings or cuttings, or otherwise directly improving conditions for colonization, establishment, and growth (Merritt and Johnson 2006). Where managed floods are too small to sufficiently disturb the site and create bare substrate, vegetation removal can be used to achieve this (Taylor et al. 1999; Cooper and Andersen 2012).

Post-Project Monitoring, Maintenance, and Adaptive Management

To assess the success of *Tamarix* removal and restoration activities in meeting a project's goals and objectives, it is critical to follow up with monitoring, evaluation, and, often, site maintenance. However, monitoring and maintenance are frequently left out of the planning process or are underbudgeted (Holmes et al. 2005). Clearly documenting the ecological outcomes (or unintended costs) of restoration projects should be prioritized in the funding, permitting, and regulatory settings at the local, state, and federal levels (Follstad Shah et al. 2007). Furthermore, strategic funding mechanisms should be established for long-term monitoring and documenting of project outcomes. Funding duration should be commensurate with the goals of monitoring.

The scope, methods, and frequency of monitoring should be decided prior to project implementation based on the project's goals and objectives and the extent to which an experimental approach is emphasized. There are many benefits to incorporating statistically sound monitoring designs, including adequate replication and controls (e.g., before and after treatment monitoring on treated sites and untreated controls). Conducting restoration projects within an experimental framework enables current and past successes and failures to inform future control and restoration projects (Zedler and Callaway 1993). Accurate assessments of control and restoration outcomes may take several years or decades (Palmer et al. 2007); short- and long-term biological and physical responses can also differ. The efficacy of *Tamarix* control may be high immediately following treatments; however, resprouting and recolonization can occur over several years (see chapter 20, this volume). That many years may be required to assess project "success" is important to consider and articulate when establishing specific project objectives and stakeholder expectations.

Site maintenance following control and restoration can be crucial for meeting project objectives, and, like monitoring, it requires advanced planning and an adequate budget (Briggs 1996; Briggs and Cornelius 1998; Briggs and Flores 2003; Sher et al. 2010). Site maintenance can include actions to help desirable vegetation become established or survive (especially in the first 2–5 years), such as irrigation (Bunting et al. 2011), reducing competition from undesirable weed species, repairing irrigation systems, replanting if necessary, and maintaining livestock fences.

An iterative process of learning from previous projects is the essence of an adaptive management framework and a key element in any restoration planning process (figure 23.1; Pastorok et al. 1997). Adaptive management is largely dependent upon rigorous monitoring and hypothesis testing to identify aspects of *Tamarix* removal and associated restoration actions that could be improved. Recommendations for adjustments or maintenance needs may be incorporated into the later implementation activities of a given project, or, if results are made broadly available to the appropriate natural resource and scientific personnel, the recommendations may benefit similar projects that have yet to be undertaken.

Conclusions and Future Research Needs

Restoration of areas that have been occupied by *Tamarix* often does not result from *Tamarix* removal alone. An understanding of site conditions is essential, including long- and short-term patterns of surface and groundwater hydrology, soils, and such key biotic factors as propagule availability, competitive interactions, and physiological requirements and tolerances of desired species. In addition to current conditions, future scenarios of climate change, water management, and land use are important considerations in the objective setting, planning, and implementation stages of restoration. In some situations, the combined effects of river regulation, dominance by exotic species, management to control exotic species, and climate change may have created novel conditions that make "restoration" to a historical natural condition or dynamic unlikely, and transformation to a novel state more likely (Hobbs et al. 2006; Seastedt et al. 2008; Cooper and Andersen 2012). These situations highlight the importance of a solid monitoring and adaptive management framework.

It is best to implement passive restoration approaches (e.g., the restoration of natural flow regimes) whenever possible, as they have greater likelihood for larger-scale and longer-term success. Choosing sites that contain nonnative species but retain natural flow regimes can be very effective. In many cases, however, conditions that favored native vegetation may have been irreparably altered (Stromberg et al. 2007b; Merritt and Poff 2010; Mortenson and Weisberg 2010). Most of these sites likely contain one or more environmental constraints to native riparian species establishment, including deep groundwater, infrequent (or absent) flooding, high soil salinity or alkalinity, and low, variable precipitation. Careful selection of suitable species is therefore critical. Greater effort and expense are required to restore native species under these conditions, and the probability of success is lower. Restoration efforts are also more likely to be successful if they include a clear articulation of the control and restoration methods to be used, as well as budgets and plans for maintenance, monitoring, and adaptive management.

Tamarix removal is being undertaken in a range of situations with a complex set of site conditions that present significant challenges (Dudley and DeLoach 2004; Merritt and Johnson 2006). Research is needed to increase our understanding of which actions are best implemented under various conditions and to improve our ability to estimate costs and predict likely benefits. Developing a better understanding of the most appropriate restoration approaches in different geomorphic and climatic settings would help during the planning and objective setting phases of restoration, and enable transfer of findings from one site to climatically and geomorphically similar ones. More work on determining desired trajectories for vegetation communities following restoration would aid in establishing the appropriate preconditions, choosing appropriate site treatments and species to restore. The environmental tolerances of many candidate taxa are relatively poorly understood; research could clarify key tolerance ranges (cf. Beauchamp et al. 2009). Whenever possible, it is crucial to include a thoughtfully designed experimental component as part of restoration projects, so that the results of particular actions can be rigorously evaluated.

It is encouraging that the current trend is away from removal of exotics with the expectation that this alone constitutes restoration. Movement toward approaches that improve techniques, guide land managers toward the most appropriate approaches in a variety of settings, and inform future projects has improved restoration projects. *Tamarix* is and will remain an important component of riparian vegetation throughout the arid western United States. Reducing the extent to which it dominates vegetation in riparian areas is an achievable goal through a combination of *Tamarix* removal and strategic restoration approaches.

Literature Cited

Allen, E. B. 1995. Restoration ecology: Limits and possibilities in arid and semiarid lands. Pages 7–15 *in* B. A. Roundy, E. D. McArthur, J. S. Haley, and D. K. Mann, editors. Proceedings: Wildland shrub and arid land restoration symposium October 19–21, 1993, Las Vegas, NV.

Allen, E. B., and M.F. Allen. 1984. Competition between plants at different successional stages: mycorrhizae as regulators. Canadian Journal of Botany 62:2625–2629.

Allen, M. F. 1991. *The Ecology of Mycorrhizae*. Cambridge University Press, Cambridge.

Allen, M.F., and E.B. Allen. 1986. Competition and mycorrhizae: Patterns and mechanisms. American Journal of Botany 73: 692.

Anderson, B. W. 1995. Salt cedar, revegetation and riparian ecosystems in the Southwest. Proceedings of the California Exotic Plant Pest Council 1995 Symposium. http://www.cal-ipc.org/symposia/archive/pdf/1995_symposium_proceedings1797.pdf.

Anderson, B. W., R. E. Russell, and R. D. Ohmart. 2004. *Riparian Revegetation: An Account of 2 Decades of Experience in the Arid Southwest*. Avvar Books, Blythe, CA.

Bateman, H.L., A. Chung-MacCoubrey, and H. L. Snell. 2008. Impact of non-native plant removal on lizards in riparian habitats in the southwestern United States. Restoration Ecology 16:180–190.

Bay, R. F., and A. A. Sher. 2008. Success of active revegetation after *Tamarix* removal in riparian ecosystems of the southwestern USA: A quantitative assessment of past restoration projects. Restoration Ecology 15:113–128.

Beauchamp, V. B., C. Walz, and P. B. Shafroth. 2009. Salinity tolerance and mycorrhizal responsiveness of native xeroriparian plants in semi-arid western USA. Applied Soil Ecology 43:175–184.

Beauchamp, V. B., J. C. Stromberg, and J. C. Stutz. 2006. Arbuscular mycorrhizal fungi associated with *Populus-Salix* stands in a semiarid riparian ecosystem. New Phytologist 2:369–380.

Beauchamp, V. B., and J. C. Stutz. 2005. Interactions between *Tamarix ramosissima* (saltcedar), *Populus fremontii* (cottonwood), and mycorrhizal fungi: Effects on seedling growth and plant species coexistence. Plant and Soil 275:221–231.

Beauchamp, V. B., and P. B. Shafroth. 2011. Floristic composition, beta diversity and nestedness of reference sites for restoration of xeroriparian areas in semi-arid western USA. Ecological Applications 21:465–476.

Bernstein, L. 1958. Salt tolerance of grasses and forage legumes. USDA Agricultural Information Bulletin No. 194. Washington, DC.

Bhattacharjee, J., J. P. Taylor, and L. M. Smith. 2006. Controlled flooding and staged draw-down for restoration of native cottonwoods in the middle Rio Grande valley, New Mexico, USA. Wetlands 26:691–702.

Briggs, M. K. 1996. Riparian ecosystem restoration in arid lands: Strategies and references. University of Arizona Press, Tucson.

Briggs, M. K., and M. L. Flores. 2003. Small-scale restoration project promotes regional restoration in the Colorado River delta: A testament to the power of tangible restoration at the community level. Southwest Hydrology 2:24–26.

Briggs, M. K., and S. Cornelius. 1998. Opportunities for ecological improvement along the lower Colorado River and delta. Wetlands 18:513–529.

Bunting, D. P., S. A. Kurc, and M. R. Grabau. 2011. Using existing agricultural infrastructure for restoration practices: Factors influencing successful establishment of *Populus fremontii* over *Tamarix ramosissima*. Journal of Arid Environments 75:851–860.

Burton P. J., and C. M. Burton. 2002. Promoting genetic diversity in the production of large quantities of native plant seed. Ecological Restoration 20:117–123.

Clewell, A., J. Rieger, and J. Munro. 2005. Society for Ecological Restoration International: Guidelines for developing and managing ecological restoration projects. [Online]. http://www.ser.org/content/guidelines_ecological_restoration.asp. Accessed November 22, 2005.

Cooper, D. J., and D. C. Andersen. 2012. Novel plant communities limit the effects of a managed flood to restore riparian forests along a large regulated river. River Research and Applications. 28:204–215.

Cooper, D. J., D. R. D'Amico, and M. L. Scott. 2003. Physiological and morphological response patterns of *Populus deltoides* to alluvial ground water pumping. Environmental Management 31:215–226.

DeWine, J. M., and D. J. Cooper. 2008. Canopy shade and the successional replacement of tamarisk by native box elder. Journal of Applied Ecology 45:505–514.

Dick-Peddie, W. A. 1993. *New Mexico Vegetation: Past, Present and Future*. University of New Mexico Press. Albuquerque.

Dixon R. M. 1990. Land imprinting for dryland revegetation and restoration. Pages 14–22 *in* J. J. Berger, editor. *Environmental Restoration: Science and Strategies for Restoring the Earth*. Island Press, Covelo, CA.

Dudley, T. L., and C. J. DeLoach. 2004. Saltcedar (*Tamarix* spp.), endangered species, and biological weed control: Can they mix? Weed Technology 18:1542–1551.

Egler, F. E. 1954. Vegetation science concepts. I. Initial floristic composition: A factor in old-field development. Vegetatio 4:412–417.

Federal Water Pollution Control Administration. 1968. Water quality criteria: Report to the National Technical Advisory Committee to the Secretary of the Interior. Federal Water Pollution Control Administration, Washington, DC.

Follstad Shah, J. J., C. N. Dahm, S. P. Gloss, and E. S. Bernhardt. 2007. River and riparian restoration in the Southwest: Results of the national river restoration science synthesis project. Restoration Ecology 15:550–562.

Friedman, J. M., M. L. Scott, and W. M. Lewis Jr. 1995. Restoration of riparian forest using irrigation, artificial disturbance, and natural seedfall. Environmental Management 19:547–557.

Gilpin, M. E. 1987. Experimental community assembly: Competition, community structure and the order of species introductions, Pages 151–162 *in* W. R Jordan III, M. E.

Gilpin, and J. D. Aber, editors. *Restoration Ecology: A Synthetic Approach to Ecological Research*. Cambridge University Press, Cambridge.

Glenn, E. P., K. Morino, P. L. Nagler, S. Murray, S. Pearlstein, K. Hultine. 2012. Roles of salt-cedar (*Tamarix* spp.) and capillary rise in salinizing a non-flooding terrace on a flow regulated desert river. Journal of Arid Environments 79:56–65.

Goodson, J. M., A. M. Gurnell, P. G. Angold, and I. P. Morrissey. 2001. Riparian seed banks: Structure, process and implications for riparian management. Progress in Physical Geography 25:310–325.

Grime, J. P. 1979. *Plant Strategies and Vegetation Processes*. John Wiley & Sons, New York, NY.

Hanson, D. E. 1991. Russian knapweed interference with corn VA mycorrhiza, western wheatgrass, and smooth brome. M.S. thesis. Colorado State University, Fort Collins.

Harms, R. S., and R. D. Hiebert. 2006. Vegetation response following invasive tamarisk (*Tamarix* spp.) removal and implications for riparian restoration. Restoration Ecology 14:461–472.

Herbel, C. H., G. H. Abernathy, C. C. Yarbrough, and D. K. Gardner. 1973. Rootplowing and seeding arid rangelands in the Southwest. Journal of Range Management 26:193–197.

Hobbs, R. J., S. Arico, J. Aronson, J. S. Baron, P. Bridgewater, V. A. Cramer, P. R. Epstein, J. J. Ewel, C. A. Klink, A. E. Lugo, D. Norton, D. Ojima, D. M. Richardson, E. W. Sanderson, F. Valldares, M. Vila, R. Zamora, and M. Zobel. 2006. Novel ecosystems: Theoretical and management aspects of the new ecological world order. Global Ecology and Biogeography 15:1–7.

Holmes, P. M., D. M. Richardson, K. J. Esler, E. T. F. Witkowski, and S. Fourie. 2005. A decision-making framework for restoring riparian zones degraded by invasive alien plants in South Africa. South African Journal of Science 101:553–564.

Hudgeons, J. L., A. E. Knutson, K. M. Heinz, C. J. DeLoach, T. L. Dudley, R. R. Pattison, J. R. Kiniry. 2007. Defoliation by introduced *Diorhabda elongata* leaf beetles (Coleoptera: Chrysomelidae) reduces carbohydrate reserves and regrowth of *Tamarix* (Tamaricaceae). Biological Control 43:213–221.

Hughes, F. M. R. 1997. Floodplain biogeomorphology. Progress in Physical Geography 21:510–529.

Hughes, F. M. R., A. Colston, and J. O. Mountford. 2005. Restoring riparian ecosystems: The challenge of accommodating variability and designing restoration trajectories. Ecology and Society 10: 12. [Online]. http://www.ecologyandsociety.org/vol10/iss1/art12/.

Hughes, F. M. R., W. M. Adams, E. Muller, C. Nilsson, K. S. Richards, N. Barsoum, H. Decamps, R. Foussadier, J. Girel, H. Guilloy, A. Hayes, M. Johansson, L. Lambs, G. Pautou, J. L. Peiry, M. Perrow, F. Vautier, and M. Winfield. 2001. The importance of different scale processes for the restoration of floodplain woodlands. Regulated Rivers: Research and Management 17:325–345.

Hultine, K. R., J. Belnap, C. vanRiper, J. R. Ehleringer, P. E. Dennison, M. E. Lee, P. L. Nagler, K. A. Snyder, S. M. Uselman, and J. B. West. 2010. Tamarisk biocontrol in the western United Sates: Ecological and societal implications. Frontiers in Ecology and the Environment 8:467–474.

Jolly, I. D., G. R. Walker, and P. J. Thorburn. 1993. Salt accumulation in semi-arid floodplain soils with implications for forest health. Journal of Hydrology 150:589–614.

Kennedy, L. J., R. L. Tiller, and J. C. Stutz. 2002. Associations between arbuscular mycorrhizal fungi and *Sporobolus wrightii* in riparian habitats in arid South-western North America. Journal of Arid Environments 50:459–475.

Kline, V. A., and E. A. Howell. 1987. Prairies. Pages 75–83 *in* W. R. Jordan III, M. E. Gilpin, and J. D. Aber, editors. *Restoration Ecology: A Synthetic Approach to Ecological Research.* Cambridge University Press, Cambridge.

Kondolf, G. M., and H. Piegay (editors). 2003. *Tools in Fluvial Geomorphology.* John Wiley and Sons, Chichester, West Sussex.

Lair, K. D., and S. L. Wynn. 2002a. Revegetation strategies and technology development for restoration of xeric *Tamarix* infestation sites. Technical Memo. No. 8220-02-04. Bureau of Reclamation Technical Service Center, Denver, CO.

Lair, K. D., and S. L. Wynn. 2002b. Revegetation strategies and technology development for restoration of xeric *Tamarix* infestation sites following fire. Technical Memo No. 8220-02-06. Bureau of Reclamation Technical Service Center, Denver, CO.

Lite, S. J., and J. C. Stromberg. 2005. Surface water and ground-water thresholds for maintaining *Populus-Salix* forests, San Pedro River, Arizona. Biological Conservation 125:153–167.

Mahoney, J. M., and S. B. Rood. 1998. Streamflow requirements for cottonwood seedling recruitment: An interactive model. Wetlands 18:634–645.

McDaniel, K. C., and J. P Taylor. 2003. Saltcedar recovery after herbicide-burn and mechanical clearing practices. Journal of Range Management 56:439–445.

McKay J. K., C. E. Christian, S. Harrison, and K. J. Rice. 2005. "How local is local?" A review of practical and conceptual issues in the genetics of restoration. Restoration Ecology 13: 432–440.

Meinzer, O. E. 1927. Plants as indicators of ground water. U.S. Geological Survey Water Supply Paper 577. Washington, DC.

Merritt, D. M., and E. E. Wohl. 2002. Processes governing hydrochory along rivers: Hydraulics, hydrology, and dispersal phenology. Ecological Applications 12:1071–1087.

Merritt, D. M., and Johnson, J.B. 2006. Vegetation aspects of the middle Rio Grande riparian restoration. Progress report, Middle Rio Grande Ecosystem Study. U.S. Forest Service, Rocky Mountain Research Station, Albuquerque, NM.

Merritt, D. M., M. L. Scott, N. L. Poff, G. T. Auble, and D. A. Lytle. 2010. Theory, methods and tools for determining environmental flows for riparian vegetation: Riparian vegetation-flow response guilds. Freshwater Biology 55:206–225.

Merritt, D. M., and N. L. Poff. 2010. Shifting dominance of riparian *Populus* and *Tamarix* along gradients of flow alteration in western North American rivers. Ecological Applications. 20:135–152.

Merritt, D. M., and P. B. Shafroth. In Press. Edaphic, salinity, and stand structural trends in chronosequences of native and non-native dominated riparian forests along the Colorado River, USA. Biological Invasions. Published online, June 20, 2012. doi:10.1007/s10530-012-0263-4.

Molles, M. C., C. S. Crawford, L. M. Ellis, H. M. Valett, and C. N. Dahm. 1998. Managed flooding for riparian ecosystem restoration. BioScience 48:749–756.

Mortenson, S.G., and P.J. Weisberg. 2010. Does river regulation increase the dominance of invasive woody species in riparian landscapes? Global Ecology and Biogeography 19:562–574.

Mortenson, S.G., P.J. Weisberg, and B.E. Ralston. 2008. Do beavers promote the invasion of non-native *Tamarix* in the Grand Canyon riparian zone? Wetlands 28:666–675.

Nagler, P. L., E. P. Glenn, C. S. Jarnevich, and P. B. Shafroth. 2011. Distribution and abundance of saltcedar and Russian olive in the western United States. Critical Reviews in Plant Sciences 30:508–523.

Nagler, P. L., O. Hinojosa-Huerta, E. P. Glenn, J. Garcia-Hernandez, R. Romo, C. Curtis, A. R. Huete, and S. G. Nelson. 2005. Regeneration of native trees in the presence of invasive saltcedar in the Colorado River delta. Conservation Biology 19:1842–1852.

Natural Resources Conservation Service. 1996. Plant materials for saline-alkali soils. NRCS Montana Plant Materials Technical Note No. 26 (rev). USDA-NRCS, Bridger, MT.

Ogle, D. 1994. Salt tolerance of plants. Plant Materials Technical Note No. 9. USDA-NRCS, Boise, ID.

Palmer, M. A., J. D. Allan, J. Meyer, and E. S. Bernhardt. 2007. River restoration in the twenty-first century: Data and experiential knowledge to inform future efforts. Restoration Ecology 15:472–481.

Pastorok, R. A., A. MacDonald, J. R. Sampson, P. Wilber, D. J. Yozzo, and J. P. Titre. 1997. An ecological decision framework for environmental restoration projects. Ecological Engineering 9:89–107.

Pattison, R. R., C.M. D'Antonio, T. L. Dudley, K. K. Allander, and B. Rice. 2011. Early impacts of biological control on canopy cover and water use of the invasive saltcedar tree (*Tamarix* spp.) in western Nevada, USA. Oecologia 165:605–616.

Paxton, E. H., T. D. Theimer, and M. K. Sogge. 2011. Tamarisk biocontrol using tamarisk beetles: Potential consequences for riparian birds in the southwestern United States. Condor 11:255–265.

Pinkney, F. C. 1992. *Revegetation and Enhancement of Riparian Communities along the Lower Colorado River*. U.S. Department of the Interior, Bureau of Reclamation, Ecological Resources Division, Denver, CO.

Quimby, P. C., C. J. DeLoach, S. A. Wineriter, J. A. Goolsby, R. Sobhian, C. D. Boyette, and H. K. Abbas. 2003. Biological control of weeds: Research by the U.S. Department of Agriculture-Agricultural Research Service, Selected case studies. Pest Management Science. 59:671–680.

Redente, E. F., and E. J. Depuit. 1988. Reclamation of drastically disturbed rangeland. Pages 559–584 *in* P. T. Tueller, editor. *Vegetation Science Applications for Rangeland Analysis and Management*. Kluwer Academic Publishers, The Netherlands.

Richards, K., J. Brasington, and F. Hughes. 2002. Geomorphic dynamics of floodplains: Ecological implications and a potential modeling strategy. Freshwater Biology 47:559–579.

Richards, L.A., editor. 1954. *Diagnosis and Improvement of Saline and Alkali Soils*. U.S. Department of Agriculture, Agricultural Research Service, Soil and Water Conservation Research Branch, Washington, DC.

Richardson, D. M., P. M. Holmes, K. J. Esler, S. M. Galatowitsch, J. C. Stromberg, S. P. Kirkman, P. Pysek, and R. J. Hobbs. 2007. Riparian vegetation: Degradation, alien plant invasions, and restoration prospects. Diversity and Distributions 13:126–139.

Roelle, J., and D. Gladwin. 1999. Establishment of woody riparian species from natural seedfall at a former gravel pit. Restoration Ecology 7:183–192.

Romney, E. M., A. Wallace, and R. B. Hunter. 1987. Pulse establishment of woody shrubs on denuded Mojave Desert land. Pages 54–57 *in* E. D. McArthur, A. Wallace, and M. Haferkamp, editors. Proceedings: Symposium on shrub ecophysiology and biotechnology. USDA Forest Service General Technical Report INT-GTR-256. Ogden, UT.

Rood, S. B., G. M. Samuelson, J. H. Braatne, C. R. Gourley, F. M. R. Hughes, and J. M. Mahoney. 2005. Managing river flows to restore floodplain forests. Frontiers in Ecology and the Environment 3:193–201.

Scott, M. L., P. B. Shafroth, and G. T. Auble. 1999. Responses of cottonwoods to alluvial water table declines. Environmental Management 23:347–358.

Seastedt, T. R., R. J. Hobbs, and K. N. Suding. 2008. Management of novel ecosystems: Are novel approaches required? Frontiers in Ecology and the Environment 10:547–553.

Shafroth, P. B., A. C. Wilcox, D. A. Lytle, J. T. Hickey, D. C. Andersen, V. B. Beauchamp, A. Hautzinger, L. E. McMullen, and A. Warner. 2010b. Ecosystem effects of environmental flows: Modeling and experimental floods in a dryland river. Freshwater Biology 55:68–85.

Shafroth, P. B., C. A. Brown, and D. M. Merritt, editors. 2010a. Saltcedar and Russian olive control demonstration act science assessment. U.S. Geological Survey Scientific Investigations Report 2009-5247. U.S. Department of the Interior, U.S. Geological Survey, Reston, VA.

Shafroth, P. B., G. T. Auble, J. C. Stromberg, and D. T. Patten. 1998. Establishment of woody riparian vegetation in relation to annual patterns of streamflow, Bill Williams River, Arizona. Wetlands 18:577–590.

Shafroth, P. B., J. C. Stromberg, and D. T. Patten. 2000. Woody riparian vegetation response to different alluvial water table regimes. Western North American Naturalist 60:66–76.

Shafroth, P. B., J. R. Cleverly, T. L. Dudley, J. P. Taylor, C. van Riper III, E. P. Weeks, and J. N. Stuart. 2005. Control of *Tamarix* in the western United States: Implications for water salvage, wildlife use, and riparian restoration. Environmental Management 35:231–246.

Shafroth, P. B., and M. K. Briggs. 2008a. Restoration ecology and invasive riparian plants: An introduction to the special section on *Tamarix* spp. in western North America. Restoration Ecology 16:94–96.

Shafroth, P.B., V.B. Beauchamp, M.K. Briggs, K. Lair, M.L. Scott, and A.A. Sher. 2008b. Planning riparian restoration in the context of *Tamarix* control in western North America. Restoration Ecology 16:97–112.

Sher, A. A., and D. L. Marshall. 2003. Competition between native and exotic floodplain tree species across water regimes and soil textures. American Journal of Botany 90:413–422.

Sher, A. A., D. L. Marshall, and J. P. Taylor. 2002. Establishment patterns of native *Populus* and *Salix* in the presence of invasive nonnative *Tamarix*. Ecological Applications 12:760–772.

Sher, A. A., K. Lair, M. DePrenger-Levin, and K. Dohrenwend. 2010. Best management practices for revegetation in the upper Colorado River basin. Denver Botanic Gardens, Denver, CO.

Smith, S. E., and J. D. Read. 1997. *Mycorrhizal Symbiosis*. Academic Press, San Diego, CA.

Stromberg, J. C. 2001. Restoration of riparian vegetation in the south-western United States: Importance of flow regimes and fluvial dynamism. Journal of Arid Environments 49:17–34.

Stromberg, J. C., D. T. Patten, and B. D. Richter. 1991. Flood flows and dynamics of Sonoran riparian forests. Rivers 2:221–235.

Stromberg, J. C., S. J. Lite, R. Marler, C. Paradzick, P. B. Shafroth, D. Shorrock, J. M. White, and M. S. White. 2007b. Altered stream-flow regimes and invasive plant species: The *Tamarix* case. Global Ecology and Biogeography 16:381–393.

Stromberg, J. C., V. B. Beauchamp, M. D. Dixon, S. Lite, and C. Paradzick. 2007a. Importance of low-flow and high-flow characteristics to restoration of riparian vegetation in arid south-western United States. Freshwater Biology 52:651–679.

Swift, C. E. 2010. Salt tolerance of various temperate zone ornamental plants. Colorado State University Cooperative Extension Service. Tri River Area, Grand Junction, Colorado. [Online]. http://www.coopext.colostate.edu/TRA/PLANTS/stable.shtml. Accessed November 22, 2011.

Tangee, K. K. and N. C. Kielen. 2002. Agricultural drainage water management in arid and semi-arid areas. FAO Irrigation and Drainage Paper 61. Food and Agriculture Organization of the United Nations. Rome.

Taylor, J. P., D. B. Wester, and L. M. Smith. 1999. Soil disturbance, flood management, and riparian woody plant establishment in the Rio Grande floodplain. Wetlands 19:372–382.

Taylor, J. P., and K. C. McDaniel. 2004. Revegetation strategies after saltcedar (*Tamarix* spp.) control in headwater, transitional, and depositional watershed areas. Weed Technology 18:1278–1282.

Titus, J. H., P. J. Titus, R. S. Nowak, and S. D. Smith. 2002. Arbuscular mycorrhizae of Mojave Desert plants. Western North American Naturalist 62:327–334.

Vandersande, M. W., E. P. Glenn, and J. L. Walworth. 2001. Tolerance of five riparian plants from the lower Colorado River to salinity drought and inundation. Journal of Arid Environments 49:147–159.

Vincent, K. R., J. M. Friedman, and E. R. Griffin. 2009. Erosional consequence of saltcedar control. Environmental Management 44:218–227.

Webb, A. A. and W. D. Erskine. 2003. A practical scientific approach to riparian vegetation rehabilitation in Australia. Journal of Environmental Management 68:329–341.

Webb, R. H., S. A. Leake, and R. M. Turner. 2007. The ribbon of green: Change in riparian vegetation in the southwestern United States. University of Arizona Press, Tucson.

Weeks, E. P., H. L. Weaver, G.S. Campbell, and B.D. Tanner. 1987. Water use by saltcedar and by replacement vegetation in the Pecos River floodplain between Acme and Artesia, New Mexico. U.S. Geological Survey Professional Paper 491–G.

24

Revegetation after Tamarisk Removal

WHAT GROWS NEXT?

Robin F. Bay

Tamarisk (*Tamarix* spp., also saltcedar) is the dominant woody species in many riparian areas of the American southwest (Brotherson and Winkel 1986; Friedman et al. 2005). Its management has been a prominent environmental and political issue since the 1970s. However, the question of what vegetation would and should be established after tamarisk removal was late to enter the public awareness and debate. For many years, the general public understanding was that tamarisk was a malicious invader actively dominating native riparian landscapes (see chapter 16, this volume). It was thought to be displacing the native cottonwood and willow trees through competition (Barrows 1998) and excessive water use (e.g., DiTomaso 1998) and providing poorer wildlife habitat than native riparian plant species (Engel-Wilson and Ohmart 1978; Ellis 1995; Bailey et al. 2001). Thus, if tamarisk was the problem, removing it would be the simple solution. With such a strong focus on control and eradication, many land managers failed to consider what would replace tamarisk after it was removed. Over the past decade this issue has become a more explicit focus of the tamarisk site restoration process, but it still lacks the emphasis it deserves.

I began research on revegetation after tamarisk removal in 2004 as a part of my graduate studies. My graduate professor and I were interested in looking at past revegetation projects (both successful and unsuccessful) and evaluating all aspects of these projects (Bay and Sher 2008). The first (unexpected) challenge to this research was that most land managers who were doing tamarisk removal were not doing revegetation. We talked to dozens of land managers throughout the desert Southwest looking for projects where tamarisk had been removed and active revegetation completed. Most of the land managers we contacted had completed the tamarisk removal step, but not the revegetation. In some cases this was due to budget constraints; in others, an assumption that tamarisk removal would be enough to restore the community. After an extensive search, we found seven land managers in Arizona, Nevada, and New Mexico who had completed a total of 28 projects that met our research criteria.

In this chapter, I outline what we and others have found regarding successful post-tamarisk revegetation. I discuss how success can be defined, measured, and modeled, and what factors contribute to success. Finally, I make a case for revegetation as not only a step in the tamarisk management process, but as a key component in the planning and design phases of these projects.

BACKGROUND

There is little debate that tamarisk is associated with many changes in riparian environments (see chapters 9 and 15, this volume). These include not only shifts in native species composition, surface and groundwater quantities, and wildlife habitat but also decreased water quality, decreased plant and animal biodiversity, increased soil salinity and pH, increased fire frequency, and stream channelization (Zouhar 2003). Recent papers and several chapters in this book discuss tamarisk's role as the driver or the passenger of the environmental changes associated with tamarisk-dominated sites (e.g., see chapters 16, 9, and 15, this volume; Stromberg et al. 2009). Regardless of causal relationships, if tamarisk-dominated sites exhibit many of these associated characteristics, then the natural regeneration of native species would be unlikely. In many cases, the water quality and quantity, along with the altered soil chemistry, are sufficiently different from a formerly cottonwood/ willow-dominated site to preclude their unassisted return. These changes cannot be ignored when considering the revegetation of tamarisk-dominated sites.

While there have been cases where native species were shown to regenerate naturally after tamarisk removal in Arizona (e.g., Briggs et al. 1994) and New Mexico (e.g., Taylor et al. 1999), a 2006 survey of natural recovery at 33 sites throughout the Southwest suggested that unassisted regeneration is unusual (Harms and Hiebert 2006). Harms and Hiebert compared mechanically cleared or burned sites to untreated sites and concluded that native communities do not recover sufficiently on their own, and that active revegetation is necessary to meet management goals for riparian ecosystem recovery. Other studies have shown that in some cases tamarisk removal followed by revegetation with native species is still not sufficient to achieve riparian ecosystem recovery (Carothers et al. 1989). Due to many of the soil and hydrological changes associated with tamarisk-dominated sites, some restoration projects will likely require additional components such as manipulation of flooding patterns, soil amendments, and long-term weed management (Carothers et al. 1989; Sher et al. 2010).

Revegetation and restoration after tamarisk removal have been attempted since the 1970s; however, much of this work was completed as an afterthought. Planning for revegetation as a primary component of a tamarisk control project did not become common until the last decade. It is still not always a key part of removal projects, but it is gaining acceptance. The National Invasive Species Council has said that "restoration is an integral component of comprehensive prevention and control programs for invasive species that may keep invasive species from causing greater environmental disturbances" (National Invasive Species Management Plan 2001). Further, research is now showing that restoring native riparian vegetation in tamarisk-dominated areas can result in economic and social benefits, including recreational, aesthetic, and land-use improvements (Zavaleta 2000).

COSTS AND BENEFITS

Studies of ecological benefits of revegetation projects usually focus on improved plant and animal diversity and habitat (Taylor and McDaniel 1998). Several researchers have shown that tamarisk seedlings are not strong competitors (Sher et al. 2000, 2002; Sher and Marshall 2003; Bhattacharjee et al. 2009). Therefore, aggressive revegetation work can prevent the reestablishment of tamarisk. Planting desirable species can also mitigate the invasion of other undesirable weedy species (Shafroth et al. 2005). Additionally, active revegetation can help to mitigate erosion problems that have followed some tamarisk removal projects (Vincent et al. 2009).

Given these benefits, it would seem that active revegetation would be an obvious component in any tamarisk management plan. However, revegetation projects can be expensive and time consuming, and proper planning is important ensure success. The past two decades have seen a variety of studies addressing riparian revegetation strategies and successes (Carothers et al. 1989; Pinkney 1992; Briggs et al. 1994; Briggs and Cornelius 1998; Taylor et al. 1999). Some studies have even have even focused specifically on revegetation after tamarisk removal (Anderson 1996; Taylor 1996; Taylor and McDaniel 1998, 2003, 2004). All these studies examined only one or a few specific locations, and all advocate basing revegetation strategies on particular ecological site characteristics. Specifically, they advocate evaluation of groundwater (depth and fluctuations), soil (salinity and texture), and water availability (flooding and/or irrigation) as the most important factors to consider when choosing revegetation strategies. Several of the unsuccessful projects evaluated by Bay and Sher (2008) could have been prevented if the planning phase had included soil analysis and/or groundwater testing and the revegetation species had been selected based on the results. Completing these tests during planning is worth the modest expense when compared to the costs of repeating a failed revegetation project.

What Makes Revegetation Successful?

DEFINING SUCCESS

The first hurdle in achieving revegetation success is to define it. The success of a project is measured by the achievement of desired goals. Defining these goals is integral to the design, implementation, and monitoring of a project (chapter 23, this volume).

Key to defining goals is determining whether the aim is restoration or simply revegetation. The difference may be subtle, but it can significantly shape the project design. Restoration goals include restoring the site to a previous state, which for many tamarisk-dominated sites means the creation of a cottonwood and willow riparian woodland. However, it also includes restoration of the historic hydrologic regime, which might not be possible in flood-controlled river systems. Revegetation is always a part of restoration, but it does not necessarily ensure that the site will be returned to a previous state. Given the many environmental changes associated with tamarisk-dominated sites, restoration may not always be possible. For instance, if the soil salinity is high and/or the water table is low, cottonwood and willow species will not survive. Selecting revegetation species that are better adapted to current site conditions could mean a successful revegetation project even if restoration of a cottonwood/willow riparian community is not achievable.

Success can be measured in many ways. A revegetation project could be successful because a desirable level of vegetation cover, species diversity, tree density, or habitat type is achieved. Success could be measured by decreased tamarisk cover, increased native plant cover or diversity, increased faunal populations or diversity, reduced stream bank erosion, or any combination of these and other measures. There is no single way to define success, but it is important to define it in the planning phase of a project to ensure that all aspects of the project will achieve the desired goal (see chapter 23, this volume)

FACTORS INFLUENCING SUCCESS

A variety of factors influence the success of riparian revegetation projects after tamarisk removal (Sher et al. 2010). These have included soil characteristics (salinity, pH, texture, and nutrients), depth to ground water, general water availability, flooding, weed management, and planting management, including species selection and fertilization. Most greenhouse and field trials that have investigated the impact of these factors have focused on soil characteristics and water availability.

Soil plays a critical role in revegetation success, providing the substrate, nutrients, and water-holding capacity for vegetation establishment. Soil chemistry, specifically salinity and alkalinity (pH), have often been shown to be higher in tamarisk-dominated sites compared to native sites (see chapter 8, this volume). Because of this, several authors have suggested that selecting species that are adapted to the soil chemistry present at the revegetation site is important for revegetation success (e.g., Carothers et al. 1989; Anderson 1996; Shafroth et al. 2005; Sher et al. 2010).

Soil texture is also likely to affect revegetation success, but its effect will again depend on the species selected. Coarser-textured soils (those high in sand) have been shown to provide a better substrate of seedling establishment of native riparian species both in the field (Friedman et al. 1995; Taylor et al. 1999; Bhattacharjee et al. 2008) and in controlled experiments (Sher and Marshall 2003). Coarse soils have better drainage and more topography for seed collection, but also have a lower water holding capacity than finer soils and are more susceptible to drying (Bhattacharjee et al. 2008).

Water availability is closely linked to soil texture, and is one of the most often-cited factors contributing to revegetation success. Native riparian species such as cottonwood and willow require moist soils to germinate (Friedman et al. 1995) and relatively shallow water tables to thrive. These species are adapted to gradually declining water tables, as naturally occurs after overbank flooding (Taylor et al 1999; Sher and Marshal 2003; Bhattacharjee et al. 2008). Rapid water table draw-down can both decrease their survival and favor tamarisk seedling establishment (Horton and Clark 2001). Additionally, extended periods without groundwater recharge can stress these native species (see chapter 9, this volume).

Flooding is the factor that can tie all of these previous elements together, and has been the focus of many tamarisk management research projects. Flooding influences not only seasonal water availability, but also soil chemistry and texture. Overbank flooding scours soils, decreasing soil salinity and pH. Flooding also redistributes soils and creates new sand bars and other bare substrate ideal for cottonwood seedling establishment (Friedman et al. 1995). When prescribed floods are timed to coincide with cottonwood and willow seedfall (as they would have been historically) they can provide the perfect conditions for recruitment and survival, including an open and moist substrate and gradual soil moisture draw-down as they recede (chapters 7 and 21, this volume). Many studies have shown that properly timed floods can improve cottonwood recruitment and survival both in field (Taylor and McDaniel 1998; Taylor et al. 1999) and controlled experiments (Levine and Stromberg 2001; Vandersande et al. 2001). There is also research to suggest that flooding is critical to the long-term success of any restoration effort in tamarisk-dominated sites (Stromberg et al. 2007) and that without flooding, the site characteristics that favor tamarisk over native species will continue to intensify (Anderson 1996).

MODELING SUCCESS

Currently, most tamarisk researchers and land managers agree about the importance of evaluating site characteristics to achieve revegetation success, and there is substantial agreement that the factors discussed earlier are the most important to evaluate. However, most studies have only focused on one or two of these factors, so it is difficult for land managers to determine the relative importance of each of these variables in achieving revegetation success in the field.

Filling this gap was the goal of our study of revegetation sites (Bay and Sher 2008). We designed this study to evaluate which site characteristics play the greatest role in revegetation success over time and across sites and regions. We wanted to identify what successful long-term revegetation projects have in common when it comes to management, soils, geography, hydrology, and climate that unsuccessful projects lack. We also wanted to see if the same factors cited as important in specific studies would prove to be important across all sites. Finally, we were interested in whether the characteristics cited by Harms and Hiebert (2006) as important in their survey of passive revegetation sites are the same in active revegetation sites. In short, by comparing results of active revegetation projects after tamarisk removal across regions (Mojave and Sonoran-Chihuahuan transitions ecoregions) and over time (18 years), we hoped to create guidelines to advise land and resource managers in the planning phases.

First, we surveyed 28 active revegetation sites without irrigation. This included 10 sites in Arizona and Nevada in the lower Colorado River watershed in the Mohave ecoregion, and 18 sites in the middle Rio Grande River watershed in the Sonoran-Chihuahuan transition ecoregion. Each site had been mechanically cleared of tamarisk, planted with native species, and had no additional mechanical tamarisk control. Each site was cleared between one and 18 years ago using a variety of removal methods. While 20 of the sites had been dense tamarisk stands prior to removal, eight had been mixed tamarisk stands with no more than 50% tamarisk cover. After tamarisk removal, 18 sites were planted with whole plants (either poles or containers, henceforth "stem-planted") of cottonwood, willow, mesquite, and/or seep willow (*Baccharis* spp.). The other ten sites were seeded with a mixture of saltbush (*Atriplex* spp.), dropseed (*Sporobolus* spp.), and other locally common, drought-tolerant species (henceforth "seeded"). Flooding had occurred since tamarisk removal on eight of the sites, where there was no damming, but not on the other 20 flood-controlled sites. In addition, we recorded geographic, climate, soil, hydrological, and management characteristics for each site, as well as vegetation cover, richness, and woody species density data.

With these data we created models to evaluate the relative importance of the various site characteristics in achieving a suite of parameters for success. We developed the models using regression tree analysis, which uses statistics to create predictive models for a single success variable with multiple predictor variables (i.e., site characteristics). The resulting "trees" branch from the "root" based on the site characteristics that best explain the variation in the success variable. For instance, the regression tree for tamarisk cover (see figure 24.1) shows that average tamarisk cover across all 28 evaluated sites is 3.14%. However, the mean tamarisk cover in the 11 sites that received more than 20.8 centimeters of annual rainfall was 0%, as opposed to the 17 sites with less rainfall where the mean cover was 5.17%. These 17 sites were further described by the presence or absence of flooding. Sites with less precipitation, but with flooding following tamarisk removal, had almost as little tamarisk cover (0.12%) as the sites with more rain. This model suggests that lower tamarisk cover can be predicted by higher water availability, with important implications for tamarisk reinvasion.

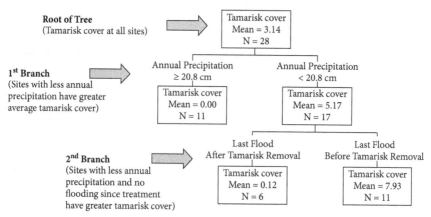

FIGURE 24.1 Regression tree to explain variability in tamarisk cover at 28 sites where tamarisk had been removed and active revegetation done.

Source: Adapted from Bay and Sher (2008).

Through the development and evaluation of dozens of these regression tree models, 15 success variables were well described by the models. For these variables, several site characteristics came out repeatedly as best predicting success (see table 24.1). Many of these variables, not surprisingly, were those that had been shown by previous studies to be the most important. Increased water availability in the form of precipitation, flooding, or nearby perennial flow was a primary descriptor for higher native vegetation cover, native perennial vegetation cover, and planted species relative density, as well as lower tamarisk and noxious species relative cover. Water availability was a factor on the regression trees for 11 of the 15 success variables reported. Soil was also key descriptive factor in nine of the regression trees with soil texture (percent sand or percent gravel) and soil salinity being of highest importance.

Management variables were also important in the regression tree models, including the revegetation method (stem-planting vs. seeding) and the size of the planting area (table 24.1). The smaller planting areas likely had increased total cover due to the increased opportunity for species other than those planted to immigrate via short-distance dispersal methods. The importance of revegetation method as a success factor suggested that on many of the sites evaluated, suitable methods were used for the given site characteristics. Either by design or by luck, appropriate species were selected, and appropriate methods were used to plant them. Additionally, the higher native perennial species richness and cover in stem-planted sites suggest that the plantings may have additional benefits over seeding, such as providing safe sites, that enhance the vegetation community beyond the obvious benefit of the planted species cover. Both total vegetation cover and planted species cover were higher in older sites according to the regression trees (table 24.1). This supports the results of Taylor et al. (2006) but contradicts those of Harms and Hiebert (2006) and those later collected by Belote et al. (2010). Harms and Hiebert (2006) and Belote et al. (2010) showed no improvement over time

TABLE 24.1

Characteristics that best described success variables in regression tree analyses of 28 sites where tamarisk was removed and active revegetation was done.

Success variable	Site characteristics in regression trees		
	1st Branch	2nd Branch	3rd Branch
Tamarisk Control			
↓ Tamarisk Cover	↑ Annual precipitation	⥮ Flooding after tamarisk removal	✘ ↑ Native species cover
↓ Tamarisk Relative Cover	↑ Annual precipitation	⥮ Flooding after tamarisk removal	▨ ↓ Soil salinity
↓ Tamarisk Relative Density	↓ Distance to perennial water	▨ ↓ Soil salinity	—
Plant Community Enhancement			
↑ Total Vegetation Cover	⊙ ↑ Time	▨ 40–60% Sand in soil ✘ Site size	▨ ↑ Gravel in soil
↑ Native Vegetation Cover	↓ Distance to perennial water	—	—
↑ Native Perennial Cover	↓ Distance to perennial water	—	—
↑ Native Relative Cover	↑ Growing season precipitation	▨ ↑ Gravel in soil	—
↑ Native Perennial Relative Cover	▨ ↑ Gravel in soil	✘ Pole planting	↑ Growing season precipitation
↑ Total Species Richness	✘ Pole planting	▨ ↓ Soil salinity	⊙ ↑ Time
↑ Native Species Richness	✘ Pole planting	⊙ ↑ Time	—
↑ Native Perennial Species Richness	✘ Pole planting	↓ Distance to perennial water	↑ Growing season precipitation
↓ Noxious Species Cover	▨ ↑ Gravel in soil	↓ Annual precipitation	—
↓ Noxious Species Relative Cover	⥮ Flooding after tamarisk removal	↑ Annual precipitation	▨ ↓ Soil pH
Planted Species Success			
↑ Planted Species Relative Cover	⊙ ↑ Time	▨ 40–60% Sand in soil	✘ Seeding
↑ Planted Species Relative Density	↑ Growing season precipitation	↑ Growing season maximum temperature	—

Water variables = ⥮

Soil variables = ▨

Management variables = ✘

Time = ⊙

in vegetation cover or diversity on passive revegetation sites. This apparent contradiction has two possible explanations. The first is that, as Harms and Hiebert (2006) suggest, active revegetation is more successful than passive revegetation. The second has to do with the statistical methods used. Harms and Hiebert (2006) used a standard linear regression analysis, and when we ran this same analysis for total vegetation cover or native vegetation cover and time since treatment, the result is very similar to theirs with very little of the variability in cover explained by time ($R^2 = 0.08$ for total cover and $R^2 = 0.10$ for native cover; see figure 24.2). Had we simply done this type of analysis, we would have concluded no improvement in cover over time, as they did. However, regression tree analysis evaluates interactions between variables. For example, the regression tree for total cover includes four variables, and while time only explains about 23% of the variability, the whole model explains 68% of the variability in total cover. Altogether, the result is a strong model for which each portion alone was not as strong. These results underscore

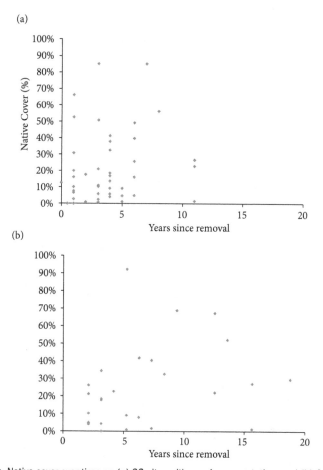

FIGURE 24.2 Native cover over time on (a) 33 sites with passive revegetation, and (b) 28 sites with active revegetation.

Sources: (a) adapted from Harms and Hiebert (2006); (b) adapted from Bay 2006.

the importance of understanding the data and applying it properly when designing a revegetation project.

LESSONS LEARNED FROM PAST PROJECTS

To evaluate the lessons that could be generalized over time and across regions,
Harms and Hiebert (2006) and Bay and Sher (2008) both did survey projects where
regional, hydrological, and soil differences were considered. Harms and Hiebert
(2006) studied natural (passive) regeneration after tamarisk removal at 33 sites
across three ecoregions: the Mojave, Colorado Plateau, and Sonoran-Chihuahuan
transition zone. Each of the sites had been treated to remove tamarisk by either
cutting or burning (followed by herbicide) between 1 and 11 years before the study
sampling occurred. The treated plots were paired with untreated control plots in the
same area. When compared to the untreated plots, treated plots showed significant
reduction in tamarisk foliar cover (82%–95%), decrease in total foliar cover, and
a 176% increase in native cover, mostly in the form of grasses and shrubs. While
the native cover and species diversity increased significantly in the Mojave region,
it did not increase significantly in the other two regions, and tamarisk cover was
not significantly lower than native cover on the treated sites. Their results show
successful restoration in that tamarisk did not return and no other exotics became
dominant. However, they also suggest that passive revegetation may be insufficient to redevelop native riparian vegetation communities, especially the trees and
shrubs that provide canopy cover.

This last finding was supported by Belote et al. (2010) who studied natural
regeneration in 13 sites in Grand Canyon National Park, cleared one to three years
earlier. While exotic species cover (of which tamarisk was the largest component)
did decrease significantly after removal, native species cover and richness were not
significantly affected by removal. However, at this site woody exotic cover prior to
removal only averaged around 10%, much lower than the cover on most tamarisk
removal sites.

Bay and Sher (2008) followed up on the idea that passive revegetation is insufficient to restore these communities. While our study was primarily designed to
evaluate site characteristics that affect success, the data can also be used to evaluate
the relative success of the various projects. The 28 sites we surveyed were in only the
Mojave and Sonoran-Chihuahuan transition ecoregions as defined by Harms and
Hiebert (2006); we were unable to find any suitable sites in the Colorado Plateau
ecoregion. Of the 28 sites surveyed, 18 could be considered successful in that they
had low tamarisk cover, increased native cover and richness, and some success
in the planting establishment. The other 10 were less successful due to either low
native cover, high tamarisk cover, or both. Across sites and in the Mojave ecoregion cover ranged from 0% native cover and 10.7% tamarisk cover at a 3 year old
seeded site near Cibola, Arizona, to 92.3% native cover and 0% tamarisk cover at
a five-year-old stem-planted site near Lake Mead, Nevada. Sonoran-Chihuahuan
transition ecoregion sites ranged from 0.1% native cover and 0.4% tamarisk cover

at a three-year-old seeded site at the Sevilleta NWR, to 52.2% native cover and 0% tamarisk cover at a 13-year-old stem-planted site near Albuquerque, New Mexico (see figure 24.3a). Of the ten less successful sites, three were in the Mojave ecoregion, and of these, two were seeded sites and one was stem planted (see figure 24.3b). Of the seven less successful sites in Chihuahuan transition ecoregion, two were seeded and five were stem-planted. The six sites with the lowest native cover (0%–4%) were seeded sites between 2 and 15 years old; however, the site with the second highest native cover (69%) was also a seeded site that was 9 years old. The apparent contradictions support my suggestion that the differences in success and failure might be due to differences in site characteristics, and that some of these site characteristics might be similar across regions and time.

FIGURE 24.3 Restoration sites where *Tamarix* had been removed and poles of native trees planted. Some were highly successful, such as this site on the Rio Grande (a), whereas at other sites poles never rooted and vegetative cover remained low, as in this one on the Colorado River (b). It should be noted that success of revegetation projects was not consistent across regions and appeared to be more dependent upon environmental variables such as water availability. *Source:* Photo courtesy of Robin Bay.

Scientists have been publishing the results of revegetation after tamarisk removal for almost 40 years now, and the revegetation literature has proliferated in the past decade (see chapter 1, this volume); however, there are still large gaps. There are likely many unpublished data sets in the hands of land managers and organizations. Tamarisk removal and revegetation projects are often funded by grants that do not include a monitoring provision or, at best, only include one or two years of monitoring. There are many projects completed by agencies or organizations that lack the funds to do long-term data collection or to collect and publish data to the standards of peer-reviewed journals. Further, land managers who have unsuccessful projects are often reluctant to publish their results, even though these results could be some of the most useful to other planners. Including more unpublished data in the literature could help to fill in the knowledge gaps, especially in areas such as the Colorado Plateau, about which very little has been published.

A Case for Planning Ahead

In the past few years, a body of literature has started to develop that is helping to guide land managers in planning and implementing successful projects, including both tamarisk control and subsequent native revegetation. Shafroth et al. (2008) published an article detailing how to plan riparian restoration projects in tamarisk-dominated areas, and there are also now best management practices guidelines for tamarisk control (Nissen et al. 2009) and revegetation after tamarisk removal (Sher et al. 2010) for the upper Colorado River region.

All these resources agree that to achieve a successful project, it is critical to follow some basic steps: defining success, setting realistic goals, evaluating site characteristics (both abiotic and biotic), planning the process from start to finish, conducting progress/success monitoring, and implementing maintenance when necessary. These steps are not unique to tamarisk control projects; as a professional ecological consultant, I try to follow all of them on every revegetation project I am involved with. However, remembering these steps is even more important for tamarisk control projects, where there are two distinct phases: removal and revegetation. For too long, land managers saw each phase as a separate project and considered the project complete when the removal was done, with revegetation as an another project that might be addressed later. The most successful tamarisk control projects will be those that are planned from start to finish and include both phases in the planning.

Conclusion

Tamarisk control projects and associated research have come a long way, and the revegetation component is finally gaining the attention it deserves. As in any field of research, the process is ongoing: there is always room for new ideas, and as much can be learned from failure as from success. Of the projects I have reviewed in this chapter, many have been successful, some have not, but all have helped to shape the body of knowledge about revegetation after tamarisk removal.

The past 40 years of tamarisk control projects have taught us that we can have success in restoring and/or revegetating tamarisk-dominated areas if we define that success realistically and plan a comprehensive implementation process. With the ever-growing number of resources available to land managers who are planning riparian restoration projects, there is little excuse for not planning projects for success. There is strong evidence in the literature to suggest that evaluating a site's soil conditions and water availability is a valuable part of the planning process. Other specific site characteristics that are important to success at any given site will be determined by the goals and objectives of the project, but projects will ultimately be the most successful if the species selected and the methods used to plant them are adapted to specific site characteristics.

Literature Cited

Anderson, B. W. 1996. Salt cedar, revegetation and riparian ecosystems in the Southwest. Pages 32–41 *in* J. Lovich, J. Randall, and M. Kelly, editors. California Exotic Pest Plant Council. Proceedings of a symposium October 6–8, 1995. Pacific Grove, CA.

Bailey, J. K., J. A. Schweitzer, and T. G. Whitham. 2001. Salt cedar negative affects biodiversity of aquatic macroinvertebrates. Wetlands 21:442–447.

Barrows, C. 1998. The case for wholesale removal. Restoration and Management Notes 16:135–139.

Bay, R. B. 2006. Success of active revegetation after *Tamarix* spp. removal in southwestern riparian ecosystems: A quantitative assessment of past restoration projects. M.S. thesis, University of Denver, CO.

Bay, R. B., and A. A. Sher. 2008.Success of active revegetation after *Tamarix* spp. removal in southwestern riparian ecosystems: A quantitative assessment of past restoration projects. Restoration Ecology 16:113–128.

Belote, R. T., L. J. Makarick, M. J. C. Kearsley, and C. L. Lauver. 2010. Tamarisk removal in Grand Canyon National Park: Changing the native-non-native relationship as a restoration goal. Ecological Restoration 28:449–459.

Bhattacharjee, J., J. P. Taylor, L. M. Smith, and D. A. Haukos. 2009. Seedling competition between native cottonwood and exotic saltcedar: Implications for restoration. Biological Invasions 11:1777–1787.

Bhattacharjee, J., J. P. Taylor, L. M. Smith, and L. E. Spence. 2008. The importance of soil characteristics in determining survival of first-year Cottonwood seedlings in altered riparian habitats. Restoration Ecology 16:563–571.

Briggs, M. K., B. A. Roundy, and W. W. Shaw. 1994. Trial and error: Assessing the effectiveness of riparian revegetation in Arizona. Restoration and Management Notes 12:160–167.

Briggs, M. K., and S. Cornelius. 1998. Opportunities for ecological improvement along the lower Colorado River and delta. Wetlands 18:513–529.

Brotherson, J. D., and V. Winkel. 1986. Habitat relationships of saltcedar (*Tamarix ramosissima*) in central Utah. Great Basin Naturalist 46:535–541.

Carothers, S. W., G. S. Mills, and R. R. Johnson. 1989. The creation and restoration of riparian habitat in southwestern arid and semi-arid regions. Pages 359–371 *in* J. A. Kusler and M. E. Kentula, editors. Wetland Creation and Restoration: The Status of the Science, Volume 1: Regional Reviews. Environmental Protection Agency, EPA/600/3-89/038.

DiTomaso, J. 1998. Impact, biology, and ecology of saltcedar (*Tamarix* spp.) in the southwestern United States. Weed Technology 12:326–336.

Ellis, L. M. 1995. Bird use of saltcedar and cottonwood vegetation in the Middle Rio Grande valley of New Mexico, U.S.A. Journal of Arid Environments 30:339–349.

Engel-Wilson, R.W., and R. D. Ohmart. 1978. Floral and attendant faunal changes on the lower Rio Grande between Fort Quitman and Presidio, Texas. Pages 139–147 *in* R. R. Johnson, J. F. McCormick, technical coordinators. Strategies for protection and management of floodplain wetlands and other riparian ecosystems, proceedings of a symposium, Callaway Gardens, Georgia. US Department of Agriculture Forest Service, Gen Tech. Rep. WO-12. Washington DC.

Friedman, J. M., G. T. Auble, P. B. Shafroth, M. L.Scott, M. F. Merigliano, M. D. Freehling, and E. R. Griffin. 2005. Dominance of non-native riparian trees in western USA. Biological Invasions 7:747–751.

Friedman, J. M., M. L. Scott, and W. M. Lewis Jr. 1995. Restoration of riparian forest using irrigation, artificial disturbance, and natural seedfall. Environmental Management 19:547–557.

Harms, R., and R. Hiebert. 2006. Vegetation response following invasive tamarisk (*Tamarix* spp.) removal and implications for riparian restoration. Restoration Ecology 14:461–472.

Horton, J. L., and J. L. Clark. 2001. Water table decline alters growth and survival of *Salix gooddingii* and *Tamarix chinensis* seedlings. Forest Ecology and Management 140:239–247.

Levine, C. M. and J. C. Stromberg. 2001. Effects of flooding on native and exotic plant seedlings: Implications for restoration south-western riparian forest by manipulating water and sediment flows. Journal of Arid Environments 49:111–131.

National Invasive Species Council. 2001. Meeting the invasive species challenge: National invasive species management plan. Washington DC.

Nissen, S., A. Sher, and A. Norton. 2009. Tamarisk best management practices in Colorado watersheds. Colorado State University, University of Denver, Colorado Department of Agriculture, and Denver Botanic Gardens.

Pinkney, F. C. 1992. Revegetation and enhancement of riparian communities along the Lower Colorado River. Bureau of Reclamation, Ecological Resources Branch, Ecological Resources Division, Denver, CO.

Shafroth, P. B., J. R. Cleverly, T. L. Dudley, J. P. Taylor, C. Van Riper III, E. P. Weeks, and J. N. Stuart. 2005. Control of *Tamarix* in the western United States: Implications for water salvage, wildlife, and riparian restoration. Environmental Management 35:231–246.

Shafroth, P. B., V. B. Beauchamp, M. K. Briggs, K. Lair, M. L. Scott, and A. A. Sher. 2008. Planning riparian restoration in the context of *Tamarix* control in western North America. Restoration Ecology 16:97–112.

Sher, A. A., and D. L. Marshall. 2003. Seedling competition between native *Populus deltoides* (Salicaceae) and exotic *Tamarix ramosissima* (Tamaricaceae) across water regimes and substrate types. American Journal of Botany 90:413–422.

Sher, A. A., D. L. Marshall and J. P. Taylor. 2002. Establishment patterns of native *Populus* and *Salix* in the presence of invasive nonnative *Tamarix*. Ecological Applications 12:760–772.

Sher, A. A., D. L. Marshall, and S. A. Gilbert. 2000. Competition between native *Populus deltoides* and invasive nonnative *Tamarix ramosissima* and the implications for reestablishing flooding disturbance. Conservation Biology 14:1744–1754.

Sher, A. A., K. Lair, M. DePrenger-Levin, and K. Dohrenwend. 2010. Best management practices for revegetation after Tamarisk removal in the Upper Colorado River Basin. Denver Botanic Gardens, Denver, CO.

Stromberg, J. C., M. K. Chew, P. L. Nagler, and E. P. Glenn. 2009. Changing perceptions of change: The role of scientists in *Tamarix* and river management. Restoration Ecology 17:177–186.

Stromberg, J. C., V. B. Beauchamp, M. D. Dixon, S. J. Lite, and C. Paradzick. 2007. Importance of low-flow and high-flow characteristics to restoration riparian vegetation along river in the arid south-western Untied States. Freshwater Biology 52:651–679.

Taylor, J. P. 1996. Saltcedar management and riparian restoration. *In* Saltcedar management and riparian restoration, proceedings of a workshop. September 17–18, 1996, Las Vegas, NV. [Online]. http://www.invasivespeciesinfo.gov/docs/news/workshopSep96/taylor.html. Retrieved October 7, 2004.

Taylor, J. P., D. B. Wester, and L. M. Smith. 1999. Soil disturbance, flood management, and riparian woody plant establishment in the Rio Grande floodplain. Wetlands 19:372–382.

Taylor, J. P., and K. C. McDaniel. 1998. Restoration of saltcedar (*Tamarix* sp.)-infested floodplains on the Bosque del Apache National Wildlife Refuge. Weed Technology 12:345–352.

Taylor, J. P., and K. C. McDaniel. 2004. Revegetation strategies after saltcedar (*Tamarix* spp.) control in headwater, transitional, and depositional watershed areas. Weed Technology 18:1278–1282.

Taylor, J. P., and K. C. McDaniel. 2003. Salt cedar control and riparian habitat restoration. Pages 1–6 *in Solutions, Technology, and a Look to the Future, Decision Makers Field Guide*. New Mexico Bureau of Geology and Mineral Resources, New Mexico Institute of Mining and Technology, Socorro.

Taylor, J. P., L. M. Smith, and D. A. Haukos. 2006. Evaluation of woody plant restoration in the middle Rio Grande: Ten years after. Wetlands 26:1151–1160.

Vandersande, M. W., E. P. Glenn, and J. L. Walworth. 2001. Tolerance of five riparian plants from the lower Colorado River to salinity drought and inundation. Journal of Arid Environments 49:147–159.

Vincent, K. R., J. M. Friedman, and E. R. Griffin. 2009. Erosional consequences of saltcedar control. Environmental Management 44:218–227.

Zavaleta, E. 2000. The economic value of controlling an invasive shrub. Ambio 29:462–467.

Zouhar, K. 2003. *Tamarix* spp. *in* Fire Effects Information System. [Online]. U.S. Forest Service, Rocky Mountain Research Station, Fire Sciences Laboratory. www.fs.fed.us/database/feis/. Retrieved June 20, 2004.

25

The Future of Tamarix in North America

Anna Sher

To make any predictions regarding the future of *Tamarix* ecology and management in North America requires that we first understand its past. Synthesizing the myriad perspectives and sometimes contradictory research on the genus *Tamarix* may at first seem a daunting task, but it is at the core intent of ecology: identifying patterns. Here, I summarize this book with an eye toward understanding how we arrived at where we are now. From this foundation, I turn to the challenge of making predictions about the future distribution and abundance of *Tamarix*, given the impacts of the biocontrol beetle and other management practices, climate change, and the evolution of the plant itself.

That a monoculture of *Tamarix* is different from historic, native plant communities of the Southwest seems abundantly clear (see table 25.1). Nevertheless, there are some ways in which this relatively new plant provides functions similar to those of willows and other shrubby vegetation in this system, including as a habitat for cup-nesting birds (see chapter 11, this volume), a source of shade for wildlife and livestock, and simply as a primary producer that fixes CO_2. It has been shown that despite the morphological and physiological differences among native and exotic trees, when overbank flows still occur, *Tamarix* stands can be similar to cottonwood stands in many respects (Stromberg 1998). Further, there are many species of animals that have adapted to use it; species abundance and richness can be very high in *Tamarix* stands (see chapters 10 and 12, this volume). However, when the dominant primary producer has changed dramatically in growth form, chemistry, and habit (see chapter 9, this volume), there will be inevitable impacts that cascade through the ecosystem. The shift to *Tamarix* has altered ecosystem processes including soil nutrient cycling (see chapter 13, this volume), fire behavior (see chapter 14, this volume), and the movement of water and sediment (see

TABLE 25.1

Summary of plant traits that differ between *Tamarix* and native riparian vegetation, i.e., cottonwoods and willows*

Traits that differ from native riparian trees	Traits that significantly overlap with native riparian trees
Phenology of flowering (onset is later)	Water use/evapotranspiration
Flowering duration (may last for weeks or months)	Capacity to shade the soil
Timing of bud-break in spring (earlier)	Leaf area index (LAI)
Seed dispersal (longer period)	
N concentrations in leaves (higher)	
C:N ratio in leaves (lower)	
Photosynthetic rates (higher)	
Branching structure (implications for habitat, sediment collection)	
Leaf structure/morphology	
Salt glands and salinity tolerance (higher)	
Wood density (burning temperature is higher)	
Flammability of leaves (higher)	
Drought resistance, including cavitation resistance (higher)	
Canopy height	
Capacity to form extremely large (i.e. dozens of hectares) monocultures	

*It is important to note that there is much more overlap between *Tamarix* and other xeric and halophytic species, which tend to be herbaceous or shrubs. Further, it should be noted that these differences are only "good" or "bad" in context of what is desirable in a particular place at a particular time. They are at once what makes the species at once both amazing and a troublesome weed.

chapter 7, this volume). One cannot have such changes in an ecosystem without a response in wildlife; species compositions of both vertebrates (see chapter 10, this volume) and invertebrates (see chapter 12, this volume) consistently differ between *Tamarix*-dominated and native-dominated stands.

Does this mean that *Tamarix* is bad for riparian ecosystems and should always be eliminated? I have been hard-pressed to find scientists or land managers who think so; the focus seems to be less on the species itself and more on its impacts in a particular place at a particular time. There are those who argue that *Tamarix*'s impacts on ecosystems have been overstated and based on "bad" science (see chapter 16, this volume) or misinterpretation of the scientific literature (Stromberg et al. 2009); clearly, there have been cases of this, but it seems unlikely that there has been any malicious intent. As scientists, we do, however, have to be fully aware of our internal biases as we design our experiments and interpret our data.

I argue that valuing the plant as universally good or bad is not only unscientific, but counterproductive to achieving goals that promote any greater good (see chapter 18, this volume). We need to be clear about falling prey to this type of thinking. The passenger-driver model (see chapters 1 and 15, this volume) can help us gain clarity about the perceptions of *Tamarix* behavior: at one end of the spectrum, the emphasis is on the ways in which *Tamarix* has been a driver of ecological change and how its presence interferes with both human and biological interests (low value); at the other, the focus is on the ways in which this plant is a passenger to anthropogenic alterations, and as such is an acceptable addition to an otherwise degraded landscape (high value). To make sense of this, we need to be fully cognizant that even when we are speaking for the interests of an endangered bird or ecosystem function for the sake of biodiversity, we are necessarily framing *Tamarix*'s value or lack thereof in terms of human interests, and that these interests will be context-dependent (see chapter 18, this volume). That is, there will always be circumstances that warrant *Tamarix* removal, and those that do not. Our role as scientists is to provide as much information as possible to inform these decision points. Regardless of whether your priorities have financial motivations or ecological ones, we must consider the possibility that all situations will not be the same, and that pre-formed ideas can get in the way of good science and management.

We are given the opportunity to confront our expectations and value judgments when we are surprised by what we see. How we respond to our biases has significant consequences for the progress of the field. Discovering that *Tamarix* was a poor competitor as a seedling was unexpected to those of us who considered it to be "aggressive" at all stages, and motivated multiple follow-up experiments to confirm the finding (Sher et al. 2000; Sher et al. 2002; Sher and Marshall 2003). We must not let our fear of losing support for riparian restoration work influence how or what data is published in the scientific literature. On the other side, we must not downplay or shy away from compelling evidence of *Tamarix*'s negative impacts for fear that the findings will fuel its widespread destruction.

Some types of *Tamarix*'s effects are easier to document (e.g., creating a particular branching structure); others, more difficult (e.g., elevating soil salinity, which may have other, concurrent causes). We should be wary of committing a type II error; that is, assuming when there is lack of solid evidence that *Tamarix* is causing problems, it means there are none. We do not want to fall into the same flawed logic as creationists, namely, that because the kind of proof they require cannot be produced, the entire theory of evolution is discredited. Despite ample evidence that *Tamarix*-dominated habitat is very different from native habitat, there are some who argue that because there have been cases in which its impacts were overstated or misunderstood, there is never a cause remove the plant. Is it dangerous to act when we have limited knowledge? Certainly, it can be. This is a challenge for all "crisis disciplines"; one must weigh the danger of the harm of acting against the harm of doing nothing (Soule 1985). We always act based on imperfect information. That said, there are real possibilities that well-intended *Tamarix* management efforts could have (even) devastating effects for some organisms (see chapter 11, this volume).

I believe that we must also acknowledge the moral component of this debate, which generally is unspoken. That is, killing a tree because it is a relative new-comer has been called "xenophytophobia" by some (Stromberg and Chew 2002), and therefore immoral. For others, it is our moral obligation to protect and restore native species and ecosystems. The idea here is that communities of organisms that have taken millennia to become structured in a particular way have infinite and intrinsic value. I admit to having bias in this direction, and to seeing the protection of native wild places as a tenet of the "normative postulates" of conservation biology that "diversity of organisms is good" (Soule 1985). It is my understanding that no native riparian tree formed the kind of monospecific thickets, miles long and miles wide, that *Tamarix* can achieve, and that such monocultures are less desirable than the complex assemblages of cottonwoods, willows, and xeric shrubs that would have been there in the past. However, as much as I may value the latter condition, I must also acknowledge that *Tamarix* removal will not always or necessarily result in that condition.

Central to the question of whether *Tamarix* should be removed is the issue of what vegetation will then fill the niche it has occupied. With careful site selection, active revegetation after *Tamarix* removal has proved successful (see chapter 24, this volume). However, there are certainly areas that will support *Tamarix* and little else; removal of the tree will result in very low vegetative cover (see figure 25.1, this volume). However, sparse vegetation may in fact be the intended outcome of *Tamarix* removal in some cases, such as where it is desirable for a river channel to be as dynamic as possible. In extremely harsh environments, it will take considerable effort and resources to establish replacement vegetation (Sher et al. 2010). It is possible that this condition represents much of the area where *Tamarix* is currently dominant, or a monoculture. A survey of 33 restoration sites

FIGURE 25.1 A *Tamarix* tree growing in Slick Rock Canyon on the lower Dolores River, under highly inhospitable conditions. No native vegetation appears able to tolerate the saline soils in this location.

Source: Photo courtesy of Rob Anderson.

where *Tamarix* was removed but no active revegetation done found highly variable responses of the plant community, from 0% to 90% native cover (Harms and Hiebert 2006). Neither was time a predictor; one site had nearly 90% native cover just three years after removal, while another had less than 5% native cover even 11 years after removal.

I have seen *Tamarix* removal sites that are no better after removal, where it was the only woody species that could grow, and without it diversity would be clearly

FIGURE 25.2 A *Tamarix* removal site on the lower Colorado River, near the Cibola National Wildlife Refuge in Arizona. There did not appear to be much chance of other trees growing, given that the site is disconnected from the river by a levee (in background). *Tamarix* is beginning to grow back.
Source: Photo courtesy of Robin Bay.

FIGURE 25.3 A *Tamarix* removal site in the upper Colorado River with successful revegetation of native grass, a project managed by Pat Arbeiter.
Source: Photo courtesy of Michelle DePrenger-Levin.

compromised (see figure 25.2). But I have also seen places that are greatly improved by removal of *Tamarix* (see figure 25.3). These have most often been locations where there has been active revegetation as described by Bay (see chapter 24, this volume) and/or overbank flows to facilitate establishment of natives as described by Dello Russo (see chapter 21, this volume) and Shafroth et al. (see chapter 23, this volume). Even for one specific desired outcome, there can be cases where removal of *Tamarix* is both good and bad; many river rafters have lamented the loss of beach and access to campsites caused by dense *Tamarix* stands, but in places where there are no other trees, rafters may value *Tamarix* as a place to tie a boat, or as a source of shade (Cole 2011). Thus, we must be very careful about making any blanket statements with regard to whom or what will benefit from *Tamarix* control.

Monoculture versus Mixed

Many of the costs associated with *Tamarix* assume monotypic stands or, at the very least, that *Tamarix* is being invasive, that is, strongly dominant. Generally speaking, plant monocultures are expected to have lower habitat value (Tilman 1982), are less aesthetically pleasing, and provide fewer ecosystem services (Elmqvist et al. 2010). This has been shown with regard to habitat for a variety of animals, where abundance or richness is greater in mixed forests than monospecific *Tamarix* stands (see chapter 10, this volume). Fire risk is likely greater in monocultures (see chapter 14, this volume). Most human values are maximized in a mixed stand of only natives, or with some *Tamarix*, rather than a *Tamarix* monoculture. The exception is with "nonbeneficial" water use, since a more complex forest that includes many layers of vegetation will be expected to use more water than one with a single canopy (see chapter 5, this volume).

In some cases, an ecosystem that includes *Tamarix* but is not completely dominated by it will have more value than a purely native stand. This was found in an analysis of bird surveys conducted at National Wildlife Refuges in Arizona, where for many species, the highest abundance was found in mixed stands with roughly equal cover of native and exotic trees, and that in *Tamarix*-dominated sites, the addition of just a few native trees greatly increased some species' abundance (van Riper et al. 2008). This paper has been used to argue both that *Tamarix* is good and bad for birds, but we need only to pay attention to the whole picture to understand what they are telling us: *Tamarix* is not always bad, but may be more likely to be bad when it is the only thing growing.

It can easily be argued that any presence of *Tamarix* trees is problematic because they are so prolific, and given the right conditions, they can become a monoculture. One habitat suitability model predicts that even with no change in hydrologic management, there are millions of hectares into which *Tamarix* could spread (Morisette et al. 2006). Given *Tamarix's* very abundant and prolonged seed production and dispersal, to prevent such a takeover will require wholesale removal wherever possible, or will it? Additional modeling will be needed to determine how much of this area would become monocultures as opposed to mixed stands (see chapter 3, this volume). Central to such models will certainly be consideration of river and stream flow regimes (see chapter 7, this volume).

Where hydrological conditions promote establishment of native vegetation, we are likely to see mixed stands of *Tamarix* and native trees, with native trees being dominant in some cases (Sprenger et al. 2002; Merritt and Poff 2010). However, given that most rivers in the United States (as in most of the developed world) are regulated, and human needs for water are ever increasing, it is not likely that natural flooding regimes will again become more common in most regions.

Regional Differences

Research on *Tamarix* ecology in North America is dominated by work done in two ecoregions, the Mojave/Sonoran (in the Lower Colorado Watershed, including Arizona and Southern California) and the northern Chihuahuan Transition (in the Rio Grande Watershed, primarily in New Mexico), with growing numbers of publications from the upper Colorado Plateau (portions of Colorado, Utah, Wyoming, Nevada, Arizona, and New Mexico; see figure 25.4). Although *Tamarix* are found in all these areas, there are differences in temperature, growing season, precipitation, soil, and other climatic and geologic features that affect the ecology among them. Further, it is likely that *Tamarix* has been in each region for different lengths of time, and has, to various degrees, adapted to the local conditions. Preliminary work suggests that distributions of genotypes differ by regions (Gaskin and Schaal 2002), likely

FIGURE 25.4 Map of southwestern United States showing major distribution of deserts. The shaded regions indicate where the most tamarisk research has been done.

reflecting the fact that *T. chinensis* was favored for planting as a streambank stabilizer in Texas and New Mexico, whereas *T. ramosissima* is more often planted as an ornamental in the Sonoran/Mojave region (see chapter 19, this volume). These differences certainly have direct effects on vegetation, and likely both direct and indirect effects on other trophic levels. For example, bird use of *Tamarix* is highly variable by region; several avian species that were found to use *Tamarix* along the Pecos River do not use *Tamarix* growing along the Colorado River, despite their presence in the range (Hunter et al. 1988). Research on how other animals use *Tamarix* has sometimes differed dramatically by region (see chapter 10, this volume). Given such variability, we must consider the influence of region when interpreting research findings, and not necessarily extrapolate these findings to represent *Tamarix* ecology overall.

The primary edaphic difference among the regions *Tamarix* occupies is water availability. The Mojave is the driest, with its growing season in the winter when the only precipitation arrives; and even then, few rivers ever run at full capacity. Although the Chihuahuan Transition can reach temperatures that are nearly as high in the summer, it has an order of magnitude more precipitation, including reliable summer rains that support a growing season from spring through fall. The upper Colorado Plateau has a wide range of conditions, but generally with precipitation greater than the Mojave and less than the Chihuahuan Transition; it has a bimodal spring and fall growing season.

We might consider the possibility that attitudes toward *Tamarix* and its control are influenced by where researchers have conducted their field work. Consider our four chapters on *Tamarix* and water use, and at least five others that touch on the subject. All authors agree that *Tamarix* water use is variable and can be equal to that of native trees that also use groundwater. They also agree that not all *Tamarix* should necessarily be removed, especially if water salvage is the primary goal. However, the authors of these chapters reach very different conclusions about the legitimacy of *Tamarix* removal. These differences seem to correlate with the location of their primary research. In general, those who have done their work in the Mojave/Sonoran (Chew, Glenn, Nagler, Sogge, and Stromberg) tend to be more accepting of the presence of *Tamarix* and embracing its current role in the ecosystem; whereas those who have concentrated on the Chihuahuan Transition zone (Cleverly, Dello Russo, Lair) seem more likely to favor removal. Californian researchers seem to be aligned with the Chihuahuan group (Dudley, Zavaleta). Those who have spent most of their professional lives in the upper Colorado River basin appear more likely to lie between the two (Shafroth, Merritt, Hultine) with some tending more toward the Chihuahuan group (Douglass, Nissen, Carlson), particularly those with California connections (Bean). Although an imperfect conceptual model,[1] this may be due at

---

[1] I base these assessments on both their writings (here and elsewhere) as well as personal conversations I had with them. There will certainly be exceptions to this pattern, and some of the authors mentioned here may protest my characterizations of their position, although I suspect most won't. For those who feel they have been incorrectly assigned, please know that I intend no harm and have the highest respect for each one of them, as evidenced by my inclusion of them in my book.

least in part to the differences in *Tamarix* behavior and response of ecosystems to *Tamarix* removal in these regions (T. Talley-Farnham, B. Hammer, and B. Maybach pers. comm.).[2]

For all regions, *Tamarix* is almost certainly a passenger of ecosystem change during the establishment phases, eventually becoming a driver depending on the extent to which the populations are monocultures. There is good reason to expect that mature *Tamarix* stands are acting as a strong driver in the Chihuahuan-transition, motivating human efforts to remove it. In this region we see some of the most dramatic populations of *Tamarix*, with impenetrable, tall thickets for many miles (see figure 25.5). *Tamarix chinensis* is the dominant species farther south in the Chihuahuan, giving way to a hybrid swarm of several genotypes in the northern area (Gaskin and Schaal 2002). *T. chinensis* is from the arid regions of China, and its hybrids are likely the epitome of the combination of drought tolerance and fast growth associated with invasive *Tamarix* (see chapter 9, this volume). Land and resource managers in this region are under extreme pressure to balance competing needs for water, and have observed first-hand the power of *Tamarix* to change fire regimes and affect wildlife habitat (see chapter 21, this volume). Therefore it is not surprising that those who have done much of their research in these impressive thickets (including myself) would be at least initially if not primarily oriented toward a driver model. It seems that populations of *Tamarix* are more likely to be sparse or mixed with natives in the Sonoran/Mojave regions (see figure 25.6), but dense monocultures do occur there (P. Shafroth and K. Hultine pers. comm.). Thus, it is difficult to determine as yet if the greater tolerance of *Tamarix* in Arizona is due to its acting less as a driver there.

The Future of Tamarix in the Context of Management

Many large-scale restoration projects are currently underway that include *Tamarix* removal as a primary component and that will affect the future of the species in its North American range. These include a coordinated effort along the Dolores River in the upper Colorado watershed by the Bureau of Land Management in cooperation with the nonprofits The Nature Conservancy, the Tamarisk Coalition, and other stakeholders. Colorado had a statewide plan to deal with *Tamarix* (CDNR

---

[2] I made my first public plea for an investigation of regional differences during my keynote address at the Tamarisk Symposium in 2007, in the context of the patterns found by Rebecca Harms and Ron Hiebert (2006), and those that were emerging in my student Robin Bay's work. The specific thesis of regional differences presented here was developed during a graduate seminar that I taught in 2009, "The Ecology of Invasive Species—Tamarix as Case Study." Two students taking the seminar, Tiffany Talley-Farnham and Bryce Hammer, wrote a carefully researched paper on this topic of regional differences, and I have used it as my primary source for this section. Beyhan Titiz Maybach also contributed significantly to the discussion and development of this thesis during the class.

(a)

(b)

FIGURE 25.5 Miles-long monoculture of mature *Tamarix* trees growing along the middle Rio Grande (a); these trees grow very densely and are difficult for even cattle or wildlife to navigate (b). *Source:* Photo courtesy of Michelle Ohrtman.

2003), as other states have done, including New Mexico, and there have been collaborative efforts for entire watersheds, such as the seven-state Colorado River Tamarisk and Russian Olive Assessment project. These efforts generally acknowledge that *Tamarix* control is only one component of a restoration plan, although increasing native plant cover is usually the focus. Such cooperative agreements may also involve tribes, corporations, and private landowners, and are predicted to continue in the future (see chapter 21, this volume; Tim Carlson, pers. comm.). At the time of printing, there are *Tamarix* control plans under way for the Verde

FIGURE 25.6 Cottonwood overstory with *Tamarix* understory, growing along the San Pedro River
Source: Photo courtesy of John L Sabo.

(Arizona) and Virgin Rivers (Utah/Arizona) as well. Where these focused projects are being applied, we can expect significant decreases in *Tamarix*.

Even though great efforts are being put forth using both chemical and mechanical control methods in these plans, neither approach will likely match the ultimate effect of the biocontrol beetle (see figure 25.7). Its range is steadily expanding, even without human assistance (see chapter 22, this volume). It is difficult to project the ongoing impact of the beetle, including how native vegetation will respond to the decrease in *Tamarix* cover. It seems probable that *Tamarix* would die most quickly from herbivory when under stress (but see chapter 22, this volume); if true, this could mean that those areas that are too salty and too dry to support other vegetation will likely see the greatest and fastest mortality by the beetle. Loss of these trees and the absence of replacement vegetation is likely to have dramatic, negative impacts on riverbank stabilization (see chapter 7, this volume) and wildlife habitat (see chapter 10, this volume). However, where active management is working in concert with beetle control, especially where hydrology is favorable, we should expect to see improvement in ecosystem function and other services associated with beetle defoliation.

Where there is no other effort to remove trees, the beetle is likely to reach some level of equilibrium with the *Tamarix*, such that neither entirely goes away. Complete control of a tree with biological control is considered difficult (McFadyen 1998). It is difficult to predict how evolution of both the beetle and *Tamarix* may affect their relationship over time, but since the beetle's generation time is an order of magnitude shorter than that of the tree suggests that it should be able to evolve at a much faster rate. The success of *Diorhabda carinulata* may also be a function of past evolution of *Tamarix* in America in the absence of a specialist herbivore but in the presence of generalists (Müller-Schärer et al. 2004).

FIGURE 25.7 Defoliation by *Diorhabda. carinulata*, the tamarisk biocontrol beetle.
Source: Photo courtesy of Andrew Norton.

The Future of *Tamarix* in the Context of Climate Change

Global climate change is a reality that will have complex and interacting effects on temperature and precipitation in the American West, and has significant implications for the future of *Tamarix*. Increased drought and warming temperatures combined with ever-expanding human demands for water certainly will favor the xeric *Tamarix* over less adapted species (see chapter 7, this volume). Warming temperatures have given us earlier springs (Cayan et al. 2001); this trend will mean a longer growing season for *Tamarix* and native species alike (Friedman et al. 2011). One model predicts that at least some western rivers will have greater than average flows in spring and winter, with drier summers than in the past (Shepherd et al. 2010). Where this is true, it could favor native species over *Tamarix*, given their earlier seed dispersal. However, a recent review of the combined effects of changes in flows, CO_2 climate, and other factors associated with global change concluded that we should expect an expansion in the range of *Tamarix* (Perry et al. 2012).

Perhaps the most tangible prediction we can make is the northward spread of *Tamarix*; currently it is scarce north of 39°N latitude, likely a function of intolerance to prolonged cold (Friedman et al. 2005). However, predicted temperature increases of 1.7°C to 5°C over the next century have been used in suitability models to predict dramatic spread of *Tamarix* in the northern range (Kerns et al. 2009). Scientists and land managers from more northern states frequently approach me with concern about managing *Tamarix* to prevent invasion. Research is being done at higher latitudes to identify conditions under which this is more likely; one such experiment determined that burning (a common management technique for managing rangeland) particularly stimulated *Tamarix* seedling establishment (Ohrtman et al. 2011).

Interacting with warming temperatures due to global warming is the capacity of *Tamarix* to tolerate cold. In an experiment that tested cold-hardiness and genetic variability across a latitudinal gradient for both *Tamarix* and cottonwoods, it was found that although cottonwoods had greater tolerance to freezing temperatures in midwinter, *Tamarix* had greater tolerance than cottonwood in spring and late fall (Friedman et al. 2011). Further, *Tamarix* showed a surprising degree of variability across latitudes, suggesting a great capacity for both plastic and evolutionary adaptation in response to climate change. In addition, both cottonwoods and *Tamarix* leafed out in the spring in direct response to temperature, but *Tamarix* was earlier—consistent with its cold hardiness at this time of year. It has been found that although warm temperatures are coming earlier in the spring, the last day of frost appears to not be changing (Inouye 2008). Given the findings above, *Tamarix* may be strongly favored over cottonwoods, which may break their dormancy and leaf out in response to earlier springs only to lose early foliage to freezing temperatures. Thus, even without climate change, we should expect northward progression of *Tamarix* as it continues to be selected for hardiness, and with climate change its progress should be faster.

Like many exotic species that succeed in their introduced range, *Tamarix* has shown a generally high capacity to evolve. Genetically, US populations are distinct from *Tamarix* in its native range, owing in part to a very high degree of hybridization in the new world (see chapter 2, this volume). Hybridization frequently confers superior fitness to its progeny ("hybrid vigor"), and certainly creates a wider gene pool with which to respond to natural selection. Thus, we can expect *Tamarix* to change over time in response to its environment, and likely faster than most native trees, given its short generation time (see chapter 9, this volume).

Future Directions for the Field

Scientific interest in *Tamarix* may have hit a peak in 2010 but does not appear to be waning (see chapter 1, this volume). Certainly, management efforts are continuing, and with them a demand for research to ever better inform field methods, especially for revegetation. I perceive a shift in the literature that is likely to increase with time, namely, that we will see more research focused on what comes after removal, including vegetation and wildlife response to the biological control beetle and other efforts currently underway. Some of the more significant gaps in understanding for *Tamarix*, signaling directions for the field (in no particular order) are:[3]

- Fine-scale regional stream inventories of *Tamarix* distribution, and suitability models that include biomass and/or abundance measures

---

[3] For additional and more detailed discussion of research gaps in *Tamarix* research, see Shafroth et al. 2010.

- Understory response to biological control of *Tamarix*
- Ecology and genetics of the biocontrol beetle, including the requirements for establishment, movement, interactions with other control techniques, interactions with other organisms, and nontarget impacts in the field
- Wildlife response to management efforts, including biological control
- Ecology of animals using *Tamarix*, especially mammals, fish, invertebrates, and herpetofauna. Particular need with regard to uncommon and specialist species. Use of *Tamarix* as migratory bird habitat.
- The relative importance of and interactions between different aspects of hydrology (groundwater, precipitation, flooding, soil moisture) for structuring riparian forests
- Extent to which managed flows can increase passive revegetation after *Tamarix* removal
- Quantifying water salvage (or lack thereof) in response to *Tamarix* removal, including testing the hypothesis that revegetation by xeric species will facilitate it. This includes having water use measurements that account for natural variation and longer term studies (see more thoughts on this in Shafroth et al. [2010]).
- The response of *Tamarix* stands to fire, particularly those defoliated by the bio-control
- The impact of *Tamarix* removal on soil properties (including salinity) and biotic communities (including mycorrhizae) with a particular emphasis on mechanistic links
- The distribution and ecological importance of dark septate endophytes (DSE) for *Tamarix*
- The ecology and invasive potential of *Tamarix aphylla*
- Response of ecosystems to *Tamarix* removal including legacy effects in the soil, longer-term monitoring, and regional variation
- Ways and mechanisms by which resource and land managers can record data in their own work, and collaborate with scientists
- GIS modeling of *Tamarix* populations and their replacement vegetation over time, including in response to climate change and effects of the bio-control and other management efforts

Research is driven by funding, and in lean times it is difficult to predict the attractiveness of *Tamarix* research to funding agencies. Much of the attention to the species was formerly motivated by concern about water (see chapter 17, this volume), a situation that appears to have run its course, given the current understanding of *Tamarix* water use (see chapter 6, this volume). Although there are conditions under which water salvage may be possible (see chapter 4, this volume), there is now enough ambiguity to diminish its power as a justification for research (see chapter 5, this volume). Future funding is more likely to be motivated by need for restoration efforts, where scientists can play a critically important role (see chapters 23 and 24, this volume). However, there are also questions that fall within the realm of pure science; we must embrace such opportunities to use *Tamarix* as a model system to advance general knowledge.

Tamarix will continue to exert its influence both biologically and intellectually in the coming century. It is a plant that has broken ecological rules and found its own way to survive in a challenging environment. In doing so, it has provided habitat for animals and ecosystem services for humans in places where no other trees would grow. It has also shown its capacity to alter its own environment, a true "ecosystem engineer" (sensu Jones et al. 1994), with many changes being inhospitable to organisms that had once thrived there. In these cases, *Tamarix* has worked in concert with humans in making dramatic modifications to the landscape. We have developed, through research and trial and error, a suite of tools we can use to manage the plant, and some of what we do in the future will necessarily be to learn how to manage these tools, such as biological control. Although there may not be consensus as to when or even if we should use these tools, it is not necessary to create an all-encompassing rule for management. Rather, as resource managers we must consider the needs of the organisms and ecosystems in each site and region and make a reasonable plan based on the best information available. And as scientists, we must continue to provide the best, most useful, and least biased information to make this possible.

Literature Cited

Cayan, D. R., M. D. Dettinger, S. A. Kammerdiener, J. M. Caprio, and D. H. Peterson. 2001. Changes in the onset of spring in the western United States. Bulletin of the American Meteorological Society 82:399–415.

CDNR. 2003. Colorado 10-year Strategic Plan for Tamarix Removal and The Coordinated Restoration of Colorado's Native Riparian Ecosystems State of Colorado. Executive Order # D 002 03. Denver, Colorado.

Cole, C. 2011. Tamarisk-eating beetle no cure. azdailysun.com, Flagstaff, Arizona.

Elmqvist, T. E., E. Maltby, T. Barker, M. Mortimer, C. Perrings, J. Aronson, R. S. de Groot, A. Fitter, G. Mace, J. Norberg, I. Sousa Pinto, I. Ring. 2010. Biodiversity, ecosystems and ecosystem services. Pages 239–242 in P. Kumar, editor. *The Economics of Ecosystems and Biodiversity: Ecological and Economic Foundations.* Earthscan, Washington, D.C.

Friedman, J. M., G. T. Auble, P. B. Shafroth, M. L. Scott, M. F. Merigliano, M. D. Freehling, and E. R. Griffin. 2005. Dominance of non-native riparian trees in western USA. Biological Invasions 7:747–751.

Friedman, J. M., J. E. Roelle, and B. S. Cade. 2011. Genetic and environmental influences on leaf phenology and cold hardiness of native and introduced riparian trees. International Journal of Biometeorology 55:775–787. [Online]. http://azdailysun.com/news/local/tamarisk-eating-beetle-no-cure/article_fa00f1f2-06f3-5da2-a22f-ed4a4808c7f6.html

Gaskin, J. F. and B. A. Schaal. 2002. Hybrid Tamarix widespread in U.S. invasion and undetected in native Asian range. PNAS 99:11256–11259.

Harms, R. S. and R. D. Hiebert. 2006. Vegetation response following invasive tamarisk (*Tamarix* spp.) removal and implications for riparian restoration. Restoration Ecology 14:461–472.

Hunter, W. C., R. D. Ohmart, and B. W. Anderson. 1988. Use of exotic saltcedar (*Tamarix chinensis*) by birds in arid riparian systems. Condor 90:113–123.

Inouye, D. W. 2008. Effects of climate change on phenology, frost damage, and floral abundance of montane wildflowers. Ecology 89:353–362.

Jones, C. G., J. H. Lawton, and M. Shachak. 1994. Organisms as ecosystem engineers. Oikos 69:373–386.

Kerns, B. K., B. J. Naylor, M. Buonopane, C. G. Parks, and B. Rogers. 2009. Modeling Tamarisk (*Tamarix* spp.) Habitat and climate change effects in the northwestern United States. Invasive Plant Science and Management 2:200–215.

McFadyen, R. E. C. 1998. Biological control of weeds. Annual Review of Entomology 43:369–393.

Merritt, D. M., and N. L. R. Poff. 2010. Shifting dominance of riparian Populus and *Tamarix* along gradients of flow alteration in western North American rivers. Ecological Applications 20:135–152.

Morisette, J. T., C. S. Jarnevich, A. Ullah, W. Cai, J. A. Pedelty, J. E. Gentle, T. J. Stohlgren, and J. L. Schnase. 2006. A tamarisk habitat suitability map for the continental United States. Frontiers in Ecology and the Environment 4:11–17.

Müller-Schärer, H., U. Schaffner, and T. Steinger. 2004. Evolution in invasive plants: implications for biological control. Trends in Ecology and Evolution 19:417–422.

Ohrtman, M. K., S. A. Clay, D. E. Clay, E. M. Mousel, and A. J. Smart. 2011. Preventing Saltcedar (*Tamarix* spp.) Seedling Establishment in the Northern Prairie Pothole Region. Invasive Plant Science and Management 4:427–436.

Perry, L. G., D. C. Andersen, L. V. Reynolds, S. M. Nelson, and P. B. Shafroth. 2012. Vulnerability of riparian ecosystems to elevated CO_2 and climate change in arid and semiarid western North America. Global Change Biology 18:821–842.

Shafroth, P. B., C. A. Brown, and D. M. Merritt. 2010. Saltcedar and Russian Olive Control Demonstration Act Science Assessment. U.S. Geological Survey Scientific Investigations Report: 2009–5247. [Online]. http://www.fort.usgs.gov/Products/Publications/pub_abstract.asp?PUBID=22895.

Shepherd, A., K. M. Gill, and S. B. Rood. 2010. Climate change and future flows of Rocky Mountain rivers: Converging forecasts from empirical trend projection and down-scaled global circulation modelling. Hydrological Processes 24:3864–3877.

Sher, A., K. Lair, M. DePrenger-Levin, and K. Dohrenwend. 2010. Best Management Practices for Revegetation in the Upper Colorado River Basin. Denver Botanic Gardens, Denver, Colorado.

Sher, A. A., and D. L. Marshall. 2003. Seedling competition between native *Populus deltoides* (Salicaceae) and exotic *Tamarix ramosissima* (Tamaricaceae) across water regimes and substrate types. American Journal of Botany 90:413–422.

Sher, A. A., D. L. Marshall, and J. P. Taylor. 2002. Establishment patterns of native *Populus* and *Salix* in the presence of invasive nonnative *Tamarix*. Ecological Applications 12:760–772.

Sher, A. A., D. L. Marshall, and S. A. Gilbert. 2000. Competition between native *Populus deltoides* and invasive *Tamarix ramosissima* and the implications for reestablishing flooding disturbance. Conservation Biology 14:1744–1754.

Soule, M. E. 1985. What is Conservation Biology? BioScience 35:727–734.

Sprenger, M. D., L. M. Smith, and J. P. Taylor. 2002. Restoration of riparian habitat using experimental flooding. Wetlands 22:49–57.

Stromberg, J., M. K. Chew, P. L. Nagler, and E. P. Glenn. 2009. Changing perceptions of change: The role of scientists in *Tamarix* and river management. Restoration Ecology 17:177–186.

Stromberg, J. C. 1998. Functional equivalency of saltcedar (*Tamarix chinensis*) and Fremont Cottonwood (*Populus fremontii*) along a free-flowing river. Wetlands 18: 675–686.

Stromberg, J. C., and M. K. Chew. 2002. Foreign visitors in riparian corridors of the American Southwest: is xenophobia justified? Pages 195–219 *in* B. Tellman, editor. *Invasive Exotic Species in the Sonoran Region*. University of Arizona Press, Tuscon.

Tilman, D. 1982. *Resource Competition and Community Structure*. Princeton University Press, Princeton, NJ.

van Riper, C., K. L. Paxton, C. O'Brien, P. B. Shafroth, and L. J. McGrath. 2008. Rethinking avian response to *Tamarix* on the lower Colorado River: A threshold hypothesis. Restoration Ecology 16:155–167.7

Glossary [1]

Acre-foot A unit of measurement equivalent to one foot deep water over 43,560 ft^2 (i.e., an acre). It is equivalent to 325,851 gallons or 1,233.5 kl.

Advection The horizontal component in the transfer of air properties. For example, the heat and water vapor content of the air at the earth's surface varies appreciably and by the wind systems these properties are transferred to other areas (Stiegeler 1976).

Adventive Not native, as in a plant that is transported to a new range either intentionally or not.

Aerodynamic *Aerodynamic Roughness*; An index of the nature of airflow near the ground surface (or in this case the vegetative canopy). A surface is aerodynamically smooth if there is a layer of air immediately above it that has laminar flow. However, in meteorological terms, nearly all surfaces are aerodynamically rough, producing turbulent flow down to the ground surface, even for the lightest winds (Stiegeler 1976).

Aggradation Sediment accumulation both in riparian vegetation and above reservoirs.

ANPP above ground net primary productivity.

APHIS Animal and Plant Health Inspection Service (a federal agency within USDA).

Arid Term used to describe a climate or habitat having a low annual rainfall of less than 250 mm with evaporation exceeding precipitation and a sparse vegetation (Lincoln èt. al 1998).

ASABE American Society of Agricultural and Biological Engineers.

ASCE American Society of Civil Engineers.

Avulsion Lateral displacement of a stream from its main channel into a new course across its floodplain. Normally it is a result of the instability caused by channel aggradation (the general accumulation of unconsolidated sediments on a surface which thereby raises its level) (Allaby and Allaby 1991).

Bowen ratio (or energy budget) Calculates evaporation as latent heat from the surface energy budget using the ratio of sensible to latent heat (Bowen ratio) derived from the ratio between atmospheric temperature and humidity gradients measured a few meters above vegetation (Shuttleworth 2008).

BR Bowen Ratio.

Cavitation Rupture of the water column in a vessel of the xylem, causing that vessel to no longer conduct water. It can be caused by evaporation within the stem, or damage to the stem.

---

[1] Adapted with permission from Appendix E "Definitions" of the Independent Peer Review of Tamarisk & Russian olive Evapotranspiration for the Colorado River Basin, as a part of the Colorado River Basin Tamarisk and Russian Olive Assessment (Tamarisk Coalition 2009). **459**

CFS Cubic feet per second.

Channel The bank-delimited length of flowing water, such as a stream or river, that may change with seasonal alterations in volume or velocity. (Also: a constructed water-way, i.e., ditch or canal.)

Channelizing The process by which a channel becomes deeper and more constrained, usually due to human constructed forms (e.g. levees) or the establishment of sediment-trapping vegetation such as *Tamarix*.

Container studies Container studies refer to that category of experiments that have grown vegetation in small containers, and are often too small to include representative amounts of soil evaporation; or that may have been placed in unnatural environments and elevated above the natural soil surface so that radiative and aerodynamic characteristics are unrepresentative of a natural environment (Allen et al. 1998).

Crop coefficient The calculated value of a given crop's evapotranspiration (ET) that, when multiplied by a reference crop's evaporation (ET_o) in similar climactic conditions, estimates that crop's evapotranspiration rate (Woodhouse 2008). One of the most basic crop coefficients (K_c) is the ratio of the ET observed for the crop studied to that observed for a reference crop under the same climactic conditions (Allen et. al 1998).

CW Cottonwood and willow (native trees that grow along rivers of the N. American southwest).

Defoliate To shed or to lose leaves; to cause a tree to lose its leaves (Durrenberger 1973).

Degradation The diminution of biological productivity or diversity (Gregorich et al. 2001).

Demonstration project Large-scale tamarisk restoration projects, which may be identified in Public Law 109–320, that will serve as research platforms to address critical tamarisk management issues. These issues include water savings, impacts to habitat and biodiversity, economics, etc.

EC Electrical conductivity.

Ecosystem services Those ways in which natural systems benefit humans, including both non-consumptive and consumptive goods and services, such as trees preventing erosion or insects pollinating crops.

Eddy covariance (also called eddy correlation) Calculates evaporation as 20- to 60-minute time averages from the correlation coefficient between fluctuations in vertical windspeed and atmospheric humidity measured at high frequency (~10Hz) at the same location, a few meters above vegetation (Shuttleworth 2008).

ESA Ecological Society of America.

ET Evapotranspiration.

ETo Reference evapotranspiration or Reference crop evapotranspiration.

EToF Fraction of reference evapotranspiration.

Evaporative demand The conditions for air to be capable of absorbing moisture. An index of this is the saturation deficit, which is the difference between the saturation vapor pressure and actual vapor pressure. If the saturation deficit is large, as in warm dry air, the gradient between the moist surface and the atmosphere will be high and so the rate of transfer will be large. With moist air the humidity gradient will be less and the rate of evaporation correspondingly smaller (Stiegeler 1976).

Evapotranspiration The combined system of vapor transfer by evaporation and transpiration from the ground surface and its vegetative layer (Stiegler 1976).

EWRI Environmental & Water Resources Institute.

Exotic A plant or animal species that is not indigenous to a region; an organism intentionally or accidentally introduced and often persisting (Peale 1996).

FAO Food and Agriculture Organization of the United Nations.

Field capacity The amount of water retained by soil once all free water has drained from it via gravity; it is expressed as the percentage of dry weight per volume of soil.

Field survey The collection of data in a "natural" environment, with the purpose of understanding the diversity or composition of a particular population or community.

FWS Fish and Wildlife Service.

Gaining stream Stream that receives groundwater. The water table is farther above the elevation of the stream's surface as distance from stream increases (Peale 1996). Also see *losing stream*.

Genotype The specific genetic makeup of an organism, often referring to the presence of particular alleles.

GSFLOW Coupled Ground-water and Surface-water FLOW model by USGS.

Halophytic *Halophyte:* a plant living in saline conditions; a plant tolerating or thriving in an alkaline soil rich in sodium and calcium salts (Lincoln et al. 1998).

Herbaceous *Herb:* a plant having stems that are not secondarily thickened and lignified; non-woody (Lincoln et al. 1998).

Herbivory A form of predation in which a heterotroph, usually an animal, consumes an autotroph, usually a plant.

Heterogeneity Heterogeneous: having a non-uniform structure or composition (Lincoln et. al 1998).

Hydrograph The seasonal variation in a river or other body of water, such as in level, velocity, or discharge.

Hydrophytic vegetation Plants that occur in very wet soils, in contrast with mesophytic vegetation that is better adapted to intermediate moisture levels and xerophytic, which is found in dry zones.

Incision The process whereby a downward-eroding stream deepens its channel or produces a narrow, steep-walled valley; esp. the downcutting of a stream during rejuvenation, whether due to relative movement (uplift) of the crust or to other cause. Also, the product of such a process, such as an incised notch or meander (Roberts and Jackson 1980).

Invasive species Legally defined as an alien species whose introduction does or is likely to cause economic or environmental harm or harm to human health. (**Executive Order 13112**). Ecologically, they are species (exotic or native) that have dramatically increased in density or range and thereby cause harm to historic patterns of ecological functioning.

LANDSAT Any of a series of satellites operated jointly by NASA and the US Geologic Service that transmit images of the Earth's surface. They are used in the scientific discipline of Remote Sensing and can be used to map vegetation, among other features.

Landscape Ecology The study of organisms in the context of spatial and temporal aspects of an environment on a grand scale, including biotic, abiotic and cultural interactions (Calow 1999).

LAI Leaf area index – The total two-dimensional area of leaf tissue per unit ground area.

Litter dispersal The loss of litter or recently fallen plant material that is only partially decomposed and in which the organs of the plant are still discernible, forming a surface layer on some soils (Lincoln et al. 1998).

Losing stream A stream in which water is being lost to the groundwater system. Ground water is deeper below stream surface as distance from stream increases (Peale 1996). Also see *gaining stream.*

Lower Colorado River Basin The Colorado River Watershed beginning at Lee's Ferry just below Glen Canyon Dam and terminating in the Gulf of California. The Lower Basin covers portions of Arizona, California, Mexico, Nevada, New Mexico, and Utah.

Lysimeter (LYS) The process of estimating ET by measuring the change in weight of an isolated, preferably undisturbed soil sample with overlying vegetation (if present), while measuring precipitation to and drainage from the sample plot (Shuttleworth 2008). Lysimeters are special containers that are placed at ground level in natural settings and where the container should be large enough to contain representative amounts of evaporation from soil and transpiration from vegetation. ET is determined from lysimeters by monitoring the change in weight of the lysimeter or by noting the change in water table elevation (Allen et al. 1998, pers. comm. 2009).

Meristem/meristematic tissue a place of rapidly dividing cells on a plant from which new tissues or organs (such as stems) can arise.

Mesic Applied to an environment that is neither extremely wet (hydric) nor extremely dry (xeric) (Allaby and Allaby 1991).

Mesic riparian fringe The transition zone between fully riparian and fully mesic vegetation communities.

Mesophytic see Hydrophytic.

Model A mathematical or conceptual framework used to explain a pattern or orientation of natural events, which may be used to make predictions.

MODIS (Moderate Resolution Imaging Spectroradiometer); an instrument that projects images from space to study the earth's surface. These images are higher resolution than LANDSAT images but are taken less frequently (see LANDSAT).

MOU Memorandum of Understanding.

Mycorrhizae fungi that form symbiotic relationships with plant roots, in which the fungi receives sugars while providing water and mineral nutrients to the plant.

Mycorrhizal inoculation The introduction of fungus that has a symbiotic relationship with the root system of a plant (Durrenberger 1973).

Nitrogen manipulation The use of nitrogen for the benefit of restoration. Nitrogen is the most abundant gas in the atmosphere and a critical constituent in the soil, and can only be used directly by a few specialized bacteria. To be of widespread value it has to be converted into the nitrate form. In nature, nitrogen is involved in cyclic changes termed the Nitrogen Cycle, the change from animal life to nitrites to nitrates to plant life to animal life (Stiegeler 1976).

NPP Net primary productivity.

NRCS Natural Resources Conservation Service.

NWR National Wildlife Refuge.

Phenology/Phenological Relating to the timing of biological events, e.g. flowering.

Photosynthetic rates The rate at which photosynthesis occurs. Photosynthesis is the biochemical process that utilizes radiant energy from sunlight to synthesize carbohydrates from carbon dioxide and water in the presence of chlorophyll (Lincoln et al. 1998).

Phreatophyte (obligate and facultative) A plant that habitually obtains its water supply from the zone of saturation either directly or through the capillary fringe (Durrenberger 1973). Obligate phreatophytes require access to groundwater at all times and life stages, whereas facultative phreatophytes access it for only a portion of their water requirements or life stages.

Plant stomatal behavior The actions of the stomata of leaves that control the loss of water vapor, or stomatal transpiration. Stomatal resistence is the property of the stomata in restricting the free exchange of carbon dioxide (CO_2) by a plant leaf; the major constraint on CO_2 uptake into the plant leaf, governed largely by the diameter of the stomatal pores. Stomatal conductance is the reciprocal of stomata resistance (Lincoln et al. 1998).

Potential evapotranspiration The theoretical baseline measurement of water loss by plants when water is unlimited.

Propagule plant tissues or organs that can be used to propagate the plant, including both seeds (for sexual reproduction) and plant parts such as stems, leaf, or root parts that are capable of growing into a new plant (for asexual or vegetative reproduction).

Reference ET An estimate of what evapotranspiration would be in a highly studied reference vegetation, that is, well-watered and actively transpiring grass of a certain height. Reference ET is [often] calculated using the Penman-Monteith equation, and expresses the energy available to evaporate water and the wind available to transport water vapor from the ground into the air, for the reference vegetation type (Woodhouse 2008).

Remote sensing estimates using energy balance Evaporation is deduced indirectly from the surface energy balance, with sensible heat calculated from the difference between air temperature and the temperature of the evaporating surface, along with an estimate of the aerodynamic exchange resistance between these two (Shuttleworth 2008).

Remote sensing using vegetation indices Remote sensing using airborne or satellite sensors (e.g., Moderate-resolution Imaging Spectroadiometer or MODIS) to measure vegetation characteristics through NDVI (Normalized Difference Vegetation Index) and EVI (Enhanced Vegetation Index) (Measuring Vegetation (NDVI&EVI)... [updated 2009]).

Restoration The process of returning a site from a disturbed or totally altered condition to a previously existing natural or altered condition. This process requires some knowledge of the type of vegetation that existed prior to modification (Peale 1996).

Rhizosphere The narrow region of soil surrounding plant roots that is directly influenced by root secretions and associated soil microorganisms.

Riparian zone Riparian zones are the interfaces between terrestrial and aquatic ecosystems. As ecotones, they encompass sharp gradients of environmental factors, ecological processes, and plant communities. Riparian zones are not easily delineated but are composed of mosaics of landforms, communities, and environments within the larger landscape (Gregory et al. 1991). Riparian lands are defined by EPA simply as areas adjacent to streams, rivers, lakes, and freshwater estuaries. BLM defines it as lands along, adjacent to, or contiguous with perennially or intermittently flowing rivers and streams, glacial potholes, and shores of lakes and reservoirs with stable water levels. Excluded are such sites as ephemeral streams or washes that do not exhibit the presence of vegetation dependent upon free water in the soil (Colorado DNR 1998).

Salinity Occurs as either the total dissolved solids (TDS) (Peale 1996) in water or as salts and minerals in the soil, available to soil moisture for dissolution. As the salinity of soil water around a plant's root system increases, greater osmotic pressure is required on the part of the plant to extract water molecules from the soil (Hem 1967). When a plant cannot generate enough osmotic pressure to separate water molecules from salt and other dissolved solids, it will succumb to drought stress and desiccation.

Sap flow (SF) The measure of plant transpiration by measuring the rate of sap flow in trunk, branches or roots, using heat as a tracer, with an estimate of the volume of wood through which flow occurs (Shuttleworth 2008).

Scintillometer A device that uses a theoretical relationship between sensible and latent heat fluxes and atmospheric scintillation introduced into a beam of electromagnetic radiation between source and detector by temperature and humidity fluctuations (Shuttleworth 2008).

SCS Soil Conservation Service, now the Natural Resource Conservation Service (NRCS).

Soil Moisture Content See *Soil water balance.*

Soil water balance The ratio of the volume of contained water in a soil compared with the entire soil volume. When a soil is fully saturated, water will drain easily into the underlying unsaturated rock. When such drainage stops, the soil still retains capillary moisture and is said to contain its field-capacity moisture content. Further drying of the soil (e.g. by evapotranspiration) creates a soil-moisture deficit, which is the amount of water which must be added to the soil to restore it to field capacity measured as a depth of precipitation (Allaby and Allaby 1991).

Stoma A small pore, moderated by guard cells, in a plant leaf or stem that allows the transfer of water vapor and gas.

Succession successive plant communities (seral stages) that follow one another over time on a given site (Peale 1996).

Systemic herbicide A pesticide that is translocated from the point of application to other tissue and organs in the plant, with the goal of killing the entire plant.

Transpiration The removal of moisture from the soil by plant roots, its translocation up the stem to the leaves, and its evaporation through the stomata (Allaby and Allaby 1991).

Upper Colorado River Basin The Colorado River Watershed beginning at its headwaters in Colorado's Rocky Mountains and extending downstream to Lee's Ferry just below Glen Canyon Dam. The Upper Basin covers portions of Arizona, Colorado, New Mexico, Utah, and Wyoming.

Upper floodplain terraces Lands within the floodplain but with a deeper water table that would normally be occupied by more xeric native vegetation.

USDA-ARS United States Department of AgricultureAgriculture Research Service.

USFS United States Forest Service.

USGS United States Geological Service.

Weed A contextually undesirable plant.

Xeric Having very little moisture; tolerating or adapted to dry conditions.

Xerophytic see Hydrophytic.

Xeroriparian Adjective describing riparian areas along ephemeral streams or dry washes in arid environments that are characterized by low moisture availability, or the vegetation that grows there.

LITERATURE CITED

Allaby A, Allaby M (eds). 1991. The Concise Oxford Dictionary of Earth Sciences. Oxford (NY): Oxford University Press.

Allen RG, Pereira LS, Raes D, Smith, M. 1998. Crop Evapotranspiration-Guidelines for Computing Crop Water Requirements, FAO Irrigation and drainage paper 56. Rome, Italy: Food and Agriculture Organization of the United Nations. ISBN 92-5-104219-5. Retrieved on 2007-11-24.

Colorado Department of Natural Resources (DNR). 1998. Native Plant Revegetation Guide for Colorado, Volume III.

Durrenberger RW. 1973. Dictionary of the Environmental Sciences. Arizona State University: National Press Books.

Gregorich EG, Turchenek LW, Carter MR, Angers DA. 2001. Soil and Environmental Science Dictionary. Canadian Society of Soil Science. USA: CRC Press.

Gregory SV, Swanson FJ, McKee WA, Cummins KW. 1991. An ecosystem perspective of riparian zones; focus on the links between land and water. Bioscience 41(8):540–551.

Hem JD. 1967. Composition of saline residues on leaves and stems of saltcedar (*Tamarisk pendantra* Pallas). Reston (VA): US Geological Survey. Professional Paper 491-C.

Lincoln R, Boxshell G, Clark P. 1998. A Dictionary of Ecology, Evolution and Systematics, Second Edition. New York (NY): Cambridge University Press.

Measuring Vegetation (NDVI & EVI) [Internet]. [updated 2009 March 4]. Greenbelt (MD): Earth Observatory, Nasa; [cited 2009 March 4]. Available from: http://earthobserva-tory.nasa.gov/Features/MeasuringVegetation/measuring_vegetation_4.php

Peale M (ed). 1996. Best Management Practices for Wetlands within Colorado State Parks. Denver (CO): Colorado State Parks.

56Plants [Internet]. [updated 2009 Jan 23]. United States Department of Agriculture, National Agricultural Library, National Invasive Species Information Center; [cited 2009 March 2]. Available from: http://www.invasivespeciesinfo.gov/plants/main.shtml.

Roberts LB, Jackson JA. (eds). 1980. Glossary of Geology, Second Edition. Falls Church (VA): American Geological Institute.

Shuttleworth WJ. 2008. Evapotranspiration Measurement Methods. University of Arizona. Southwest Hydrology January/February: 22–23.

Stiegeler SE (ed). 1976. A Dictionary of Earth Sciences. New York (NY):PICA PRESS; Pan Books Ltd.

Woodhouse B. 2008. Approaches to ET Measurement. University of Arizona. Southwest Hydrology January/February: 20–21.

INDEX

467